Studies in Computational Intelligence

Volume 1085

The series "Studies in Computational Intelligence" (SCI) publishes new developments and advances in the various areas of computational intelligence—quickly and with a high quality. The intent is to cover the theory, applications, and design methods of computational intelligence, as embedded in the fields of engineering, computer science, physics and life sciences, as well as the methodologies behind them. The series contains monographs, lecture notes and edited volumes in computational intelligence spanning the areas of neural networks, connectionist systems, genetic algorithms, evolutionary computation, artificial intelligence, cellular automata, self-organizing systems, soft computing, fuzzy systems, and hybrid intelligent systems. Of particular value to both the contributors and the readership are the short publication timeframe and the world-wide distribution, which enable both wide and rapid dissemination of research output.

Indexed by SCOPUS, DBLP, WTI Frankfurt eG, zbMATH, SCImago.

All books published in the series are submitted for consideration in Web of Science.

Rajiv Pandey · Nidhi Srivastava ·
Neeraj Kumar Singh · Kanishka Tyagi
Editors

Quantum Computing: A Shift from Bits to Qubits

Springer

Editors
Rajiv Pandey
Amity Institute of Information Technology
Amity University Uttar Pradesh
Lucknow, India

Nidhi Srivastava
Amity Institute of Information Technology
Amity University Uttar Pradesh
Lucknow, India

Neeraj Kumar Singh
INPT-ENSEEIHT/IRIT
University of Toulouse
Toulouse, France

Kanishka Tyagi
Aptiv Advance Engineering Center
Agoura Hills, CA, USA

ISSN 1860-949X ISSN 1860-9503 (electronic)
Studies in Computational Intelligence
ISBN 978-981-19-9529-3 ISBN 978-981-19-9530-9 (eBook)
https://doi.org/10.1007/978-981-19-9530-9

This Springer imprint is published by the registered company Springer Nature Singapore Pte Ltd.
The registered company address is: 152 Beach Road, #21-01/04 Gateway East, Singapore 189721, Singapore

Preface

Physicists and computer scientists came up in the 1970s and early 1980s with the idea of a device based on quantum mechanics. In October 2019, scientists at Google reported their "quantum supremacy" with their Sycamore chip. It led a race among researchers to push the boundaries to usher us into a new quantum computing revolution. Companies like D-Wave commercialized quantum computing, which further fuelled rapid innovation. All these and more events proved the fact that quantum computers can perform the given task in a substantially shorter duration as compared to the classical computer.

Quantum computers work on quantum bits known as Qubit. Qubits are the backbone behind the working of the quantum computer. For every Qubit, two possible values allow performing most calculations simultaneously. A Qubit can be anything that exhibits quantum behavior. Quantum computers, with the use of superposition and entanglement, can store a vast amount of information which is not possible now. This superposition of Qubit being 0 or 1 is a powerful feature that can be exploited in many valuable ways.

Quantum computers can be of help in many ways spanning across domains. It can simulate and analyze molecules for drug development and materials design, optimizing supply chain logistics, traffic flow, etc. One more advantage of this is that it is almost impossible to crack quantum encryption, which is based on the laws of physics rather than mathematics which most of the encryption algorithms are based on in classical computers.

Although quantum computing was introduced a few decades back, many barriers must be overcome before it can be used to its full potential. The coordination between the qubits gets disturbed if there is environmental disruption like temperature variations, mechanical vibrations, stray electromagnetic fields, etc. Thus this reduces the system's reliability. Thus there is a need to isolate these machines, so there is no disturbance. Error-correction routines can also correct this. For each Qubit, at least five error-correcting qubits are needed. However, this increases the complexity and also the cost of the computers.

The government and private sector are spending vast amounts of money to exploit these features of quantum physics in computing. If used to its full potential, quantum computing can bring revolution; still, lots of work needs to be done in this field.

We received a total of 48 chapters, and through a peer-reviewed process, we narrowed out 23 chapters into five parts. The book includes chapters dedicated to an introduction to quantum computing, the work behind these systems, quantum algorithms, quantum communications, quantum cryptography, and practical applications of quantum computing.

The book shall help the user understand the theory, concepts, working principles, simulation tools, and future direction of quantum computing. Our focus in the book is to detail quantum computing, a contemporary subject of research gaining popularity both in academia and industry. The benefits of quantum technology can be reaped in several ways. The book should answer the thriving question: how and when can quantum technologies be considered?

It will be dedicated to both theoreticians and practitioners, allowing them to get experience in applying quantum computing technology backed by solid theoretical foundations. It could be used as a supplementary tutorial in academic quantum technologies/computing courses, primarily by graduate students and application engineers. Quantum computing is a regular topic in significant signal processing, high-performance computing, machine learning, and other advanced technical conferences. This book may be a reference manual for many research articles.

Part—Scientific Theory for Quantum: This part details the foundational principles of quantum theory from the lens of physics as well as computer science.

- *Quantification of Correlations in Quantum States by Cristian Susa* presents a discussion on the quantification of correlations in quantum states. This chapter describes a strategy to quantify the genuine total correlations at any size of multipartite quantum states. Moreover, their approach to the given quantum state of any number of systems is compared with the set of tensor product states in any order.
- *From Quantum Mechanics to Quantum Computing by Pooja Srivastava, Anushtup Mishra, and Yogesh Kumar Srivastava,* the basics of quantum mechanics were discussed by the authors. The concepts explained are Quantum state, Hilbert space, Qubits, Composite systems and entanglement, mixed states, and density matrix. Finally, the authors have talked about decoherence which is a considerable challenge to building a practical quantum computer.
- *Phase Space Quantization I: Geometrical Ideas by Carlos Alberto Alcalde and Kanishka Tyagi takes* Radar theory as an example problem to explain classical phase space and exhibit quantum features of the Heisenberg group. The chapter shows that using the quantization of the spin ½ particle arising in deformation theory, a star product on the Bloch sphere in geometrical data can be understood.
- *Phase Space Quantization II: Statistical Ideas by Carlos Alberto Alcalde and Kanishka Tyagi* reviews the origin of information in signal theory as a gateway to understanding quantum information theory. Concepts of operational physics and

Lie group representation theory are applied to the study of phase space measurements. Quantum Signals are described by the SU(2) group on the Poincare sphere introduced in the deformation theory approach by the star exponential.

- *Efficient Quantum Circuit for Karatsuba Multiplier by James Selsiya M., Kalaiarasi M., Venkatasubramani V. R., and Rajaram S.* present work on quantum circuits. Since the number of qubits is an essential parameter in a quantum circuit design, the authors have given two novel designs for GF(2n) multiplier. To implement the design flow, they used a quantum programming studio, and to evaluate quantum gates, they used a quantum computer emulator.

Part—Quantum Computing: Building Concepts: This part talks about concept, analysis, and quantum entanglement in detail.

- *Quantum Concepts, by Manjula Gandhi S., Gayathri Devi S., Sathya C. and Vani K. H., and Kiruthika K.* demonstrates basic concepts by implementing them on IBM's Quantum Experience– Qiskit, which is an open-source framework to write and execute quantum programs. The authors have explained and created quantum entanglement on Qiskit and represented the output of entanglement in different forms.
- *Evolutionary Analysis: Classical Bits to Quantum Qubits by Rajiv Pandey, Pratibha Maurya, Guru Dev Singh, and Mohd. Sarfaraz Faiyaz presents* a comparative analysis between classical and quantum computers. Finally, they explain different quantum gates.
- *Non-silicon Computing With Quantum Superposition and Quantum Entanglement Using Qubits by N. Vidhya, V. Seethalakshmi, and S. Suganyadevi* emphasizes the advantage of quantum features to overcome the miniaturization issues of classical computing. They explain the importance of reversible computation and the impact of quantum tunneling.

Part—Quantum Algorithms–Theory and Applications: The focus of this part is the theoretical aspects of quantum computing along with industrial applications.

- *A Reversible Hybrid Architecture for Multilayer Memory Cells in Quantum Dot Cellular Automata with Minimized Area and Less Delay contributed by Suparba Tapna, Debarka Mukhopadhyay, and Kisalaya Chakrabarti*, describes Quantum Dot Cellular Automata (QCA), which are among the latest advancements in the plan of computerized nanocircuits. The authors propose a three-level design for lessening the utilization of energy. It also diminishes the number of cells and consumption territory in a QCA-based memory.
- *Quantum Neural Network for Image Classification Using TensorFlow Quantum by Arun Pandian J. and Kanchanadevi K.* deals with Quantum neural networks. The authors show a practical demonstration of quantum neural networks on benchmarking datasets. The authors explain the steps for data preparation and building a quantum neural network. Finally, they compare the performance of the Quantum Neural network with the classical neural network.
- *Quantum Network Architecture and Its Topology by Supriyo Banerjee, Biswajit Maiti, and Banani Saha* deal with Quantum Key Distribution (QKD). The authors

discuss different types of network communications regarding QKD. The quantum network's current state and development have also been covered, wherein the authors have shown the different networks, QKD network type, QKD protocol, and the topologies used.

- *Quantum Computing Enabled Machine Learning for an Enhanced Model Training by Jayesh Soni, Nagarajan Prabakar, and Himanshu Upadhyay* presents Quantum machine learning. The authors cover the concepts of quantum computers—qubits, superposition, entanglement, and algorithms. The authors present an implementation for the breast cancer classification problem using the traditional machine learning algorithms on traditional computers and QSVM trained on IBM Quantum Computer.
- *Numerical Modeling of the Major Temporal Arcade Using a Quantum Genetic Algorithm by Jose Soto-Alvarez, Ivan Cruz-Aceves, Arturo Hernandez-Aguirre, Martha Alicia Hernandez-Gonzalez, and Luis Miguel Lopez-Montero* presents an automatic MTA segmentation and numerical modeling based on spline curves and Quantum genetic algorithm (QGA). Modeling of MTA is useful for computer-aided diagnosis in ophthalmology.
- *Entangled Quantum Neural Network by Qinxue Meng, Jiarun Zhang, Zhao Li, Ming Li, and Lin Cui* provides a new framework called Entangled Quantum Neural Network, EQNN. They have conducted experiments on three UCI datasets. EQNN was used for measurement processes of the entangled states to replace specific layers of the MLP and was found to outperform the baseline algorithms.

Part—Quantum Simulation Tools and Demonstrations: The subject matters of this part are the industry tools and software deployed for doing hands-on quantum computing.

- *Exploring IBM Quantum Experience by Gayathri Devi S., Manjula Gandhi S., Chandia S., and Boobalaragavan P.* describes each step for navigating IBM Quantum Experience. They then cover the applications of teleportation and quantum algorithms. Finally, they conclude the chapter by listing the various research fields where IBM Q Experience has played an important part.
- *Quantum Programming on Azure Quantum—An Open Source Tool for Quantum Developers by Kumar Prateek and Soumyadev Maity* emphasizes Microsoft's cloud-based quantum computing as a service, Azure Quantum. The authors have written this chapter keeping in mind a developer who is new to quantum and needs to develop their skill to access quantum computers. The chapter in detail describes Azure Quantum, Azure Quantum workspace, quantum programs, and ways to develop quantum software.
- *Survey of Open Source Tools/Industry Tools to develop Quantum Software by Dhaval Mehta, Amol Ranadive, Jigna Bhupendra and Rajiv Pandey* focuses on the basic functionality of the tools for developing applications/solving real-life problems using Quantum Computing. They explain Azure Quantum and IBM Quantum. In the end, the authors list the various open-source software, their small description, programming language used, license of usages, and supported OS for developing the algorithms.

- *Simulating Quantum Principles: Qiskit Versus Cirq by Rajiv Pandey, Pratibha Maurya, Guru Dev Singh, and Mohd. Sarfaraz Faiyaz* depicts the concepts of superposition, entanglement, and interference through programming on IBM's Composer, Qiskit, and Google's Cirq.

Part—Future Direction and Applications: This part mainly focuses on what's next and coming areas of applications of Quantum.

- *Quantum Machine Learning in Prediction of Breast Cancer by Jigna Prajapati, Himanshu Paliwal, Bhupendra Prajapati, Rajiv Pandey, and Surovi Saikia* discusses the various Quantum Machine learning techniques like Quantum Neural networks, Dimensionality Reduction Algorithms, Support Vector Machines, etc., which are useful in the prediction of Breast cancer and breast cancer type.
- *Understanding of Argon Fluid Sensor Using Single Quantum Well Through K-P model: A Bio-Medical Application Using Semiconductor Based Quantum Structure by Gopinath Palai, Nitin Tripathy and Biswaranjan Panda* investigates the argon concentration in their fluids with three layers of silicon based on quantum well structure. The mathematics of the research relies on the configuration of the proposed structure which is derived through the Kronig–Penny model. The authors conclude that one can determine the concentration of argon fluid by knowing the output potential.
- *A Study On Quantum Cryptography And Its Need by Vandani Verma* discusses quantum cryptography. They explain the concept of cryptology. The chapter talks about BB84 and Ekert 91 protocol and the detection of eavesdroppers in quantum key distribution. Finally, it discusses quantum-masked authentication protocols.
- *Evolution of Quantum Machine Learning and An Attempt of Its Application for SDN Intrusion Detection by Aakash Shinde and Shailesh Bendale* provides a general overview of major concepts regarding Quantum Computation and QML. The chapter illustrates the information regarding the flow of experimentation, methods considered while experimenting, tools utilized, the process followed, and findings regarding the implementation of QML algorithms on the Software Defined Network (SDN) dataset.
- *Implications of Deep Circuits in Improving Quality of Quantum Question Answering by Pragya Katyayan and Nisheeth Joshi* deals with SelQA (Selection-based Question Answering) dataset and performed question classification using quantum support vector machine (QSVM) and variational quantum classifier (VQC). The results show an improvement in a Question Answering system with the classification results.

We would like to extend our gratitude to all the reviewers who have evaluated the submissions and helped improve the ones that have been selected for inclusion in this volume. Thank you for your hard work and support.

Lucknow, India	Dr. Rajiv Pandey
Lucknow, India	Dr. Nidhi Srivastava
Toulouse, France	Dr. Neeraj Kumar Singh
Agoura Hills, CA, USA	Dr. Kanishka Tyagi

Contents

Scientific Theory for Quantum

Quantification of Correlations in Quantum States 3
Cristian E. Susa-Quintero

From Quantum Mechanics to Quantum Computing 15
Pooja Srivastava, Anushtup Mishra, and Yogesh K. Srivastava

Phase Space Quantization I: Geometrical Ideas 31
Carlos Alberto Alcalde and Kanishka Tyagi

Phase Space Quantization II: Statistical Ideas 53
Carlos Alberto Alcalde and Kanishka Tyagi

Efficient Quantum Circuit for Karatsuba Multiplier 79
M. James Selsiya, M. Kalaiarasi, S. Rajaram,
and V. R. Venkatasubramani

Quantum Computing: Building Concepts

Quantum Concepts 99
S. Manjula Gandhi, S. Gayathri Devi, K. Sathya, K. H. Vani,
and K. Kiruthika

Evolutionary Analysis: Classical Bits to Quantum Qubits 115
Rajiv Pandey, Pratibha Maurya, Guru Dev Singh,
and Mohd. Sarfaraz Faiyaz

Non-silicon Computing with Quantum Superposition Entanglement Using Qubits 131
N. Vidhya, V. Seethalakshmi, and S. Suganyadevi

Quantum Algorithms–Theory and Applications

A Reversible Hybrid Architecture for Multilayer Memory Cell in Quantum-Dot Cellular Automata with Minimized Area and Less Delay 153
Suparba Tapna, Debarka Mukhopadhyay, and Kisalaya Chakrabarti

Quantum Neural Network for Image Classification Using TensorFlow Quantum 171
J. Arun Pandian and K. Kanchanadevi

Quantum Network Architecture and Its Topology 183
Supriyo Banerjee, Biswajit Maiti, and Banaini Saha

Quantum Computing-Enabled Machine Learning for an Enhanced Model Training Approach 201
Jayesh Soni, Nagarajan Prabakar, and Himanshu Upadhyay

Numerical Modeling of the Major Temporal Arcade Using a Quantum Genetic Algorithm 217
Jose A. Soto-Alvarez, Ivan Cruz-Aceves, Arturo Hernandez-Aguirre, Martha A. Hernandez-Gonzalez, and Luis M. Lopez-Montero

Entangled Quantum Neural Network 245
Qinxue Meng, Jiarun Zhang, Zhao Li, Ming Li, and Lin Cui

Quantum Simulation Tools and Demonstrations

Exploring IBM Quantum Experience 265
S. Gayathri Devi, S. Manjula Gandhi, S. Chandia, and P. Boobalaragavan

Quantum Programming on Azure Quantum—An Open Source Tool for Quantum Developers 283
Kumar Prateek and Soumyadev Maity

Survey of Open-Source Tools/Industry Tools to Develop Quantum Software 311
Dhaval Mehta, Amol Ranadive, Jigna B. Prajapati, and Rajiv Pandey

Simulating Quantum Principles: Qiskit Versus Cirq 333
Rajiv Pandey, Pratibha Maurya, Guru Dev Singh, and Mohd. Sarfaraz Faiyaz

Future Direction and Applications

Quantum Machine Learning in Prediction of Breast Cancer 351
Jigna B. Prajapati, Himanshu Paliwal, Bhupendra G. Prajapati, Surovi Saikia, and Rajiv Pandey

Understanding of Argon Fluid Sensor Using Single Quantum Well Through K-P Model: A Bio-medical Application Using Semiconductor Based Quantum Structure 383
Gopinath Palai, Nitin Tripathy, Biswaranjan Panda, and Chandra Sekhar Mishra

A Study on Quantum Cryptography and Its Need 407
Vandani Verma

Evolution of Quantum Machine Learning and an Attempt of Its Application for SDN Intrusion Detection 437
Aakash R. Shinde and Shailesh P. Bendale

Implications of Deep Circuits in Improving Quality of Quantum Question Answering 457
Pragya Katyayan and Nisheeth Joshi

Editors and Contributors

About the Editors

Dr. Rajiv Pandey Senior Member IEEE is a Faculty at Amity Institute of Information Technology, Amity University, Uttar Pradesh, Lucknow Campus India. He has edited various volumes on AI/ML in Edge Computing and Semantic IoT, published by Springer and Elsevier. He possesses a diverse background experience of around 40 years to include 15 years of Industry and 20 years of academics. His research interests include contemporary technologies as Quantum computing, Blockchain and cryptocurrencies, Semantic Web Provenance, Cloud and Big Data, and Data Analytics. He has published more than 50 research papers in Scopus, and other science indexed journals of repute. He has been Session chairs, technical committee member for various IEEE and Elsevier conferences. He has been awarded by DST, Government of India during IISF and is on technical committees of various government and private universities. He is intellectually involved in supervising doctorate research scholars Three scholars have been awarded Ph.D. and three are on the verge of submitting the final thesis. He is also an active contributor in professional bodies like IEEE, IET, and Lucknow Management Association. He has filed patents and received Grants form AICTE India.

Dr. Nidhi Srivastava is currently working as Assistant professor at Amity Institute of Information Technology, Amity University, Uttar Pradesh, Lucknow Campus, India. She has more than 15 years of teaching experience. Dr. Nidhi has a rich and diverse experience in academia. Her research interests include Human–Computer Interaction, Cloud Computing, Semantic Web, and Artificial Intelligence. She has more than 30 publications in various international/national journals and conferences. She has also edited a book on Semantic Web published in book series Studies in Computational Intelligence, Springer. She is Reviewer and Editorial Board Member of several international journals. She has been nominated as Member of Technical Committee and Organizing Committee of many international/national conferences. She is also Member of many international and national bodies like CSI, IAENG, IDES, SDIWC, etc.

Dr. Neeraj Kumar Singh is an Associate Professor in computer science at INPT-ENSEEIHT and member of the ACADIE team at IRIT. Before joining INPT, Dr. Singh worked as a research fellow and team leader at the Centre for Software Certification (McSCert), McMaster University, Canada. He worked as a research associate in the Department of Computer Science at University of York, UK. He also worked as a research scientist at the INRIA Nancy Grand Est Centre, France, where he has received his Ph.D. in computer science. He leads his research in the area of theory and practice of rigorous software engineering and formal methods to design and implementation of safe, secure and dependable critical systems. He has authored/edited two books and one conference proceeding. He published more than 80+ peer-reviewed research articles in well-known journals, books and international conferences. He has been involved in many scientific activities. He is also involved in several research projects on formal methods and system engineering as project leader and as scientific coordinator.

Dr. Kanishka Tyagi is a senior machine learning scientist at Aptiv advances engineering center, California. He received his Bachelor's Degree in Electrical Engineering in 2008 from Pantnagar, India. Later he worked as a Research Associate at the Department of Electrical Engineering, Indian Institute of Technology, Kanpur, with Dr. P. K. Kalra. He received his Master's and Doctoral degree with Dr. Michael Manry in the Department of Electrical Engineering at The University of Texas at Arlington in 2012 and 2017. Prior to Aptiv, he worked at Siemens research, interned in machine learning groups at The MathWorks and Google Research. He has worked as a visiting researcher at Ajou University and Seoul National University. His research interests are radar machine learning, neural networks and hardware machine learning. He received the 2007 and 2011 IEEE CIS Outstanding Student Paper Travel Grant Award and 2013 IEEE CIS Walter Karplus Summer Research Grant award. Dr. Tyagi has taught four undergraduate/graduate courses on Big Data, soft computing and machine learning. Dr. Tyagi is a IEEE senior member, member of IEEE-CIS industrial-academic committee and IEEE standards committee on Explainable AI. He currently serves as an associate editor for IEEE Transaction on Neural Network and Learning Systems and IEEE Transactions on Artificial Intelligence. Dr. Tyagi has published over 30+ papers and filed 17 U.S. patents and trade secrets.

Contributors

Carlos Alberto Alcalde Technical Director for Radar SW, NXP Semiconductors, Grenoble, Auvergne-Rhône-Alpes, France

J. Arun Pandian School of Information Technology and Engineering, Vellore Institute of Technology, Vellore, India

Supriyo Banerjee Kalyani Government Engineering College, Kalyani, West Bengal, India

Shailesh P. Bendale Head of Department (Computer Engineering), NBN Sinhgad School of Engineering, Maharashtra, Pune, India

P. Boobalaragavan Department of Computing, Coimbatore Institute of Technology, Coimbatore, Tamil Nadu, India

Kisalaya Chakrabarti Haldia Institute of Technology, Haldia, India

S. Chandia Department of Computing, Coimbatore Institute of Technology, Coimbatore, Tamil Nadu, India

Ivan Cruz-Aceves CONACYT - Centro de Investigación en Matemáticas (CIMAT), Guanajuato, Gto, México

Lin Cui College of Information Engineering, Suzhou University, Suzhou, China

Mohd. Sarfaraz Faiyaz Amity School of Engineering and Technology, Amity University Uttar Pradesh, Lucknow, India

S. Gayathri Devi Department of Computing, Coimbatore Institute of Technology, Coimbatore, Tamil Nadu, India

Arturo Hernandez-Aguirre Centro de Investigación en Matemáticas (CIMAT), Guanajuato, Gto, México

Martha A. Hernandez-Gonzalez Unidad Médica de Alta Especialidad (UMAE) - Hospital de Especialidades No.1. Centro Médico Nacional del Bajio, IMSS, León, Gto, Mexico

Nisheeth Joshi Department of Computer Science, Banasthali Vidyapith, Banasthali, Rajasthan, India

M. Kalaiarasi ECE Department, Thiagarajar College of Engineering, (An Autonomous Institution Affiliated to Anna University Chennai), Madurai, India

K. Kanchanadevi Department of Computer Science and Engineering, Vel Tech Rangarajan Dr. Sagunthala R&D Institute of Science and Technology, Chennai, India

Pragya Katyayan Department of Computer Science, Banasthali Vidyapith, Banasthali, Rajasthan, India

K. Kiruthika Department of Computing, Coimbatore Institute of Technology, Tamil Nadu, Coimbatore, India

Ming Li The Key Laboratory of Intelligent Education Technology and Application of Zhejiang Province, Zhejiang Normal University, Jinhua, China

Zhao Li Zhejiang University, Link2Do Technology Ltd., Hangzhou, Zhejiang, China

Luis M. Lopez-Montero Unidad Médica de Alta Especialidad (UMAE) - Hospital de Especialidades No.1. Centro Médico Nacional del Bajio, IMSS, León, Gto, Mexico

Biswajit Maiti Maulana Azad College, Kolkata, West Bengal, India

Soumyadev Maity Department of Information Technology, Indian Institute of Information Technology Allahabad, Prayagraj, India

S. Manjula Gandhi Department of Computing, Coimbatore Institute of Technology, Coimbatore, Tamil Nadu, India

Pratibha Maurya Amity Institute of Information Technology, Amity University Uttar Pradesh, Noida, India

Dhaval Mehta School of Engineering and Technology, Navrachana University, Vadodara, Gujarat, India

Qinxue Meng College of Information Engineering, Suzhou University, Suzhou, China

Anushtup Mishra Amity School of Applied Sciences, Amity University Uttar Pradesh, Lucknow, India

Debarka Mukhopadhyay Christ (Deemed to be University), Bengaluru, India

Gopinath Palai Gandhi Institute for Technological Advancement (GITA), Bhubaneswar, Odisha, India

Himanshu Paliwal Shree S K Patel College of Pharmaceutical Education and Research, Ganpat University, Mahesana, Gujarat, India

Biswaranjan Panda IISER Berhampur, Berhampur, Odisha, India

Rajiv Pandey Amity Institute of Information Technology (AIIT), Amity University, Lucknow Campus, Uttar Pradesh, India;
Amity University, Noida, Delhi NCR, India

Nagarajan Prabakar Knight Foundation School of Computing and Information Sciences, Florida International University, Miami, FL, USA

Bhupendra G. Prajapati Shree S K Patel College of Pharmaceutical Education and Research, Ganpat University, Mahesana, Gujarat, India

Jigna B. Prajapati Acharya Motibhai Patel Institute of Computer Studies, Ganpat University, Mahesana, Gujarat, India

Kumar Prateek Department of Information Technology, Indian Institute of Information Technology Allahabad, Prayagraj, India

S. Rajaram ECE Department, Thiagarajar College of Engineering, (An Autonomous Institution Affiliated to Anna University Chennai), Madurai, India

Amol Ranadive School of Business and Law, Navrachana University, Vadodara, Gujarat, India

Banaini Saha University of Calcutta, Kolkata, West Bengal, India

Surovi Saikia Translation Research Laboratory, Department of Biotechnology, Bharathiar University, Coimbatore, Tamil Nadu, India

K. Sathya Department of Computing, Coimbatore Institute of Technology, Tamil Nadu, Coimbatore, India

V. Seethalakshmi Department of ECE, KPR Institute of Engineering and Technology, Coimbatore, India

Chandra Sekhar Mishra Gandhi Institute for Technological Advancement (GITA), Bhubaneswar, Odisha, India

M. James Selsiya ECE Department, Thiagarajar College of Engineering, (An Autonomous Institution Affiliated to Anna University Chennai), Madurai, India

Aakash R. Shinde MS in Quantum Information Sciences, University of Southern California, Los Angeles, CA, USA

Guru Dev Singh Amity Institute of Information Technology, Amity University Uttar Pradesh, Lucknow, India

Jayesh Soni Knight Foundation School of Computing and Information Sciences, Florida International University, Miami, FL, USA

Jose A. Soto-Alvarez Centro de Investigación en Matemáticas (CIMAT), Guanajuato, Gto, México

Pooja Srivastava Amity School of Applied Sciences, Amity University Uttar Pradesh, Lucknow, India

Yogesh K. Srivastava National Institute of Science Education and Research (NISER), Jatni, Khurda, Odisha, India;
Homi Bhabha National Institute, Training School Complex, Mumbai, India

S. Suganyadevi Department of ECE, KPR Institute of Engineering and Technology, Coimbatore, India

Cristian E. Susa-Quintero Department of Physics and Electronics, University of Córdoba, Montería, Colombia

Suparba Tapna Brainware University, Barasat, India

Nitin Tripathy IISER Berhampur, Berhampur, Odisha, India

Kanishka Tyagi Senior Machine Learning Scientist, Aptiv Advance Research Center, Agoura Hills, CA, USA

Himanshu Upadhyay Applied Research Center, Florida International University, Miami, FL, USA

K. H. Vani Department of Computing, Coimbatore Institute of Technology, Tamil Nadu, Coimbatore, India

V. R. Venkatasubramani ECE Department, Thiagarajar College of Engineering, (An Autonomous Institution Affiliated to Anna University Chennai), Madurai, India

Vandani Verma Amity Institute of Applied Sciences, Amity University-125, Noida, India

N. Vidhya Department of ECE, KPR Institute of Engineering and Technology, Coimbatore, India

Jiarun Zhang University of California San Diego, San Diego, USA

Scientific Theory for Quantum

Quantification of Correlations in Quantum States

Cristian E. Susa-Quintero

Abstract Non-classical correlations such as entanglement and quantum discord are essential concepts in the context of quantum technologies nowadays. A plethora of both theoretical and experimental advances on this sort of correlation can be found in the literature. However, it is worth pointing out that new insights on their quantification, detection and application, to name a few, continue to be very welcome. In this chapter, we present a discussion on the quantification of correlations in quantum states. For doing so, we discuss how to quantify the so-called quantum discord in the light of a recently proposed measure inspired by the resource theory for coherence and operationally well-defined in the context of parameter estimation. On the other hand, we also comment on the problem of identifying genuine multipartite correlations in many-body systems. In this case, we discuss a proposal to quantify genuine total (classical plus quantum) correlations at different order $2 \leq k \leq N$.

1 Introduction

Since the seminal paper from Einstein, Podolsky and Rosen more than 90 years ago [1], quantum entanglement has become the cornerstone of the second quantum revolution we are living in today. However, at the beginning of the nineteenth century, a new class of non-classicality more general than entanglement arose. Olliver and Zurek [2] as well as Henderson and Vedral [3] discovered independently the notion of quantum discord. Despite the great progress in the definition, quantification, interpretation and detection of quantum discord ([4, 5]), there still exist open questions that allow contributions on this quantum property. For instance, new definitions of discord with a well-motivated operational interpretation and quantification of genuine multipartite correlations are two important subjects of active research interest that we discuss from our own perspective. In particular, after a brief comment on entanglement in Sect. 2, we present in Sect. 3 an operational definition of quantum

C. E. Susa-Quintero (✉)
Department of Physics and Electronics, University of Córdoba, 230002 Montería, Colombia
e-mail: cristiansusa@correo.unicordoba.edu.co

R. Pandey et al. (eds.), *Quantum Computing: A Shift from Bits to Qubits*,
Studies in Computational Intelligence 1085,
https://doi.org/10.1007/978-981-19-9530-9_1

discord that we recently proposed within the context of quantum metrology and that is related to local coherence [6]. Then, we deal with the multipartite scenario in Sect. 4 by discussing a way of quantifying genuine multipartite correlations [7]. We do not give a particular definition for quantum discord, but we consider the quantification of total (classical + quantum) correlations. Section 5 is devoted to the conclusion and remarks.

2 Quantum States and Entanglement

The quantum state of a physical system $\mathcal{S}$ with Hilbert space $\mathcal{H}_S^d$ of dimension d can be represented by a density operator:

$$\rho_S \equiv \rho = \sum_i p_i |\psi_i\rangle\langle\psi_i| , \tag{1}$$

where probabilities p_i satisfy $\sum_i p_i = 1$. We say that ρ describes a statistical mixture of pure states. From Eq. (1), it is clear that ρ is pure if and only if just one probability equals unity $p_k = 1$ and $p_{i\neq k} = 0$. Then, a pure state can be written as $\rho = |\psi\rangle\langle\psi|$, for any vector state $|\psi\rangle \in \mathcal{H}_S^d$.

For composite systems,[1] the total Hilbert space is structured by the tensor product of the Hilbert spaces of the subsystems:

$$\mathcal{H}_S^{\Pi_{i=1}^N d_i} = \mathcal{H}_{S_1}^{d_1} \otimes \mathcal{H}_{S_2}^{d_2} \otimes \cdots \otimes \mathcal{H}_{S_N}^{d_N} .$$

Quantum mechanics evidenced that counterintuitive phenomena occur when composing quantum systems. In the bipartite case, the superposition principle and quantum coherence are responsible for *magic* and *useful* phenomena such as the well-known entanglement nowadays. The simplest scenario to understand this is given by the two-level (qubit) system. Making use of the computational basis $\{|0\rangle, |1\rangle\}$ for any qubit, the so-called Bell's entangled states lead:

$$|\Phi^{\pm}\rangle = \frac{1}{\sqrt{2}}(|00\rangle \pm |11\rangle), \quad |\Psi^{\pm}\rangle = \frac{1}{\sqrt{2}}(|01\rangle \pm |10\rangle). \tag{2}$$

Entangled states are not able to be written in terms of the subsystem states, i.e., we say that ρ is entangled if $\rho \neq \rho_{[1]} \otimes \rho_{[2]}$, where $\rho_{[i]} \equiv \rho_S$ stands for a state of the subsystem i. Definition and quantification of entanglement have been arguably investigated with many contributions from both theoretical and experimental viewpoints. As entanglement is not the main topic of this chapter, we encourage interested

[1] A system with two parts (subsystems) is usually referred to as a *bipartite* system. Along the chapter, we make use of the term *multipartite* to refer to many-body systems (generally with more than two parts).

readers to extend the knowledge about entanglement with further literature (e.g.—but not limited to—[8–10]).

From a different perspective, quantum correlations more general than entanglement can also emerge in quantum systems. Perhaps the most general way of quantumness known so far is the so-called quantum discord [2, 3]. This kind of non-classicality has been exploited for more than the last two decades such that a plethora of quantum discord quantifiers have been developed and discussed [5]. More recently, quantum discord, entanglement and other quantum properties like coherence have all been described in the same framework of *quantum resource theory* (see, e.g., [11]). It is remarkable that in spite of the big progress on quantum correlations (entanglement included), new insights may still arise, for instance, in the context of multipartite systems.

3 Quantum Correlations Beyond Entanglement as Dephasing Sensitivity

Since the concept of quantum discord arose, many strategies have been developed to quantify and operationally interpret this form of quantumness (the reader is encouraged to read for instance [4, 5, 12] for historical as well as technical issues). In this section, we discuss a recently proposed quantifier of discord-like correlations by testing the sensitivity of quantum bipartite states to local dephasing [6].

Considering a finite-dimensional system, the quantum coherence with respect to a basis $\{|i\rangle\}$ in a state described by a density matrix ρ may be given in terms of the off-diagonal elements $\rho_{ij}, i \neq j$, for example, the distance of the state of interest to the closest diagonal state. Quantifying coherence in the framework of quantum resource theory has been formalized [13].

In the scenario of parameter estimation, we have proposed a measure of coherence as the ability of the system to act as a useful probe to estimate a dephasing channel, in other words, how sensitive a quantum system is to a dephasing process. Let Φ_i^p be a completely positive trace-preserving (CPTP) map modelling the decoherence process:

$$\begin{aligned} \Phi_i^p(\rho) &:= \rho \mapsto \rho_p = p\rho_d + (1-p)\rho, \qquad 0 < p < 1 \\ \rho_d &= \sum_i \langle i|\,\rho\,|i\rangle\,|i\rangle\,\langle i|\,, \end{aligned} \tag{3}$$

where ρ_d is the matrix given by the diagonal entries of the state with respect to the basis $\{|i\rangle\}$.

The success of the task of estimating the parameter p may be quantified by the Quantum Fisher Information (QFI)

$$F(\rho, \Phi_i^p) = \mathrm{Tr}\left[\rho_p L_p^2\right] = \mathrm{Tr}\left[\partial_p \rho_p L_p\right], \tag{4}$$

where $\partial_p \rho_p = \frac{L_p\rho_p+\rho_p L_p}{2}$ is the Lyapunov matrix equation for the symmetric logarithmic derivative (SLD) L_p. The inverse of the QFI determines the lower bound of the variance for an estimator of p according to the Cramér-Rao bound, $\mathrm{Var}(O_p) \geq [\mu F(\rho, \Phi_i^p)]^{-1}$, where the estimator O_p satisfies $\mathrm{Tr}(O_p \rho_p) = p$ and μ is the number of copies of ρ_p used for estimating the parameter [14]. In the eigenbasis of $\rho_p = \sum_n \lambda_n |\psi_n\rangle\langle\psi_n|$, the solution for the SLD is

$$L_p = 2\sum_{nm} \frac{\langle\psi_m|\, \partial_p\rho_p \,|\psi_n\rangle}{\lambda_m + \lambda_n} |\psi_m\rangle\langle\psi_n|\,, \qquad \lambda_m + \lambda_n \neq 0, \tag{5}$$

and the QFI reduces to

$$F(\rho, \Phi_i^p) \equiv F(\rho, \rho_p) = 2\sum_{nm} \frac{\left|\langle\psi_m|\, \partial_p\rho_p \,|\psi_n\rangle\right|^2}{\lambda_m + \lambda_n}. \tag{6}$$

We have proved in [6] that the QFI given by Eq. (6) is a measure of coherence with respect to the basis $\{|i\rangle\}$, and for any value of the parameter $p \in (0, 1)$. According to the accepted criteria for a measure of coherence [15], $F(\rho, \Phi_i^p)$ satisfies the following properties: (i) Faithfulness: $F(\rho, \Phi_i^p) = 0$ if and only if $\rho = \rho_d$. (ii) Monotonicity under strictly incoherent operations (SIO): $F(\mathcal{E}(\rho), \Phi_i^p) \leq F(\rho, \Phi_i^p)$ for any trace-preserving SIO $\mathcal{E}(\rho) = \sum_{k,j,i} c_{k,j} c_{k,i}^* \langle j|\, \rho \,|i\rangle\, |f_k(j)\rangle\langle f_k(i)|$.[2] And (iii) Convexity: for any ensemble $\{p_k, \rho_k\}$ it turns out $F(\sum_k p_k\rho_k, \Phi_i^p) \leq \sum_k p_k F(\rho_k, \Phi_i^p)$.

Let us discuss the case of a qubit with a density operator:

$$\rho = \begin{pmatrix} \rho_{00} & \rho_{01} \\ \rho_{10} & \rho_{11} \end{pmatrix},$$

where $\rho_{00} + \rho_{11} = 1$ and $\rho_{10} = \rho_{01}^*$. Considering the decoherence process with respect to the computational basis $\{|0\rangle, |1\rangle\}$, known as phase flip channel $\rho_p = (1-p)\rho + p\sigma_z\rho\sigma_z$, our measure of coherence reads:

$$F(\rho, \rho_p) = \frac{4|\rho_{01}|^2}{1 - (1-2p)^2|\rho_{01}|^2/\rho_{00}\rho_{11}}. \tag{7}$$

Figure 1 shows the behaviour of our proposed measure of coherence for the qubit case Eq. (7): (main plot) F in terms of the off-diagonal element ρ_{01}; (inset) maximal coherence (given for pure states) in terms of the dephasing parameter p for the maximum value $|\rho_{01}| = 1/2$. It is worth noting that the divergence for the case $p = 0$ is due to the sudden change of rank for the pure state density operator introduced by

[2] The SIO is assumed to have the Kraus representation $\mathcal{E}(\rho) = \sum_k K_k \rho K_k^\dagger$ with $K_k^\dagger = \sum_i c_{k,i}^* |i\rangle\langle f_k(i)|$, and $f_k(i)$ any permutation of the set $\{0, 1, \ldots, d-1\}$. It turns out that $\langle l|\, K_k\rho K_k^\dagger \,|l\rangle = \sum_{j,i} c_{k,j} c_{k,i}^* \langle j|\, \rho \,|i\rangle$ with vanishing off-diagonal elements as $K_k^\dagger |j\rangle = \sum_i c_{k,i}^* |i\rangle \langle f_k(i)|j\rangle$ has only one nonzero term for a given k.

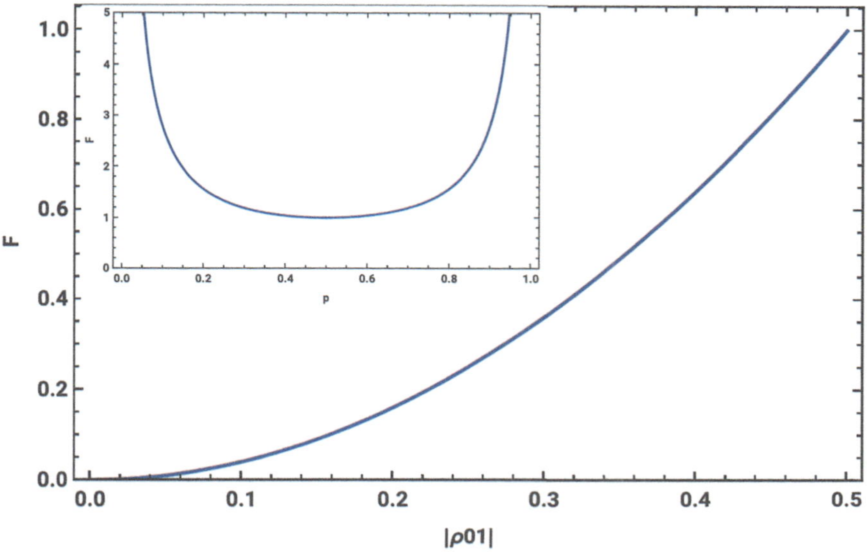

Fig. 1 (Main) Measure of coherence for $p = 1/2$ in terms of the off-diagonal element ρ_{01}. (Inset) Behaviour of F for pure states as a function of p. It is worth noting that each value of p defines a measure of coherence

the dephasing channel. On the other hand, the case $p = 1$ immediately reduces to $F = 0$ as the initial state is diagonal on the basis $\{|i\rangle\}$ (Eq. (3)).

As an application of the measure of coherence proposed in Eq. (6), we present an interpretation of the non-classicality of composite quantum systems as the sensitivity of a quantum state to local dephasing. For a bipartite system in the state ρ_{AB}, we have proposed a quantifier of quantum discord as follows:

$$D(\rho, \Phi_p^A) = \min_U F(\rho, \Phi_p^U \otimes I), \tag{8}$$

where the local dephasing acting on subsystem A is given on the basis $\{U\,|i\rangle\}$, and the minimum is taken over all the possible bases. In [6], we proved that

1. Faithfulness: $D(\rho, \Phi_p^A) = 0$ if and only if $\rho = \sum_i p_i\,|i\rangle\langle i| \otimes \rho_B^{(i)}$.
2. Monotonicity under local operations on the unperturbed subsystem B: $D(\mathcal{E}_B(\rho), \Phi_p^A) \leq D(\rho, \Phi_p^A)$, for any CPTP map $\mathcal{E}_B$.
3. Entanglement monotonicity for pure states: $\sum_i p_i D(|\psi_i\rangle\langle\psi_i|\,, \Phi_p^A) \leq D(|\psi\rangle\langle\psi|\,, \Phi_p^A)$, for a pure state $|\psi\rangle$ that can be transformed to an ensemble $\{p_i;\,|\psi_i\rangle\}$.

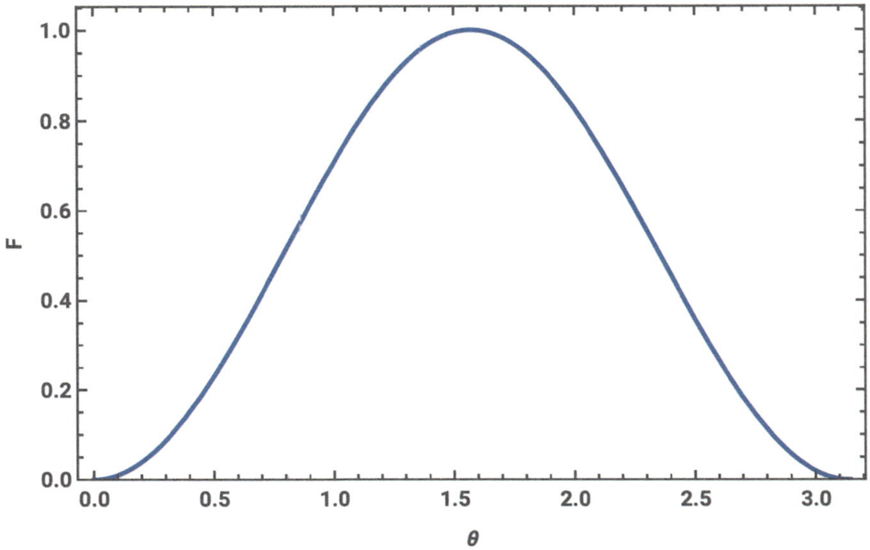

Fig. 2 Measure of quantum discord D for $p = 1/2$ in terms of the state parameter $\theta \in [0, \pi]$

4. Monotonicity under extendible commutativity-preserving operations (ECPO): [3] $D(\mathcal{E}_A(\rho), \Phi_p^A) \leq D(\rho, \Phi_p^A)$, for any ECPO $\mathcal{E}_A$ on subsystem A.

Let us now discuss a bit our measure of quantum correlations for the case of two qubits. For the set of partially entangled states $|\theta\rangle = \cos(\theta/2)\,|00\rangle + \sin(\theta/2)\,|11\rangle$, our measure of quantum discord is shown in Fig. 2 as a function of θ for $p = 1/2$. As expected, and according to criterion 3, F is an entanglement monotone for pure states. F achieves its maximum value for $\theta = \pi/2$ when the state $|\theta\rangle = |\Phi^+\rangle$.

The extension of this quantum discord quantifier to multipartite systems could be an interesting matter of future research.

[3] In the context of quantum resource theories, the free operations for quantum discord is still subject to debate. The accepted set so far is the set of commutativity-preserving operations (CPO) [5]; $[\mathcal{E}(\rho), \mathcal{E}(\sigma)] = 0$, for any CPO $\mathcal{E}$ and any density operators such that $[\rho, \sigma] = 0$. In [6], we introduced the extension ECPO as operations defined as follows: for any extension ρ_{SA} of ρ_S throughout an ancillary system A; there exists an operation $\mathcal{F}_{SA}(\rho_{SA})$ such that $\mathcal{E}_S = \mathrm{Tr}_A[\mathcal{F}_{SA}]$. We say that $\mathcal{F}_{SA}(\rho_{SA})$ is free for any free $\mathcal{E}_S(\rho_S)$.

4 Genuine Multipartite Total Correlations

We devote this section to discuss a proposed strategy to quantify the genuine[4] total correlations at any size of multipartite quantum states [7]. In short, our approach compares the given quantum state of N systems with the set of tensor product states at any order, i.e., among any $1 \leq k \leq N$ parts. In this sense, classical and quantum correlations are both caught by our measure of genuine multipartite correlations.

To define the correlations of order higher than k, $2 \leq k \leq N-1$, for an N-partite state represented by the density operator ρ_N, we firstly write the set of tensor products describing up to k-partite clusters

$$\underline{\mathcal{P}}_k = \left\{ \sigma_N = \bigotimes_{i=1}^{m} \sigma_{k_i};\ \forall k_i : \sum_{i=1}^{m} k_i = N,\ k = \max\{k_i\} \right\} \tag{9}$$

where σ_{k_i} is the quantum state of a k_i-partite cluster. It is worth noting that $\underline{\mathcal{P}}_k$ satisfies the hierarchy

$$\underline{\mathcal{P}}_1 \subset \underline{\mathcal{P}}_2 \subset \cdots \subset \underline{\mathcal{P}}_{N-1} \subset \underline{\mathcal{P}}_N \tag{10}$$

where $\underline{\mathcal{P}}_N$ gives the whole Hilbert space of the N-partite system. The hierarchy in Eq. (10) can be sequentially constructed as follows.

For $k = 1$, we have

$$\underline{\mathcal{P}}_1 = \left\{ \sigma = \bigotimes_{i=1}^{N} \sigma_{k_i \,|\, \max\{k_i\}=1} \equiv \sigma_{[1]} \otimes \sigma_{[2]} \otimes \cdots \otimes \sigma_{[N]} \right\}, \tag{11}$$

where $\sigma_{[i]}$ is a state of the i-th single part. Hence, $\underline{\mathcal{P}}_1$ represents the set of tensor product states of the N single subsystems.

Now, for $k = 2$: $\underline{\mathcal{P}}_2 = \underline{\mathcal{P}}_1 \cup \mathcal{B}_2$, where

$$\mathcal{B}_2 = \left\{ \sigma = \bigotimes_{i=1}^{m} \sigma_{k_i \,|\, \max\{k_i\}=2} \otimes \bigotimes_{j=1}^{n} \sigma_{k_j \,|\, \max\{k_j\}=1} : \sum_{i=1}^{m} k_i + \sum_{j=1}^{n} k_j = N \right\}, \tag{12}$$

leads to the set of tensor product states with at least one irreducible two-partite state. A state $\sigma_{k_i \,|\, \max\{k_i\}=2}$ must be understood as any $\sigma_2 \neq \sigma_{[i]} \otimes \sigma_{[j]}$ for arbitrary single subsystems i, j.

Following a similar structure for higher orders k, it is straightforward to see that $\underline{\mathcal{P}}_k = \underline{\mathcal{P}}_{k-1} \cup \mathcal{B}_k$, where $\mathcal{B}_k$ is the set of tensor product states with at least one joint (irreducible) k-partite state

---

[4] A genuine multipartite correlation is understood as a joint correlation that cannot be reduced to properties of the composing parts. For instance, a genuine tripartite correlation cannot be described in terms of either two-partite correlations or single-part properties.

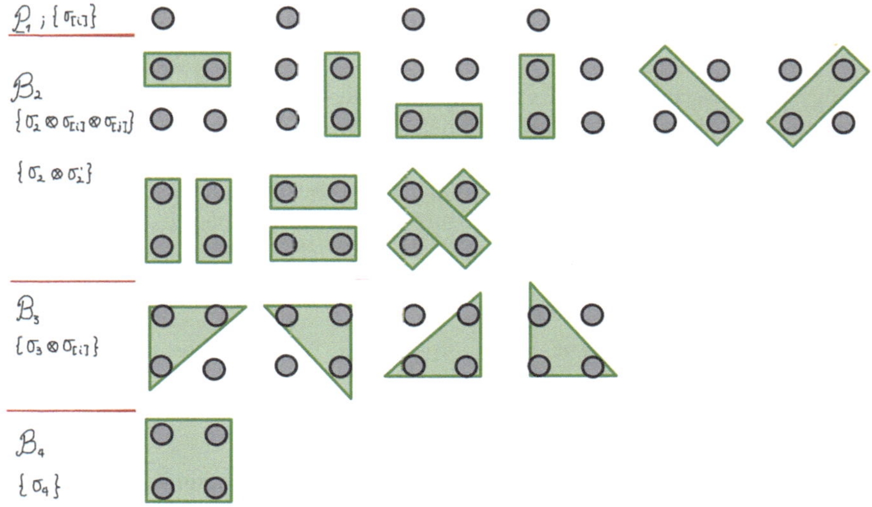

Fig. 3 Hierarchy scheme of sets $\underline{\mathcal{P}}_k$ for $N = 4$. Circles represent single systems $\mathcal{S}_i$. Rectangles identify irreducible two-partite states, while triangles represent irreducible correlations among three subsystems. The biggest set also contains irreducible four-partite states here represented by the big square

$$\mathcal{B}_k = \left\{ \sigma = \bigotimes_{i=1}^{m} \sigma_{k_i \,|\, \max\{k_i\}=k} \otimes \bigotimes_{j=1}^{n} \sigma_{k_j \,|\, \max\{k_j\}<k} \;:\; \sum_{i=1}^{m} k_i + \sum_{j=1}^{n} k_j = N \right\}, \quad (13)$$

In Fig. 3, a particular example of the hierarchy of $\underline{\mathcal{P}}_k$ is shown for $N = 4$. From Eq. (11), we have the set of tensor product states of the form $\{\sigma_{[1]} \otimes \sigma_{[2]} \otimes \sigma_{[3]} \otimes \sigma_{[4]}\}$. Thus, for constructing $\underline{\mathcal{P}}_2$ we just add the set of all irreducible two-partite states (Eq. (12)): $\{\sigma_2 \otimes \sigma_{[i]} \otimes \sigma_{[j]}\} \cup \{\sigma_2 \otimes \sigma_2'\}$. Here, σ_2 must be understood as a joint irreducible state between any two subsystems: $\sigma_2 \equiv \sigma_{[ij]} \neq \sigma_{[i]} \otimes \sigma_{[j]}$. These pairwise states are represented by rectangles in Fig. 3. In a similar way, the set $\mathcal{B}_3$ (connections represented by triangles) contains all the states of the form $\{\sigma_3 \otimes \sigma_{[i]}\}$ with $\sigma_3 \equiv \sigma_{[ijk]}$ which the set $\underline{\mathcal{P}}_3$ is constructed with. Finally, the biggest set in this example also contains irreducible four-partite states $\{\sigma_4 \equiv \sigma_{[1234]}\}$.

From an information-theoretic perspective, the correlations higher than order k can be defined as the distance of ρ_N with respect to the set $\underline{\mathcal{P}}_k$ [7]:

$$D_k(\rho_N) = \min_{\sigma_N \in \underline{\mathcal{P}}_k} D(\rho_N, \sigma_N), \quad (14)$$

where $D(\rho, \sigma)$ is an appropriate distance measure; $D(\rho, \sigma) \geq 0$ with equality if and only if $\rho = \sigma$ and $D(\mathcal{E}(\rho), \mathcal{E}(\sigma)) \leq D(\rho, \sigma)$ for any CPTP map $\mathcal{E}$ and arbitrary states ρ, σ. Genuine k-partite correlations can then be defined as

$$G_k(\rho_N) = D_{k-1}(\rho_N) - D_k(\rho_N). \tag{15}$$

As in Sect. 3 for quantum discord, a measure of genuine multipartite correlations must satisfy some accepted criteria that we resume as follows [16] (here, we write the criteria for Eq. (14) but it must be clear that they are also satisfied by Eq. (15)):

1. Faithfulness: $D_k(\rho_N) = 0$ if and only if $\rho_N \in \underline{\mathcal{P}}_k$. We then say that ρ_N does not have k-partite correlations.
2. Adding a disjoint n-partite system does not create multipartite correlations higher than order n: $D_n(\rho_N \otimes \rho_n) = D_n(\rho_N) = 0$ for any ρ_n.
3. Local CPTP maps cannot create correlations higher than any order k: $D_k(\mathcal{E}(\rho_N)) \leq D_k(\rho_N)$ for any local CPTP $\mathcal{E}(\rho_N) = [\mathcal{E}_1 \otimes \mathcal{E}_2 \otimes \cdots \otimes \mathcal{E}_N](\rho_N)$.
4. Partial trace over n subsystems cannot increase correlations higher than order $k < N - n$; $D_k(\rho_{N-n}) \leq D_k(\rho_N)$ for any $\rho_{N-n} = \mathrm{Tr}_n(\rho_N)$.
5. Distilling n subsystems by fine-graining does not create correlations higher than order $k + n$; $D_{k+n}(\rho_{N+n}) = D_k(\rho_N) = 0$.

Definitions in Eqs. (14) and (15) imply an optimization procedure that can be difficult to deal with depending on the choice of $D(\rho_N, \sigma_N)$. We have shown in [7] that by employing the relative entropy between quantum states, for which the closest set of tensor product states is constructed by the tensor product of the marginals of ρ_N, both Eqs. (14) and (15) reduce to nice analytical formulas.

Coming back to the case of qubits, let us consider a system of N qubits in a noisy state of the form $\rho_{\mathrm{GHZ}} = \frac{p}{2^N} I + (1-p)\,|\mathrm{GHZ}\rangle\langle\mathrm{GHZ}|$, where $|\mathrm{GHZ}\rangle = \frac{1}{\sqrt{2}}\left(|0_{[1]}0_{[2]}\cdots 0_{[N]}\rangle + |1_{[1]}1_{[2]}\cdots 1_{[N]}\rangle\right)$ is the well-known Greenberger-Horne-Zeilinger (GHZ) state [17]. ρ_{GHZ} is invariant under subsystem permutations. Making use of the relative entropy $S(\rho||\sigma) = -S(\rho) - \mathrm{Tr}\,(\rho \log \sigma)$ for arbitrary states ρ and σ, where $S(\rho) = -\,\mathrm{Tr}(\rho \log \rho)$ is the von Neumann entropy, our measure of multipartite correlations Eq. (14) reduces to

$$D_k(\rho_N) = \lfloor N/k \rfloor S(\rho_k) + (1 - \delta_{N \bmod k, 0}) S(\rho_{N \bmod k}) - S(\rho_N), \tag{16}$$

where $\lfloor\cdot\rfloor$ stands for the floor integer part and N mod k is the remainder of N when divided by k. Hence, Eq. (16) allows to easily compute the genuine multipartite correlations in Eq. (15). In Fig. 4, the behaviour of some genuine multipartite correlations for the case of $N = 127$ is shown: (left panel) $G_3(\rho_{\mathrm{GHZ}})$ (thick), $G_4(\rho_{\mathrm{GHZ}})$ (dashed), $G_5(\rho_{\mathrm{GHZ}})$ (dot-dashed) and $G_6(\rho_{\mathrm{GHZ}})$ (dotted). On the other hand, (right panel) $G_{126}(\rho_{\mathrm{GHZ}})$ (thick), $G_{125}(\rho_{\mathrm{GHZ}})$ (dashed), $G_{124}(\rho_{\mathrm{GHZ}})$ (dot-dashed) and $G_{123}(\rho_{\mathrm{GHZ}})$ (dotted). Note that genuine correlations of small order achieve higher values than those of orders close to N. This is understood by remembering that the considered state has genuine N-partite correlations, thus, the *distance* of this state with respect to the different sets $\underline{\mathcal{P}}_k$ in general increases when $k \to 1$.

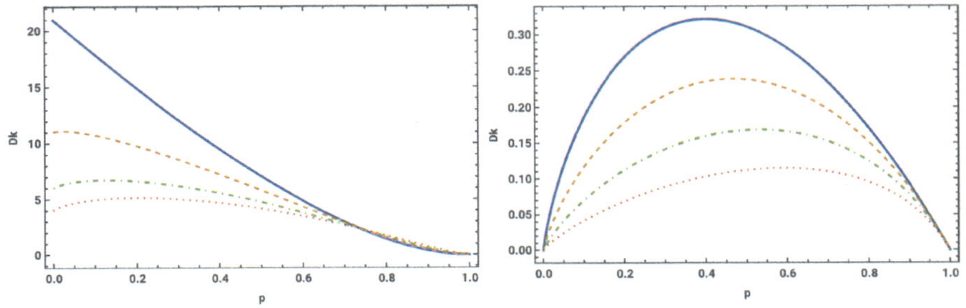

Fig. 4 Genuine multipartite correlations of order: (left) $k = 3, 4, 5, 6$ and (right) $k = 123, 124, 125, 126$ for a system of 127 particles in a GHZ state mixed with white noise. Noisy parameter $0 \leq p \leq 1$

5 Conclusion and Remarks

We have discussed a couple of recent contributions: (i) an operationally motivated definition of quantum discord which is interpreted as the sensitivity of quantum states to be locally dephased. An interesting further work on this could be to propose an extension of our discord measure to the case of multipartite systems. (ii) An information-theoretic measure of genuine multipartite total correlations that allows us to quantify not only genuine multipartite correlations but also their pattern complexity in terms of the size of the system. An interesting further work, related to contribution (i), could be to developing a measure of genuine multipartite purely quantum correlations.

Acknowledgements I thank all my colleagues and friends who have been part of the development of these projects. In particular, I would like to thank Davide Girolami for his hospitality and all the knowledge shared with me. I acknowledge partial support from University of Córdoba.

References

1. A. Einstein, B. Podolsky, N. Rosen, Can quantum-mechanical description of physical reality be considered complete? Phys. Rev. **47**, 777–780 (1935)
2. H. Ollivier, W.H. Zurek, Quantum discord: A measure of the quantumness of correlations. Phys. Rev. Lett. **88**, 017901 (2001)
3. L. Henderson, V. Vedral, Classical, quantum and total correlations. J. Phys. A: Math. Gen. **34**(35), 6899–6905 (2001)
4. K. Modi, A. Brodutch, H. Cable, T. Paterek, V. Vedral, The classical-quantum boundary for correlations: discord and related measures. Rev. Mod. Phys. **84**, 1655–1707 (2012)
5. G. Adesso, T.R. Bromley, M. Cianciaruso, Measures and applications of quantum correlations. J. Phys. A: Math. Theoret. **49**(47), 473001 (2016)
6. B. Yadin, P. Bogaert, C.E. Susa, D. Girolami, Coherence and quantum correlations measure sensitivity to dephasing channels. Phys. Rev. A **99**, 012329 (2019). (Jan)

7. D. Girolami, T. Tufarelli, C.E. Susa, Quantifying genuine multipartite correlations and their pattern complexity. Phys. Rev. Lett. **119**, 140505 (2017). (Oct)
8. R. Horodecki, P. Horodecki, M. Horodecki, K. Horodecki, Quantum entanglement. Rev. Mod. Phys. **81**, 865–942 (2009). (Jun)
9. S. Haddadi, M. Bohloul, A brief overview of bipartite and multipartite entanglement measures. Int. J. Theor. Phys. **57**, 3912–3916 (2018). (Dec)
10. M. Navascués, E. Wolfe, D. Rosset, A. Pozas-Kerstjens, Genuine network multipartite entanglement. Phys. Rev. Lett. **125**, 240505 (2020). (Dec)
11. E. Chitambar, G. Gour, Quantum resource theories. Rev. Mod. Phys. **91**, 025001 (2019). (Apr)
12. A. Bera, T. Das, D. Sadhukhan, S.S. Roy, A. Sen(De), U. Sen, Quantum Discord Its allies: Rev. Recent Progr. **81**(2), 024001 (2017)
13. A. Streltsov, G. Adesso, M.B. Plenio, Colloquium: quantum coherence as a resource. Rev. Mod. Phys. **89**, 041003 (2017). (Oct)
14. G.A. Matteo, Paris. Quantum estimation for quantum technology. Int. J. Quantum Inform. **07**(supp01), 125–137 (2009)
15. T. Baumgratz, M. Cramer, M.B. Plenio, Quantifying coherence. Phys. Rev. Lett. **113**, 140401 (2014). (Sep)
16. C.H. Bennett, A. Grudka, M. Horodecki, P. Horodecki, R. Horodecki, Postulates for measures of genuine multipartite correlations. Phys. Rev. A **83**, 012312 (2011). (Jan)
17. D.M. Greenberger, M.A. Horne, A. Zeilinger, *Going Beyond Bell's Theorem* (Springer, Netherlands, 1989), pp. 69–72

From Quantum Mechanics to Quantum Computing

Pooja Srivastava, Anushtup Mishra, and Yogesh K. Srivastava

Abstract Quantum computing is an upcoming technology which has the potential to revolutionize our lives with its unprecedented computating powers. A quantum computer works on the principles of Quantum Mechanics, a branch of Physics. Quantum mechanics is the universal theory of nature which governs the microworld and underlies all the physical phenomena. In this chapter, we outline how the basic quantum concepts like probability amplitudes, linear superposition principle, entanglement, etc can be harnessed to perform computations. We discuss the basic concepts of quantum mechanics, starting with Hilbert space methods and describe how quantum bits or qubits can be used.

Keywords Quantum mechanics · Linear superposition principle · Entanglement

1 Introduction

> "Quantum mechanics, that mysterious, confusing discipline, which none of us really understands, but which we know how to use."
>
> –Murray Gell-Mann

We live in a macroscopic world which is governed by the laws of classical physics. However, to get acquainted with the microscopic world we need to drive into another

P. Srivastava (✉) · A. Mishra
Amity School of Applied Sciences, Amity University Uttar Pradesh,
Malhore Railway Station Road, Lucknow 226010, India
e-mail: psrivastava5@lko.amity.edu

Y. K. Srivastava
National Institute of Science Education and Research (NISER),
Bhubaneswar, P.O., 752050 Jatni, Khurda, Odisha, India
e-mail: yogeshs@niser.ac.in

Homi Bhabha National Institute, Training School Complex,
Anushakti Nagar, 400085 Mumbai, India

R. Pandey et al. (eds.), *Quantum Computing: A Shift from Bits to Qubits*,
Studies in Computational Intelligence 1085,
https://doi.org/10.1007/978-981-19-9530-9_2

world, which is tiny and more alien than anything we can intuitively understand: the subatomic world of quantum mechanics. Quantum Mechanics (QM) is a theory provides us with the formalism to describe the microscopic laws which govern our universe at the atomic and subatomic scale. In addition to the explanatory and predictive power, this theory has also been found very useful in various applications, for example, LASER, integrated circuits etc. It is the foundational theory of quantum chemistry, quantum field theory, quantum technology, and many others. One of the most exciting application of quantum mechanics, with profound but yet to be realised applications, is the field of quantum computing. Quantum computing generalises the notions of computer science to take into account the laws of quantum mechanics. The study of quantum computing comes under quantum information science. Quantum information science is an interdisciplinary field that seeks to understand the analysis, processing, and transmission of information using laws of quantum mechanics.

In this chapter, we discuss the basics of quantum mechanics, with special attention to those ideas which underlie the quantum computing. Since it's not possible to discuss all of quantum mechanics, even those parts relevant to quantum computing in one chapter, we have been very selective in choice of topics as well as in their treatment. Quantum computing applications mostly require systems with finite number of degrees of freedom and so we have restricted ourselves to such systems (or in quantum mechanical terminology, to systems with finite dimensional Hilbert spaces). There is not enough space to thoroughly discuss any particular topic and so we have assembled various notions and formulae which are useful in applications of quantum mechanics to computing, as discussed in the rest of the book. More details about any particular topic discussed here can be obtained from the references listed at the end. Given the introductory nature of this chapter, references are to books rather than to the original literature.

2 Quantum State

One of the most mysterious concept in Quantum Mechanics is that of quantum state. To understand it, let's first go back a step and recall how we specify state in classical physics. In classical mechanics, we can specify the state of a particle at a given time by giving it's position and velocity (as in Lagrangian Mechanics) or it's position and momentum (as in Hamiltonian Mechanics). We can represent this in phase space as $(x(t), p(t))$ at a give time t. Every dynamical variable Ω in classical mechanics is a function of x, p i.e $\Omega(x, p)$. An important feature of classical physics is that if a particle is in a state given by x, p then the measurement of a dynamical variable Ω will give $\Omega(x, p)$. The state of the system will remain unaffected by the act of measurement. We will see later that this is not true in quantum mechanics.

Since the position and momentum in classical physics are continuous variables, this is an example of a state labelled by a continuous variable. The Hamilton or Lagrange equations of motion, then specify how the classical state of the system changes with time. For example, in phase space Hamilton equations are

$\dot{x} = \frac{\partial H}{\partial p}$, $\dot{p} = -\frac{\partial H}{\partial x}$ for a single particle described by the classical Hamiltonian $H(x, p)$. We have described a classical system of one particle in one dimension but it can be straightforwardly generalised to any number of particles in any number of dimensions. For example, a system of N particles in 3 dimensions is described by a phase space of $6N$ dimensions.

We will now briefly describe some general features of the quantum state and how it differs from the classical state. We will elaborate on these features in later sections. So one shouldn't worry if some ideas are not fully clear.

The quantum description differs from the classical description mentioned above in two crucial ways. Firstly, the quantum state $\psi(t)$ lives in an abstract space, called Hilbert space, which is not directly connected to physical space. Secondly, given a state $\psi(t)$, quantum mechanics provides *statistical* predictions about the outcomes of all observations on the system described by $\psi(t)$. In the Hilbert space, quantum state is abstractly represented, using Dirac's notation, as $|\psi(t)\rangle$. The quantum state can be concretely represented in various ways- for example, by a function $\psi(x, t)$, called configuration space wavefunction or by a function $\psi(p, t)$ called momentum space wavefunction. But phase space as a whole can not be used to represent quantum state i.e. we can not write a function on phase space $\psi(x, p, t)$. This is related to the uncertainty principle as we will discuss later. Schrödinger's equation describes how the quantum state changes with time. In the abstract Hilbert space, the Schrödinger equation is given by $i\hbar\frac{d}{dt}|\Psi(t)\rangle = \hat{H}|\Psi(t)\rangle$. Here $\hat{H}$ is the quantum Hamiltonian and $\hbar$ is the Planck constant. This will be discussed in more detail later on.

3 Hilbert Space in Quantum Mechanics

In this section, we get into the basic arena where the quantum states live i.e Hilbert space. We will assume a basic familiarity with the elementary ideas of linear algebra, particularly the basics of linear vector spaces and matrix theory. Since our subject of interest is quantum mechanics, we will use the notations used by physicists (called Dirac's bra and ket notation) instead of the usual mathematical notation.

A Hilbert space is a complex linear vector space with inner product. Let's discuss the Hilbert space terminology a bit. Recall that a linear vector space $\mathcal{V}$ over the field of complex numbers is a collection of objects $|\psi_1\rangle, |\psi_2\rangle, ...,$ called vectors such that one can form complex linear combinations of the form $c_1|\psi_1\rangle + c_2|\psi_2\rangle$ where c_1, c_2 are complex numbers and $|\psi_1\rangle, |\psi_2\rangle$ are vectors. What this means is that one can add vectors and multiply vectors by complex numbers (not just two but an arbitrary number i.e. we can form the vector $|\Psi\rangle = \sum_{n=1}^{N} c_n|\psi_n\rangle$ for any N). Inner product on a Hilbert space is an operation which takes two vectors, say $|v\rangle, |w\rangle$ and gives a complex number, denoted by $\langle w|v\rangle$. Inner product is same as the usual dot product of vector algebra except that since we now have a complex vector space, inner product satisfies $\langle w|v\rangle = (\langle v|w\rangle)^*$ under complex conjugation operation $*$. It's useful to consider inner product between two vectors in an alternative (but equivalent) way. For each vector $|v\rangle$ (called ket vector) in the Hilbert space, we can define an element in the dual space, denoted by $\langle v|$ and called a bra vector. To each ket $|v\rangle$ there corresponds a unique bra vector $\langle v|$ and vice-versa. The set of all bra vectors is the dual space and we say that the action of bra vector $\langle w|$ on ket vector $|v\rangle$ gives the inner product $\langle w|v\rangle$.

The dimension of a vector space is the maximum number of linearly independent vectors and a set of maximum number of linearly independent vectors is called a basis. In quantum mechanics, it turns out to be very convenient to choose an orthonormal basis i.e if $|\phi_i\rangle$, for $i = 1, ..N$ form a basis then $\langle\phi_i|\phi_j\rangle = \delta_{ij}$. Here Kronecker delta $\delta_{ij} = 1$ for $i = j$ and 0 otherwise. Any arbitrary vector $|\Psi\rangle$ in an N dimensional Hilbert space can be written as a linear combination of the basis vectors $|\Psi\rangle = \sum_{n=1}^{N} c_n|\phi_n\rangle$. Using orthonormality, we can easily see that $c_m = \langle\phi_m|\Psi\rangle$. Using this in the expansion of $|\Psi\rangle$, we get $|\Psi\rangle = \sum_{n=1}^{N}\langle\phi_n|\Psi\rangle|\phi_n\rangle$. Since this is true for any state $|\Psi\rangle$, we can read off a useful relation, called completeness relation or resolution of identity, $\sum_{n=1}^{N}|\phi_n\rangle\langle\phi_n| = I$ where I is the unit operator which gives $I|v\rangle = |v\rangle$ for any vector $|v\rangle$.

Operators A linear operator A in Hilbert space is a mapping from Hilbert space to itself which transforms a ket $|\psi\rangle$ to another ket $|\psi'\rangle$ and which is linear i.e $|\psi'\rangle = A|\psi\rangle$ and $A\left(\sum_i c_i|\phi_i\rangle\right) = \sum_i c_i A|\psi_i\rangle$. Due to linearity, the action of an operator on any general state is completely determined by the action of the operator on the basis states. Since the action of the operator on a basis element $|\phi_i\rangle$ is another vector in the same Hilbert space and that resultant vector can be expanded in terms of the same basis, we can write $A|\phi_j\rangle = \sum_i A_{ij}|\phi_i\rangle$. Matrix with entries A_{ij} is said to give the matrix representation of the operator in the given basis. It can be easily checked that $A = \sum_{i,j} A_{ij}|\phi_i\rangle\langle\phi_j|$ or equivalently $A_{ji} = \langle\phi_j|A|\phi_i\rangle$. An equivalent way to get the same representation is to use the completeness relation twice:

$$A = IAI = \sum_{i,j}|\phi_j\rangle\langle\phi_j|A|\phi_i\rangle\langle\phi_i| = \sum_{i,j}\langle\phi_j|A|\phi_i\rangle|\phi_j\rangle\langle\phi_i| = \sum_{i,j} A_{ji}|\phi_j\rangle\langle\phi_i| \tag{1}$$

A nice way to define operators in a Hilbert space is using the outer product $|\phi\rangle\langle\psi|$ for a ket $\phi\rangle$ and a bra vector $\langle\psi|$. This defines an operator because acting on any ket $|v\rangle$ it gives another ket $\langle\psi|v\rangle|\phi\rangle$, just like an operator. Completeness relation $\sum_{n=1}^{N}|\phi_n\rangle\langle\phi_n| = I$ can be thought of as outer product representation of the identity operator. Simplest operator is the identity operator whose matrix representation is just the identity matrix. We will see later that physical quantities correspond to Hermitian operators or equivalently Hermitian matrices (satisfying $A = A^\dagger$ where hermitian conjugation $A^\dagger = (A^T)^*$ is the combination of complex conjugation and transpose operations) and since all eigenvalues of a Hermitian matrix are real, observed value of any physical quantity is a real number. All the eigenvectors of a Hermitian matrix form (or can be made into) an orthonormal set and hence can serve as basis.

Many times, we will use the operator and it's corresponding matrix interchangeably. For most practical computations, one works in a given basis and with corresponding matrix representation of operators. Equations containing operators can equivalently be written as matrix equations. Similar to the trace $Tr(A_{ij}) = \sum_i A_{ii}$

for the matrix representation, we can define the trace of an operator as $Tr(A) = \sum_k \langle k|A|k\rangle$.Trace will be of use to use later when we discuss density matrices in quantum mechanics.

Tensor Product Till now, we have considered only a single Hilbert space which, as we will see, corresponds to a single particle or a single quantum mechanical system. For discussing multiparticle systems or systems with many parts, we need to consider tensor product of Hilbert spaces.

If H_1 and H_2 are Hilbert spaces of dimension m, n respectively then their tensor product or direct product $H_1 \otimes H_2$ is a Hilbert space of dimension mn. If $|\phi_i\rangle$, with $i = 1, ..m$, is a basis for H_1 and $|\psi_j\rangle$, with $j = 1, ..n$, is a basis for H_2 then $|\phi_i\rangle \otimes |\psi_j\rangle$ forms a basis for the tensor product space. If A and B are linear operators in H_1 and H_2 respectively then we can define a linear operator $A \otimes B$ on the tensor product space which acts as

$$(A \otimes B)(|\phi\rangle \otimes |\psi\rangle) = A|\phi\rangle \otimes B|\psi\rangle \tag{2}$$

where $|\phi\rangle$, $|\psi\rangle$ belong to H_1, H_2 respectively.

Connection between the mathematics of Hilbert spaces and Quantum Mechanics is through the following postulates:

Postulate 1 Any isolated physical system has an associated Hilbert space (sometimes called State Space). The quantum state of the system is completely described, at any given time, by a (unit normalised vector) called state vector $|\psi(t)\rangle$, with $\langle\psi(t)|\psi(t)\rangle = 1$, which belongs to this State space. Quantum state contains all the possible information about the system.

In physical terms, the possibility of making complex linear combinations of vectors in Hilbert space is called superposition principle and it lies at the heart of counterintuitive nature of quantum mechanics. It tells us that if we have two quantum states $|\psi\rangle$ and $|\phi\rangle$ then $|\chi\rangle = c_1|\psi\rangle + c_2|\phi\rangle$ is also a quantum state. We say that the system is in a superposed state with amplitude c_1 for $|\psi\rangle$ and c_2 for $|\phi\rangle$.

Postulate 2 Observable quantities correspond to Hermitian operators on the Hilbert space. The measurement of an observable A when the system is in a state $|\Psi(t)\rangle$ corresponds to the action of the operator on the state. Using the set of eigenvectors $|\phi_n\rangle$ of A as basis, we can write $A|\Psi(t)\rangle = \sum_n c_n|\phi_n\rangle$ where $c_n = \langle\phi_n|A|\Psi(t)\rangle$.

3.1 *Time Evolution*

A very important question is how does a quantum state changes with time?

Postulate 3 The time evolution of a *closed* quantum system is effected by a unitary operator U , with $U^\dagger = U^{-1}$ i.e. state vector for a system at times t_1 and t_2 are related as $|\Psi(t_2)\rangle = U(t_2, t_1)|\Psi(t_1)\rangle$ with $U(t_1, t_1) = I$.

A standard result in linear algebra is that a unitary operator can always be written in terms of a Hermitian operator as $U = e^{iK}$ where K is a Hermitian operator (i.e. $K^\dagger = K$). For systems with continuous time evolution, it's conventional to write the operator K as $K = -\frac{\hat{H}(t_2-t_1)}{\hbar}$ where the Hermitian operator $\hat{H}$ is known as the quantum Hamiltonian operator. We can write the quantum time evolution as a differential equation, known as Schrödinger time-dependent equation, by considering infinitesimal time evolution i.e. $t_2 - t_1 = \Delta t$ with $O(\Delta t)^2$ terms neglected. Then we get $|\Psi(t+\Delta t)\rangle = (I - \frac{i\Delta t}{\hbar}\hat{H}|\Psi(t)\rangle$. Taking the limit $\Delta t \to 0$, we get the Schrödinger time-dependent equation $i\hbar\frac{d}{dt}|\Psi(t)\rangle = \hat{H}|\Psi(t)\rangle$.

3.2 Quantum Mechanics and Measurements

Till now, we have not said anything about how the quantum mechanical formalism is related to what we observe when we perform any measurement except mentioning that quantum mechanics gives, in general, statistical predictions. The connection is given by

Postulate 4 If the state of the system is an eigenstate $|\phi_n\rangle$ of an operator $\hat{A}$ corresponding to an observable A then on measuring the observable A, we are certain to get the corresponding eigenvalue c_n where $\hat{A}|\phi_n\rangle = c_n|\phi_n\rangle$. When the system is in a general but normalized state $|\psi\rangle$, the *probability* of getting the eigenvalue c_n on measuring observable A is given by

$$P_n(c_n) = |\langle\phi_n|\psi\rangle|^2 = |c_n|^2. \tag{3}$$

Since quantum mechanics gives statistical predictions i.e. probability distribution for different values of an observable, we need to repeat the measurement on a quantum ensemble. Normalization of the state vector ensures that all probabilities add up to one.

A quantum ensemble is a large number (say M) of copies of the system, each in the same state $|\psi\rangle$. If repeated measurements for some observable A are made on this ensemble then the system will be one of the eigenvalues of the observable with probabilities, as $M \to \infty$), given by the Eq. 3 above. As mentioned above, a single measurement of an observable A always gives one of the eigenvalues of the operator A, say c_n. But measurements of an ensemble (all systems in the ensemble are in a normalized state $|\psi\rangle$ give different eigenvalues with probabilities as given above. So, like in a statistical system, we can define an average value or expectation value $\langle A\rangle$ of the observable A as follows:

$$\langle A\rangle = \langle\psi|\hat{A}|\psi\rangle = \sum_{n,m}\langle\psi|\phi_m\rangle\langle\phi_m|\hat{A}|\phi_n\rangle\langle\phi_n|\psi\rangle = \sum_n c_n|\langle\phi_n|\psi\rangle|^2 = \sum_n c_n P_n(c_n) \tag{4}$$

In any system which is described in terms of probability distribution, apart from the average, we can also describe fluctuations about the mean using standard deviation. In quantum mechanics, it is called uncertainty and for any operator $\hat{A}$ when the system is in a state $|\psi\rangle$ is defined as $\Delta A = \sqrt{\langle\psi|(\hat{A} - \langle\hat{A}\rangle)^2|\psi\rangle}$. Even in absence of detailed knowledge of the state, expectation value and uncertainty provide a good description of the state. One should keep in mind that the result of any measurement of an observable A will be one of the eigenvalues and statistical measures like average and uncertainty correspond to measurements on the whole ensemble.

After a measurement corresponding to an observable A the state of the system collapses to an eigenstate of A and any repeat measurement immediately will confirm that the system is in that eigenstate. There is no state which is simultaneously an eigenstate of all observables. So if we perform a measurement of observable A first (and hence collapse it to an eigenstate of A) and then a measurement of observable B the result will, in general, be a probability distribution over the eigenvalues of B. On the other hand if the orders are reversed then the system will be in an eigenstate of B after first measurement and will be a statistical distribution over the eigenvalues of A on measuring A. Therefore, unless the given state is simultaneously an eigenstate of both A and B, we will get different results depending on the order and hence the measured values can not be used to characterise the state. A theorem in linear algebra says that two matrices/operators can have simultaneous eigenstates if and only if they are commuting. Hence two observables can have simultaneous eigenstates if and only if they are commuting. Among all observables we can pick a set of mutually commuting operators (called complete set of commuting observables) and use their eigenvalues to label the states. In such a simultaneous eigenstate, uncertainty is zero for all the commuting observables. The situation is different for the non-commuting operators, say C and D. Since no state is a simultaneous eigenstate, both will have non-zero uncertainty in general and generalised Heisenberg uncertainty principle relates the two. It can be shown that

$$\Delta(C)\Delta(D) \geq \frac{|\langle\psi|[C, D]|\psi\rangle|}{2} \tag{5}$$

where $[C, D] = CD - DC$ is the commutator of two observables C, D and uncertainties are evaluated in the same state $|\psi\rangle$. An example of non-commuting observables is position and momentum of a particle and hence we can't write quantum state in whole of phase space but need to use position space and momentum space separately. Other examples are various components of angular momenta, e.g. L_x, L_y, L_z and in a system with spin, the spin components etc.

4 Qubits and Two State systems

The simplest quantum mechanical system, sometimes called Qubit, has a two dimensional State space. If $|0\rangle, |1\rangle$ form an orthonormal basis in the state space then an arbitrary state vector can be written as $|\Psi(t)\rangle = C_1(t)|0\rangle + C_2(t)|1\rangle$ where C_1, C_2 are complex numbers. Two state systems can be realized in many different ways, either occurring naturally or made artificially. For example, if we only consider spin degrees of freedom for an electron i.e. ignoring spatial degrees of freedom, we get a two state system corresponding to spin up and spin down states. Basis states $|0\rangle, |1\rangle$ are sometimes, also called computational basis states in quantum computing applications.

Qubits are the analogues of bits (unit of classical information) and the qubits serve as the building blocks of quantum computers. Possibility of superposing base states $|0\rangle, |1\rangle$ distinguishes classical information from quantum information. Classical information is in terms of bits, which can be in two possible states, labelled 0 and 1 and any classical information can be in represented as a string of bits, with the number of bits required increasing with the amount of information to be stored. Similarly a qubit can also be in two states $|0\rangle, |1\rangle$ but unlike classical bit, it can also be in superposition of the two base states. For a set of N qubits, the Hilbert space has dimension 2^N and an arbitrary superposition requires 2^N complex numbers or 2^{N+1} real numbers. There is one constraint due to the requirement of normalization. Also, any quantum state, can be multiplied by a constant phase factor ($e^{i\phi}$) without changing the probabilities. Hence we have to remove two real numbers from the counting, thus requiring $2^{N+1} - 2$ real numbers to specify an arbitrary superposition in an N qubit system. This is the amount of classical bits which are required to store the information in an N qubit system.

As mentioned earlier, physical observables are represented by a Hermitian operator/matrix. A general 2×2 Hermitian matrix can be written as

$$\begin{bmatrix} q+z & x-iy \\ x+iy & q-z \end{bmatrix} = q\begin{bmatrix} 1 & 0 \\ 0 & 1 \end{bmatrix} + x\begin{bmatrix} 0 & 1 \\ 1 & 0 \end{bmatrix} + y\begin{bmatrix} 0 & -i \\ i & 0 \end{bmatrix} + z\begin{bmatrix} 1 & 0 \\ 0 & -1 \end{bmatrix} = qI + x\sigma_x + y\sigma_y + z\sigma_z \tag{6}$$

where $\sigma_x, \sigma_y, \sigma_z$ are called Pauli spin matrices and I is the identity matrix. Pauli matrices and identity matrix thus provide a basis for the vector space of 2×2 Hermitian matrices. Thus any observable for a two state system can be written in terms of Pauli matrices. One prominent observable is the quantum Hamiltonian which governs the time-evolution of the two state quantum system and it can be written in above basis. The Schrödinger equation $i\hbar\frac{d}{dt}|\Psi(t)\rangle = \hat{H}|\Psi(t)\rangle$ in the basis $|0\rangle, |1\rangle$ can be written as a matrix equation

$$i\hbar\frac{d}{dt}C_i(t) = \sum_j H_{ij}C_j(t) \tag{7}$$

where both i, j take on the values 1 and 2 and $\langle i|H|j\rangle = H_{ij} = H^*_{ji}$. As mentioned earlier

$$|\Psi(t)\rangle = \langle 0|\Psi(t)\rangle|0\rangle + \langle 1|\Psi(t)\rangle|1\rangle = C_1(t)|0\rangle + C_2(t)|1\rangle \tag{8}$$

This matrix equation can be solved in a standard way. For the case where the elements of the Hamiltonian matrix H_{ij} are time-independent, the corresponding first order matrix differential equation can be easily solved by first transforming to the basis of eigenvectors, solving the Schrödinger equation there and then transforming back to the original basis. Same procedure can be applied to solve a general N state system though finding eigenvalues and eigenvectors for a general $N \times N$ Hamiltonian matrix can, in general, be done only numerically.

5 Composite Systems and Entanglement

For multiparticle system or systems made of several parts i.e. composite systems, we construct the Hilbert space for the full system using the following postulate:

Postulate 5 The Hilbert space of a composite physical system is the tensor product of the Hilbert spaces of it's components. If the component systems for an n component system are in states $|\psi_i\rangle$, with $i = 1, ..n$ then the total system is in a state given by $|\psi_1\rangle \otimes |\psi_2\rangle \otimes ....|\psi_n\rangle$.

Consider a system W which we divide into two subsystems A and B. The above postulate says that the Hilbert space of the system W is given by the tensor product $H_W = H_A \otimes H_B$. As discussed earlier, the tensor product structure implies that if $|a_i\rangle$ and $|b_j\rangle$ are basis for the subsystems A, B then any state in the composite system W can be written as $|\Psi_W\rangle = \sum_{i,j} c_{ij}|a_i\rangle \otimes |b_j\rangle$. For notational simplicity, we will sometimes write $|a_i\rangle \otimes |b_j\rangle$ as $|a_i b_j\rangle$.

We say that the two subsystems are in an entangled state if it is *not* possible to assign a separate states to the subsystems i.e. if it is not possible to write $|\Psi_W\rangle = |\psi_A\rangle \otimes |\psi_B\rangle$.

A simple example of entanglement is given by a system W consisting of two qubits A and B. The total 4 dimensional Hilbert space of the two qubit system is spanned by the basis elements $|00\rangle, |01\rangle, |10\rangle, |11\rangle$ where we are using the simplified notation $|i\rangle \otimes |j\rangle = |ij\rangle$ for the basis elements. We can construct a superposed state in the 2 qubit Hilbert space which is given by $|\Psi_W\rangle = \frac{1}{\sqrt{2}}(|01\rangle + |10\rangle)$. One can check that it is *not* possible to write this state as a tensor product of a state of first qubit and a state of second qubit i.e $|\Psi_W\rangle \neq |\psi_A\rangle \otimes |\psi_B\rangle$. We say that the total system is in an entangled state $|\Psi_W\rangle$ or that the two subsystems qubit A and qubit B, are entangled. On the other hand, a state $|\Phi_W\rangle = \frac{1}{2}(|00\rangle + |01\rangle + |10\rangle + |11\rangle)$ is not an entangled state since it can be written as $|\Phi_W\rangle = \left(\frac{1}{\sqrt{2}}(|0\rangle + |1\rangle) \otimes \frac{1}{\sqrt{2}}(|0\rangle + |1\rangle)\right)$.

Unfortunately there is no straightforward way of determining if a given state is entangled or not. There is no operator whose expectation value gives the degree of entanglement and entanglement is not associated with any observable. But it's important for quantum computing and manipulation of quantum information. An important difference between classical and quantum computers is that in classical computers, the bits are independent of each other. The state of a bit is not influenced by the state of any other bit but in quantum computers the qubits can be entangled with each other, which means that they become part of one large quantum state together. Initially unentangled qubits can become entangled due to quantum time-evolution.

One difference between classical mechanics and quantum mechanics is that a classical (made of classical states) source of information, say a book, doesn't change or get degraded on repeated readings or measurements. On the other hand, a quantum source of information, say something made out of qubits, gets irrevocably changed by measurement or reading. As an example, consider a normalized qubit $|\psi\rangle = a|0\rangle + b|1\rangle$ where a, b. are complex numbers. To 'read' it we would like to measure a, b (subject to normalization condition). Unfortunately there is no measurement which can give us a, b directly . As the Postulate 4 above describes, we will only get base states $|0\rangle$ or $1\rangle$ with probabilities $|a|^2$ and $|b|^2$ respectively. We can not repeat the measurement (to get the full probability distribution) on the same qubit since the act of measurement has 'collapsed' the state to either $|0\rangle$ or $|1\rangle$ according to the outcome of the measurement. So we can informally say that the quantum information, unlike classical information, is 'destroyed' by 'reading' it.

One possible way to evade the above quantum mechanical limitation would be to make many copies of the unknown qubit before the measurement. Measuring on all the copies would give us the probability distribution for $|a|^2$, $|b|^2$ to any desired accuracy. Unfortunately, there is a theorem in quantum mechanics, called No Cloning Theorem, which says that it's impossible to copy a quantum state without destroying the original. The impossibility of making a quantum Xerox machine follows directly from the linearity of quantum mechanics and was pointed out by Wootters, Zurek [7] and Dieks [8] in 1982. To see how this happens, let's assume that there *can* exist a device which can duplicate a given quantum state. We want a machine which takes the given quantum state $|\psi\rangle$ (this is the 'original' and another state $|C\rangle$ (this is like the blank paper in ordinary Xerox machine) as inputs and gives the output state $|\psi\rangle \otimes |\psi\rangle$ (this is like the original plus copy in an ordinary Xerox machine). If this purported machine works as claimed for any given state then it will do $|\psi_1\rangle \otimes |C\rangle \rightarrow |\psi_1\rangle \otimes |\psi_1\rangle$ and $|\psi_2\rangle \otimes |C\rangle \rightarrow |\psi_2\rangle \otimes |\psi_2\rangle$ for two input states $|\psi_1\rangle, |\psi_2\rangle$. What happens if we input a linear combination $|\psi\rangle = a|\psi_1\rangle + b|\psi_2\rangle$ to this machine? Assuming that the device is also governed by Schrödinger equation (which is a linear equation), we get $|\psi\rangle \otimes |C\rangle \rightarrow a|\psi_1\rangle \otimes |\psi_1\rangle + b|\psi_2\rangle \otimes |\psi_2\rangle$. But this is *not* a copy $|\psi\rangle \otimes |\psi\rangle = ((a|\psi_1\rangle + b|\psi_2\rangle) \otimes (a|\psi_1\rangle + b|\psi_2\rangle))$ of the superposed state $|\psi\rangle$ as we wanted. For example, for a 2-qubit system, if we input $a|0\rangle + b|1\rangle$ in the quantum Xerox machine, we will get $a|0\rangle \otimes |0\rangle + b|1\rangle \otimes |1\rangle$ instead of true copy which would have been $((a|0\rangle + b|1\rangle) \otimes a|0\rangle + b|1\rangle) = a^2|0\rangle \otimes |0\rangle + b^2|1\rangle \otimes |1\rangle + ab(|0\rangle \otimes |1\rangle + |1\rangle \otimes |0\rangle)$.

So quantum information is very fragile in a sense and not easily transferable. This means that various ingenious schemes have to be devised to transfer quantum information. For example, using entangled states and some classical states, it is possible to transfer (or quantum teleport [9]) quantum information.

5.1 *Multiple Qubits*

A two qubit state has a 4 dimensional Hilbert space and it's (computational) basis states can be written as $|00\rangle$, $|01\rangle$, $|10\rangle$, $|11\rangle$ using the notation $|xx\rangle = |x\rangle \otimes |x\rangle$ for any state $|x\rangle$. For a state with two classical bits we would have had 00, 01, 10, 11 as possible states. But now, due to the possibility of superposition of states, a general 2 qubit state can be written as

$$|\Psi(t)\rangle = c_{00}(t)|00\rangle + c_{01}(t)|01\rangle + c_{10}(t)|10\rangle + c_{11}(t)|11\rangle \tag{9}$$

Each coefficient c_{ij} gives the probability $|c_{ij}|^2$ of finding the system in the corresponding basis state. Normalization condition states that sum of all such probabilities equal 1. With multiple qubits, we can measure a subset of qubits, e.g. only the first qubit in a 2 qubit system. In this case, we get 0 with probability $|c_{00}|^2 + |c_{01}|^2$ and the state after such a measurement is the partially collapsed state

$$|\Psi'(t)\rangle = \frac{c_{00}(t)|00\rangle + c_{01}(t)|01\rangle}{\sqrt{|c_{00}|^2 + |c_{01}|^2}} \tag{10}$$

An alternative orthonormal basis for 2 qubit Hilbert space is given by so called Bell states or EPR pairs. These can be written in terms of computational basis states as

$$|\psi_1\rangle = \frac{|00\rangle + |11\rangle}{\sqrt{2}}, \quad |\psi_2\rangle = \frac{|00\rangle - |11\rangle}{\sqrt{2}}, \quad |\psi_3\rangle = \frac{|10\rangle + |01\rangle}{\sqrt{2}}, \quad |\psi_4\rangle = \frac{|01\rangle - |10\rangle}{\sqrt{2}}$$

This basis is quite useful in quantum computing and in studies on the foundations of quantum mechanics.

6 Mixed States and Density Matrix

Till now we have considered so called pure states only, in which quantum ensemble contains all the members in the same state $|\psi\rangle$. A mixed quantum state corresponds to a probabilistic mixture of pure states and is described by a density matrix/operator. If there are k pure states, represented by kets $|i\rangle$, with $i = 1, 2, ..k$, in the ensemble and there is a probability p_i that a system randomly picked out of the ensemble is in the state $|i\rangle$ then the system is said to be in a mixed state described by the density operator $\rho = \sum_i p_i |i\rangle\langle i|$. A system in a pure state $|\psi\rangle$ has a density operator of the form $\rho = |\psi\rangle\langle\psi|$ i.e. all p_i except one is zero. We can easily see that the density operator can be written as $\rho = \sum_i p_i \rho_i$ where $\rho_i = |i\rangle\langle i|$ is the density operator corresponding to pure state $|i\rangle$. In this way it's easy to see how mixed states can be thought of as mixture of pure states.

Mixed states can arise if we don't know what is the state in which has been prepared and so we have to assign a probability distribution to possible states. For example, if a beam of electrons is coming out of some source without any selection then we can assign(in absence of any other information) a probability $1/2$ of having spin up and a probability $1/2$ of having spin down. These probabilities are *not* the quantum probabilities but normal probabilities of the kind we have in statistical mechanics. Another example of a mixed state is unpolarized light (e.g. light coming directly out of sun or a bulb). Unpolarized light (or electron spin mentioned above) can not be represented as any pure state and has to be represented by a density operator of the form $\rho = \frac{1}{2}|RCP\rangle + \frac{1}{2}|LCP\rangle$ where LCP, RCP states correspond to left circularly polarized or right circularly polarized light respectively. The corresponding density matrix is

$$\rho_{mixed} = \frac{1}{2}\begin{bmatrix} 1 & 0 \\ 0 & 1 \end{bmatrix} \tag{11}$$

On the other hand, light in the $|LCP\rangle$ or $|RCP\rangle$ or any linear combination of these two is in a pure state. Vertically polarized light(VP) , for example, can be represented as a superposition $|VP\rangle = \frac{1}{\sqrt{2}}(|RCP\rangle + |LCP\rangle)$. It's density matrix is given by

$$\rho_{pure} = |VP\rangle\langle VP| = \frac{1}{2}\begin{bmatrix} 1 & 1 \\ 1 & 1 \end{bmatrix} \tag{12}$$

Another way that mixed states can arise is when we have a system which is entangled with another system, e.g., when the given system is part of a composite system that includes entangled states and we don't have the information about the other part(s). Description of mixed states arising this way is given by reduced density operator. Consider a system W made of two subsystems A and B. If the density matrix for the total system is given by ρ^W then the reduced density operator for system A is given by $\rho^A = tr_B(\rho^W)$. Here tr_B is an operation called partial trace which is defined as

$$tr_B(|a_1\rangle\langle a_2| \otimes |b_1\rangle\langle b_2|) = tr(|b_1\rangle\langle b_2|)|a_1\rangle\langle a_2| \tag{13}$$

Here $|a_1\rangle$, $|a_2\rangle$ are any two vectors in the Hilbert space of subsystem A and $|b_1\rangle$, $|b_2\rangle$ are any two vectors in the Hilbert space of subsystem B. On the RHS, we have the usual trace operation on subsystem B and so $tr(|b_1\rangle\langle b_2|) = \langle b_2|b_1\rangle$.

As mentioned above, density operator can describe both pure and mixed states. A simple criteria to decide whether a given density matrix describes pure or mixed state is to calculate trace of it's square i.e. $Tr(\rho^2)$. Pure states always have $Tr(\rho^2) = 1$ (as can be easily checked by using $\rho = |\psi\rangle\langle\psi|$) while mixed states have $tr(\rho^2) < 1$. This can be easily checked for the case of polarized and unpolarized light density matrices given above. It can easily be established using the definition of the density operator that $Tr(\rho) = 1$ and $\rho^\dagger = \rho$. The expectation value of an observable A in a mixed ensemble consisting of pure states $|\psi_j\rangle$ with probabilities p_j can be written in terms of the corresponding density matrix as

$$\langle A\rangle = \sum_j p_j\langle\psi_j|A|\psi_j\rangle = \sum_j p_j tr(|\psi_j\rangle\langle\psi_j|A) = tr(\rho A) \tag{14}$$

7 Measurement Problem and Quantum Decoherence

Quantum mechanics is a very successful theory for describing the behaviour of microscopic world. But how does our macroscopic, approximately classical world emerges out of this microscopic description is still a question which has not been fully answered yet. In this section, we briefly discuss how measurements in quantum mechanics, using a classical apparatus leads to some puzzling questions.

As we have mentioned earlier also, linear superposition principle is the most enigmatic aspect of quantum mechanics and it leads to some of the most counterintuitive features of quantum mechanics. This linearity is maintained by the time-evolution due to Schrödinger equation and in absence of any measurement, system continues to evolve linearly. If we have a state which is a superposition of eigenstates of some observable then it will evolve while maintaining the same superposition. But when we perform a measurement of the observable A, the Postulate 4 says that the system, initially in a superposed state, $|\psi\rangle = \sum_i c_i|\phi_i\rangle$, irreversibly reduces to one of the eigenstates, say $|\phi_i\rangle$. This mysterious interaction with the measuring apparatus is called state reduction or wavefunction collapse i.e. $|\psi\rangle \rightarrow |\phi_i\rangle$. In the standard presentation of quantum mechanics, the measurement apparatus is treated as a macroscopic, classical system and wavefunction collapse is a postulate. This dichotomy between the description of a system using quantum mechanics and treating the measuring apparatus as a classical system (which is not allowed to be in a superposed state) is known as quantum measurement problem.

It's very puzzling because we know that the matter is ultimately made of atoms or some other microscopic constituents which obey quantum mechanics. Thus, in principle, a macroscopic detector which is also made of these should also be described by some (possibly very complicated) quantum state. If $|\psi(0)\rangle$ is the initial state of the system (for concreteness, we take it to be a qubit $a|0\rangle + b|1\rangle$) and the apparatus is described, at time $t = 0$, by a state vector $\phi(0)$ then the total system (qubit + apparatus) can be treated as a composite system and described, at time $t = 0$, by a total state vector $|\Psi(0)\rangle = |\psi(0)\rangle \otimes |\phi(0)\rangle$. Time evolution of the total system will still be described by a unitary operator $|\Psi(t)\rangle = U(t, 0)|\Psi(0)\rangle$ where Schrödinger equation now involves total Hamiltonian, including the interaction between the system (qubit) and the apparatus. As a consequence of this interaction, the total state vector evolves as

$$|\Psi(0) \rightarrow |\Psi(t) = a|0\rangle \otimes |\phi_0\rangle + b|1\rangle \otimes |\phi_1\rangle \tag{15}$$

Here $|\phi_0\rangle, |\phi_1\rangle$ are the states of the apparatus, showing states $|0\rangle, |1\rangle$ respectively. For example, qubit is maybe represented as spin states of an electron while $|\phi_0\rangle, |\phi_1\rangle$ may be the pointer in the apparatus showing electron spin up and down respectively. Important point to note is that the total state vector in the Eq. 15 is still a superposition of two states (system and apparatus states are entangled now). The puzzle is how does one go from this superposed state to the situation described by the Postulate 4 i.e. in terms of projected state $a|0\rangle \otimes \phi_0\rangle$ *or* $b|1\rangle \otimes \phi_1\rangle$? Does it happen when a human observes the apparatus? Strictly applying quantum mechanics though, one can consider a combined qubit + apparatus + human system and write a superposed state for this system also. The notion of applying the idea of linear superposition to macroscopic objects like humans or even detectors seems preposterous (Schrödinger's proposed experiment with cat dramatically exemplifies this).

The questions mentioned above have been discussed a lot in the literature even though a consensus solution hasn't emerged yet. One approach, which doesn't answer all questions but throws new light on these aspects of quantum mechanics, is the appreciation of the phenomenon of quantum decoherence.

The idea of quantum decoherence is that if a quantum system is closed i.e. perfectly isolated, it will maintain linear superpositions indefinitely, as expected. But such a closed system can never be investigated as any measurement is a form of interaction from outside. In fact, any system we study is, to some extent, an open system that interacts with an environment and due to this interaction, quantum coherence (maintenance of definite phase relations in a linear superposition) gets lost. This process is called quantum decoherence and even though the total system (given system + environment) can still be thought of in a coherent, superposed state, the state of the system alone becomes a mixed state given by classical probabilities. Schematically, we should write the total system (qubit + apparatus + environment) state vector as

$$|\Psi(0)\rangle \rightarrow |\Psi(t)\rangle = a|0\rangle \otimes |\phi_0, \chi_0\rangle + b|1\rangle \otimes |\phi_1, \chi_1\rangle \tag{16}$$

where the variables χ_0, χ_1 describe, symbolically, different environments. Since apparatus is a macroscopic object, it will interact with the environment. Now comes the crux of the decoherence. Because there are a huge number of environment variables and they are uncorrelated, the averaging over the environment variables washes out any interference terms (or cross terms like $\langle\phi_0, \chi_0|\phi_1, \chi_1\rangle \rightarrow 0$). As an example, we can consider a system where qubit corresponds to the spin of the electron. If we calculate following expectation values of the Pauli sigma matrices

$$\langle\Psi(t)|\sigma_x|\Psi(t)\rangle = a^*b\langle\phi_0, \chi_0|\phi_1, \chi_1\rangle + cc \tag{17}$$

$$\langle\Psi(t)|\sigma_z|\Psi(t)\rangle = |a|^2\langle\phi_0, \chi_0|\phi_0, \chi_0\rangle - |b|^2\langle\phi_1, \chi_1|\phi_1, \chi_1\rangle \tag{18}$$

Here, *cc* in the first equation denotes complex conjugation of the first term. Using the normalization conditions $\langle\phi_0, \chi_0|\phi_0, \chi_0\rangle = 1 = \langle\phi_1, \chi_1|\phi_1, \chi_1\rangle$ and the decoherence condition $\langle\phi_0, \chi_0|\phi_1, \chi_1\rangle \rightarrow 0$, we get the result that $\langle\Psi(t)|\sigma_x|\Psi(t)\rangle \rightarrow 0$ and $\langle\Psi(t)|\sigma_z|\Psi(t)\rangle \rightarrow |a|^2 - |b|^2$.

What this means is that, as a result of interaction with the environment, we get the same results for expectation values as we would have gotten if there was no superposed state for the qubit but there was a mixture of two separate states $|0\rangle$ and $|1\rangle$ which come out of a box with probabilities $|a|^2$ and $|b|^2$ respectively.

Decoherence, thus, provides a mechanism for the transition from superposition to a mixed state even though the total state of the system and environment combined is still the entangled state. Decoherence makes it look like as if a state reduction has taken place and since we can never completely isolate a system from it's environment, for all practical purposes, we can just use the Postulate 4. Even though we have discussed decoherence using state vectors, one can also use the density matrix for the total system and then trace over the environment degrees of freedom to get the reduced density matrix for the system. Both approaches give the same results. Decoherence happens on a very small time scale for macroscopic objects unless special care is taken to maintain the coherence of the superpositions.

7.1 Decoherence and Quantum Computing

Controlling Quantum Decoherence due to the environment presents a huge challenge to building a practical quantum computer. As we discussed earlier, the main advantage of a qubit (or any quantum system) over classical bits is the possibility of having superposition of the basis states and interference between them. This greatly increases the power of quantum computers over the classical computers. But decoherence, as we discussed above, converts superpositions into mixed states and prevents the undisturbed evolution of quantum systems.

Preventing decoherence usually means isolating the system from it's environment as much as possible. In practice, this usually requires cooling the system to ultra-low temperatures, trying to keep it as close to zero kelvin as possible (using superconducting circuits, for example), avoiding external radiations like cosmic rays etc. It's a huge challenge to keep the qubits in a coherent state, at least till the time when the quantum computation is running.

8 Conclusion

In this chapter, we have tried to give a concise discussion of quantum mechanical formalism, focussing on those aspects which are of relevance to quantum computing applications. In quantum computing applications, there are two main approaches to using the quantum mechanics. As a subfield of quantum information science, quantum computers generalize the notions of classical computer science like algorithms, Turing machine etc. For this arena, the formalism of quantum mechanism appears in it's abstract form and we don't need to worry about the particular realisations and details of the systems used to implement quantum algorithms and quantum circuits.

That is why we have mostly focussed on the basics of quantum formalism. On the other hand, for the actual building of quantum computers, one needs to work with some particular system and the details do matter. Promising candidates for physical realizations of quantum computers differ in the ways in which qubit is physically realized. Examples include superconducting junctions, molecular magnets, trapped ions, optical lattices etc. Of course, these systems differ a lot in their physics and we haven't discussed any of these systems in this chapter.

References

1. N. Zettlie, *Quantum Mechnics Concepts and Applications* (Wiley India Pvt Ltd, New Delhi, 2022)
2. F. Arute, K. Arya, Quantum supremacy using a programmable superconducting processor. Nature **574**, 505–510 (2019)
3. M.L. Nielsen, Chuang, *Quantum Computation and Quantum Information* (Cambridge University Press, 2010)
4. S. Gasiorowicz, *Quantum Physics*, 3rd edn. (Wiley, 2005)
5. D.J. Griffiths, *Introduction to Quantum Mechanics*, 2nd edn. (Pearson Prentice Hall, 2005)
6. R.P. Feynman, *The Feynman Lectures on Physics, Vol. III: The New Millennium Edition: Quantum Mechanics Basic Books* (2011)
7. W. Wootters, W. Zurek, A single quantum cannot be cloned. Nature **299**, 802–803 (1982). https://doi.org/10.1038/299802a0
8. D. Dieks, Phys. Lett. A **92**, 271–272 (1982)
9. C.H. Bennett, G. Brassard, C. Crépeau, R. Jozsa, A. Peres, W.K. Wootters, Teleporting an unknown quantum state via dual classical and Einstein-Podolsky-Rosen channels. Phys. Rev. Lett. **70**(13), 1895–1899 (1993)

Phase Space Quantization I: Geometrical Ideas

Carlos Alberto Alcalde and Kanishka Tyagi

Abstract This chapter introduces quantum mechanics in phase space. One of the guiding principles is the belief that the physical states of a system can be parametrized by finite dimensional smooth manifolds. Our motivating example is radar theory. Radar measurements take place on classical phase space and exhibit quantum features such as operator representations of the Heisenberg group. Classical physics is described by algebraic and geometrical structures in symplectic manifolds that serve as the arena of classical as well as quantum measurements. We treat the Kahler case as a special kind of symplectic manifold that paves the way to a presentation of quantum mechanics by deformation theory. Deformation quantization proceeds via the introduction of a star product, an associative but not commutative product between functions in phase space. The founders of this approach "suggest quantization be understood as a deformation of the structure of the algebra of classical observables rather than a radical change in the nature of the observables." We show how the quantization of the spin 1/2 particle arises in deformation theory: it can be understood by a star product on the Bloch sphere autonomously, through the geometrical data. Hilbert space methods are introduced in this approach to make contact with conventional approaches.

Keywords Radar theory · Statistical manifolds · Deformation quantization · Koopman representation · Co-adjoint orbit · RKHS · Functions of positive type

C. A. Alcalde
Technical Director for Radar SW, NXP Semiconductors, Grenoble, Auvergne-Rhône-Alpes, France

K. Tyagi (✉)
Senior Machine Learning Scientist, Aptiv Advance Research Center, Agoura Hills, CA, USA
e-mail: kanishka.tyagi@aptiv.com

R. Pandey et al. (eds.), *Quantum Computing: A Shift from Bits to Qubits*,
Studies in Computational Intelligence 1085,
https://doi.org/10.1007/978-981-19-9530-9_3

1 Introduction

In their standard formulations, classical and quantum theories look very different. A unifying arena to both is the concept phase space. Abstractly it is the set that contains all the relevant variables needed to describe our physical theories such as position and momentum. It is modeled by a finite dimensional symplectic manifold. While we can study classical physics without making any reference to measurements of physical quantities, in quantum mechanics one introduces statistical concepts such as expectation values from the start. In order to provide a transition toward quantum mechanics we consider the problem of electromagnetic illumination of an object by a radar sensor as example of a classical measurement of phase space parameters.

The idea to associate phase space functions to quantum mechanical operators and vice versa is as old as quantum mechanics [1]. In one of the early contributions to quantization, Moyal [2] found the phase space function that corresponds to the commutator of two operators, thereby discovering the bracket that bears his name. In the 70s, M. Flato, A. Lichnerowicz, and D. Sternheimer related the Moyal bracket to deformations of Poisson brackets. Furthermore, they sought an autonomous formulation of quantum mechanics as a manifestation of a deformation of the classical algebra of observables [3]. Their work and the ideas behind their program are given in two remarkable papers [4, 5]. This work as well of that of many others [6–11] find their conceptual origin there. The basic principle of deformation quantization is to view quantum mechanics as arising from a deformation of the classical algebra of functions in phase space obtained via the introduction of a new product, the star product. Consult [12] for a pedagogical introduction to the subject.

Quantization in phase space can also be understood by Feynman's method. Phase space path integrals in $\mathbb{R}^{2n}$ were related to the so-called star exponential in [13]. While providing many insights and although they are used widely, the formulation of Feynman path integrals remains heuristic. By defining autonomously a star product in some Kahler manifolds, path integrals appear naturally in the construction of the star exponential for those spaces. Our philosophy is that defining path integrals via star products may pave the way toward making them rigorous. Though not expounded here, it is worth mentioning that this is another application of deformation quantization.

The plan of the article is as follows. Radar theory is used as a concrete example of measurements of macroscopic objects in phase space. The purpose of this section is to serve as a motivating example but it can be skipped on first reading. Before considering quantum measurements, the radar signal is the mathematical analog of the wavefunction in quantum mechanics and can be seen to generate functions on symplectic manifolds by group transformations of the signal. This is the reverse of the polarization step in geometric quantization where one starts with functions in phase space and reduces the number of dimensions to produce spaces where the irreducible representations live. While this is of historical importance, recent developments of POVM quantization [14, 15] suggest along with the deformation quantization program that symplectic geometrical data contains all that is needed to describe generic quantization schemes.

To achieve this understanding, in Sect. 3, we review the setting of classical mechanics on symplectic manifolds, with emphasis on Hamiltonian group actions. In Sect. 4, we specialize to Kahler manifolds that are complex manifolds of great interest in modern applications [16, 17]. We consider the situation a group G acting via holomorphic automorphism. The Kahler data allow us to construct locally Hamiltonian functions that produce Lie algebra homomorphisms with the Lie algebra of the group G via Poisson brackets. We call these Hamiltonians preferred observables. Section 5 starts by describing how quantum mechanics is defined by the introduction of a star product: a deformation of the point-wise product in the classical algebra of observables. When the star product is G-invariant, construction of the star exponential establishes a lien with group representation theory as explained by Fronsdal [18]. The preferred observables allow us to define a star product algebra geometrically. Quantization is thereby performed using only the geometrical Kahler data without the introduction of a Hilbert space. Furthermore, associativity of the product defines quantization for all observables in the algebra. This is reminiscent of Feynman path integrals in that it defines quantization by using the classical action and the sum-over-all-paths postulate. The paper concludes with applying these concepts to quantization of the Bloch sphere.

2 Phase Space Radar Measurements

A concrete example of measurements of phase space parameters is radar theory [19]. An electromagnetic signal $\varphi(t)$ is transmitted for a time $t \in [0, T] \subset \mathbb{R}$ within a measurement interval T and receives signals reflected from objects in the scene. Each received signal (attenuated and with additive noise) is assumed to be a replica of the transmit signal modulated by a Doppler frequency $x \in \mathbb{R}$ and a time delay $y \in \mathbb{R}$. Signals are added linearly at the radar receiver.

The purpose of radar signal processing is to retrieve information about the scene from the received signal. A popular technique is to estimate the key parameters x and y from the received signal $s(t)$ modeling it as a transformed version of the transmit signal

$$\varphi_{xy}(t) = e^{ixt}\varphi(t - y), \text{ with } \varphi(t) = e^{-i\gamma t^2}. \tag{1}$$

This transmit signal φ, with chirp rate γ, is a time limited electromagnetic wave with a finite peak power and therefore a square integrable function. The processing of the signal is referred to as the matched filter or Windowed Fourier transform.

$$V_\varphi : L^2(\mathbb{R}) \to L^2(\mathbb{R}^2), \quad s \mapsto \langle \overline{\varphi_{xy}}, s \rangle_{L^2(\mathbb{R})}. \tag{2}$$

The map V_φ is obviously not surjective. The diagonal matrix element ($\varphi = s$) is called the radar ambiguity function. The ambiguity function lives in a reproducing kernel Hilbert space inside $L^2(\mathbb{R}^2)$. As such, it is a two-dimensional signal and the notation $\Psi(\alpha) \stackrel{\text{def}}{=} \Psi(x, y) = (V_\varphi)(x, y)$ is sometimes used. Unfortunately, this notation does not make explicit its dependence on the original signal $\varphi \in L^2(\mathbb{R})$. This map also goes by the name of the short-time Fourier transform or the windowed Fourier transform. The following properties of the windowed Fourier transform are important for the present work:

- Isometry:

$$\langle V_\varphi, f_1, V_\varphi, f_2\rangle_{L^2(\mathbb{R})} = \|\varphi\|^2\langle f_1, f_2\rangle_{L^2(\mathbb{R})}. \tag{3}$$

- Resolution of the identity:

$$\int_{\mathbb{R}^2} \overline{\varphi_{\alpha\beta}(t)}\langle\varphi_{\alpha\beta}, \cdot\rangle \frac{d\alpha d\beta}{2\pi} = \hat{I} \stackrel{\text{def}}{=} \text{The Identity Operator in } L^2(\mathbb{R}). \tag{4}$$

- Reproducing Kernel Hilbert space condition: Let

$$K(a, b, \alpha, \beta) = \langle\varphi_{ab}, \varphi_{\alpha,\beta}\rangle_{L^2(\mathbb{R})} = \int_R \overline{\varphi_{ab}(t)}\varphi_{\alpha\beta}(t)dt, \tag{5}$$

then for all function in the range of V_φ, the reproducing property holds

$$f(\alpha, \beta) = \int_{\mathbb{R}^2} K(a, b, \alpha, \beta) f(a, b)\frac{d\alpha d\beta}{2\pi}. \tag{6}$$

The windowed Fourier transform can be extended to the Schwartz space S (infinitely differentiable rapidly decreasing functions), and its dual S' (tempered distributions).

Remarks

1. The above properties hold leaving the mother wavelet φ unspecified. The range of V_φ is a proper subspace of $L^2(\mathbb{R}^2)$.
2. Infinitesimal time translations and modulations are generated by the operators of position and momentum of quantum mechanics: for all $\varphi \in L^2(\mathbb{R})$ define (without worrying about domains of operators and letting $\hbar = 1$ for the moment)

$$\hat{P}\varphi = i\frac{d}{dt}\varphi \tag{7}$$

and

$$\hat{Q}\varphi = t\varphi. \tag{8}$$

One has that the one-parameter groups generated by this operators become

$$e^{i\beta\hat{P}}\varphi(t) = \varphi(t - \beta) \tag{9}$$

$$e^{i\alpha\hat{Q}}\varphi(t) = e^{i\alpha t}\varphi(t). \tag{10}$$

In 1926 Heisenberg discovered that momentum and position did not commute in quantum mechanics:

$$[\hat{P}, \hat{Q}] \stackrel{\text{def}}{=} \hat{P}\hat{Q} - \hat{Q}\hat{P} = \hat{I}. \tag{11}$$

Shortly after, Dirac pointed out that this expression had an analog in classical physics:

$$\{p, q\} = 1, \tag{12}$$

where{, } denotes the Poisson bracket of two functions of phase space (p, q):

$$\{f_1, f_2\} = \frac{\partial f_1}{\partial p}\frac{\partial f_2}{\partial q} - \frac{\partial f_1}{\partial q}\frac{\partial f_2}{\partial p}. \tag{13}$$

In fact the passage of coordinates (p, q) in phase space to operators $\hat{P}$, $\hat{Q}$ on a Hilbert Space is what physicists call quantization. Modulations and time translations form a group called the Heisenberg group and the expression of its group elements in the space of square integrable functions described above is related to the so-called the Schroedinger representation. Notice that the coordinates (α, β) as they appear above are not most naturally identified with phase space coordinates (p, q) but in a dual relationship to them.

3 Symplectic Geometry

So far geometrical concepts have been absent. In radar theory and signal analysis, the idea of phase space of classical mechanics appears with momentum and position replaced by time and frequency. Both in the initial design as in the problem of detection, estimation of these phase space parameters enters naturally. Although interpreted differently, in quantum theory as in signal analysis, these estimates can be thought of taking the expectation values of operators.

The appropriate phase space to consider for the time-frequency problem (as well as conventional quantum mechanics) is the two-dimensional plane $\mathbb{R}^2$. Natural geometric quantities to consider are length and area, both already present if we consider complex multiplication in the plane as the real and imaginary parts. Remark that the area spanned by two vectors (x_1,y_1) and (x_2,y_2) is given by $(x_2y_1 - x_2y_1)/2$. This is the extra factor in the definition of the Heisenberg group.

Area-preserving transformations in the plane are called symplectic transformations (they are called canonical in the physics literature). It is well known [20] that infinitesimal symplectic transformations are generated by a smooth functions H, called the Hamiltonians (an example is the energy in time-independent systems). Using (p, q) for phase space coordinates this means that the vector field which is the infinitesimal generator of the symplectic transformation can be written with the aid of the Poisson bracket as

$$\xi_H = \{H, \cdot\} = \frac{\partial H}{\partial p}\frac{\partial}{\partial q} - \frac{\partial H}{\partial q}\frac{\partial}{\partial p}. \tag{14}$$

To those unfamiliar with this notation for a vector, it means the components in the **p**- and in the **q**-directions are given by $\frac{\partial H}{\partial p}$ and by -$\frac{\partial H}{\partial q}$, respectively. The equations of motion are given by

$$\dot{q} = \frac{\partial H}{\partial p} \text{ and } \dot{p} = -\frac{\partial H}{\partial q}. \tag{15}$$

Hence to a differential geometer, classical physics is finding solutions to flows of symplectic transformations. Solving the harmonic oscillator in this formulation gives its geometric meaning. The Hamiltonian is $H = \frac{1}{2}\, p^2 + \frac{1}{2}\, q^2$ (in units where mass and frequency are one). The equation of motion can be written in vector form as

$$\frac{d}{dt}\begin{bmatrix} p \\ q \end{bmatrix} = \begin{bmatrix} 0 & -1 \\ 1 & 0 \end{bmatrix}\begin{bmatrix} p \\ q \end{bmatrix}. \tag{16}$$

Given initial data (p_o, q_o) the solution is

$$\begin{bmatrix} p \\ q \end{bmatrix} = e^{t\begin{bmatrix} 0 & -1 \\ 1 & 0 \end{bmatrix}}\begin{bmatrix} p_o \\ q_o \end{bmatrix} = \begin{bmatrix} cos(t) & -sin(t) \\ sin(t) & cos(t) \end{bmatrix}\begin{bmatrix} p_o \\ q_o \end{bmatrix}. \tag{17}$$

This shows how the Hamiltonian of the harmonic oscillator generates rotations in classical phase space. Among the symplectic transformations, the orientation-preserving linear coordinate transformations form a group isomorphic to the matrix group of two by two real matrices of determinant one. This group is referred to as $SL(2, \mathbb{R})$. The infinitesimal generators are the matrices

$$H = \begin{bmatrix} 0 & -1 \\ 1 & 0 \end{bmatrix} \quad X = \begin{bmatrix} 1 & 0 \\ 0 & -1 \end{bmatrix} \quad Y = \begin{bmatrix} 0 & 1 \\ 1 & 0 \end{bmatrix}. \tag{18}$$

They form a Lie algebra under commutator bracket:

$$[Y, X] = 2H \;,\; [H, Y] = -2X \;,\; [H, X] = 2Y. \tag{19}$$

In phase space, we generate symplectic transformations by the action of Hamiltonians under the Poisson bracket. Defining the Hamiltonians (overloading the use of H, X, and Y)

$$H = \frac{1}{2}(p^2 + q^2) \;,\; X = pq \;,\; Y = \frac{1}{2}(p^2 - q^2). \tag{20}$$

We see we obtain the same Lie Algebra under Poisson bracket:

$$\{Y, X\} = 2H \;, \{H, Y\} = -2X \;, \{H, X\} = 2Y. \tag{21}$$

In this example, some important geometric and algebraic ideas are shown:

- Infinitesimal transformations in phase space can be generated by functions under the Poisson bracket (in symplectic geometry it is called a moment map). In fact, these functions provide a realization of the abstract commutator bracket, under the Poisson bracket in the space of phase space functions.
- Exponentiation of the infinitesimal symmetry gives rise to a group action, in this case the rotation group $O(2)$.

Poisson brackets arise on more general manifolds that $\mathbb{R}^2$, in particular, they always exist for symplectic manifolds. For geometers, a rich source of symplectic manifolds arises from the orbits of a group in the dual of its Lie algebra. This construction is important for signal theory because it relates the geometry of phase space with representation theory.

Let G be a general Lie group, $\mathfrak{g}$ the Lie algebra of G is the vector space tangent of $\mathbb{G}$ at the identity element and let $\mathfrak{g}^*$ denote the dual vector space of $\mathfrak{g}$.

The group G acts on itself by conjugation: $x \mapsto gxg^{-1}$. This is an automorphism that fixes the identity element, and therefore the differential of this action is a map $Ad_g \colon \mathfrak{g} \to \mathfrak{g}$.

This map of the Lie group G on its Lie algebra $\mathfrak{g}$ is called the adjoint representation. The dual representation on the vector space dual to g is called the co-adjoint representation and denoted by $Ad_g^* \colon \mathfrak{g}^\star \to \mathfrak{g}^*$.

It is defined by

$$\langle Ad_g^* \xi, X \rangle = \langle \xi, Ad_{g^{-1}}, X \rangle. \tag{22}$$

The importance of the co-adjoint orbit construction is the correspondence of orbits of a Lie group G in the dual $\mathfrak{g}^\star$ of its Lie algebra and its irreducible representations [21]. In particular, for the Heisenberg group, one has that the irreducible representations of the Heisenberg group $\mathbb{H}$ are in one-to-one correspondence with the orbits of $\mathbb{H}$ on the vector space $\mathfrak{h}^\star$ (the dual of its Lie algebra $\mathfrak{h}$). Consider the following assumptions:

- For a function $f \in L^2(\mathbb{H}) \cap L^1(\mathbb{H})$, let $\mathscr{F}^2(f)$ be the usual Fourier transform on the orbit that identifies $\mathfrak{h}$ and $\mathfrak{h}^\star$ (restricted to the orbit).
- Let $d\mu(\xi) = \frac{dpdq}{2\pi\lambda}$ be the Liouville measure on the orbit (p, q, λ).
- Let $\pi(f)$ be the integrated representation of the Heisenberg group defined in the radar section.

Kirillov's theorem for the Heisenberg group in two dimensions is the following statement:

There is a bijection between the co-adjoint orbits and the equivalent classes of unitary irreducible representations of $\mathbb{H}$. The Orbit O_λ is associated to a representation π_λ by the character formula [22, 23]:

$$Tr\pi_\lambda(f) = \int_{\mathcal{O}_\lambda} \mathcal{F}^2(f)\, d\mu(\xi). \tag{23}$$

To see how this program works, consider the Lie algebra $\mathfrak{h}$ of $\mathbb{H}$ realized as the three-dimensional vector space of upper triangular matrices (isomorphic to $\mathbb{R}^3$). Take as a basis the matrices

$$P = \begin{bmatrix} 0\,1\,0 \\ 0\,0\,0 \\ 0\,0\,0 \end{bmatrix}, Q = \begin{bmatrix} 0\,0\,0 \\ 0\,0\,1 \\ 0\,0\,0 \end{bmatrix}, Z = \begin{bmatrix} 0\,0\,1 \\ 0\,0\,0 \\ 0\,0\,0 \end{bmatrix}$$

A generic vector X can be written as

$$X = \begin{bmatrix} 0 & \alpha & \gamma \\ 0 & 0 & \beta \\ 0 & 0 & 0 \end{bmatrix}. \tag{24}$$

The Lie algebra structure of $\mathfrak{h}$ is given by the commutator bracket of its elements. Here we have $[P, Q] = Z$ and all others zero.

Lie algebra coordinates of $\mathfrak{h}$ can be used as coordinates for the Lie group $\mathbb{H}$ via the exponential map $exp : \mathfrak{h} \to \mathbb{H}$ so that the group coordinates of an element $e^X \in \mathbb{H}$ are (α, β, γ).

The following identity holds between three arbitrary group elements

$$e^{X_1} e^{X_2} = e^{\frac{1}{2}[X_1, X_2]} e^{X_1+X_2} \overset{\text{def}}{=} e^{X_3}. \tag{25}$$

This gives the group law of the Heisenberg group as in the previous section

$$\alpha_3 = \alpha_1 + \alpha_2, \beta_3 = \beta_1 + \beta_2, \gamma_3 = \gamma_1 + \gamma_2 + \frac{1}{2}(\alpha_1\beta_2 - \alpha_2\beta_1). \tag{26}$$

The adjoint action is

$$Ad_g X = Ad_{\alpha,\beta,\gamma} X = gXg^{-1} = g \begin{bmatrix} 0\; a\; c \\ 0\; 0\; b \\ 0\; 0\; 0 \end{bmatrix} g^{-1} = \begin{bmatrix} 0\; a & c - \alpha b + \beta a \\ 0\; 0 & b \\ 0\; 0 & 0. \end{bmatrix} \tag{27}$$

In this realization, the dual of the Lie algebra $\mathfrak{h}^\star$ is identified with the lower triangular matrices. A generic element can be written as

$$\begin{bmatrix} 0 & 0 & 0 \\ p & 0 & 0 \\ \lambda & q & 0. \end{bmatrix} \tag{28}$$

The pairing $\langle Y, X\rangle$ is given by the trace, which is simply the inner product of the two vectors:

$$\langle Y, X\rangle = tr(YX) = pa + qb + \lambda c. \tag{29}$$

The co-adjoint action can be found by duality and straight computation to be given by

$$Ad^*_{(\alpha,\beta,\gamma)}(p, q, \lambda) = Ad^*_g(p, q, \lambda) = (p + \beta\lambda, q - \alpha\lambda, \lambda). \tag{30}$$

The orbits of $\mathbb{H}$ can be seen to be planes isomorphic to $\mathbb{R}^2$ for each value of $\lambda \neq 0$. Let $g = (\alpha, \beta, \gamma) \in \mathbb{H}$ be generated by $a = (P, Q, Z) \in \mathfrak{h}^\star$. The vector fields of right translation X^R_a are

$$X^R_P = \frac{\partial}{\partial\alpha} - \frac{\beta}{2}\frac{\partial}{\partial\gamma}\,,\ X^R_Q = \frac{\partial}{\partial\beta} - \frac{\alpha}{2}\frac{\partial}{\partial\gamma}\,,\ X^R_Z = \frac{\partial}{\partial\gamma}. \tag{31}$$

While the vector fields of left translation X^L_a are

$$X^L_P = -\frac{\partial}{\partial\alpha} - \frac{\beta}{2}\frac{\partial}{\partial\gamma},\ X^L_Q = \frac{\partial}{\partial\beta} + \frac{\alpha}{2}\frac{\partial}{\partial\gamma},\ X^L_Z = \frac{\partial}{\partial\gamma}. \tag{32}$$

The commutator bracket of X^R_P and X^R_Q is

$$[X^R_P, X^R_Q] = X^R_Z = \frac{\partial}{\partial\gamma}. \tag{33}$$

This is the infinitesimal character which is explained in the following remarks:

1. The coordinate λ (in dual relationship with the group parameter γ) parameterizes the orbits that stack under planes of "height" λ. Kirillov theorem gives a geometrical meaning to the representation theory of the Heisenberg group by giving a picture of the irreducible representations. Each "lives" in a two-dimensional plane isomorphic to $\mathbb{R}^2$.
2. The orbit method labels in phase space the irreducible representations of the group. We could have studied the action of the Heisenberg group on phase space with the method described previously. Consider the simple Poisson bracket of the coordinate functions as was mentioned earlier:

$$\{p, q\} = 1\ \text{, all others zero.} \tag{34}$$

It is a realization in phase space of the abstract Heisenberg Lie algebra $\mathfrak{h}$:

$$[P, Q] = Z\ \text{, all others zero.} \tag{35}$$

Now consider the vector field defined by the Hamiltonian $H = -\lambda\,(\alpha p + \beta q)$:

$$\xi_H = \{H, \cdot\} = \frac{\partial H}{\partial p}\frac{\partial}{\partial q} - \frac{\partial H}{\partial q}\frac{\partial}{\partial p} = -\alpha\lambda\frac{\partial}{\partial q} + \beta\lambda\frac{\partial}{\partial p}. \tag{36}$$

The equations of motion can easily be integrated to give the co-adjoint action of the Heisenberg group $\mathbb{H}$ as was found before.

4 Symplectic Geometry and Holomorphic Automorphisms

Kahler manifolds are both Riemannian and symplectic and of great importance in many fields like information theory [24] and quantum computation [25].

Let M be an n-dimensional complex manifold and let G denote the group of holomorphic transformations of M. Following Calabi [26], it is useful to think of M together with its conjugate manifold $\overline{M}$. The group G acts on a chart $U \bigotimes \overline{U}$ of $M \bigotimes \overline{M}$ via holomorphic transformations on M and anti-holomorphic ones on $\overline{M}$ in a local coordinate system given by $(z, \overline{v}) = (z_1, \ldots, z_n, \overline{v}_1, \ldots, \overline{v}_n)$. We recover M itself by simply considering the real analytic submanifold obtained from $M \bigotimes \overline{M}$ by taking the diagonal set of points. Real analytic objects in M can be thought of having local coordinates $(z, \overline{z})$ of a chart U without any confusion. From now on let us assume that M is a Kahler manifold. To say that M is Kahler implies the existence of a real function $\Phi(z, \overline{z})$ on a chart U called the Kahler potential of the metric

$$ds^2 = g_{j\overline{k}} dz^j d\overline{z}^k \tag{37}$$

given by

$$g_{j\overline{k}} = \frac{\partial^2 \Phi}{\partial z_j \partial \overline{z}_k}. \tag{38}$$

Any other Kahlerian potential on U differs from Φ by the sum of a holomorphic and an anti-holomorphic function.

As a real manifold, M is symplectic of real dimension $2n$ with a codes two form ω of type $(1, 1)$. Compatibility with the complex structure requires that it be written as

$$\omega = g_{j\overline{k}} dz^j \wedge d\overline{z}^k \sqrt{-1}. \tag{39}$$

Now recall how classical mechanics is developed on a symplectic manifold (M, ω). Aided by the non-degeneracy of ω one sets a vector bundle isomorphism $S : TM \to TM^*$ between the real tangent and cotangent bundles defined by the interior product with ω:

$$S(X) = i(X)\omega \in TM^* \text{ for } X \in TM$$

The vector field $X_f = S^{-1}(df)$ is called a Hamiltonian vector field and f a Hamiltonian function. The Poisson bracket of two functions f_1 and f_2 is defined in a coordinate-free way by

$$\{f_1, f_2\} = -\omega(X_{f_1}, X_{f_2}). \tag{40}$$

In the local coordinates of our Kahlerian manifold M, it takes the form

$$\{f_1, f_2\} = \omega^{j\bar{k}} \left(\frac{\partial f_1}{\partial z_j} \frac{\partial f_2}{\partial \bar{z}_k} - \frac{\partial f_1}{\partial \bar{z}_j} \frac{\partial f_2}{\partial z_k} \right), \tag{41}$$

where $g_{j\bar{k}}$ and $\omega^{j\bar{k}}$ are the inverses of each other. One may describe classical mechanics in this setting by thinking of a point in (M, ω) as the dynamical state of our physical system in phase space. Fixing a Hamiltonian H, the motion of our dynamical system is given, by definition, by an integral curve $\lambda(t)$ of the Hamiltonian vector field X_H, with the parameter t interpreted as the time. One can make a slight generalization by considering a differentiable family f_t of elements of $C^\infty(M)$ satisfying

$$\frac{df_t}{dt} = \{H, f_t\}. \tag{42}$$

This can be considered the fundamental coordinate-free equation of classical mechanics. Indeed, if one lets f_t be the usual coordinate functions one obtains (in a chart) the usual Hamilton equations. The point here is that one can consider the functions in phase space as the fundamental objects of the theory. This way of thinking lies at the core of the passage to quantum theory via star products.

Classical symmetries are incorporated in the language of symplectic geometry via a Lie group action on M that here we take to be holomorphic. We associate to an element X of the Lie algebra $\mathfrak{g}$ of G a vector field in M, also denoted by X, by

$$(X_f)(p) = \frac{d}{dt} f(e^{tX} \circ p)|_{t=0} \tag{43}$$

for an arbitrary $f \in C^\infty(M)$ and $p \in M$. Such an action is said to be symplectic if it leaves the symplectic form ω invariant. The action is to be Hamiltonian if in addition the associated vector field X is Hamiltonian. Given a Hamiltonian action it is natural to consider the linear function that associates to an element X of the Lie algebra $\mathcal{G}$, the value of the corresponding Hamiltonian at each point of M. One then defines the moment map $J : M \to \mathfrak{g}^\star$ by assigning to each point of M the linear function defined above. As is well known, Hamiltonian vector fields only determine a Hamiltonian function up to a co-cycle in Lie algebra cohomology. It will be useful to consider the following.

Definition 1 A holomorphic vector field on a chart U of a Kahlerian manifold M is a complex vector field Z of type $(1, 0)$ such that $Z(f)$ is a holomorphic function for every holomorphic function f. An anti-holomorphic vector field is defined analogously.

Since we are considering holomorphic automorphisms, the fundamental two form ω is invariant under the G-action. In particular the one-parameter subgroup generated by a Lie Algebra element X satisfies $\mathcal{L}_X \omega = 0$, which means that X is locally Hamiltonian. This is equivalent in the Kahler case to say that X is a Killing field of the Kahler metric. In our situation, X is defined as the infinitesimal action

$$X = \frac{d}{dt}\left(e^{tX} \circ z, \overline{e^{tX} \circ z}\right)|_{t=0}, \tag{44}$$

where the decomposition $X = Z + \overline{Z}$ is such that $Z \in TM^{(1,0)}$ is holomorphic and $\overline{Z} \in TM^{(1,0)}$ is anti-holomorphic. On the local coordinates of a chart U the fundamental vector field is given by

$$X = \xi^i(z)\frac{\partial}{\partial z_i} + \overline{\xi^i}(\overline{z})\frac{\partial}{\partial \overline{z}_i}.$$

But X being locally Hamiltonian implies the existence of a local function H such that

$$dH = -i(x)\omega \tag{45}$$

Putting these facts together we have that on chart U, H must satisfy independently

$$\omega^{i\overline{j}}\frac{\partial H}{\partial \overline{z}_j} = \xi^i(z)$$

and

$$\omega^{i\overline{j}}\frac{\partial H}{\partial z_j} = \overline{\xi^i}(\overline{z}).$$

Knowledge of a Kahler potential Φ on a chart U will determine a solution to these local equations up to the addition of constants. Indeed, because we know how a given Kahler potential Φ may differ another one on a chart U, its Lie derivative with respect to X will have the form

$$\mathcal{L}_X\Phi = Z(\Phi) + \overline{Z}(\Phi) = \varphi(z) + \overline{\varphi}(\overline{z}). \tag{46}$$

Here ϕ and $\overline{\phi}$ are holomorphic and anti-holomorphic functions on U, generally not complex conjugates of one another unless the G action is real. Armed with this data on U the solution to the above equations is found to be given by the following.

Lemma 1 *A Hamiltonian H for the vector field $X = Z + \overline{Z}$ described above is given by either of the following local expressions:*

$$H = 2i(Z)\alpha - \varphi(z) \tag{47}$$

$$= 2i(\overline{Z})\alpha - \overline{\varphi}(\overline{z}) \tag{48}$$

$$= i(X)\alpha + \frac{1}{2}\left(\overline{\varphi}(\overline{z}) - \varphi(z)\right) \tag{49}$$

where α is the canonical one form determined in chart by $d\alpha = \omega$ which in terms of a Kahler potential Φ becomes

$$\alpha = \frac{1}{2}\left(\partial\Phi - \overline{\partial}\Phi\right)$$

Proof The equivalence of the first two equations follows from

$$2i(Z)\alpha - 2i(\overline{Z})\alpha = Z(\Phi) + \overline{Z}(\Phi) = \varphi(z) + \overline{\varphi}(\overline{z})$$

so that

$$2i(Z)\alpha - \varphi(z) = 2i(\overline{Z})\alpha + \overline{\varphi}(\overline{z})$$

The third expression for H is just the average of the first two. To verify that H is a Hamiltonian for X in U consider

$$\begin{aligned} dH &= d \circ i(X)\alpha + \frac{1}{2}\left(\overline{\partial}\Phi - \partial\Phi\right) \\ &= \mathcal{L}_X\alpha - i(X) \circ d\alpha + \frac{1}{2}\left(\overline{\partial}\Phi - \partial\Phi\right) \end{aligned}$$

But the Lie derivative of α itself is

$$\mathcal{L}_X\alpha = \frac{1}{2}\left(\overline{\partial}\Phi - \partial\Phi\right)$$

so that

$$dH = -i(X) \circ d\alpha = -i(X)\omega$$

and the Lemma is proven. □

Up to the addition of constants, these Hamiltonians (under the Poisson bracket) provide us with a Lie algebra homomorphism with the Lie algebra $\mathfrak{g}$ of G. We shall refer to them as preferred observables.

Corollary 1 *If $\mathcal{L}_X\alpha = 0$, the Hamiltonians of interest can be simplified to*

$$H = 2i(Z)\alpha \tag{50}$$

$$= 2i(\overline{Z})\alpha \tag{51}$$

$$= i(X)\alpha \tag{52}$$

5 Deformation Quantization

Let (M, ω) denote a Kahler manifold, which we interpret here as the phase space of a physical system. The associative algebra $N = (C^\infty(M), \cdot)$, defined on $C^\infty(M)$ by the usual product of functions, is thought of as the classical algebra of observables. As is well known, $C^\infty(M)$ possesses also a Lie algebra structure provided by Poisson bracket multiplication. The main idea behind deformation quantization is to arrive a quantum mechanics via a formal deformation (in the sense of [27]) of these algebraic structures. Let $E(N, \hbar)$ be the space of formal functions of $\hbar \in \mathbb{C}$ with coefficients in N. We shall call $\hbar$ the deformation parameter. An element F of $E(N, \hbar)$ can be thought as a series in $\hbar$ with coefficients in $C^\infty(M)$. Consider the bilinear map $E(N, \hbar) \bigotimes E(N, \hbar) \to E(N, \hbar)$ given by

$$(F, G) \to F \star G = F \cdot G + \sum_{n=1}^{\infty} C_n(F, G)\hbar^n. \tag{53}$$

This map is said to define a star product if, for all, $F, G, H \in E(N, \hbar)$, $\lambda \in \mathbb{C}$, we have

(1) Associativity, i.e., $(F \star G) \star H = F \star (G \star H)$.
(2) $C_1(F, G) - C_1(G, F) = \{F, G\}$, the Poisson bracket of F and G.
(3) $C_n(\lambda, F) = C_n(F, \lambda) = 0$ for all n.

Now consider an action of a Lie group G by holomorphic symplectomorphisms as above. The moment map described above gives us (modulo constants) a set of functions that close under the Poisson bracket to the Lie algebra $\mathfrak{g}$ of G. We would like to preserve this structure upon quantization. We distinguish these functions with the name preferred observables.

Definition 2 A star product is said to be G-invariant if it satisfies

$$a \star F - F \star a = i\hbar\{a, F\} \tag{54}$$

for all $a \in \mathfrak{g}$, $F \in E(N, \hbar)$.

This assumption carries profound consequences as was first shown by Fronsdal [18]. One starts by introducing the star exponential of an observable H (not necessarily a preferred observable):

$$\text{Exp}(tH) = \sum_{n=0}^{\infty} \frac{1}{n!} \left(\frac{t}{i\hbar}\right)^n (H*)^n, \tag{55}$$

where

$$(H*)^n \stackrel{\text{def}}{=} \underbrace{H * H \ldots * H}_{n-times}.$$

That is, $\text{Exp}(tH)$ is the exponential function of H using the deformed product. We will assume that $\text{Exp}(tH)$ converges to a function or a distribution on M and has a Fourier expansion of the form

$$\text{Exp}(tH) = \sum_{\lambda \in I} \Pi_\lambda\, e^{\left(\frac{-it}{\hbar} E_\lambda\right)}. \tag{56}$$

We call I (a sequence in $\mathbb{C}$), the spectrum of H, E_λ an eigenvalue of H, and Π_λ the projector associated with λ.

By obtaining the star exponential of a Hamiltonian, one has effectively a full independent formulation of Quantum Mechanics as was shown in Bayen et al. [4, 5]. In a nutshell, one obtains the analog of the spectral decomposition of a Unitary operator by considering the Fourier transform in t, of $\mathrm{Exp}(tH)$, thereby obtaining the spectrum associated with H as well as the eigen-projectors in one decomposition. Many examples of this have been given. We see therefore deformation quantization as an independent method of quantization. If desired one can make contact with Hilbert space methods using the concept of star polarizations as was developed by Fronsdal [18]. Also, deformation quantization arises naturally using coherent states [7]. Both approaches can be seen to imply the other. Here we would like to emphasize the relation of deformation quantization with Feynman integrals. Indeed, as we shall see in the next section, the star exponential of an observable is directly related to the phase space path integral as it is usually defined.

Remark 1 One has several methods to construct the star exponential, instead of trying to perform the sum directly [5, 6]. Here we would like to point out that the star exponential can be obtained by solving the differential equation

$$i\hbar\frac{\partial}{\partial t}\mathrm{Exp}(tH) = H \star \mathrm{Exp}(tH) \tag{57}$$

plus boundary conditions depending on the order of the differential equation obtained.

For a second-order equation, we have

$$\mathrm{Exp}(tH)|_{t=0} = 1 \tag{58}$$

and

$$i\hbar\frac{\partial}{\partial t}\mathrm{Exp}(tH)|_{t=0} = H, \tag{59}$$

which follow from the definition of $\mathrm{Exp}(tH)$. One way to establish this autonomous formulation is to consider the geometric data: a Kahler manifold (M, ω) and the action of a holomorphic automorphism group G. Together with the moment maps described above this is the pre-quantum data of the geometric quantization program. It turns out that it is all that is needed.

We will call $Lie_{\mathfrak{g}} \subset N$ the algebra of preferred observables with basis $L_k (k = 1, \ldots, dim\mathfrak{g})$. By a star polynomial one means an expression involving $\mathbb{C}$, $Lie_{\mathfrak{g}}$ and the operations of addition and star multiplication.

Proposition 1 *Let (M, ω) be a Kahler manifold and let $Lie_{\mathfrak{g}}$ denote the algebra of preferred observables constructed above. Let*

$$a \star f = a \cdot f + i\hbar Z_a(f) \tag{60}$$

$$f \star a = a \cdot f - i\hbar \overline{Z}_a(f) \tag{61}$$

for all $a \in Lie_{\mathfrak{g}}$, $f \in E(N, \hbar)$. Then we have a $\mathfrak{g}$-invariant star product on star polynomials of $Lie_{\mathfrak{g}}$.

Proof It is not hard to show that $a \star f$ and $f \star a$ satisfy the definition of a star product. In fact, it is $\mathfrak{g}$-invariant since

$$a \star f - f \star a = i\hbar\left(Z_a(f) + \overline{Z}_a(f)\right) = i\hbar\{a, f\}. \tag{62}$$

The only non-trivial thing to check is associativity. We need to verify that $(a \star f) \star b = a \star (f \star b)$ for all $a, b \in Lie_{\mathfrak{g}}$, $f \in E(N, \hbar)$. We have

$$\begin{aligned}(a \star f) \star b &= (a \cdot f + i\hbar Z_a(f)) \star b \\ &= a \cdot f \cdot b + i\hbar\left(bZ_a(f) - \overline{Z}_b(a \cdot f)\right) + \hbar^2 \overline{Z}_b(Z_a(f))\end{aligned} \tag{63}$$

and

$$\begin{aligned}a \star (f \star b) &= a \star (b \cdot f + i\hbar \overline{Z}_b(f)) \\ &= a \cdot f \cdot b + i\hbar\left(Z_a(b \cdot f) - a\overline{Z}_b(f)\right) + \hbar^2 Z_a(\overline{Z}_b(f)).\end{aligned} \tag{64}$$

Comparing the last two equations we see that the terms of order $\hbar$ are equal since $Z_a(b) = -\overline{Z}_b(a)$. The equality of terms of order $\hbar^2$ follows from the fact that holomorphic and anti-holomorphic vector fields commute. Star polynomials are defined inductively. In terms of a basis, monomials are defined by

$$L_{i_1} \star \ldots \star L_{i_n} = L_{i_1} \star \left(L_{i_2} \star \ldots \star L_{i_n}\right) \tag{65}$$

and the star product between them as

$$\left(L_i \star \ldots \star L_j\right) \star (L_m \star \ldots \star L_n) = L_i \star \ldots \star L_j \star L_m \star \ldots \star L_n. \tag{66}$$

It is easy to see that this procedure is internally consistent. □

Example 1 ($M = \mathbb{R}^{2n} = \mathbb{C}^n$) Let $(z, \overline{z}) = (z_1, \ldots, z_n, \overline{z}_1, \ldots, \overline{z}_n)$ be canonical coordinates on M. The symplectic form ω can be obtained from the Kahler potential $\Phi = z\overline{z}$. The preferred observables are z_i, $\overline{z_j}$ and $z_i\overline{z}_j$. They close to the oscillator algebra of dimension $2n$. In this case, we have a closed expression for the star product of any two functions $F, G \in C^\infty(M) = N$. Let us define

$$(F \star G)(z, \overline{z}) = \sum_{k=0}^{\infty} \frac{\hbar^k}{k!} C^k(F, G), \tag{67}$$

where

$$C^k(F,G) = \sum_{i=1}^{n} \left(\frac{\partial^I}{\partial z_i} \frac{\partial^{II}}{\partial \overline{z_i}} \right)^k (F,G) \tag{68}$$

and the I (resp., II) operator applies to the first (resp., second) function.

One can verify that this is a star product on $N = C^\infty(M)$ in the sense given above. In particular, the star products of the preferred observables with any arbitrary function have the form of equations (60) and (61). Consider, for example, the Hamiltonian of n-uncoupled harmonic oscillators: $H = \sum_{k=1}^{n} \alpha_k z_k \overline{z}_k$ where the α_k are real numbers. We have

$$(H \star F)(z, \overline{z}) = HF + \hbar \sum_{k=1}^{n} \alpha_k \overline{z}_k \frac{\partial F}{\partial \overline{z}_k}.$$

To obtain the star exponential of H, we solve equation (57) under initial condition (58). The solution can be easily found to be

$$\mathrm{Exp}(tH) = \exp(\sum_{k=1}^{n} \frac{1}{\hbar}(e^{-i\alpha_k t} - 1)\overline{z}_k z_k). \tag{69}$$

This result can also be obtained by a purely combinatorial calculation from the definition of the star exponential as was done in [11] for the one-dimensional case. In either case, one obtains the spectrum (here the positive integers) and the quantum states directly from the Fourier expansion in t. (see [4–6] for other examples). We would like to point out that in this example we have that

$$\mathrm{Exp}(tH) = e^{\frac{z\overline{z}}{\hbar}} U(z, \overline{z}, t), \tag{70}$$

where $U(z, \overline{z}, t)$ denotes the kernel of the evolution operator $e^{\frac{-i}{\hbar}H}$ [28]. Usually such a kernel is expressed in terms of a Feynman path integral. In the case of the harmonic oscillator, Feynman's path integral is determined by the evaluation of the action along the classical path. We shall see that one can introduce star products autonomously for homogenous Kahler manifolds to recover analogous results. That is, the preferred observables will have an exact saddle-point approximation. Similarly, remark that the star exponential of H is still a Gaussian, and therefore the steepest descent approximation of the integral over M of $\mathrm{Exp}(tH)$ is exact. In the case of compact symplectic manifolds, this exactness generalizes via the Duistermaat–Heckman localization formula.

That is, f is a good observable if $X_f = Z_f + \overline{Z}_f$ with Z_f holomorphic and $\overline{Z}_f$ anti-holomorphic. It is not hard to verify that good observables close under the Poisson bracket to an algebra of good observables. In many situations, the algebra of good observables will be large enough to serve as a basis for the smooth functions of M. This will be the case of complex symmetric spaces.

Proposition 2 *Let M be a Kahler manifold. Let $\mathcal{G}$ denote an algebra of good observables. Let*

$$a \star f = a \cdot f + i\hbar Z_a(f) \tag{71}$$

$$f \star a = a \cdot f - i\hbar \overline{Z}_a(f) \tag{72}$$

for all $a \in \mathcal{G}$, $f \in E(N, \hbar)$. This defines a $\mathcal{G}$-invariant star product on star polynomials of $\mathcal{G}$.

It is instructive to see what a good observable looks like in local coordinates. If H is a good observable,

$$Z_H = \omega^{i\bar{j}} \frac{\partial H}{\partial \overline{z}_j} \frac{\partial}{\partial z_i}$$

and

$$\overline{Z}_H = \omega^{i\bar{j}} \frac{\partial H}{\partial z_i} \frac{\partial}{\partial \overline{z}_j}$$

are holomorphic and anti-holomorphic vector fields. This means that locally, there exist holomorphic and anti-holomorphic functions such that

$$\omega^{i\bar{j}} \frac{\partial H}{\partial \overline{z}_j} = h_i(z)$$

and

$$\omega^{i\bar{j}} \frac{\partial H}{\partial z_i} = \overline{h}_i(\overline{z}),$$

where $\overline{h}_i$ is the complex conjugate of h_i if H is a real function. Using the Kahler potential Φ, we can give a local expression for H:

$$\begin{aligned} H &= \frac{\partial \Phi}{\partial z_i} h_i(z) + h_0(z) \\ &= \frac{\partial \Phi}{\partial \overline{z}_i} \overline{h}_i(\overline{z}) + \overline{h}_0(\overline{z}) \end{aligned}$$

$$H = \frac{1}{2}\left(\frac{\partial \Phi}{\partial z_i} h_i(z) + \frac{\partial \Phi}{\partial \overline{z}_i} \overline{h}_i(\overline{z}) \right) + \frac{1}{2}\left(h_0(z) + \overline{h}_0(\overline{z}) \right), \tag{73}$$

where the locally holomorphic and anti-holomorphic functions:$h_0(z)$ and $\overline{h}_0(\overline{z})$ are such as to make the equality between the first two expressions hold.

We can apply our proposition to the following situation. Consider the co-adjoint orbits of a complex semisimple Lie group G on the dual of its Lie algebra $\mathcal{G}^*$. Let G_λ be the stabilizer of a point $\lambda \in \mathcal{G}^*$. The co-adjoint orbit of G passing through λ is isomorphic to the homogeneous space G/G_λ. We have that $M = G/G_\lambda$ is a Kahlerian manifold and G acts on M holomorphically ([29]). Examples of

M are bounded symmetric domains and projective spaces. Interpreting M as a co-adjoint orbit, we have via the moment map a set of preferred observables related to the Lie algebra generators. On the other hand, since the G-action is holomorphic these observables will be good.

The simplest non-trivial example is that of $M = S^2$. Using z as the local canonical coordinate (obtained by stereographic projection) we have a holomorphic action of $SL(2, C)$ given by

$$z \mapsto gz = \frac{\alpha z + \gamma}{\beta z + \delta},$$

where

$$g = \begin{bmatrix} \alpha & \gamma \\ \beta & \delta \end{bmatrix}.$$

The Kahler potential is usually given by $\Phi = \ln(1 + z\bar{z})$, which yields the metric

$$g(z, \bar{z}) = \frac{\partial^2 \Phi}{\partial z \partial \bar{z}} = \frac{1}{(1 + z\bar{z})^2}.$$

The Poisson bracket of two functions f and g is

$$\{f, g\} = (1 + z\bar{z})^2 \left(\frac{\partial f}{\partial z} \frac{\partial g}{\partial \bar{z}} - \frac{\partial f}{\partial \bar{z}} \frac{\partial g}{\partial z} \right).$$

It is easy to verify that the vector fields associated with the one- parameter subgroups are Hamiltonian and that we have a set of good observables L_i, $i = 1, 2, 3$:

$$L_1 = \frac{z + \bar{z}}{1 + z\bar{z}}, \quad L_2 = \frac{z - \bar{z}}{1 + z\bar{z}}, \quad \text{and} \quad L_3 = \frac{z\bar{z}}{1 + z\bar{z}}.$$

So far we have obtained a deformation of the polynomial algebra of observables given by the above formulas. No restriction on the value of $\hbar$ is needed. When constructing the star exponential, however, restrictions on the value of $\hbar$ can occur if we impose single-valuedness to obtain a reasonable physical interpretation. This is similar to the quantization of coupling constants in the path integral quantization of multi-valued actions. Indeed, while star products yield an autonomous method of quantization, the construction of the star exponential is intimately related to path integration, as it has been pointed out in [6, 13].

6 Hilbert Space Methods

So far we have found a path integral in S^2 while looking for a general formula for the star exponential of a given Hamiltonian. We should remark that in an independent approach, both the star exponential and the path integral have separately an autonomous meaning. The correspondence with conventional Hilbert space quantum mechanics can be found by a method similar to polarization of geometric quantization [6]. Let $F \in C^\infty(W)$ denote the solution space of the following equation:

$$\Psi * a = \lambda(a)\Psi \tag{74}$$

for all $a \in g_o$, where g_o is a sub-algebra of g and $a \mapsto \lambda(a)$ is an infinitesimal character of g_o.

Associativity of the star product guarantees the stability of F under the transformation $\Psi \mapsto h * \Psi$, $h \in C^\infty(W)$. This construction works where polarization in the sense of geometric quantization might not. A slight variation of this construction can be given for SU(2) as follows. Using equations (30) and (31), one can determine the star product of the Lie algebra generators with arbitrary functions in phase space.

We found that it is useful to locally parameterize in terms of x and $\overline{x}$. If $\overrightarrow{L}$ $= 2l(\sin\theta\cos\phi, \sin\theta\sin\phi, \cos\theta)$, let $z = 2l\cot\frac{\theta}{2}e^{i\phi}$, the stereographic projection, and obtain for the $\overline{L} \sim \xi$ the following expressions:

$$L_+ = \frac{1}{2}(L_1 + iL_2) = 2l\left(\frac{z}{1+z\overline{z}}\right) \tag{75}$$

$$L_- = \frac{1}{2}(L_1 - iL_2) = 2l\left(\frac{\overline{z}}{1+z\overline{z}}\right) \tag{76}$$

$$L_o = L_3 = 2l\left(\frac{z\overline{z}-1}{1+z\overline{z}}\right). \tag{77}$$

Then we have for an arbitrary function:

$$L_+ * f = L_+ f - z^2\partial_z f \tag{78}$$

$$L_- * f = L_- f - \partial_z f \tag{79}$$

$$L_o * f = L_o f - 2z\partial_z f \tag{80}$$

$$f * L_+ = L_+ f + \partial_z f \tag{81}$$

$$f * L_- = L_- f - \overline{z}^2\partial_z f \tag{82}$$

$$f * L_o = L_o f + 2z\partial_z f. \tag{83}$$

We now consider as our space F those functions in the image of the star exponential that satisfy the following condition:

$$\Phi * L_+ = 0 \tag{84}$$

the solutions have the form

$$\Psi = \left(\frac{1}{1+z\overline{z}}\right)^{2l} \psi(z). \tag{85}$$

It is not hard to show that one can define operators $\hat{L}_i = \hat{L}_+, \hat{L}_-, \hat{L}_o$ by

$$L_i * \Psi = \left(\frac{1}{1+z\overline{z}}\right)^{2l} (\hat{L}_i\psi)(z) \tag{86}$$

$$\hat{L}_+\psi(z) =< z|\hat{L}_+|\psi >= (2lz - z^2\partial_z)\psi(z) \tag{87}$$

$$\hat{L}_-\psi(z) =< z|\hat{L}_-|\psi >= \partial_z\psi(z) \tag{88}$$

$$\hat{L}_o\psi(z) =< z|\hat{L}_o|\psi >= (2z\partial_z - 2l)\psi(z). \tag{89}$$

We establish therefore a relation between operators and functions

$$f(z, \overline{z}) = \langle z|\hat{F}\ |z\rangle \tag{90}$$

and, in particular,

$$\mathrm{Exp}(tH) = \langle z|e^{\frac{i}{\hbar}\hat{H}}\ |z\rangle \tag{91}$$

relating the star exponential to the definition of the holomorphic path integral.

7 Conclusions and Discussions

Motivated by phase space measurements in radar theory, classical physics in symplectic manifolds was reviewed with emphasis in Hamiltonian group actions. The special case of classical dynamics in Kahler manifolds is considered. Quantum theory is introduced autonomously using only geometrical data by the introduction of star products on preferred observables. This can be applied to bounded symmetric domains, productive spaces, and Grassmann manifolds. The example of the Bloch sphere is constructed explicitly and its connection with Hilbert approaches is given for comparison.

References

1. H. Weyl, *The Theory of Groups and Quantum Mechanics* (Courier Corporation, 1950)
2. J.E. Moyal, Quantum mechanics as a statistical theory, in *Mathematical Proceedings of the Cambridge Philosophical Society*, vol. 45 (Cambridge University Press, 1949), pp. 99–124
3. M. Flato, D. Sternheimer, harmonic analysis and representations of Semisimple Lie Groups (1980)
4. F. Bayen, M. Flato, C. Fronsdal, A. Lichnerowicz, D. Sternheimer, Deformation theory and quantization. I. Deformations of symplectic structures. Ann. Phys. **111**, 61–110 (1978)
5. F. Bayen, M. Flato, C. Fronsdal, A. Lichnerowicz, D. Sternheimer, Deformation theory and quantization. II. Physical applications. Ann. Phys. **111**, 111–151 (1978)
6. C. Alcalde, Star quantization on the cylinder. J. Math. Phys. **31**, 2672–2682 (1990)
7. S. Gutt, Cohomology associated with the Poisson Lie algebra. Lett. Math. Phys. **3**, 297–310 (1979)
8. C. Moreno, *-products on some Kähler manifolds. Lett. Math. Phys. **11**, 361–372 (1986)
9. C. Moreno, Invariant star products and representations of compact semisimple Lie groups. Lett. Math. Phys. **12**, 217–229 (1986)
10. C. Moreno, Geodesic symmetries and invariant star products on Kähler symmetric spaces. Lett. Math. Phys. **13**, 245–257 (1987)
11. R. Berger, Algebraic quantization. Lett. Math. Phys. **17**, 275–283 (1989)
12. A.C. Hirshfeld, P. Henselder, Deformation quantization in the teaching of quantum mechanics. Am. J. Phys. **70**, 537–547 (2002)
13. P. Sharan, Star-product representation of path integrals. Phys. Rev. D **20**, 414 (1979)
14. L. Ioos, D. Kazhdan, L. Polterovich, Berezin–Toeplitz quantization and the least unsharpness principle, in *International Mathematics Research Notices* (2021)
15. H. Bergeron, E.M. Curado, J.P. Gazeau, L.M. Rodrigues, Weyl-Heisenberg integral quantization (s): a compendium (2017). arXiv preprint arXiv:1703.08443
16. K. Fujii, Introduction to Grassmann manifolds and quantum computation. J. Appl. Math. **2**, 371–405 (2002)
17. A.V. Gorokhov, Coherent states, Kähler manifolds, and quantum dynamics, in *Laser Physics, Photonic Technologies, and Molecular Modeling*, vol. 12193 (SPIE, 2022), pp. 161–164
18. C. Fronsdal, Some ideas about quantization. Rep. Math. Phys. **15**, 111–145 (1979)
19. R.E. Blahut, W. Miller, C.H. Wilcox et al., *Radar and sonar: part I, in The IMA Volumes in Mathematics and its Applications*, vol. 32 (Springer, New York, 1991)
20. V. Guillemin, S. Sternberg, *Symplectic Techniques in Physics* (Cambridge University Press, 1990)
21. A.A. Kirillov, *Lectures on the Orbit Method*, vol. 64 (American Mathematical Society, 2004)
22. L.P. Rothschild, for Nilpotent Lie Groups. Math. Z. **140**, 63–65 (1974)
23. H. Zahir, Produits STAR et représentation des groupes de Lie. PhD thesis, Université Paul Verlaine-Metz (1991)
24. M. Molitor, Exponential families, Kähler geometry and quantum mechanics. J. Geom. Phys. **70**, 54–80 (2013)
25. M.R. Kibler, Quantum information and quantum computing: an overview and some mathematical aspects (2018). arXiv preprint arXiv:1811.08499
26. E. Calabi, Isometric imbedding of complex manifolds. Ann. Math. 1–23 (1953)
27. M. Gestenhaber, Deformation theory of algebraic structures. Ann. Math. **79**, 59 (1964)
28. L. Faddeev, A. Slavnov, Gauge fields: introduction to quantum theory (1980)
29. A.W. Knapp, Representation theory of semisimple groups, an overview based on examples (1986)

Phase Space Quantization II: Statistical Ideas

Carlos Alberto Alcalde and Kanishka Tyagi

To the memory of ***Moshe Flato***

Abstract The purpose of this paper is to introduce statistical states through signals used as the probes of physical systems. Phase space is the arena of both quantum and classical mechanics. This work reviews the origin of information in signal theory as a gateway to understanding quantum information theory. Concepts of operational physics and Lie group representation theory are applied to the study of phase space measurements. A statistical model of Hamiltonian dynamics in the Koopman representation is a Hilbert bundle over a symplectic manifold. Bochner's theorem and its relationship with group characters pave the way to statistical observables in classical physics through an extension of the Fourier transform. The natural extension to non-commutative signal theory gives rise to representations of dynamical groups like in quantum mechanics. Two methods of quantization are discussed with emphasis on the appearance of statistical states: Souriau's functions of positive type and the more general scheme of star product quantization. Quantum Signals are described by states arising from representations of the SU(2) group on the Poincare sphere that are introduced in the deformation theory approach by the star exponential.

Keywords Radar theory · Statistical manifolds · Deformation quantization · Koopman representation · Co-adjoint orbit · RKHS · Functions of positive type

C. A. Alcalde
Technical Director for Radar SW, NXP Semiconductors, Grenoble, Auvergne-Rhône-Alpes, France

K. Tyagi (✉)
Senior Machine Learning Scientist, Aptiv Advance Research Center, Agoura Hills, CA, USA
e-mail: kanishka.tyagi@aptiv.com

R. Pandey et al. (eds.), *Quantum Computing: A Shift from Bits to Qubits*,
Studies in Computational Intelligence 1085,
https://doi.org/10.1007/978-981-19-9530-9_4

1 Introduction

The operational approach to physics [1, 2] suggests elevating statistical considerations to a fundamental principle relating the states of physical systems and the observations we make of them. Hence it is natural for the geometry of information to appear alongside the symplectic geometry of classical physics and geometrical approaches to quantum theory. Phase space can be understood as the set that contains the variables that describe a physical system. Classically, the concept of state is associated with the initial conditions. It is assumed to have the structure of a symplectic manifold X. Any probability distribution on X is called a state. Mathematically, this involves considering a measurable space $(X, \mathscr{B})$ and observations as realizations of random variables on X. A statistical manifold is defined as the set all probability distributions on X. Signals are elements of L^2 Hilbert spaces and signal processing aims to extract information from the registration of values from stochastic processes, often referred to as a time series. The appearance of Hilbert spaces is reminiscent of quantum theory. There are indeed formal similarities between the two theories. It is therefore important to keep in mind their obvious differences. In signal theory measurements, there are two distinct physical systems. One is the object being observed by the signal and the other is the signal itself when received by a sensor. The wavefunction of quantum theory and the so-called the waveform of classical signal theory are not the same objects at all in spite of having similar mathematical descriptions. Most importantly, measurements in standard signal processing do not alter what is observed. There is no wave function collapse as the observables remain in the realm of classical physics.

The interpretation of signal theory as a statistical model requires a comparison with the more familiar statistical models of classical mechanics and quantum theory. Because classical dynamics is not usually presented as a statistical theory, it is relevant to consider probability distributions in the abstract setting of C-star algebras. The appearance of characters and measure algebras are by no means superfluous mathematical abstractions but necessary elements for the passage to quantum mechanics and the understanding of signal theory.

In the usual presentation of quantum theory, states are not represented by measures in phase space but instead by rays in Hilbert space. More generally, states are density matrices. One approach to facilitate the classical to quantum transition is to construct a Hilbert space over phase space, in the so-called the prequantization step [3]. Full quantization requires that we provide an irreducible representation of the Lie Algebra of observables, the quantum operators. In signal theory, the states of the signal are also rays in Hilbert space but since measurements themselves do not change the state system, the observables must remain commutative. The Koopman operators [4] are well suited for such a description. In our approach, the space of signals, a well-defined object of study, is identified with the Koopman representation space.

In a nutshell, signal theory is composed of classical observables and signal states responding to the classical dynamics in Hilbert space. Abstractly, signal theory forms a Hilbert bundle over the space of motions. The resemblance of quantum mechanics and signal theory is in the mathematical description of the wavefunction on the one hand and the waveform of the signal on the other. In particular:

- Signals are in general complex waves obeying the superposition principle.
- Dynamics in phase space is represented in the space of signals by complex amplitudes like in quantum mechanics.

The localization of phase space points within the resolution of the sensor and the non-commutativity of the Lie groups describing the dynamics have an impact on the statistical descriptions and the choice of probability distributions.

These issues will not be tackled here directly but considerations from the theory of coherent states will be relevant [5]. Further insights of the theory can be found through geometric approaches to quantum mechanics. For this purpose, functions of positive type as reproducing kernels of signals and deformation quantization are briefly introduced leaving the developments and applications for future work [6]. This paper is organized as follows.

- Section 1: Radar signals in phase space.
- Section 2: The algebraic representation of classical mechanics and the introduction of statistical states in the stochastic phase space.
- Section 3: The Koopman representation and the statistical model of signals on phase space.
- Section 4: Non-commutative auto-correlation of signals and quantum mechanical states.
- Section 5: Deformation Quantization of the Sphere.

2 Radar Signals and Quantum Wave Functions

Radar signal processing introduces the idea of probes (initial states), observables (the radar environment), and final states as the registration of values. It is in this sense that it resembles the quantum measurement problem with the important difference that the radar signal is an electromagnetic wave that interacts but is not associated with the object like in quantum mechanics. It is a classical measurement since it does not change the nature of the observable. Radar is an indirect measurement of an environment and estimation of phase space parameters. In particular, it calculates the range and range-rate from the sensor to the radar targets. It will useful to keep in mind the following properties of the radar problem as an example of a statistical model of signal theory in phase space.

- A probe (the transmit signal) defines the resolution and probability of detection of the radar sensor.
- The ambiguity function or wavelet transform, is an inner product of the original probe with its return to the radar sensor (the receive signal). If the received signal is modeled by a Lie group transformation, it is a function in an homogeneous space.

- For the Heisenberg group, a two-dimensional Fourier transform of the ambiguity function is called the Wigner distribution and it defines its variables from the corresponding co-adjoint orbit.
- The interaction between the transmit and the receive signals makes it possible for the information to be extracted from the environment.

2.1 The Heisenberg Group

It has been recognized that radar signal processing fits well within the representation theory of the Heisenberg group [7]. The Heisenberg group $\mathbb{H}$ is a nilpotent Lie group isomorphic to $\mathbb{R}^3 = \mathbb{R}^2 \otimes \mathbb{R} \approx \mathbb{C} \otimes \mathbb{R}$ as a manifold. In the parametrization of a generic element $g \in \mathbb{H}$ given by (x, y, s) the group multiplication is

$$g_1 g_2 = (x_1, y_1, s_1)(x_2, y_2, s_2) \stackrel{\text{def}}{=} \left(x_1 + x_2, y_1 + y_2, s_1 + s_2 + \frac{x_1 y_2 - x_2 y_1}{2} \right) \quad (1)$$

The group identity is the origin $(0, 0, 0)$ and the inverse element $g^{-1} = (-x, -y, -s)$. The infinite-dimensional representations that we are interested in signal theory are classified by a non-zero real number λ: For $\varphi \in L^2(\mathbb{R})$ define the operator $\pi_\lambda(x, y, s)$

$$\pi_\lambda(x, y, s)\varphi(t) \stackrel{\text{def}}{=} e^{i\lambda s} e^{i\lambda \frac{xy}{2}} e^{i\lambda x t} \varphi(t + y) \quad (2)$$

This representation of the Heisenberg group is related to the wave equation formulation of quantum theory, hence it is the most appropriate that goes by the name of Schroedinger representation. It is a representation in the sense that for any two group elements g_1, g_2 we have

$$\pi_\lambda(g_1)\pi_\lambda(g_2) = \pi_\lambda(g_1 \cdot g_2) \quad (3)$$

The Schroedinger representation is unitary, irreducible (no invariant L^2 subspaces in $\mathbb{R}$) and by the Stone von-Neumann theorem, any other unitary irreducible representation $\tilde{\pi}$ there are unitary intertwining operators U such that $\tilde{\pi} = U^{-1}\pi_\lambda U$ for some non-zero λ.

The infinitesimal generators of the group, are the elements of the Heisenberg Lie Algebra $\mathfrak{h}$ usually denoted by P, Q and I and the Lie Algebra multiplications are represented by commutators or by Poisson brackets depending if operators or functions in phase space are used. This is of course Dirac's paradigm.

The so-called integrated representation of the Heisenberg group can be considered the definition of the Fourier transform of a function of the group $f \in L^2(\mathbb{H}) \cap L^1(\mathbb{H})$ as an operator valued integral [8]

$$\pi(f) \stackrel{\text{def}}{=} \int_{\mathbb{H}} f(g)\pi_\lambda(g)dg, \quad \text{with } g \in \mathbb{H} \text{ and } dg \text{ the } \mathbb{H}\text{-invariant Haar measure.} \tag{4}$$

Fixing on an irreducible representation λ, the Heisenberg group Fourier transform is

$$\pi(f) = \int\int_{\mathbb{R}^2} f_\lambda(x, y)\pi_\lambda(x, y)\, dx\, dy, \quad \text{where } \pi_\lambda(x, y) \stackrel{\text{def}}{=} \pi_\lambda(x, y, 0). \tag{5}$$

For $\varphi \in L^2(\mathbb{R})$ the operator $\pi_\lambda(x, y, s)$ is the fundamental representation of the theory that bears the name of Schrödinger.

$$\pi_\lambda(x, y)\varphi(t) \stackrel{\text{def}}{=} e^{i\lambda\frac{xy}{2}}\varphi_{xy}(t) \tag{6}$$

The action of $\pi_\lambda(x, y)$ induces the windowed Fourier transform.

$$V_\varphi : L^2(\mathbb{R}) \to L^2(\mathbb{R}^2), \quad s \mapsto \langle\overline{\varphi_{xy}}, s\rangle_{L^2(\mathbb{R})} \tag{7}$$

Following Thangavelu [8] consider $\pi(f)$ as a Hilbert-Schmidt operator on $\varphi \in L^2(\mathbb{R})$ in

$$(\pi(f)\varphi)(u) = \int_{\mathbb{R}} K_f(u, v)\varphi(v)\, dv \tag{8}$$

$$K_f(u, v) = \int_{\mathbb{R}} f_\lambda(x, u - v)\, e^{i\lambda x(\frac{u+v}{2})}\, dx. \tag{9}$$

The Kernel of the Operator $\widehat{F}$ acting on $L^2(\mathbb{R})$ is a the result of partial transform of the function in the Lie group $f_\lambda(x, y)$ in the first variable. In the theory of pseudo differential operators, f_λ is called the covariant symbol or spreading function. In terms of the Kernel K_f,

$$K_f(u, v) \equiv \langle u, \widehat{F}v\rangle \equiv \langle u|\widehat{F}|v\rangle, \tag{10}$$

the contravariant symbol $F(p, q)$ of $\widehat{F}$ is defined by

$$F_\lambda(p, q) = \int_{\mathbb{R}} K_f(q + \tfrac{t}{2}, q - \tfrac{t}{2})\, e^{i\lambda pt}\frac{dt}{2\pi\lambda} \tag{11}$$

This is the Weyl transform discovered in 1927 as an attempt to map operators to functions in phase space. The Wigner and Windowed Fourier transform are contravariant and covariant symbols of the rank-one operator $\Phi = \varphi_1\langle\varphi_2, \cdot\rangle$ and are Fourier transforms of each other. The Fourier transform of V_φ is the Wigner transform $W_\varphi = \mathscr{F}^2(V_\varphi)$.

Alongside with the symbols of an operator, let us consider the matrix element of the operator π_λ of the Schroedinger representation between two signals ψ and $\varphi \in L^2(\mathbb{R})$. It is the windowed Fourier transform once again:

$$\langle \pi_\lambda(\alpha, \beta)\varphi, \psi \rangle = \int_{\mathbb{R}} e^{i\lambda\frac{\alpha\beta}{2}} e^{-i\lambda\alpha t}\overline{\varphi(t-\beta)}\psi(t)dt \tag{12}$$

$$= \int_{\mathbb{R}} e^{-i\lambda\alpha u}\overline{\varphi(u-\beta/2)}\psi(u+\beta/2)du \tag{13}$$

$$= e^{i\lambda\frac{\alpha\beta}{2}}(V_\varphi\psi)(\alpha\beta) \tag{14}$$

The windowed Fourier transform can thus be seen as the covariant symbol of a rank-one operator and the matrix element of the operator defining the Schroedinger representation.

The matrix element of an operator expressed through auxiliary signals is:

$$\langle \hat{F}\varphi, \psi \rangle = \iint_{\mathbb{R}^2} K_f(x, y)\overline{\varphi(x)}\psi(y)dxdy \tag{15}$$

In terms of the Wigner transform it can be written as

$$\langle \hat{F}\varphi, \psi \rangle = \iiint_{\mathbb{R}^3} F_\lambda(p, \frac{x+y}{2})e^{i\lambda p(x-y)}\overline{\varphi(x)}\psi(y)dxdydp \tag{16}$$

$$\langle \hat{F}\varphi, \psi \rangle = \iint_{\mathbb{R}^2} F_\lambda(p, q)(W_\varphi\psi)(p, q)dpdq \tag{17}$$

While in terms of the Ambiguity function (windowed Fourier transform) is:

$$\langle \hat{F}\varphi, \psi \rangle = \iiint_{\mathbb{R}^2} f_\lambda(x, y)\langle \pi_\lambda(x, y)\varphi, \psi \rangle dxdy \tag{18}$$

$$= \iint_{\mathbb{R}^2} f_\lambda(x, y)e^{i\lambda\frac{xy}{2}}(V_\varphi\psi)(x, y)dxdy \tag{19}$$

2.2 Statistical Interpretation

In both quantum mechanics and signal theory, matrix elements like the ambiguity function, carry initial statistical data. The above representations can be interpreted as the match of the observable matrix element with statistical information provided by the signals. In a nutshell, the "generic" representation $f_\lambda(x, y)$ responds through the prior probability amplitude given by the ambiguity function $(V_\varphi\psi)$. The Fourier dual of the same concepts are the phase space function $F_\lambda(p, q)$ and the Wigner transform of the signals $(W_\varphi\psi)$, with the difference being the symplectic manifold with coordinates (p, q) as opposed to the (homogeneous) group parameters x, y.

We see that the ambiguity function is invariant under the central element $(0, 0, s)$. This is an example of the general situation of a Lie group G and a subgroup H producing a function on G/H, which is $(V_\varphi)(x, y)$ for the Heisenberg group. The two dimensional Fourier transform lives on a co-adjoint orbit of $\mathbb{H}$ so its coordinates are labeled by (p, q). Hence one writes the Wigner transform $W = W_\varphi(p, q)$. The parameter γ parametrizes the probe used in the measurement and is related to the radar resolution and probability of detection. The choice $\gamma = \frac{1}{2}$ is special. In that case V_φ is proportional to an analytic signal so one can write the image of V_φ as holomorphic functions $\psi(z)$. In this situation one uncovers the function analytic properties of signal theory and corresponding holomorphic coherent states in quantum theory.

The integrated group representation implies that on the functions in the group is given by group convolution which is in general non-commutative:

$$\pi(\tilde{f}_1)\pi(\tilde{f}_2) = \pi(\tilde{f}_1 o \tilde{f}_2) \tag{20}$$

The phase space counterpart of the group convolution product is the star product $F_1 \star F_2$ in the co-adjoint orbit. Since matrix elements like the ambiguity function are related to the states, it could be expected that in phase space the Wigner function would correspond to the analogous statistical state.

The reason Wigner did not succeed to find probability distributions related to quantum mechanics states (other than violating the uncertainty principle) is that the Wigner function is the image of rank-one-projectors that do not satisfy positivity in general. However, probability distributions can be obtained from Wigner functions if we consider that:

- Wigner functions are positive if generated by Gaussian probes.
- If $\Omega(p, q)$ is any pointwise non-negative phase space function (Wigner function or not), then the Abelian convolution $\Omega \circ W$, is a positive Wigner function for any Wigner function W.
- If $\Omega(p, q)$ and W are Wigner functions, then $\Omega \circ W$ is always non-negative (albeit possibly not a Wigner function), while $\Omega \star W$ will always be a Wigner function (albeit possibly not positive).
- Wigner functions are distributions in phase space roughly concentrated on the duration and bandwidth of the signal, they do not describe the phase space localization of the radar target.

Hilbert-Schmidt operators form a vector space (they can be considered a Hilbert space if the operators are Trace class and an operator algebra, which means that they can be multiplied like matrices) but phase space signal analysis is limited by the fact that operators need to act on signals to be defined in specific domains and the fact that by choosing an irreducible representation we have halved the phase space variables. The geometry of phase space is less than clear. It is the aim of this paper to introduce the star product formalism which can be used to multiply functions in phase is a complete analogy with operators thereby allowing for geometric insights and allows further generalizations.

The elements of radar signal theory are classical dynamics and the representation theory of electromagnetic signals reflecting from a classical object [9]. When (as a practical approximation) one describes the radar target using Hamiltonian mechanics, its dynamics get encoded in the reflected electromagnetic wave like the wavefunction in phase space quantum mechanics.

The appearance of non-commutativity in radar theory has an elementary physical explanation. Measurement of the velocity of an object does not commute with the measurement of the range to the object. This is simply a consequence of both being functions of time in general. In case of short time radar probes (short in time with respect to the movement of the object), the velocity does not change in during the measurement and both the speed and the range to the object can be measured simultaneously in this approximation. In that situation, radar signal processing can thought of being generated by an Abelian group and commutative Fourier analysis holds. Otherwise, like is the case of quantum mechanics in general, we shall speak of non-commutative signals or probes.

2.3 Stationary Stochastic Processes

Measurements in signal theory are random variables through additive noise and a statistical description is natural given the finite resolution of any system. In other words, the signal probe provides the phase space of the object with a fixed size Borel family turning it into a Probability space of signal densities. From the point of view of ergodic theory, the time evolution of the objects is described by the group of automorphisms of phase space and after interaction with the probe, this information gets registered in the signal.

One approach is to consider the stochastic process to be defined by transition probabilities, objects that like Markov kernels have a double life as probability measures in phase space and measurable functions at any fixed time. The following mathematical structure is useful for the understanding of the data recording of signals. Let $\Omega \subseteq \mathbb{R}^N$ be an open set. A function $\Phi : \Omega \to \mathbb{C}$ is said to be of positive type if it is continuous and for every set of M points $x_k \in \Omega$, the M by M matrix Φ defined by $\Phi_{kl} = \Phi(x_k - x_l)$ is positive definite. Signals, ordered in time are often described as random vectors and the following definitions are usual in:

- The vector $\{x_1, x_2, ..., x_M\}$ of **realizations** of random variables $X_1, X_2, ..., X_M$ is called a time series if it is composed of samples of a signal $\varphi(t) \in L(\mathbb{R}^2)$ over a common probability distribution.
- A stochastic process is said to be stationary it is invariant under the Lie group of time translations: $(U_t\varphi)(u) = \varphi(u - t)$
- $R(t) = \langle \varphi, U_t\varphi \rangle_{L^2(\mathbb{R})} = (\overline{\varphi} \circ \varphi)(t)$ is a function of positive type.
- In finite dimensions, the matrix R build of correlations of the type $\overline{x_k}x_l$ is called the covariance matrix and serves as an estimate of the true covariance (see definition below).
- When the stochastic process is Gaussian, the matrix R^{-1} is the Fisher matrix.

The case of non-commutative Lie groups is generalized by Jorgensen [10] as follows. Let G be a locally compact Lie group, and let $(\Gamma, \mathscr{B}(\Gamma))$ be a probability space on a manifold Γ. A stochastic L^2 process is a system of random variables labeled by points in $G : \{X_g, g \in G\}$. Covariance is defined through the expectation value: $r_X(a, b) = \mathbb{E}(X_a X_b)$. Assuming zero mean, a stationary process is defined by the condition

$$r_X(ga, gb) = r_X(a, b). \tag{21}$$

If follows that $\mathbb{E}(X_a X_b) = r_X(a^{-1}b)$.

A stochastic process is said to be stationary if the covariance is of positive type. That is if and only if any realization of stochastic process creates positive definite matrices in finite dimensional subspaces. An example of non-commutative covariance is the ambiguity function of radar theory described in the previous section. Functions of positive type will be denoted by $\mathscr{P}(G)$ and will remain relevant in the remaining sections.

3 Statistical Models

3.1 *The Statistical Model of Classical Mechanics*

The axiomatic notion of a statistical model of a physical theory starts with a given set S, whose elements are called states, and another set $C(S)$ considered the observables. For an arbitrary $F \in C(S)$ there will be a set M_F of outcomes. For an arbitrary state $s \in S$ and an observable F there is a probability distribution μ_s^F on M_F called the probability distribution of the observable F in the state s.

An abstract C-star algebra characterization is a good starting point to realize these axioms in classical physics. The act of observation is not supposed to change the physical states which leads to consider an algebra of commuting observables $C(X)$ over a space of states X. Furthermore, commutative C-star algebras allow us to study topological spaces X directly through the algebraic properties of the associated complex-valued function space $C(X)$. If X is compact and Hausdorff one can reconstruct the space X itself via $C(X)$ invoking Gelfand theory. Let $A = C(X)$ be a commutative algebra (under point multiplication) and consider its character space, they are non-zero homomorphisms that satisfy

$$\omega(ab) = \omega(a)\omega(b), \quad \text{with} \quad \omega : A \to \mathbb{C} \tag{22}$$

collected as a set $\Delta(A) = \{$ all the characters of A $\}$. The map $X \to \Delta(C(X))$ that sends $\xi \in X$ to $\omega_\xi \in \Delta(C(X))$ is simply an evaluation map of an element $u \in C(X)$

$$\omega_\xi(u) \equiv u(\xi) \in \mathbb{C} \tag{23}$$

Gelfand proved that this map is an isomorphism such that $X \cong \Delta(C(X))$. A similar reasoning works for $L^1(\mathbb{R})$ since $L^1(\mathbb{R})$ is a Banach algebra under convolution:

$$(u \circ v)(x) = \int u(x-y)v(y)\,dy \tag{24}$$

For each $x \in \mathbb{R}$ define a character via the Fourier transform $\mathscr{F}(u)$:

$$\omega_x(\mathscr{F}(u)) \equiv (\mathscr{F}(u))(x) = \int u(t)e^{-i\langle x,t\rangle}\,dt \tag{25}$$

This is the map that diagonalizes the Abelian convolution product $\mathscr{F}(u \circ v) = \mathscr{F}(u)\mathscr{F}(v)$ pointwise.

This algebraic setting can be applied to classical mechanics: The physical interpretation which starts by considering a bounded region $\Gamma \subseteq X$ as the phase space of our physical system (To illustrate the main idea it enough to consider the simplest situation):

- Points in Γ describe the state of our system. Example: $(q, p) \in \Gamma$
- Observables are smooth functions on Γ. Example: Coordinate functions $(Q, P) \in C(\Gamma)$
- The outcomes of measurements are point evaluations of specific observables. Example: (Q_0, P_0)
- $C(\Gamma)$ and $C(\Delta)$ are Fourier transform pairs related by the translation invariance of phase space.

The Gelfand isomorphism of commutative C-star algebras makes all these physically relevant ideas degenerate in the sense that the distinction of a state, an observable, and the registration of its values are blurred. From this (Aristotelian) point of view, we can describe observations without having to do the experiments. Operationally, one considers any physical system after it is prepared (calibrated) and allowed to interact with an environment resulting in measurement and the registration of data. The outcome space is defined within the resolution of our instruments. The statistical description comes from the necessity to correlate our observations with the states of our physical system. In the case of classical physics, phase space is defined as a smooth manifold X that contains the complete set of variables that describe our physical system. The engineering application of measure theory is to consider a Borel σ-field of subsets $\mathscr{B}(X)$ as resolution cells of a physical system. Any positive measure or probability distribution on X can be considered a classical state.

3.2 The Stochastic Phase Space

Consider the following heuristics to motivate the understanding of phase space functions on stochastic spaces. Lebesgue integration builds simple functions from a measurable space $(X, \mathscr{B})$

$$F(\xi) = \sum_{k \in I} F(k)\chi_k(\xi), \tag{26}$$

with

$$\chi_K(\xi) = \begin{cases} 1, & \text{if } \xi \in \Delta_k \\ 0, & \text{otherwise} \end{cases} \tag{27}$$

where

$$\Delta_k \in \mathscr{B}, \quad \text{and} \quad \Delta_k \cap \Delta_l = \emptyset \quad \text{if} \quad k \neq l. \tag{28}$$

A careful limit process with suitable approximate identities allows one to write in the sense of distributions, the measurement $k \to \xi$

$$F(\xi) = \int_X F(\xi')\delta(\xi - \xi')\, d\xi' \tag{29}$$

When the measurable space is turned into a probability space by the introduction of a probability measure: $(X, \mathscr{B}, \rho)$ one can write

$$F_\rho(\xi) = \int_X F(\xi')\rho(\xi - \xi')\, d\xi' = (F \circ \rho)(\xi) \tag{30}$$

A Lebesgue-integrable bounded function F (built up from indicator functions like above) is a member of the function space $L^1(X)$. Let $\mathscr{M}(\Gamma)$ be the set all complex-valued measures on X, that is of the type $\rho d\mu$. We have an inclusion $L^1(X) \subseteq \mathscr{M}(X)$. Probability measures are then associated with the positive $L^1(X)$ functions as probability densities. The set of such probabilities distributions on $(X, \mathscr{B}(X))$ is called the statistical manifold S.

Consistent with the approach above, Prugovecki [11], enhances the phase space X (more properly, the space of motions) with a statistical Manifold S and the interpretation is that an element ρ in S describes the probability of reading $\xi \equiv (p, q)$ of an experiment that has been prepared within $\Delta \in \mathscr{B}(X)$:

$$\mu_\xi(\Delta) = \int_\Delta \rho_\xi(\xi')\, d\mu(\xi') \tag{31}$$

When X possesses an Abelian group symmetry, the measure $d\mu$ is the Haar translationally invariant measure and one should interpret it as the (complete ignorance) uniform probability distribution. In this situation we shall write $G = X$ to emphasize the group property and re-write the equation above for any bounded function f whether positive or not.

$$\mu_f(\Delta) = \int_\Delta f(g)\, d\mu(g) \tag{32}$$

It is well known that the characters $\chi(g)$ of an Abelian group G generate the dual group $\widehat{G}$ and one can associate to an $L^1(G)$ its Gelfand transform: comparison with the definition of the Fourier transform of $L^1(\mathbb{R})$ given previously shows that both transforms coincide in this context:

$$\widehat{f}(\chi) = \int f(g)\chi(g^{-1})\, d\mu(g) \tag{33}$$

These notions come together if one considers the functions of positive type $\mathscr{P}(G)$. These are functions in the group whose matrix entries are represented by Hermitian positive definite matrices in every dimension. For more details consult Folland [12]. The Gelfand isomorphism still holds with respect to density weighted Fourier transform [13]. This isomorphism is now between elements that measure algebra and a reproducing kernels which are precisely the functions of positive type. In other words, general integrable functions are related to complex measures while the subset of functions of positive type gives us real probability distributions that we can associate with classical states. This is a consequence of Bochner's theorem that relates functions of positive type (reproducing kernels) and probability measures via the Fourier transform of Abelian groups. In the notation above, this means that if $\Phi \in \mathscr{P}(G) \Rightarrow \widehat{\Phi}$ is the probability measure on $\widehat{G}$ given by $\mu_{\widehat{\Phi}}$.

Introducing statistical distributions through the measure algebra is an elegant way to re-write a consistent algebraic classical model. Nevertheless, probability distributions appear as an afterthought and are not direct a consequence of the theory. The passage to a statistical description renders the values of the observables fuzzy or non-sharp in the POVM sense (see discussion below). When projection valued measures are utilized one returns to the usual (sharp) description of trajectories in phase space

4 The Statistical Model of Signals in Phase Space

The statistical model of signals exploits the duality between states and observables and is equivalent to the Koopman representation of classical mechanics in Hilbert space. In classical physics, probability distributions were originally introduced in the study of the kinetic theory of gases. In signal theory, statistical considerations appear due to the precision of our instruments and the practical notion of a pure state. A pure state like in quantum mechanics refers to a rank-one projection but the determination of the purity of a state depends on the data collected and the instrument itself. These considerations are epistemological and refer to the signal only.

An abstract dynamical system consists of a measurable space (M, B, μ) and a one-parameter dynamical group $\mathrm{Aut}(M, B, \mu)$ of measure-preserving transformations of M into itself. This generalizes classical mechanics where M is the constant energy surface, T_t is Hamiltonian evolution and μ is the Liouville measure. For more details see [14, 15].

Let $T \in \mathrm{Aut}(M, B, \mu)$, $\xi \in M$ and define

$$(U_T \Psi)(\xi) = \left(\frac{d(\mu(T^{-1}(\xi))}{d\mu}\right)^{-1/2} (\Psi(T^{-1}(\xi)) \tag{34}$$

The mapping $T \mapsto U_T$ defines a group homomorphism from $\mathrm{Aut}(M, B, \mu)$ into the group of unitary operators on the Hilbert space $L^2(M, \mu, \mathscr{H})$, where $\mathscr{H}$ is an auxiliary Hilbert space interpreted here to be the space of signals.

This model for signals in phase space contains a representation of C_0 [14], the space of functions of compact support which we give the interpretation of the classical observables. We ask these functions form an commutative C-star algebra since classical dynamics evolves the states deterministically, the observables cannot change the state. In the usual presentation of Koopman, Hilbert space vectors are somewhat auxiliary: they are useful to understand general (not necessarily Hamiltonian) dynamical systems. In this incarnation, the Koopman wavefunction is the probe and the question becomes how well it represents the classical state itself.

An example of such representation is obtained given any positive measure μ on M and the Hilbert space $L^2(M, \mu) \equiv L^2(M, \mu, \mathbb{C})$. Define

$$\pi_\mu(f)\Psi = f\Psi, \quad \forall f \in C_0(M), \Psi \in L^2(M, \mu). \tag{35}$$

More generally, signal theory in phase space is not a scalar theory but has a natural Hilbert bundle structure.

Vectors in $L^2(M, \mu, \mathscr{H})$ are sections Ψ : that is, Ψ is an assignment $\alpha \in M \to \mathscr{H}$ such that the $\mathbb{C}$-valued function $\langle \Psi_1(x), \Psi_2(x)\rangle_{\mathscr{H}}$ is measurable for every $\Psi_1, \Psi_2 \in \mathscr{H}$. The two-dimensional notation $\alpha \equiv (x, y)$ is used in this section.

$$\int_M \|\Psi(\alpha)\|^2_{\mathscr{H}} \mathbf{d}\mu < \infty \tag{36}$$

$L^2(M, \mu, \mathscr{H})$ with the completion of the inner product $\int \langle \Psi_1(\alpha), \Psi_2(\alpha)\rangle_{\mathscr{H}} \mathbf{d}\mu(\alpha)$ becomes a Hilbert space of its own right.

For each $f \in L^2(M, \mu)$, operators M_f form by pointwise multiplication by $f(\alpha) \in L^2(M, \mu, \mathscr{H})$, the multiplier algebra. Expectation values in signal theory are given like in statistical mechanics by averages of the observables against probability densities.

Koopman observables form a commutative Von-Neumann algebra so they remain diagonal but an irreducible representation of the Lie algebra of observables is not required here. Pre-quantum operators suffice here to induce the dynamics in the

Koopman Hilbert space. In data science observables appear as weak measurements (inner products) that can be observed directly from the data or given a priori. It is natural to build models to explain the data from the known invariances of the system. In the Koopman representation, as in signal theory, the phase space probability density $\rho(\alpha)$ is associated with $\| \Psi(\alpha) \|^2_{\mathscr{H}}$.

Positive Operator Valued Measures (POVM) tie in the various approaches by replacing the characteristic functions implicit in the Koopman representation [16]. Projection Valued (PV) measures are defined as:

$$(P(\Delta)\Psi)(\alpha) = \chi_\Delta(\alpha)\Psi(\alpha), \quad \forall \Psi \in L^2(M) \tag{37}$$

A commutative POVM can always be expressed as the probability average of PV measures. For arbitrary $\Delta \in B$ define

$$(a(\Delta)\Psi)(\alpha) = (\chi_\Delta \circ \rho)(\alpha)\Psi(\alpha), \quad \forall \Psi \in L^2(M), \text{ and } \rho \text{ a probability density} \tag{38}$$

Measurements in signal theory are random variables on a Probability space $(\mathbb{R}^2, B, \rho)$. Equivalently they appear as POVM on the Von-Neumann algebra $L^2(\mathbb{R}^2, \sqrt{\rho})$ in the Koopman Representation.

An classical observable $F(\xi)$ on phase space $\mathbb{R}^2$ associated to the probability density ρ is defined as

$$\begin{aligned} F_\rho(\xi) &= (\rho \circ F)(\xi) &(39)\\ &= \int_{\mathbb{R}^2} F(\xi')\rho(\xi - \xi')\, d\xi' &(40)\\ &= \mathscr{F}^{-1}(\mathscr{F}(F)\mathscr{F}(\rho)) &(41)\\ &= \int_{\mathbb{R}^2} \tilde{f}(\alpha) r(\alpha) e^{i\langle \alpha, \xi \rangle}\, d\xi &(42) \end{aligned}$$

where $F : F \leftrightarrow \tilde{f}$ and $\rho \leftrightarrow r$ are Fourier transform pairs. We are led to the following:

Definition 1 The statistical observable $F_\rho(\xi)$ is defined as the adapted Fourier transform

$$F_\rho(\xi) = \int_{\mathbb{R}^2} \tilde{f}(\alpha)\, E_\xi(\alpha^{-1})\, d\alpha \tag{43}$$

with

$$E_\xi(\alpha^{-1}) \stackrel{\text{def}}{=} r(\alpha) e^{i\langle \alpha, \xi \rangle} \tag{44}$$

The commutative algebraic structure of classical observables, pointwise multiplication translates to convolution in signal theory:

$$(F_\rho \cdot G_\rho)(\xi) = \int_{\mathbb{R}^2} (\tilde{f} \circ \tilde{g})(\alpha) E_\xi(\alpha^{-1})\, d\xi \tag{45}$$

The Poisson bracket structure does not translate very well due to the fact that random variables do not have smooth trajectories. This is a hint that a statistical description exists on the space of motions instead of phase space itself. The passage to quantum theory is carried by replacing the pointwise product with a star product that is in general not commutative. The signal itself (which is of a completely different nature than the classical physical system) can evolve through non-commutative evolution.

In practical applications, one can use exponential coordinates and $r(\alpha) = V_\varphi(g = e^\alpha)$ is the auto-correlation of a stochastic process generated by a group transformation. The probability density in phase space is

$$\rho_\varphi(\xi) = \int_{\mathbb{R}^2} E_\xi(\alpha^{-1})\, d\alpha = \int_{\mathbb{R}^2} r_\varphi(\alpha) e^{i\langle\alpha,\xi\rangle}\, d\alpha = \int_{\mathbb{R}^2} \langle\varphi, \pi(e^\alpha)\varphi\rangle e^{i\langle\alpha,\xi\rangle}\, d\alpha \quad (46)$$

where π_φ is a Unitary group representation of the Lie group G, and φ is the signal that generates a stationary stochastic process. It is useful to compare what happens in the case of the non-commutative group like Heisenberg:

$$W_\varphi(p, q) = \int_{\mathbb{R}^2} V_\varphi(x, y) e^{i\lambda(px+qy)} \frac{dpdq}{2\pi\lambda} = \int_{\mathbb{R}^2} \langle\varphi, \pi_\lambda(x, y)\varphi\rangle e^{i\lambda(px+qy)} \frac{dpdq}{2\pi\lambda} \quad (47)$$

The non-commutativity of the correlation can make the Wigner function in the left-hand side of the equation non-positive but it should not be considered a failure of the theory but a feature that requires proper interpretation.

5 Non-commutative Signal Theory

The statistical manifold in signal theory is generated by probability amplitudes like in quantum mechanics. These phase space wave functions do not need to produce irreducible spaces for the observables but rather are parametrized by symplectic manifolds like in the prequantization step of geometric quantization. In order to relate with the introduction of probabilities through auto-correlations of stochastic processes, we discuss the idea of introducing quantum statistical states through functions of positive type. Deformation quantization is briefly reviewed next. The introduction of a star product is the most general geometric quantization of phase space observables as a deformation of the usual pointwise product of functions.

5.1 Some Ideas from Quantum Statistical Models

Whenever phase space points can be assigned to rays in Hilbert space (equivalently to rank-one projection operators), we have a one-to-one relationship between classical (non-statistical) and pure signal states. This correspondence is also at the core of the spectral theorem and the concept of irreducible representations. Introducing probability distributions in classical physics involves the introduction of a duality with a statistical manifold, the space of motions or a linear manifold like Hilbert space.

As long as the algebra of observables is commutative (realized by independent random variables), the associated invariant group is Abelian and there is an analogous correspondence between probability measures and functions of positive type (auto-correlations). It is remarkable that a complex amplitude such as an auto-correlation would yield a positive real measure by Fourier transform but this a consequence of positivity of eigenvalues of Hermitian matrices and the existence of enough characters to represent the space. In the case of correlated observables, such as position and momentum, functions of the positive type still represent the states in the sense of probability amplitudes of quantum theory in phase space.

In the usual presentation of classical mechanics, phase space is a symplectic manifold M, and a statistical state on M is a probability measure on M, whereas observables are smooth functions on M. Souriau [3] dualizes classical physics and makes the important distinction that the statistical states should be built on the space of motions $\hat{M}$.

$\hat{M}$ as the space of dynamical solutions, is parametrized by points (q_0, p_0), the (arbitrary) initial conditions. Souriau considers $\hat{M}$, the sample space and defines the statistical state as a probability law on $\hat{M}$. It is a symplectic manifold that lives naturally in a co-adjoint orbit of a canonical transformation in the case of Hamiltonian mechanics.

In *Interpretation géométrique des états quantiques*, Souriau [3] defines quantum states through functions of positive type. He is guided by the following facts:

- In classical statistical mechanics, Bochner's theorem relates probability measures to functions of positive type via the Gelfand isomorphism which is a linear relation between functions and their characters.
- Functions of positive type generate Hilbert spaces via the GNS construction so conventional quantum mechanics is always at hand.

The duality inherent in the above developments is natural in deformation quantization. "The idea is to isolate the principal components of the theory, equations of motion, and initial conditions from each other and to associate the former with the observables and the latter with the states" [17]. Star products are introduced as a deformation of the algebra of the observables. Trajectories in the evolution space are abandoned from the start and an algebra of classical functions on the co-adjoint orbit is built with the new product replacing the Hilbert algebra of operators of conventional quantum theory. Quantum states are introduced by the analogy of Gibbs states in deformation quantization, the star exponential. The introduction of a star product represents a full quantization and is totally general but a class of star products will be shown to be related to functions of positive type.

5.2 Reproducing Kernels and Functions of Positive Type

Signals can be associated to the symplectic manifold of motions and behave like quantum wavefunctions of phase space quantization. They form a symplectic manifold of their own, but our interest here is to interpret them in the sense of Souriau [18]. Every quantity in signal theory is epistemic in the sense of Caticha [19] and auto-correlations (weak values) allow one to construct positive operators directly from the data. Here Caticha's distinction [20] between ontic and epistemic physical theories comes in handy. Because variables are purely epistemic they only tell us about the probe or signal and in contradistinction to some quantum theories there is no pretension it could be otherwise.

Signal measurements can be understood as the result of linear transformations of the signal. For each $\alpha \in M$, let T_α be a bounded operator on the Hilbert space $L^2(M, \mu, \mathscr{H})$. If $\langle T_\alpha \varphi_1, \varphi_2 \rangle$ is a measurable function for all $\varphi_1, \varphi_2 \in \mathscr{H}$ (assuming all Hilbert spaces are isomorphic) $\{T_\alpha : \alpha \in M\}$ forms a field of operators. We can think of these operators abstractly and give them a specific representations after the signal is recorded. This is in essence what is meant by signal processing. Covariance operators are examples to keep in mind. The analysis of such operators has been carried by Jean Pierre Gazeau and others [16, 21–23]. A useful basis to study such operators in signal theory are functions of positive type [12].

If Φ is a function of positive type on a Lie group G, there is an unitary representation U of G on a Hilbert space $\mathscr{V}$ such that

$$\Phi = \langle U(g)\varphi, \varphi \rangle, \text{ for some } \varphi \in \mathscr{V}. \tag{48}$$

Functions of positive type form a positive cone and the extremes of the cone are the pure states that parameterize the irreducible representations [12]. Also, functions of positive type are reproducing kernels that project to invariant sectors in the space of square integrable functions of Lie groups [22].

Let Φ be a Function of Positive type on G, Define the space

$$\mathscr{H}(\Phi) = \{f \circ \Phi : \forall f \in L^2(G)\}. \tag{49}$$

Let $\Phi(f) \equiv f \circ \Phi$, and define an inner product on $\mathscr{H}(\Phi)$ by

$$\langle \Phi(f), \Phi(g) \rangle = \int_G \Phi(y^{-1}x)\, \bar{f}(x)\, g(y)\, dx\, dy \geq 0 \tag{50}$$

The completion of $\Phi(f)$ under this norm turns $\mathscr{H}(\Phi)$ into a RKHS with kernel $\Phi(x^{-1}y)$.

The construction can be reversed.

Let $\mathscr{H}$ be a RKHS of $L^2(G) \Rightarrow$ There is a $\Phi \in \mathscr{H}$ such that $\psi = \psi \circ \Phi, \forall \psi \in \mathscr{H}$.

- $f(x) = (f \circ \Phi)(x) = \langle f, L(x)\Phi \rangle$, where L is the left translation in the group.
- The mapping $f \mapsto f \circ \Phi$ is a continuous projection from $L^2(G)$ onto $\mathscr{H}$.

For locally compact Abelian groups if Φ is a normalized function of positive type $\Rightarrow$ $\hat{\Phi}$ is a probability measure. In signal theory, functions of positive type can be thought as basis functions for covariance matrices, and by Fourier transform to probabilities in homogeneous spaces. Reproducing Kernels and their coherent states have a rich geometry related to the Fisher metric [5, 24].

5.3 Star Products in Co-adjoint Orbits

Spherical functions of positive type are known to parametrize the set of irreducible representations [12]. In this context they appear as functions in homogeneous spaces that are matrix elements of unitary representations induced from subgroups. The dual of this construction is Kirillov's orbit method.

The importance of the co-adjoint orbit construction is the correspondence of orbits of a Lie group G in the dual $\mathfrak{g}^\star$ of its Lie algebra and its irreducible representations [25]. In particular for the Heisenberg group on has that the irreducible representations of the Heisenberg group $\mathbb{H}$ are in one-to-one correspondence with the orbits of $\mathbb{H}$ on the vector space $\mathfrak{h}^\star$ (the dual of its Lie Algebra $\mathfrak{h}$). Consider the following assumptions:

- For a function $f \in L^2(\mathbb{H}) \cap L^1(\mathbb{H})$, let $\mathscr{F}^2(f)$ be the usual Fourier transform on the orbit that identifies $\mathfrak{h}$ and $\mathfrak{h}^\star$ (restricted to the orbit).
- Let $d\mu(\xi) = \frac{dpdq}{2\pi\lambda}$ be the Liouville measure on the orbit (p, q, λ).
- Let $\pi(f)$ be the integrated representation of the Heisenberg group.

Kirillov's theorem for the Heisenberg group in two dimensions is the following statement:

There is a bijection between the co-adjoint orbits and the equivalent classes of unitary irreducible representations of $\mathbb{H}$. The Orbit O_λ is associated with a representation π_λ by the character formula [26, 27] :

$$Tr\pi_\lambda(f) = \int_{O_\lambda} \mathscr{F}^2(f)\, d\mu(\xi). \tag{51}$$

Remark that once a bi-invariant Riemann metric on a Lie group G is defined, left translations are the geodesics. Furthermore, every geodesic on Lie Group can be written this way. This makes the co-adjoint orbit method a very useful tool to study symmetries and dualities.

Recall the co-adjoint representation of a Lie group G in the dual of its Lie Algebra is

$$\langle K(g)F, X\rangle \equiv \langle F, Ad(g^{-1})X\rangle \tag{52}$$

We follow the construction of Kirillov [25] closely. On the cotangent bundle $T^*(G)$ of a Lie Group G, the bundle is trivial $T^*(G) \cong G \times g^*$ and the co-vector $K(g)F$ corresponds a pair (g, F).

Remark that $T^*(G)$ is itself a group with multiplication

$$(g_1, F_1)(g_2, F_2) = (g_1 g_2, K(g_2)^{-1} F_1 + F_2) \tag{53}$$

Hence,$T^*(G)$ is the Lie group manifold of the semidirect product $G \ltimes g^*$.

Since $T^*(G)$ is a Lie Group, the tangent bundle $T(T^*(G))$ is also trivial and

$$T(T^*(G)) \cong (G \times g^*) \times (g \oplus g^*) \tag{54}$$

The group action of $G \times G$ on G by right and left shifts $g \mapsto xgy^{-1}$ extends to a Hamiltonian action on $T^*(G)$ and can be considered as multiplication on the group manifold $T^*(G)$:

$$(x, y)(g, F) = (xgy^{-1}, K(y)F) \tag{55}$$

The usefulness of this construction is the ability to perform symplectic reduction by right, left, or combined actions and in particular the symplectic planes where the solution of our Hamiltonian equations lie. In the deformation quantization picture, the orbits of $\mathbb{H}$ on the dual of its Lie algebra $\mathfrak{h}^\star$ are determined by fixing the value of the identity element in the Lie Algebra. The identity element is the Casimir element (the group Nilpotent). The adjustment of functions in the group to the orbit and imposing group invariance is a natural way to define the star product. Details are given in [28].

A basic tool of the theory is the formal exponential in star multiplication

$$\text{Exp}(tH) = \sum_{n=0}^{\infty} \frac{1}{n!} \left(\frac{t}{i\hbar}\right)^n (H\star)^n$$

where

$$(H\star)^n \stackrel{\textbf{def}}{=} \underbrace{H \star H \star \ldots \star H}_{n-times}$$

The Star exponential can be used to define star products and is the Fourier dual of Feynman path integrals [29, 30].

Let G be the Lie group related to a Lie Algebra $\mathfrak{g}$ by the exponential map. Let the group elements e^a, e^b and e^c, associated with Lie Algebra elements a, b and c, be related by the Baker-Campbell-Hausdorff formula (BCH): $e^a e^b = e^c$.

This realization of the Lie Algebra $\mathfrak{g}$ in phase space allows us to work with phase space functions (moment maps) a, b and c under the Poisson bracket. These functions naturally live on the dual $\mathfrak{g}^\star$of the Lie Algebra. More properly the evaluation of such functions should be written as $\langle J(a), \beta \rangle$, where $J : M \to \mathfrak{g}^\star$ and $\beta \in \mathfrak{g}$. Denote simply by M a co-adjoint orbit (the space of motions).

As a consequence, the star exponential of the corresponding Lie Algebra elements satisfies a similar relationship:

$$\text{Exp}(a) \star \text{Exp}(b) = \text{Exp}(c) \tag{56}$$

This relationship allows one to define a star product on $\mathfrak{g}^\star$, or by being parametrized by the Lie Algebra coordinates we can consider $\text{Exp}(a)$ as an element of the group G itself. Define $\mathbb{E}(e^a) = \text{Exp}(a)$ and write $\mathbb{E}_\xi(g)$ for the value of the function at $\xi \in M$ (a point in the orbit). Hence we consider $\mathbb{E}_\xi(g)$ as a function on $M \times G$. By construction it has the following covariance property:

$$\mathbb{E}_{K(g)\xi}(x) = \mathbb{E}_\xi(g^{-1}xg) \quad \forall g, x \in G,$$

and where $K(g)$ denotes the co-adjoint action. The following construction based on left and right invariant vector fields works as the definition of covariant star products and a method for their construction. The strategy starts by obtaining the star exponential directly via the solution of differential equations. Let $\mathfrak{g}$ be the Lie Algebra of G, $\{L_A\} \in \mathfrak{g}$ and $Q = g_{AB}L_AL_B$ be an element of the center of the enveloping algebra. Thinking of M as an orbit of G on $\mathfrak{g}^\star$, Q_M the restriction of Q to the orbit shall be a constant function on M. Let $l(a)$ and $r(a)$ denote the vector fields associated with left and right translations on G. One finds by differentiation that

$$a * \text{Exp} = -i\hbar\, l(L^A)\,\text{Exp}, \tag{57}$$

$$\text{Exp} * a = i\hbar\, r(L^A)\,\text{Exp}. \tag{58}$$

It follows that $g_{AB}l(L^A)l(L^B)\,\text{Exp} = -Q_M\,\text{Exp}$.

These are equations that define infinitesimal characters so the Exp function can be seen to be related to functions of positive type and hence to probability measures. One interpretation of the star product in signal theory is that it intertwines distributions in G/H and in the co-adjoint orbit.

Definition 2 The quantum statistical observable is defined by the adapted Fourier transform between functions in $L^2(G)$ and functions in the dual of the Lie Algebra via the star exponential:

$$F(\xi) = Exp_\xi(\tilde{f}) = \int_G \tilde{f}(x)\, E_\xi(x^{-1})\, dx \tag{59}$$

where dx is the Haar measure.

The most general h-invariant solution is given explicitly by

$$E_\xi(e^a) = \text{Exp}_\xi(a) = \Omega(a)e^{\frac{i}{\hbar}\langle a,x\rangle} \tag{60}$$

Where $a \sim \{x_1, x_2\}$ are exponential coordinates in the Heisenberg group homogeneous space and ξ is the corresponding point in the orbit $\mathfrak{h}^\star$.

The arbitrariness of the function $\Omega(x_1, x_2)$ reflects the ambiguity of choices of various orderings that can be considered. They are called Cohen distributions in time frequency analysis. For each choice one gets a different star product or different statistical state. The generalization of [34, 35] at the end of Sect. 4 as functions of positive type in the Heisenberg group is new and leads directly to positive measures. A natural family is that of matrix elements of the metaplectic representation: $\Omega(x_1, x_2) = \langle \varphi, \pi_\lambda(x_1, x_2)\varphi \rangle$. They turn out to be Gaussian exponentials that include the Husimi distribution. These do not generate the most general star products but they parameterize states generated by canonical transformations. They hint to the relevance of symplectic geometry rather than Riemannian geometry in the study statistical manifolds in phase space.

5.4 Star Quantization in the Bloch Sphere

When $G = SU(2)$, co-adjoint orbits are fized by a real positive number R.

$$\vec{L} \cdot \vec{L} = R^2 \tag{61}$$

Here L^A denotes a basis for the SU(2) Lie algebra, thus phase is the sphere S^2. It is convenient to parameterize $SU(2)$ by identifying it with $S^3 : x = \{x_o, \vec{x}\}$ with $x_o^2 + \vec{x} \cdot \vec{x} = 1$. The vector fields of left and right translations on $SU(2)$ are given by $SO(4)$ rotations of S^3.

$$l(L^A) = (x_A \partial_o - x_o \partial_A) - (x \wedge \partial)_A \tag{62}$$

$$-r(L^A) = (x_A \partial_o - x_o \partial_A) - (x \wedge \partial)_A \tag{63}$$

where $(x \wedge \partial)_A = \epsilon_{ABC} x_B \partial_C$ General expressions for star exponential satisfy the $SU(2)$ Laplacian.

$$\sum_A l(L^A) l(L^A) Exp = -2l(2l+2) Exp \tag{64}$$

Regular solutions can only be encountered for $2l$ being an integer. In general R and l are unrelated.

Our objective is to look for solutions that are spherical functions the SU(2) Laplacian in G, which can be considered as a Laplace operator $\widetilde{G} = G \times G$, where we demand that functions $F(x, y)$ with $x, y \in G$, satisfy it in the two variables. Let G_o denote the diagonal subgroup of $\widetilde{G}$. We can think of G as the homogeneous space $\widetilde{G}/G_o$ with the identification $\underset{\sim}{(x, y)} \mapsto yx^{-1}$.

Under this map functions on $\widetilde{G}/G_o$ are identified with central functions in G since functions on $\widetilde{G}/G_o$ satisfy

$$F(gx, gy) = F(x, y) \tag{65}$$

for all $g, x, y \in G$. This function now satisfies the radial Laplace operator of homogeneous space $\widetilde{G}/G_o$. Interpreted as functions of G, because of their centrality they have to be propositional to $\chi_l(yx^{-1})$, where χ_l is an irreducible character of G. This result was verified explicit by Fronsdal for the case of SU(2). In fact this construction is known to identify characters of irreducible representations by simply letting $y = e$ (the identity). Our purpose, however is to obtain a function $\mathbb{E}_t(x^{-1})$ of $W \times G$. to this end we can use the onto map

$$G/T \times T \mapsto G, \tag{66}$$

$$\xi, D(t) \mapsto g(\xi)^{-1} D(t) g(\xi) = y \tag{67}$$

where T denotes a maximal torus on G and $g(\xi)$ is any element of gT. Alternatively, given $y \in G$, we can always diagonalize it as above in a way that is unique up to permutations of the eigenvalues. for the case of $SU(2), T = U(1)$ the map is realized by sending

$$(y_o, \vec{y}) \text{ to } (\cos t, \sin t \frac{\vec{L}}{R}) \tag{68}$$

where y^μ denote the SU(2) parameters of the group element y thought of as a point in S^3. That is, if $\vec{L} = (\sin\theta\cos\phi, \sin\theta\cos\phi, \cos\theta)$, then

$$D(t) = \begin{bmatrix} e^{it} & 0 \\ 0 & e^{-it} \end{bmatrix} \tag{69}$$

$$g(\xi) = \begin{bmatrix} \cos\frac{\theta}{2} & i\sin\frac{\theta}{2}e^{i\phi} \\ i\sin\frac{\theta}{2}e^{-i\phi} & \cos\frac{\theta}{2} \end{bmatrix} h \tag{70}$$

with an arbitrary element $h \in U(1)$ that projects to the space G/T.

The reason that G/T can be identified with the co-adjoint orbit on $T^*(G)$, taken here as the phase space W , can be seen as follows. We can start by observing that the co-adjoint action of G on W is transitive. Let ξ_o denote a fixed point on W. Next consider all $h \in G$ such that $K(h)\xi_o$. Such elements form a subgroup of G called the stabilizer group ($h \in H$), of ξ_o. For any compact Lie group G , H can always be taken to be a maximal torus. If a transformation $K(g)$ carries ξ_o to another point ξ on W, then all other transformations $K(\tilde{g})$ say, carrying ξ_o to ξ have the form $\tilde{g} = gh$ where $h \in H$. We have therefore a one-to-one correspondence between points on W and points in G/H. Furthermore, the co-adjoint action of an arbitrary element $\tilde{g}$, turns into left multiplication of the corresponding coset member such that the following diagram commutes.

$$\begin{array}{ccc} \xi & \xrightarrow{h} & g(\xi) \\ \downarrow{\scriptstyle K(\tilde{g})} & & \downarrow{\scriptstyle \tilde{g}} \\ K(\tilde{g})\xi & \longrightarrow & \tilde{g}g(\xi) \end{array} \tag{71}$$

Based on the arguments just presented we can construct a function $\Phi^l_\xi(x^{-1})$ on $W \times G$ based on the known $SU(2)$ characters such that

$$\Phi^l_\xi(x^{-1}) \propto \chi_l(yx^{-1}) = \chi_l(g(\xi)^{-1}D(t)g(\xi)x^{-1} \tag{72}$$

Finally we can define the star exponential for SU(2) by integrating the U(1) dependence to arrive on the coset which is the sphere.

$$E^l_\xi(x^{-1}) = \int_T e^{ilt}\chi_l(g(\xi)^{-1}D(t)g(\xi)x^{-1}dt \tag{73}$$

In the general case the factor e^{ilt} is replaced by a character of the corresponding maximal torus T. It can be verified that $E^1_\xi(x^{-1})$ as defined above is a general star exponential for $SU(2)$.

Writing $\chi_l(g) = \sum_m d^l_{mm}(g)$ where d^l_{mn} denote the matrix elements of the irreducible representations of $SU(2)$, we can do the integral to obtain explicit expressions for Exp. The height weight choice coincides with the Kahler orbit.

$$E^1_\xi(x^{-1}) = d^l_{ll}(g(\xi)x^{-1}g(\xi)^{-1}) \tag{74}$$

$$E^l_\xi(x^{-1}) = (x_o + i\frac{\overline{x}.\overline{L}}{R})^{2l} \tag{75}$$

With our choice of star exponential from Eq. (11) it follows that $R = 2l$ making connection with Kirillov's method of co-adjoint orbits where the orbits are in one-to-one correspondence with irreducible representations. Having the orbits parameterized by the integers we can describe the functions in phase space given by the adapted Fourier Transform

$$f^l(\xi) = Exp_\xi(\tilde{f}) = \int_G \tilde{f}(g)E^l_\xi(g^{-1}) \tag{76}$$

For each l, the dimension of the image (restricted to the $2l$ sphere)

$$A_l = \{Exp_\xi(\tilde{f}) \in C^\infty(G)\} \tag{77}$$

is $(2l + 1)^2$ and it can be shown that Eq. (23) defines an invertible adapted Fourier transform from $L^2(G)$ to $\bigoplus_l A_l$

6 Summary Conclusion and Future Work

Signal theory is presented as a new example of a statistical model in addition to the usual classical mechanical and quantum theoretical models. Probability distributions are introduced through complex amplitudes over symplectic manifolds in phase space quantum mechanics. This is also the case in signal theory where the signal itself can be considered as a type of square root of a probability measure. As long we can restrict ourselves to commuting observables, information geometry in phase space is equivalent via Bochner's theorem to the weak values of observables which turn out to be auto-correlation functions of positive type. Signal amplitudes produced by Hamiltonian dynamics that are non-commutative in the Poisson bracket sense generate quantum probability distributions but remain classical measurements as they describe the probe to a physical system and have nothing to say of the object themselves.

Our main contribution consists on our interpretation of phase space signals as probability amplitudes of quantum theory. Our presentation of classical states and observables through matrix elements in signal theory is original. Furthermore, we show that the Abelian Fourier duality of positive measures and auto-correlations is still holds in quantum theory through functions of positive type and the star exponential. We emphasize the strong relationship between statistical models and geometrical quantum states introduced by Souriau as a generalization of Bochner's theorem. The usefulness of deformation quantization as statistical models in information geometry will be expounded in our future work [6]:

- Observables and States are related through probability distributions. In the presence of group symmetries a relationship between Lagrangian and Hamiltonian formulations and the symplectic duality of a homogeneous space G/H and its co-adjoint orbit appears through the concept of yokes in information geometry [31].
- Passage from the signal space back to the space of movements is useful for comparing the signal probabilities in Hilbert space with probability distributions in the space of motions as a generalization of Lie thermodynamics. This analogy goes further if one considers that general quantization schemes can be introduced via star products. It can be shown that the choice of operator ordering in quantization is equivalent to the introduction of prior probabilities in signal theory.
- Group theoretical coherent states are related to deformation quantization and the Cartan decomposition. A star product depends on the point of observation between the initial and final states. One obtains a group theoretical interpretation of dual connections [32] that can be understood by the concept of star product equivalence. Furthermore if the Fisher matrix is approximated from the data, it generates maximum likelihood coherent states.

Finally, one should point out that deformation quantization is a vast subject but not so well known outside of mathematics. Here is a very incomplete list of contributors from the physics side [29, 33–35].

References

1. A.S. Holevo, *Probabilistic and Statistical Aspects of Quantum Theory*, vol. 1 (Springer Science & Business Media, 2011)
2. P. Busch, M. Grabowski, P.J. Lahti, *Operational Quantum Physics*, vol. 31 (Springer Science & Business Media, 1997)
3. J.M. Souriau, Interpretation geometrique des etats quantiques, in *Differential Geometrical Methods in Mathematical Physics* (Springer, 1977), pp. 76–96
4. T. Eisner, B. Farkas, M. Haase, R. Nagel, *Operator Theoretic Aspects of Ergodic Theory*, vol. 272 (Springer, 2015)
5. C. Alcalde, Coherent state signal processing, in *Preparation*
6. C. Alcalde, Radar theory and quantum mechanics, in *Preparation*
7. R.E. Blahut, W. Miller, C.H. Wilcox et al., Radar and sonar: part I, in *The IMA Volumes in Mathematics and its Applications*, vol. 32 (Springer, New York, 1991)
8. S. Thangavelu, *An Introduction to the Uncertainty Principle: Hardy's Theorem on Lie Groups*, vol. 217 (Springer Science & Business Media, Boston, 2004)
9. F. Innovative, Barbaresco, tools for radar signal processing based on Cartan's geometry of SPD matrices and information geometry, in *IEEE Radar Conference* (IEEE, 2008), pp. 1–6
10. P. Jorgensen, S. Pedersen, F. Tian, *Extensions of Positive Definite Functions*. Lecture Notes in Mathematics, vol. 2160 (Springer, Switzerland, 2016)
11. E. Prugovecki, *Quantum Mechanics in Hilbert Space*, 2nd edn. (Dover, New York, 1981)
12. G.B. Folland, *A Course in Abstract Harmonic Analysis*, vol. 29 (CRC Press, Boca Raton, 1995)
13. W. Rudin, Measure algebras on abelian groups. Bull. Am. Math. Soc. **65**, 227–247. (American Mathematical Society, 1959)
14. V. Sunder, Notes on the imprimitivity theorem, in *Analysis, Geometry and Probability. Texts and Readings in Mathematics*. ed. by R. Bhatia, vol. 10 (Hindustan Book Agency, Gurgaon, 1996), pp. 299–321
15. B. Misra, I. Prigogine, M. Courbage, From deterministic dynamics to probabilistic descriptions, in *Physica A: Statistical Mechanics and its Applications*, vol. 98 (Elsevier, 1979), pp. 1–26
16. S.T. Ali, J.P. Antoine, J.P. Gazeau et al., *Coherent States, Wavelets and Their Generalizations*, vol. 3 (Springer, 2000)
17. F. Bayen, M. Flato, C. Fronsdal, A. Lichnerowicz, D. Sternheimer, Deformation theory and quantization. I. Deformations of symplectic structures. Ann. Phys. **111**, 61–110 (1978)
18. J.M. Souriau, *Des principes geometriques pour la mecanique quantique, in La Mécanique Analytique de Lagrange et son héritage* (Collège de France, Paris, 1988), pp.27–29
19. A. Caticha, Entropic dynamics: quantum mechanics from entropy and information geometry. Ann. Phys. **531**, 1700408 (2019)
20. K. Vanslette, A. Caticha, Quantum measurement and weak values in entropic dynamics, in *AIP Conference Proceedings*, vol. 1853 (AIP Publishing LLC, 2017), p. 090003
21. B. Moran, S. Howard, D. Cochran, Positive-operator-valued measures: a general setting for frames, in *Excursions in Harmonic Analysis, Volume 2: The February Fourier Talks at the Norbert Wiener Center*, ed. by T.D. Andrews, R. Balan, J.J. Benedetto, W. Czaja, K.A. Okoudjou (Birkhäuser Boston, Boston, 2013), pp. 49–64. https://doi.org/10.1007/978-0-8176-8379-5_4
22. J.G. Christensen, G. Olafsson, Sampling in spaces of bandlimited functions on commutative spaces, in *Excursions in Harmonic Analysis*, vol. 1 (Springer, 2013), pp. 35–69
23. J.P. Gazeau, B. Heller, Positive-operator valued measure (POVM) quantization. Axioms **4**, 1–29 (2014)
24. M. Spera, On some geometric aspects of coherent states, in *Coherent States and Their Applications* (Springer, 2018), pp. 157–172
25. A.A. Kirillov, *Lectures on the Orbit Method*, vol. 64 (American Mathematical Society, 2004)
26. L.P. Rothschild, For Nilpotent Lie Groups. Math. Z. **140**, 63–65 (1974)
27. H. Zahir, Produits STAR et représentation des groupes de Lie. PhD thesis, Université Paul Verlaine-Metz (1991)

28. C. Fronsdal, Some ideas about quantization. Rep. Math. Phys. **15**, 111–145 (1979)
29. C. Alcalde, Star quantization on the cylinder. J. Math. Phys. **31**, 2672–2682 (1990)
30. P. Sharan, Star-product representation of path integrals. Phys. Rev. D **20**, 414 (1979)
31. O.E. Barndorff-Nielsen, P.E. Jupp, Statistics, yokes and symplectic geometry. Annales de la Faculté des sciences de Toulouse: Mathématiques **6**, 389–427 (1997)
32. P. Perrone, Dual connections and holonomy (2015). arXiv preprint arXiv:1511.07737
33. R. Fioresi, M. Lledó, A comparison between star products on regular orbits of compact Lie groups. J. Phys. A: Math. Gen. **35**, 5687 (2002)
34. G. Dito, Moyal star-product on a hilbert space, in *Topics in Mathematical Physics, General Relativity and Cosmology in Honor of Jerzy Plebanski* (World Scientific, 2006), pp. 137–146
35. M. Cahen, S. Gutt, Wavelet transform and* exponential, in *Conference Series-Institute of Physics*, vol. 173 (Institute of Physics, Philadelphia, 1999, 2003), pp. 855–862

Efficient Quantum Circuit for Karatsuba Multiplier

M. James Selsiya, M. Kalaiarasi, S. Rajaram, and V. R. Venkatasubramani

Abstract The fundamental element of quantum computing is the quantum circuit. An efficient quantum circuit saves quantum hardware resources by reducing the number of gates without increasing the number of qubits. Quantum circuits with many qubits are very difficult to realize. Thus, the number of qubits is an important parameter in a quantum circuit design. Using reversible logic in quantum circuits has many advantages such as diminishing power consumption, reducing heat propagation and decreasing quantum cost, ancilla inputs, and garbage outputs that lead to increased performance of quantum computers. Quantum circuits for arithmetic operations such as addition, subtraction, and multiplication are required in the implementation of quantum circuits for many quantum algorithms in this area. In this article two novel designs for GF(2^n) multiplier using Karatsuba algorithm have been proposed that have been proved to have an improvement in qubits, garbage outputs, and ancilla inputs when it comes to comparison with recent research that have been done concerning this field. Bennett's garbage removal strategy with the SWAP gate is used to remove garbage output from existing works in order to establish a fair comparison to existing work.

Keywords Ancilla input · Garbage output · Karatsuba multiplier · QISKIT · Quantum circuit · Qubit

M. J. Selsiya (✉) · M. Kalaiarasi · S. Rajaram · V. R. Venkatasubramani
ECE Department, Thiagarajar College of Engineering, (An Autonomous Institution Affiliated to Anna University Chennai), Madurai, India
e-mail: jamesselsiyaraj@gmail.com

S. Rajaram
e-mail: rajaram_siva@tce.edu

R. Pandey et al. (eds.), *Quantum Computing: A Shift from Bits to Qubits*,
Studies in Computational Intelligence 1085,
https://doi.org/10.1007/978-981-19-9530-9_5

1 Introduction

1.1 Scope

In circuits-based VLSI, power consumption is a considerable factor. The amount of generated heat from the simple circuit is very small, but in complex large integrated circuits, this amount of heat increases. The problem of energy loss is caused by bit loss and it could be solved by introducing reversible circuits. Using reversible Logic, power consumption, and heat dissipation can be minimized. One of the important applications in quantum computing is reversible logic. So, there is a need to implement operations in a quantum computer.

A Quantum Computer uses a different fundamental unit of information, called the Qubit. Qubits are great because of their properties. It has two properties, the ability to simultaneously be in multiple states is called 'superposition' and 'Entanglement' is referred to Changing the state of one of the qubits will instantaneously change the state of the other one.

In quantum circuits, the main parameters to be minimized are qubit count, ancilla input and garbage output. The number of primitive reversible quantum gates termed as the quantum cost is also an important parameter. Ancilla inputs are nothing but constant inputs present in quantum circuits. Garbage outputs are the output which is not directly generated from the input qubit.

1.2 Research Contributions

1.2.1 Optimized Implementation of Binary Field Multiplication

The binary field polynomial multiplication is optimized with the-state-of-art Karatsuba multiplication, which reduced the total number of Toffoli and CNOT gates. The quantum circuit for the multiplication of two polynomial based on Karatsuba algorithm is optimized by utilizing Modified bennet's garbage removal scheme. Modification is done by changing the usage of CNOT gate into SWAP gate and RST in Bennet garbage removal method, the number of CNOT gate is optimized. The initialized qubit gets used in the following operation, which reduces the total number of qubits required for the multiplication operation.

We made use of a quantum computer emulator to evaluate quantum gates. The proposed technique was evaluated using the well-known quantum programming studio framework. The Quantum Programming Studio is a web-based graphical user interface designed to allow users to construct quantum algorithms and obtain results by simulating directly in the browser or by executing on real quantum computers. The framework includes a quantum resource estimator and a quantum computer compiler. This is important for accurate evaluation. We compare the number of Toffoli gates, CNOT gates, and qubit.

1.3 Process Overview

In quantum logic, there are several quantum gates that existed like Feynman, Peres, Toffoli gates, etc. as shown in Table 1. Which are used to implement quantum circuits. The 2×2 Feynman gate is also called as CNOT gate. The CNOT gate has a similar behavior to the XOR logic gate with some extra information to make it reversible. If the bit on the control line is 1, invert the bit on the target line is the operation of the Fenman gate. Peres, Toffoli, and Fredkin gates are 3×3 quantum gates. By using these quantum gates any digital circuits like flip flops, encoder, decoder, and the basic arithmetic operations such as addition, subtraction, and multiplication can be implemented in quantum computers to reduce power and heat losses.

Table 1 Basic quantum gates

Name of the gates	Gate symbol	Matrix
Pauli-X (X) (NOT)		$\begin{bmatrix} 0 & 1 \\ 1 & 0 \end{bmatrix}$
Pauli-Y (Y)	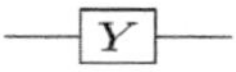	$\begin{bmatrix} 0 & -i \\ i & 0 \end{bmatrix}$
Pauli-Z (Z)	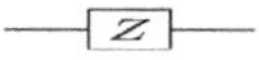	$\begin{bmatrix} 1 & 0 \\ 0 & -1 \end{bmatrix}$
Hadamard (H)	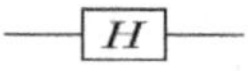	$1/\sqrt{2}\begin{bmatrix} 1 & 1 \\ 1 & -1 \end{bmatrix}$
Controlled Not (CNOT, CX)	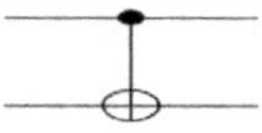	$\begin{bmatrix} 1000 \\ 0100 \\ 0001 \\ 0010 \end{bmatrix}$
Toffoli (CCNOT, CCX, TOFF)	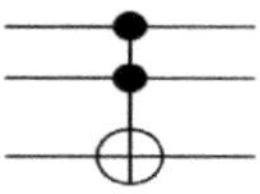	$\begin{bmatrix} 10000000 \\ 01000000 \\ 00100000 \\ 00010000 \\ 00001000 \\ 00000100 \\ 00000001 \\ 00000010 \end{bmatrix}$

Very important arithmetic function is multiplication, and it serves as a building block for multiple complex functions. For the fast and efficient multiplication our Karatsuba-based quantum multiplier circuit is used and also Karatsuba algorithm allows to reduce qubits, ancilla input, and circuit size from $O(n^2)$ to $O(n^{\log_2 3})$ (where n is the bit-size of the numbers to be multiplied). The Karatsuba algorithm is one of the most extensively utilized. The Karatsuba algorithm divides a single n-bit multiplication operation into three n/2-bit operations. Although Karatsuba multiplication necessitates multiple additional addition operations (i.e., exclusive-or), the method greatly reduces multiplication complexity. The computation complexity in quantum circuits is reduced by using the Karatsuba algorithm in quantum computing. Many multiplication algorithms have been utilized by researchers, including the partial product addition and shifting method [2], the Tree-based multiplication technique [3], and the Shift add multiplication algorithm [1]. The usage of reversible full adders and reversible half adders to implement partial product addition significantly increases the overheads in terms of qubits, ancilla inputs, and garbage outputs, as stated in Tables 8 and 9. An ideal quantum circuit should be made garbage less in nature. We can apply Bennett's garbage removal scheme to make designs garbage less. We presented a quantum multiplier based on the Karatsuba algorithm and tricks to solve these drawbacks. When compared to prior reversible quantum multiplier designs [1–3], this will provide a significant improvement in terms of qubit count and garbage output.

In this work, we present the simple quantum circuit design for GF (2^n) Karatsuba multiplier with less qubit cost, ancilla inputs and No garbage output cost. Quantum programming studio is used to implement the design flow. We made use of a quantum computer emulator to evaluate quantum gates. The proposed technique was evaluated using the well-known "quantum programming studio" framework. The Quantum Programming Studio is a web-based graphical user interface designed to allow users to construct quantum algorithms and obtain results by simulating directly in browser or by executing on real quantum computers. The framework includes a quantum resource estimator and a quantum computer compiler. This is important for accurate evaluation. It is web-based quantum programming IDE and simulator built on top of this package. Circuit can be executed on real quantum computer directly from the UI. Quantum-circuit is open-source quantum circuit simulator implemented in JavaScript. Smoothly runs 20+ qubit simulations in browser or at server (node.js). We can use it to execute quantum simulations in a JavaScript programmer. Open QASM and Quil can both import circuits. We are able to export circuits. To Open QASM, pyQuil, Quil, Qiskit, Cirq, TensorFlow Quantum, QSharp, and QuEST, so that it can be used for quantum programming language conversion. Also, Circuit drawing can be exported to SVG vector image.

1.4 Background

1.4.1 Quantum Gates

The fault-tolerant Clifford + T gate set is used in fault-tolerant quantum circuit design. Table 1 shows the gates that make up the Clifford + T gate set. The design of fault tolerant implementations of quantum circuits is gaining the attention of researchers because physical quantum computers are prone to noise errors [8–10]. The quantum multiplier circuit proposed in this work is composed of the quantum NOT gate, Feynman (CNOT) gate, CCNOT gate, and SWAP gate. A quantum gate (or quantum logic gate) is a fundamental quantum circuit that operates on a small number of qubits in quantum computing. Quantum gates use them in the same way as traditional logic circuits use normal digital logic gates. Unitary matrices are used to represent quantum logic.

Quantum computers employ qubits to measure information. Qubits have two important properties: superposition and entanglement. The ability to be in numerous states at once is known as superposition. Entanglement occurs when the state of one of the qubits changes in a predictable manner when the state of the other is changed. This happens despite the fact that they are separated by huge distances. Qubits vary from classical bits in that they can be in both states at the same time (where |0 represents state 0 and |1 represents state 1), whereas classical bits can only be either 0 or 1. The most frequent quantum gates, on the other hand, work with one or two qubit spaces, similar to how classical logic gates work with one or two bits.

A gate which operates on two qubits is called a Controlled-NOT Gate. It is also called as CNOT gate shown in Table 1. If the bit on the control line is 1, invert the bit on the target line. The CNOT gate has a similar behavior to the XOR gate with some extra information to make it reversible.

A gate which operates on three qubits is called a controlled-controlled NOT (CCNOT) gate shown in Table 1. If the bits on both of the control lines are 1, then the target bit is inverted. If the target is fixed as zero then this gate will be act like AND gate.

1.5 Organization of the Paper

This paper is organized as follows. The methods for designing a quantum Karatsuba multiplier circuit are presented in Sect. 2. In part 3, the proposed quantum circuit for the GF (2^n) Karatsuba multiplier is designed. Section 4 contains the Methodology to Remove Garbage Outputs from Quantum polynomial Multiplication Circuit Designs. Results and discussion are there in Sect. 5. The conclusion is found in Sect. 6.

2 Methodology to Design Quantum Karatsuba Multiplier Circuit

Karatsuba [7] developed the first algorithm for multiplying polynomials of size n digits with a complexity lower than that of the traditional approach i.e., $O(n^{\log_2(3)})$, which can be proved easily by using a recurrence relation, as described below.

Consider the multiplication of two n-bit polynomials X and Y. After three steps, the Karatsuba multiplication method yields the coefficient of the product Z. For multiplying two polynomials in the polynomial ring $\mathbb{F}_2$. Assume that $n = 2^N$, let be binary polynomials of degree $\leq n-1$.

$$X = \sum_{k=0}^{n-1} X_k t^k \quad Y = \sum_{k=0}^{n-1} Y_k t^k \quad Z = \sum_{k=0}^{n-1} Z_k t^k \in \mathbb{F}_2[t]$$

We built the Karatsuba multiplication algorithm for $X = X[1]\, t^{n/2} + X[0]$, $Y = Y[1]\, t^{n/2} + Y[0]$ are polynomials of degree at most n − 1. The coefficients of input polynomials are X[1], X [0], Y[1], and Y [0]. The mathematical flow of the Karatsuba algorithm is visualized in Fig. 1.

Following the three stages outlined below, the Karatsuba multiplication method yields the coefficient of product Z.

$X \cdot Y = (X[1]\, t^{n/2} + X[0]) \cdot (Y[1]\, t^{n/2} + Y[0]) = C[2]\, t^{n} + (C[1] + C[2] + C[0])\, t^{n/2} + C[0] = Z.$

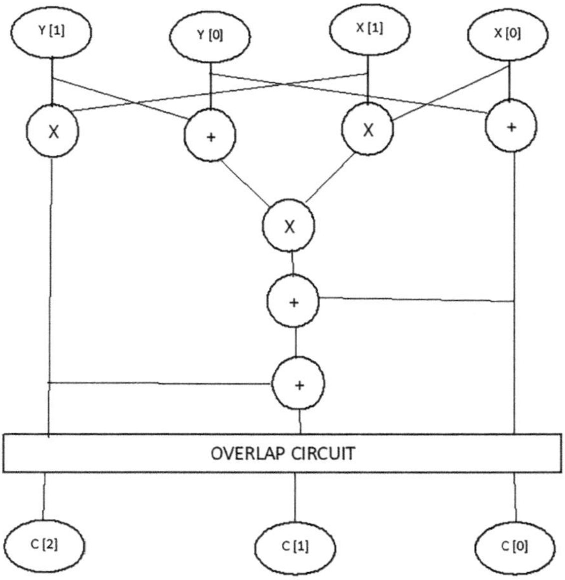

Fig. 1 Block diagram representation of classical Karatsuba multiplier

Step 1: We divide the given polynomials coefficient bits into halves, for n bit polynomials, divide its bits by n/2.

X = X [1] t n/2 + X [0], Y = Y [1] t n/2 + Y [0]. Where X[1], X [0], Y[1], and Y [0] are polynomial coefficients of degree at most n-1.

Step 2: Add leftmost to rightmost bits of given n bit polynomials i.e. (X [0] + X[1]) (Y [0] + Y[1]).

Step 3:

C [2] = X [1] Y[1],
C [1] = ((X [0] + X [1]) (Y [0] + Y [1]) + C [2] + C [0]),
C [0] = X [0] Y [0].

The product of X and Y is then obtained and given the designations C[0], C[1], and C[2]. We define C[1] in a particular manner Using Karatsuba as,

C [1] = ((X [0] + X [1]) (Y [0] + Y [1]) + C [0] + C [2]) = (X [0] Y [0] + X [0] Y [1] + X [1] Y [0] + X [1] Y [1] + X [1] Y [1] + X [0] Y [0]) = X [0] Y [1] + X [1] Y [0]. i.e., using the already existing terms C [2] and C [0] to eliminate unneeded terms in C [1].

Now, C [0], C [1], and C [2] are products of polynomials of degree $\leq$ n$-$1, and by applying this process recursively, we find X $\cdot$Y with a total of $O(n^{\log_2(3)})$ additions and multiplications in $\mathbb{F}_2$.

3 Design of Proposed Quantum Circuit for GF (2^n) Reversible Karatsuba Multiplier

3.1 Quantum Circuits for Karatsuba-Based Polynomial Multiplication

The Karatsuba algorithm is used to calculate the multiplication in the proposed quantum multiplication circuit. The proposed methodology is generic and can design a quantum multiplier circuit of any size. The steps involved in the proposed methodology are presented for finding the multiplication of the polynomials GF (2^3) with degree n$-$1.

In comparison to the standard method, which requires 9 multiplications and 4 additions for polynomials of degree 2, the Karatsuba method requires 6 multiplications and 12 additions. The binary field polynomial multiplication is optimized with Karatsuba multiplication, which reduced the total number of multiplications. This will reduce the number of times CNOT and CCNOT gate are used to design quantum circuit.

The conventional Karatsuba algorithm flow has already been observed in Fig. 1. The product of two polynomials X and Y with 2° can be calculated using the equation below,

$$(X[2]x^2 + X[1]x + X[0])(Y[2]x^2 + Y[1]x + Y[0]) = (X[2]Y[2])x^4 + ((X[1] + X[2])(Y[1] + Y[2]) - X[1]Y[1] - X[2]Y[2])x^3 + ((X[2] + X[0])(Y[2] + Y[0]) - X[0]Y[0] + X[1]Y[1] - X[2]Y[2])x^2 + ((X[1] + X[0])(Y[1] + Y[0]) - X[0]Y[0] - X[1]Y[1])x + X[0]Y[0]$$

the quantum circuit for 3×3 Karatsuba multiplier for the input combination $(X[2]x^2 + X[1]x + X[0])(Y[2]x^2 + Y[1]x + Y[0])$ is shown in Fig. 2. The first step in creating a quantum circuit for $X = (X[2]x^2 + X[1]x + X[0]) \cdot Y = (Y[2]x^2 + Y[1]x + Y[0])$ using the Karatsuba approach is to assign input polynomials to the appropriate qubits. The input coefficients X [0], X [1], X [2] are assigned to qubits q0, q1, and q2, respectively, while the input coefficients Y [0], Y [1], Y [2] are assigned to qubits q3, q4, and q5. According to the circuit which is shown in Fig. 2. We will perform the multiplication process. At the end of the computation, the quantum register $|Z\rangle$ will have $|X \cdot Y\rangle$.

A gate which operates on three qubits is called a controlled-controlled NOT (CCNOT) gate shown in Table 1. If the bits on both of the control lines is 1, then the

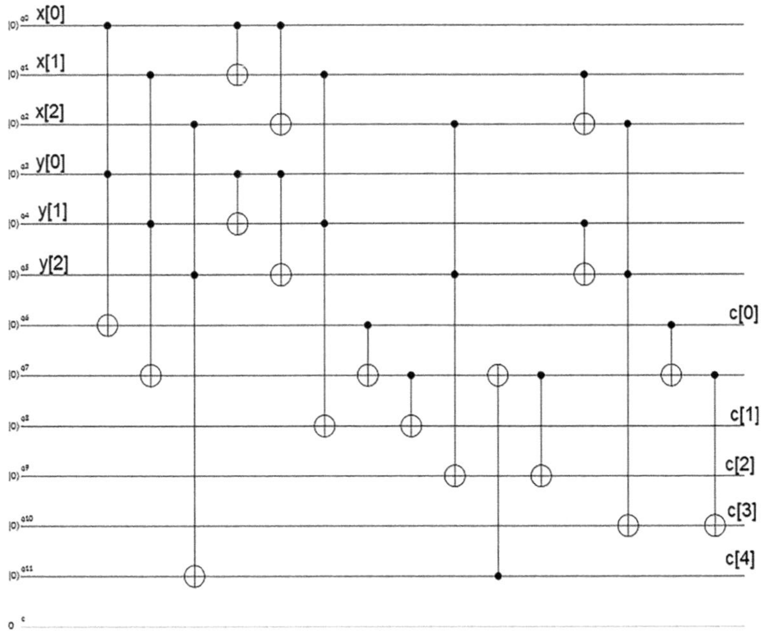

Fig. 2 Quantum circuit for Karatsuba algorithm-based multiplier for X·Y ∈ GF (2^3)

Table 2 Combination of X and Y during X · Y computation on multiplication of polynomials GF (2^3) with 2°

C [0]	X [0] Y [0]
C [1]	((X [1] + X [0]) (Y [1] + Y [0])−C [0]−X [1] Y [1])
C [2]	((X [2] + X [0]) (Y [2] + Y [0])−C [0] + X [1]Y[1] −C [4])
C [3]	((X [1] + X [1]) (Y [1] + Y [2])−C [0]−C [4])
C [4]	X [2] Y [2]

target bit is inverted. If the target is fixed as zero then this gate will be act like AND gate.

Then, on the qubits q6, q7, q8, q9, q10, and q11, we use the CCNOT gate to perform the operations X [2] Y [2], X [1] Y [1], and X [0] Y [0] by setting the target of the CCNOT gate to zero. They are called constant inputs because they are always in the state zero.

Then, to create a circuit to get the product C [0] …C [4], we use CNOT gates alone to perform the following operations: (X [1] + X [0]), (Y [1] + Y [0]), (X [2] + X [0]), (Y [2] + Y [0]), (X [1] + X [2]), (X [1] + Y [2]), (X [1] + Y [2]). These are the building blocks for the Karatsuba algorithm. We got the product coefficient Represented in Table 2 when we used Karatsuba instead of others. Using the already existing basic terms X [1]Y [1], C [1] and C [4] and the CNOT gate is used to ignore the coefficients that are not in the final result. As a result, the subtraction symbol in Table 2 is designed using CNOT in the quantum circuit for Fig. 2. Table 3 lists the outputs from the quantum circuit depicted in Fig. 2 for various combinations of input polynomials of 2°.

Because of the entire circuit may be built with only CNOT and CCNOT gates, the circuit design costs are kept to a minimum. Finally, the measurement tools will be placed in the positions of qubits q6, q8, q9, q10, and q11 in order to measure the appropriate outputs and display the probability of receiving zero and one. Figure 2 depicts the suggested design approach for a reversible quantum Karatsuba multiplier.

3.2 Quantum Circuit for GF (2^4) Karatsuba Multiplier

We obtained the Karatsuba multiplication of two input polynomials $Z = X \cdot Y$ where X = (X [3]x^3 + X [2]x^2 + X [1]x + X [0]) and Y = (Y [3]y^3 + Y [2]y^2 + Y [1]y + Y [0]) as shown in below.

$$(X[3]x^3 + X[2]x^2 + X[1]x + X[0])(Y[3]y^3 + Y[2]y^2 + Y[1]y + Y[0]) = X[3]Y[3]x^6 + ((X[2] + X[3])(Y[2] + Y[3]) - X[2]Y[2] - X[3]Y[3])x^5 + ((X[1] + X[3])(Y[1] + Y[3]) - X[1]Y[1] - X[3]Y[3] + X[2]Y[2])x^4 + ((X[0] + X[3])(Y[0] + Y[3])(X[1] + X[2])(Y[1] + Y[2]) - X[0]Y[0] - X[1]Y[1] - X[2]Y[2] - X[3]Y[3])x^3 + ((X[0] + Y[2])(Y[0] + Y[2]) - X[0]Y[0] - X[2]Y[2])x^2 + ((X[0] + X[1])(Y[0] + Y[1]) - X[0]Y[0] - X[1]Y[1])x + X[0]Y[0].$$

Table 3 (X [2]x^2 + X [1]x + X [0]) · (Y [2]x^2 + Y [1]x + Y [0]) = (C [4] x^4 + C [3] x^3 + C [2] x^2 + C [1] x + C [0])

S. no	|X⟩ (X [2]x^2 + X [1]x + X [0])	|Y⟩ (Y [2]x^2 + Y [1]x + Y [0])	C [4]	C [3]	C [2]	C [1]	C [0]	|Z⟩
1.	x^2+x+1	x^2+x+1	1	0	1	0	1	10101
2.	x^2+x+1	x^2+1	1	1	0	1	0	11010
3.	x^2+x+1	x^2+x	1	0	0	1	0	10010
4.	x^2+x+1	$x+1$	0	1	0	0	1	01001
5.	x^2+x+1	x	0	1	1	1	0	01110
6.	x^2+1	x^2+x+1	1	1	0	1	1	11011
7.	x^2+1	x^2+1	1	0	0	0	1	10001
8.	x^2+1	x^2+x	1	1	1	1	0	11110
9.	x^2+1	$x+1$	0	1	1	1	1	01111
10.	x^2+1	x	0	1	0	1	0	01010
11.	x^2+x	x^2+x+1	1	0	0	1	0	10010
12	x^2+x	x^2+1	1	1	1	1	0	11110
13.	x^2+x	x^2+x	1	0	1	0	0	10100
14.	x^2+x	$x+1$	0	1	0	1	0	01010
15.	x^2+x	x	0	1	1	0	0	01100
. . . . .	. . . . .	. . . . .	. . . . .	. . . . .	. . . . .	. . . . .	. . . . .	. . . . .

In comparison to the traditional approach, which uses 9 multiplications and 4 additions, this method requires 6 multiplications and 12 adds. For clarity, the coefficients of product Z are represented by corresponding C [x]. Even though the GF (2^3) Karatsuba multiplier and the gates flow introduced in the first section of this research were employed, the proposed GF (2^4) Karatsuba multiplier circuit is developed using basic quantum mechanics gates and can be considered a unique 4×4 multiplier module.

Quantum circuit for GF (2^4) Karatsuba multiplier is created and shown in Fig. 3, the coefficients of the inputs X and Y are provided on the wires labeled X [0] …X [3] and Y [0] …Y [3], respectively. The coefficients of the product X · Y in GF_{16} is available on the wires labeled C [0] …C [4]. We got the product coefficient like in Table 4 when we used Karatsuba instead of others. We used 6 CCNOT gates and 26 CNOT gates to create a GF (2^4) Karatsuba multiplier. As a result, we need six constant inputs, also known as ancilla inputs.

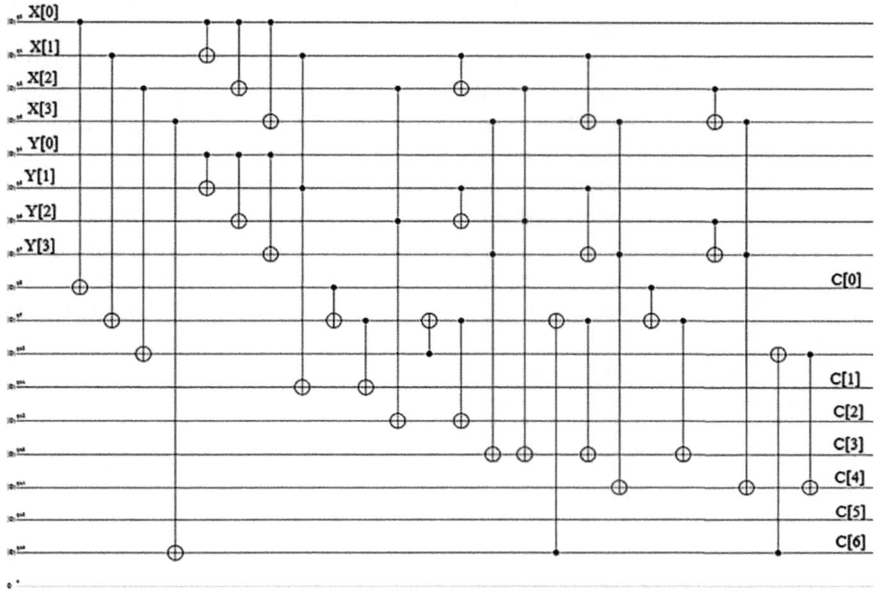

Fig. 3 Quantum circuit for Karatsuba algorithm-based multiplier for X·Y ∈ GF (2^4)

Table 4 Combination of X and Y during X · Y computation on multiplication of polynomials GF (2^4) with degree 3

C [0]	X [0] Y [0]
C [1]	((X [1] + X [0]) (Y [1] + Y [0])−C [0]−X [1] Y [1])
C [2]	((X [0] + Y [2]) (Y [0] + Y [2])−X [0] Y [0]−X [2] Y [2])
C [3]	((X [0] + X [3]) (Y [0] + Y [3]) (X [1] + X [2]) (Y [1] + Y [2])−C [0]−X [1]Y [1]−X [2]Y[2]−X [3]Y[3])
C [4]	((X [1] + X [3]) (Y [1] + Y [3])−X [1] Y [1]−C [4] + X [2] Y [2])
C [5]	((X [2] + X [3]) (Y [2] + Y [3])−X [2] Y [2]−C [4])
C [4]	X [3]Y [3]

4 Methodology to Remove Garbage Outputs from Quantum Polynomial Multiplication Circuit Designs

In this Section we presented existing quantum polynomial multiplication circuits that are garbage less in nature. However, other recent works, such as the design [2, 3] that show promise in the terms of quantum gate count suffer from significant ancilla and garbage output overhead. An ideal quantum circuit should be made garbage less in nature. We can apply Modified Bennett's garbage [6] removal scheme with swap gate to make designs garbage less.

Consider the multiplication of two n bit numbers X and Y stored in quantum registers |X⟩ and |Y⟩ by a design. At the end of computation, the quantum registers |X⟩ and |Y⟩ will keep the values X and Y respectively. Further, consider quantum registers |P⟩, |G⟩ that are each initialized to 0.

At the end of computation, |P⟩ will have the product of X and Y, |G⟩ will contain the garbage outputs. Bennett's garbage removal scheme removes the garbage outputs by applying the logical reverse of the original design to the quantum registers |X⟩, |Y⟩, |P⟩ and |G⟩. Therefore, at the end of computation, the quantum registers |X⟩ and |Y⟩ will keep the values X and Y respectively, while quantum registers will be restored to the value 0.

The steps of the modified Bennett's garbage removal scheme are explained below. The methodology is generic and can be used on any quantum circuit that has garbage outputs. An illustrative example of the methodology for a multiplication circuit with garbage outputs is also shown. Figure 5 illustrates Steps 1 through 3. The GF (2^3) polynomial multiplication circuit and its logical reverse are labeled in Fig. 5 with a box respectively (Fig. 4).

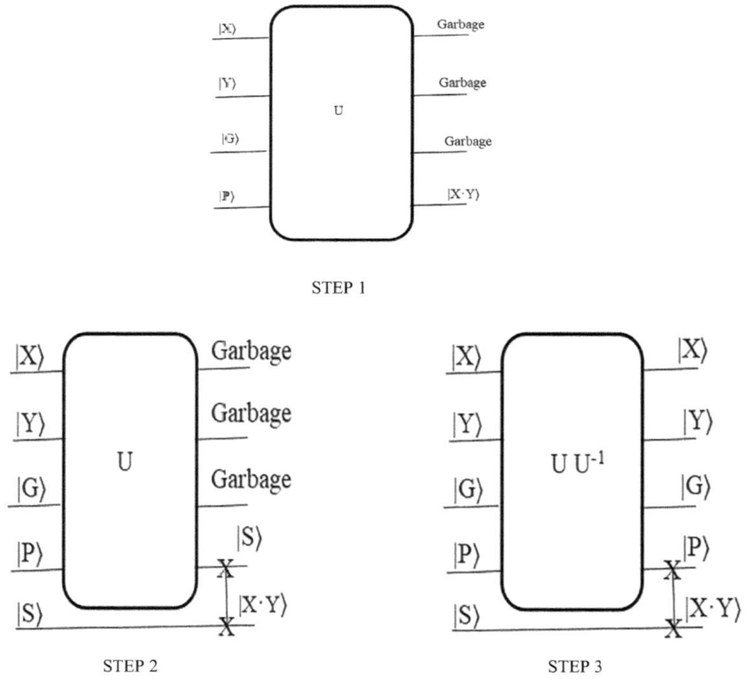

Fig. 4 Generation of the garbage less quantum multiplication circuit: Steps 1–3

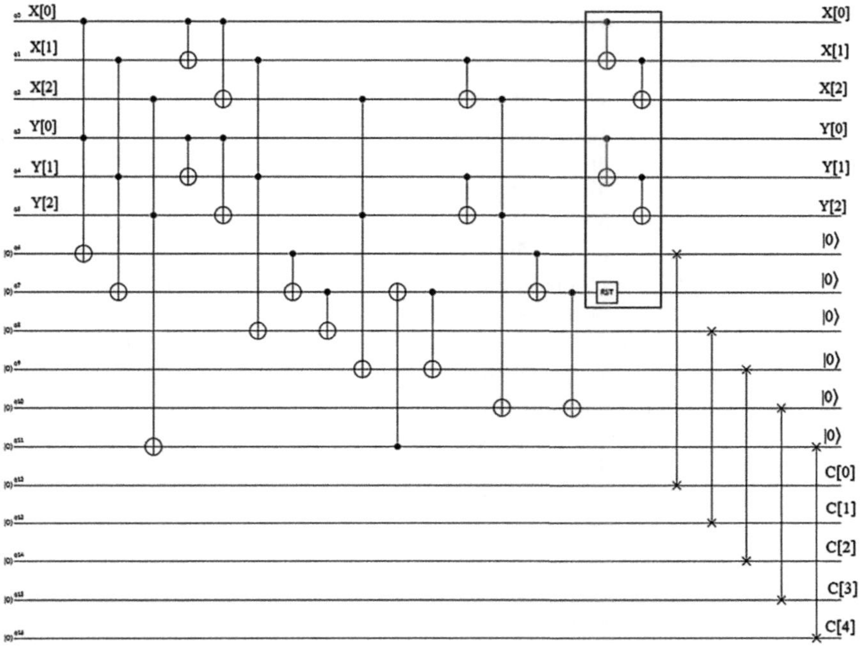

Fig. 5 Quantum circuit for Karatsuba algorithm-based multiplier for X·Y ∈ GF (2^3) before bennet's scheme (Zero garbage)

Step 1: At quantum registers |X⟩, |Y⟩, |P⟩ and |G⟩ apply the quantum multiplication circuit such that the registers |X⟩ and |Y⟩ will maintain the same value, location |P⟩ will hold the product and location |G⟩ will contain the garbage outputs.

Step 2: For i = 0: 1: 2* n−1

At locations $|P_i\rangle$ and $|S_i\rangle$ apply a SWAP gate such that location $|S_i\rangle$ is transformed to the value in location $|P_i\rangle$.

Step 3: At quantum registers |X⟩, |Y⟩, |G⟩ and |P⟩ apply the logical reverse with RST gate of the quantum multiplication circuit |X⟩ and |Y⟩ will maintain the same value, locations |P⟩ and |G⟩ will each be restored to the value 0.

This methodology, while able to remove garbage outputs from a quantum multiplication design but it will add qubit to the design and increase the total number of gate count that is shown in Table 7.

5 Results and Discussion

The binary field polynomial multiplication is optimized with the state of art Karatsuba multiplication, which reduced the total number of Toffoli and CNOT gates. The quantum circuit for the multiplication of two polynomial's coefficients based on Karatsuba algorithm is optimized by utilizing bennet's garbage removal scheme. By changing the usage of SWAP gate and RST into Bennet garbage removal method, the number of CNOT gate is optimized which is shown in Sect. 4. The initialized qubit gets used in the following operation, which reduces the total number of qubits required for the multiplication operation.

We made use of a quantum computer emulator to evaluate quantum gates. The proposed technique was evaluated using the well-known "quantum programming studio" framework. The Quantum Programming Studio is a web-based graphical user interface designed to allow users to construct quantum algorithms and obtain results by simulating directly in browser or by executing on real quantum computers. The framework includes a quantum resource estimator and a quantum computer compiler. This is important for accurate evaluation.

In Table 5, both classical and quantum results for Karatsuba algorithm-based GF (2^3) multipliers are calculated and compared. By comparing classically produced results to our quantum output, we were able to verify our quantum circuit. Both classical and quantum mathematics yielded the same outcomes.

In simulation using IBMQ lab, a histogram plot is used to visualize the output. Due to the entangle nature of qubits, the x axis displays the probability to get output state while the y axis shows the 2^n states that are simulated throughout the simulation. This histogram plot can take from using IBMQ lab. Figure 6 shows histogram plots for particular input i.e. [1 x^2 + 1 x + 1) (1 y^2 + 1 y + 1)] for Karatsuba algorithm-based quantum circuit as shown in Fig. 2 respectively. Similarly, the histogram plot for all the combinations of inputs in GF (2^3) which is shown in Table 3 is taken and verified.

Table 6 reveals the obtained results from the proposed GF (2^n) Karatsuba multiplier circuits. The number of gates utilized in the GF (2^3) quantum Karatsuba multiplication circuit is 18 gates, according to Fig. 2 where there are 6 CCNOT gates and

Table 5 Comparing the results of conventional and quantum circuits for X·Y ∈ GF (2^3)

S.no	X (X [2] X [1] X [0])	Y (Y [2] Y [1] Y [0])	Classical Z = X·Y	Quantum	
				| X·Y⟩	Probability
1	111	111	10101	|10101⟩	1
2	101	101	10001	|10001⟩	1
3	100	101	10100	|10100⟩	1
4	001	101	00101	|00101⟩	1
5	010	010	00100	|00100⟩	1

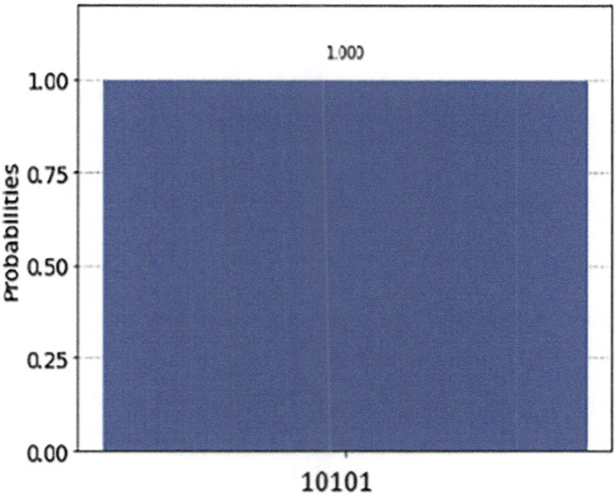

Fig. 6 Sample histogram plot for X·Y ∈ GF (2^3) Quantum circuit with input [111] * [111]

12 CNOT gates, total of 11 qubits were utilized in the design. In general, the quantum IBM launcher uses some processors for operations that need more than 32 qubits.

The GF (2^3) Karatsuba multiplier in our suggested architecture requires just 11 qubits, which is less than the 32 qubits. In addition, our design structure has six ancilla inputs and ten garbage outputs, which is fewer than the other survey participants. As a result, our proposed approach saves time during the simulation.

By the application of bennet's scheme, we can design more efficient garbage less circuit for GF (2^3) Karatsuba multiplier-based quantum circuit with 28 gates includes SWAP and reversible circuit which is denoted in Table 7.

From Table 6 we analyze that the number of gates used for GF (2^4), (X = (X [3]x^3 + X [2]x^2 + X [1]x + X [0]) and Y = (Y [3]y^3 + Y [2]y^2 + Y [1]y + Y [0]) quantum multiplication circuit is 32 gates and the number of qubits used for the

Table 6 A comparison of proposed gf (2^n) quantum reversible Karatsuba multiplier before bennet's scheme

Quantum circuit with garbage	No of gates (Quantum gate cost)	No of qubits	Ancilla inputs	Garbage outputs
Gf (2^3)	18	11	6	10
Gf (2^4)	32	16	6	'14

Table 7 A comparison of proposed GF (2^n) quantum reversible Karatsuba multiplier after bennet's scheme

Quantum circuit without garbage	No of gates (Quantum gate cost)	No of qubits	Ancilla inputs	Garbage outputs
Gf (2^3)	28	16	11	0
Gf (2^4)	46	23	13	0

Table 8 Qubit cost comparison of quantum multiplier circuit

Gf (2^4) Quantum multipliers	Algorithms used	No of Qubits Or Qubit cost
[2]	Partial product addition and shifting algorithm	52
[3]	Tree based multiplication technique	28
Our work	Karatsuba algorithm with bennet's scheme	23

Table 9 Garbage output cost comparison of quantum multiplier circuit

Gf (2^4) Quantum multipliers	Algorithms used	Garbage output
[2]	Partial product addition and shifting algorithm	22
[3]	Tree based multiplication technique	16
[13]	Karatsuba algorithm with reduction	17
[14]	Karatsuba algorithm	14
Our work	Karatsuba algorithm with bennet's scheme	0

quantum multiplication is 16 qubits. And also, our design structure consists totally 6 ancilla inputs and 14 garbage output which is less than others in survey. By the application of bennet's scheme, we can design more efficient garbage less circuit for GF (2^4) Karatsuba multiplier-based quantum circuit with 46 gates includes SWAP and reversible circuit which is denoted in Table 7.

We now illustrate the comparison of the proposed quantum polynomial multiplication circuit to the existing work. Existing works that target reversible computing (such as [15, 16]) are not considered in this comparison because they suffer from high garbage output costs. Works that present quantum multiplication circuits in the quantum Fourier transform domain are not considered in this comparison because they suffer from prohibitively high quantum gate costs.

Researchers have developed many multiplication algorithms such as Partial product addition and shifting algorithm [2], Tree based multiplication technique [3] and Shift add multiplication algorithm [1]. The use of reversible full adders and reversible half adders to perform the addition of partial products significantly increases the overheads in terms of qubit cost and garbage output cost, shown in Tables 8 and 9. Works that present multiplication circuits in the quantum Fourier transform domain such as [11, 12] suffer from prohibitively high Clifford + T quantum gate costs. The design of quantum multiplication circuits has received considerable attention in the literature. However, most works target reversible computing and suffer from high garbage output costs [2, 3, 13, 14]. To overcome these disadvantages, we have proposed Karatsuba and bennet's scheme-based quantum polynomial multiplier for GF(2^n). This will give a significant improvement in terms of qubit cost and garbage output cost compared to the existing designs of reversible quantum multiplier designs [2, 3, 13, 14].

The number of qubits is calculated by summing the number of qubits required for the inputs and the ancilla. We have calculated ancilla assuming that the quantum multiplication circuits would be utilized to multiply two polynomials in GF (2^n). Table 8 illustrates the qubit cost of our design versus the works of Babu [3] and Kotiyal et al. [2] for values of n = 4 for GF(2^n). We observed that the proposed design technique achieves improvement. The experimental results in Table 8 show that our proposed work has a reduced qubit cost of 55.76% when compared to [2] and 17.85% when compare to [3]. The experimental results in Table 9 show that our proposed work has a reduced garbage output cost of 100% when compared to [2, 3, 13, 14].

6 Conclusion

In this work, we present a new design of a quantum multiplier circuit. we have presented an efficient reversible quantum circuit for GF (2^3) and GF (2^4) multipliers using Karatsuba and modified bennet's algorithm with CNOT and CCNOT Gates optimized in terms of garbage output cost and qubit cost. The generated quantum Karatsuba multiplier circuits are optimized for quantum gate cost. Further, the proposed quantum polynomial multiplication circuit does not produce garbage outputs. Our proposed design has zero overhead in terms of garbage output cost. The proposed design also requires fewer qubits than the current state of the art. The proposed quantum multiplier circuit has been formally verified. The proposed quantum multiplier circuit could form a crucial component in the quantum hardware implementations of scientific algorithms where qubits and garbage cost are of primary concern. We conclude that the proposed quantum Karatsuba multiplier for GF (2^3) and GF (2^4) is shown to be superior to the existing designs in terms of qubit cost and garbage output cost. We conclude that the proposed quantum polynomial multiplication circuit will find applications in quantum computing that requires less qubit cost where quantum garbage output cost is a primary concern. Construction of various other mathematical functions and algorithms with our proposed Karatsuba multiplier as a base is future work.

References

1. E. Munoz-coreas, H. thapliyal, Quantum circuit design of a t-count optimized integer multiplier. IEEE Trans. Comput. **68** (5), (2020)
2. S. Kotiyal, H. Thapliyal, N. Ranganathan, Circuit for reversible quantum multiplier based on binary tree optimizing ancilla and garbage bits, *in 2014 27th International conference on VLSI design and 2014 13th International conference on embedded systems*, (2014), pp. 545–550
3. H. M. H. Babu, Cost-efficient design of a quantum multiplier—accumulator unit. Quantum Inf. Process. **16**(1), (2016), Art. No. 30

4. S. Dutta, Y. Tavva, D. Bhattacharjee, A. Chattopadhyay, Efficient quantum circuits for square-root and inverse square-root, in *2020 33rd International conference on vlsi design and 2020 19th International conference on embedded systems (vlsid)*, (2020), pp. 55–60. https://doi.org/10.1109/vlsid49098.2020.00027
5. P. N. Singh, S. Aarthi, Quantum circuits—an application in qiskit-python, in *2021 third international conference on intelligent communication technologies and virtual mobile networks (icicv)*, (2021), pp. 661–667
6. L. Kowada, Reversible Karatsuba's algorithm. Article in J. Univers. Comput. Sci, (2006)
7. A. Karatsuba, Y. Ofman, Multiplication of multidigit numbers on automata, in *Soviet physics—doklady*, vol. 7, issue no. 7 (1963), pp. 595–596
8. P. Webster, S. D. Bartlett, D. Poulin, Reducing the overhead for quantum computation when noise is biased. Phys. Rev. A. **92**, (2015). Art. No. 062309. [online]. Available https://doi.org/10.1103/physreva.92.062309
9. X. Zhou, D. W. Leung, I. L. Chuang, Methodology for quantum logic gate construction. Phys. Rev. A. **62**, (2000). Art. No. 052316. [online]. Available https://doi.org/10.1103/physreva.62.052316
10. I. Polian, A. G. Fowler, Design automation challenges for scalable quantum architectures, in 2015 52nd ACM/EDAC/IEEE design automation conference (DAC), (2015), pp. 1–6
11. L. Ruiz-Perez, J. C. Garcia-Escartin, Quantum arithmetic with the quantum fourier transform. Quantum. inf. Process. **16** (6), (2017). Art. No. 152. [online]. Available https://doi.org/10.1007/s11128-017-1603-1
12. G. Florio, D. Picca, Quantum implementation of elementary arithmetic operations. eprint arxiv:quant-Ph/0403048, (2004).[online]. Available arxiv.org/abs/quant-ph/0403048
13. S. Kepley, R. Steinwandt, Quantum circuits for F2n-multiplication with subquadratic gate count. Quantum. Inf. Process. **14**(7), 2373–2386 (2015). https://doi.org/10.1007/s11128-015-0993-1
14. K. Jang, S. Choi, H. Kwon, Z. Hu, H. Seo, Impact of optimized operations $A \cdot B, A \cdot C$ for binary field inversion on quantum computers. ed. by I. You (Springer Nature Switzerland AG, 2020), WISA 2020, LNCS 12583, pp. 154–166. https://doi.org/10.1007/978-3-030-65299-9_12
15. M. Haghparast, M. Mohammadi, K. Navi, M. Eshghi, Optimized reversible multiplier circuit. J. Circuits Syst. Comput. **18**, (02), 311–323 (2009). [Online]. Available https://doi.org/10.1142/S0218126609005083.
16. E. P. A. Akbar, M. Haghparast, K. Navi, Novel design of a fast reversibleWallace signmultiplier circuit in nanotechnology. Microelectron. J. **42** (8), 973–981 (2011). [Online]. Available http://www.sciencedirect.com/science/article/pii/S0026269211001194
17. Z.G. Wang, S.J. Wei, G.L. Long, A quantum circuit design of AES requiring fewer quantum qubits and gate operations. Front. Phys. **17**, 41501 (2022). https://doi.org/10.1007/s11467-021-1141-2
18. S. Immareddy, A. Sundaramoorthy, 2022. A survey paper on design and implementation of multipliers for digital system applications. Artif. Intell. Rev.1–29 (2022)

Quantum Computing: Building Concepts

Quantum Concepts

S. Manjula Gandhi, S. Gayathri Devi, K. Sathya, K. H. Vani, and K. Kiruthika

Abstract This chapter will help in understanding the concepts of representing quantum states, superposition states, quantum gates, quantum entanglement, and various output visualizations, along with hands-on code using IBM's Quantum Experience Qiskit, an open-source framework to write and execute quantum programs. Classical computers work on bits whereas quantum computers work on quantum bits or qubits. A state in a qubit can be represented as $|0\rangle$, $|1\rangle$ or both in $|0\rangle$ and $|1\rangle$ at the same time, called a superposition state. The advantage of storing a qubit in a superposition state is one can perform computation on all values at the same time, which leads to quantum parallelism. Qubits can be represented using two-dimensional vectors and there are three computational bases called Z basis, X basis, and Y basis. The next part of the chapter will discuss the various quantum gates—single qubit and multi qubit quantum gates. Output of these gates when applied to various states such as $|0\rangle$, $|1\rangle$, $|+\rangle$, $|-\rangle$, $|i\rangle$, $-|i\rangle$ are discussed. The last part of this chapter will discuss the code to generate Bell states also called as quantum entanglement, the output of quantum entanglement using various visualizations supported by Qiskit such as qsphere, state_city, state_hinton, bloch sphere, etc. are explored.

Keywords Qubit · Quantum state · Superposition · Parallelism · Entanglement · Bell state · IBM quantum experience · Simulators · Real quantum device · Qiskit

1 Introduction

Quantum computers [1] perform calculations using qubit (quantum bit), which can represent both 1 and 0 at the same time. The power of quantum computers increases exponentially in proportion to the number of quantum bits or qubits [2]. In this chapter we will discuss the representation of qubits, quantum gates, quantum parallelism, quantum entanglement, and output visualization. Qiskit [3] supports output in

S. Manjula Gandhi (✉) · S. Gayathri Devi · K. Sathya · K. H. Vani · K. Kiruthika
Department of Computing, Coimbatore Institute of Technology, Tamil Nadu, Coimbatore, India
e-mail: manjulagandhi@cit.edu.in

R. Pandey et al. (eds.), *Quantum Computing: A Shift from Bits to Qubits*,
Studies in Computational Intelligence 1085,
https://doi.org/10.1007/978-981-19-9530-9_6

various visualization forms such as state vector representation, qsphere representation, histogram, and bloch sphere. The concepts will be discussed along with Hands-on code using IBM's Quantum Experience [4]—Qiskit, an open-source framework to write and execute quantum programs.

Quantum programs are divided into two phases, setting up the circuit called the Build Phase and executing the circuit using simulators or quantum real devices called the Execute Phase.

The following steps are followed during Build Phase:

1. Passing required number of quantum bits and classical bits into the system (circuit).
2. Applying a specified combination of gates (which make up an algorithm) to those qubits.
3. Measuring the quantum bits and storing the results in classical bits. (In 0 and 1 s).

Simulators are quicker and produce 'perfect' results, Qiskit supports various Backend simulators such as State vector, Unitary, and QASM, whereas real quantum computers are 'noisy', meaning that they don't always produce the right results, so run experiments multiple times to determine the correct outcome.

2 Representing Quantum States

Single qubit states can be written down generally as given in Eq. (1)

$$\sqrt{1-p}\,\big|0> + \; e^{i\varphi}\sqrt{p}\,\big|1> \tag{1}$$

Here, p is the probability that a measurement of the state in the computational basis $\{|0\rangle,\ |1\rangle\}$ will have the outcome 1, and ϕ is the phase between the two computational basis states. Single qubit gates [5] can be applied to manipulate this quantum state by changing either p, ϕ or both.

Let's begin by creating a single qubit quantum circuit; we can do this in Qiskit [6] using the code given in Block 1.

```
#Block 1
from qiskit import *
from qiskit import QuantumCircuit, QuantumRegister
q=QuantumRegister(1)
mycircuit = QuantumCircuit(q)
mycircuit.draw('mpl')
```

By default, when we create a quantum circuit of N qubits, all the qubits are initialized in the state $|0\rangle$. Qiskit code to create state $|\psi\rangle = |0\rangle$ and executing the program using state vector simulator is given in Block 2. The output displayed is the state vector representation of state |0> .

```
#Block 2
#First Quantum program to create a state |Ψ⟩=|0⟩
from qiskit import QuantumRegister, QuantumCircuit, Aer, execute

my_simulator = Aer.get_backend(name='statevector_simulator')

q = QuantumRegister(1)
hello_qubit = QuantumCircuit(q)

job = execute(hello_qubit, my_simulator)
result = job.result()
result.get_statevector()
```

```
array([1.+0.j, 0.+0.j])
```

We have imported QuantumRegister, QuantumCircuit, Aer, execute [7] in the code in Block 2. These imports help us to build quantum circuits, execute them and measure the results.

QuantumRegister—A class that holds qubits.

QuantumCircuit—A class that helps to send "instructions" for the quantum system.

Aer—A class that supports classical simulator backends.

Execute—A function used to run quantum algorithms.

my_simulator—this is a variable created for the purpose of storing classical simulator.

QuantumCircuit () takes two arguments, the first parameter is the number of quantum bits and the second parameter is the number of classical bits (which is optional also).

In the example code in Block 1 and Block 2, a quantum circuit with one quantum register has been created. The last three lines of code in Block 2, helps in executing the circuit.

The instructions of our Quantum Circuit go through two more classes namely *job* and *result* before finally displaying the output [8].

3 Quantum Gates

There are three computational bases Z basis, X basis, and Y basis [9] as shown in Fig. 1.

The eigen states of the Z gate $\{ |0\rangle,\ 1\rangle\}$ form the Z basis. The eigen states of X gate $\{ |+\rangle, |-\rangle\}$ form the X basis as given in Eq. (2). The eigen states of Y gate $\{ |+i\rangle, |-i\rangle\}$ form the Y basis, as given in Eq. (3).

$$|+\rangle = \frac{(0\rangle + |1\rangle)}{\sqrt{2}}\ and\ |-\rangle = \frac{(|0\rangle - |1\rangle)}{\sqrt{2}} \tag{2}$$

$$|+i\rangle = \frac{(|0\rangle + i|1\rangle)}{\sqrt{2}}\ and\ |-i\rangle = \frac{(|0\rangle - i|1\rangle)}{\sqrt{2}} \tag{3}$$

Quantum gates [10] play an important role in quantum computing. Table 1 depicts the various single and multi qubit quantum gates along with the functionality, matrix representation, and qiskit function names.

The code in Block 3 creates a superposition state of one qubit. Our qubit has a 50% chance of being in the state $|0\rangle$ or 50% chance of being in the state $|1\rangle$. The code in Block 4 creates a superposition state of two qubits, so we have 25% chance of being in state $|00\rangle$, $|01\rangle$, $|10\rangle$, $|11\rangle$.

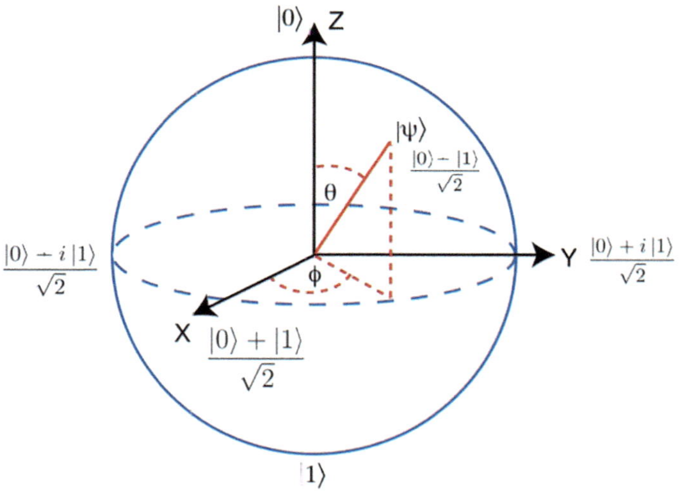

Fig. 1 Computational bases

Table 1 Single qubit and multi qubit quantum gates

Gate	Functionality	Matrix	Qiskit (Function name)
I	Leaves the state of a qubit unchanged	$\begin{bmatrix} 1 & 0 \\ 0 & 1 \end{bmatrix}$	Id
X	Flips the state of a qubit behave like classical NOT gate	$\begin{bmatrix} 0 & 1 \\ 1 & 0 \end{bmatrix}$	X
Hadamard	Creates a superposition state of qubits	$1/2\begin{bmatrix} 1 & 1 \\ 1 & -1 \end{bmatrix}$	H
Y	Performs rotation by π, around the y-axis of the bloch sphere	$\begin{bmatrix} 0 & -i \\ i & 0 \end{bmatrix}$	Y
Z	Performs rotation by π, around the z-axis of the bloch sphere	$\begin{bmatrix} 1 & 0 \\ 0 & -1 \end{bmatrix}$	Z
S	Performs rotation by $\pi/2$, around the z-axis of the bloch sphere	$\begin{bmatrix} 1 & 0 \\ 0 & e^{i\Pi/2} \end{bmatrix}$	S
T	Performs rotation by $\pi/4$, around the z-axis of the bloch sphere	$\begin{bmatrix} 1 & 0 \\ 0 & e^{i\Pi/4} \end{bmatrix}$	T
CNOT	Two qubit quantum gate, control qubit, and target qubit. If the state of control qubit is $\|1\rangle$ the target qubit gets flipped, otherwise it remains unaltered	$\begin{bmatrix} 1 & 0 & 0 & 0 \\ 0 & 1 & 0 & 0 \\ 0 & 0 & 0 & 1 \\ 0 & 0 & 1 & 0 \end{bmatrix}$	CX

```
#Block 3
# Code to create a superposition state
from qiskit import QuantumRegister, QuantumCircuit, Aer, execute
import numpy as np
import math as m
my_simulator = Aer.get_backend(name='statevector_simulator')


q = QuantumRegister(1)

H_circuit = QuantumCircuit(q)

H_circuit.h(q[0])

job = execute(H_circuit, my_simulator)
result = job.result()
result.get_statevector()
```

```
array([0.70710678+0.j, 0.70710678+0.j])
```

```
#Block 4
# Code to create a superposition state of two qubits
q = QuantumRegister(2)
H_circuit = QuantumCircuit(q)

H_circuit.h(q[0])
H_circuit.h(q[1])
job = execute(H_circuit, my_simulator)
result = job.result()
result.get_statevector()
```

```
array([0.5+0.j, 0.5+0.j, 0.5+0.j, 0.5+0.j])
```

We have created a superposition state from two qubits, by applying a hadamard gate on each of them, a combined state of the four combinations $|00\rangle$, $|01\rangle$, $|10\rangle$, $|11\rangle$ with a 25% probability.

3.1 Some Important Results

Qiskit supports various output visualizations such as state vector, QSphere, Bloch Sphere and Histogram representation [11]. Table 2 depicts the fundamental qubit states such as $|0\rangle$, $|1\rangle$, $|+\rangle$, $|-\rangle$, $|i\rangle$, $-|i\rangle$ in various representations along with the quantum circuit for generating the qubit state.

Figure 2 applies quantum gates X, Y, Z, and S on various inputs such as $|0\rangle$, $|1\rangle$, $|+\rangle$, $|-\rangle$, $|i\rangle$, $|-i\rangle$ and shows the outputs generated.

Table 2 Representing quantum states

State	Circuit diagram	State vector	Q-sphere	Bloch sphere	Histogram
|0>		[1 + 0j, 0 + 0j]	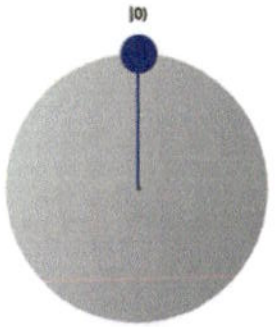	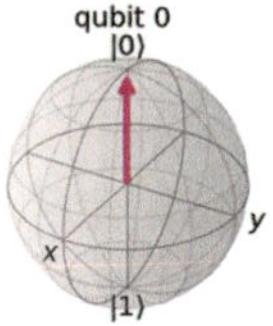	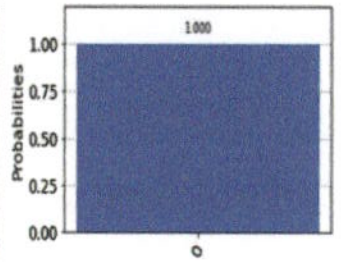
|1>	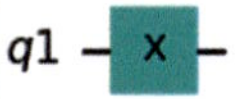	[0 + 0j, 1 + 0j]	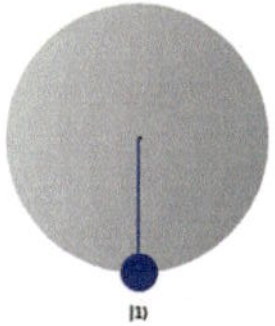	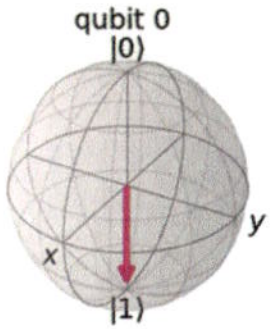	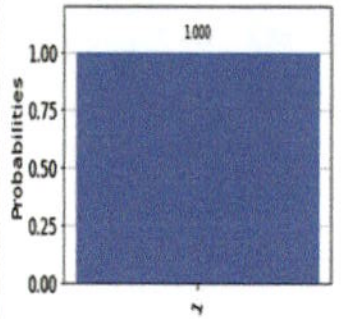
|+>	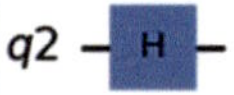	[0.707 + 0j, 0.707 + 0j]	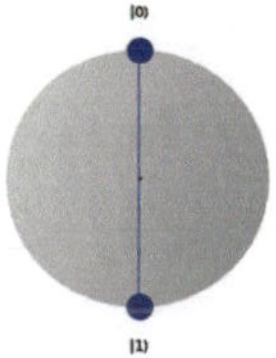	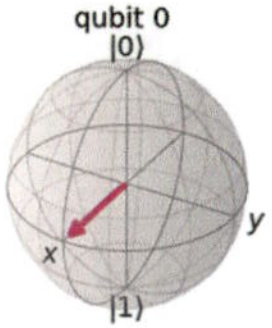	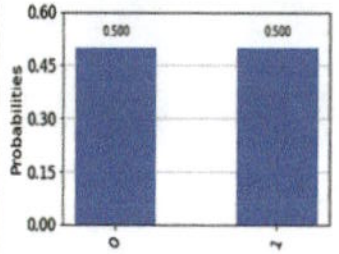

(continued)

Table 2 (continued)

State	Circuit diagram	State vector	Q-sphere	Bloch sphere	Histogram
|–>	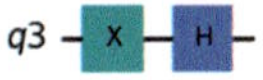	[0.707 + 0j, -0.707 + 0j]	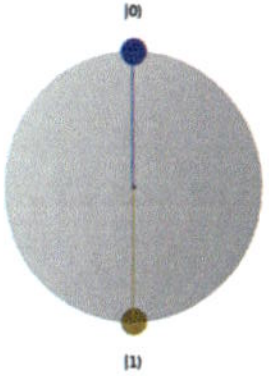	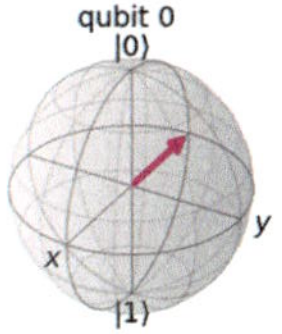	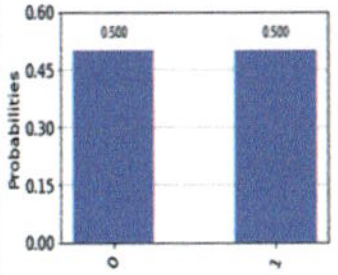
|i>	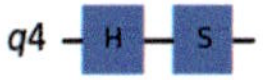	[0.707 + 0j, 0 + 0.707j]	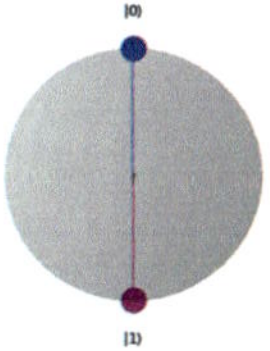	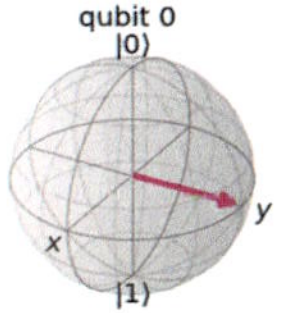	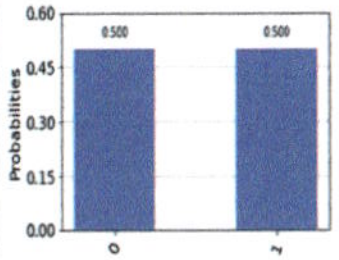
|–i>		[0.707 + 0j, 0–0.707j]	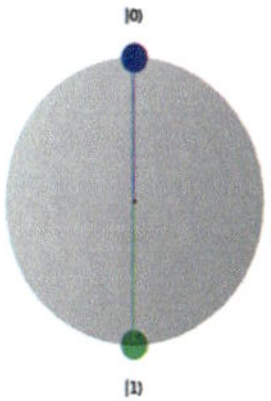	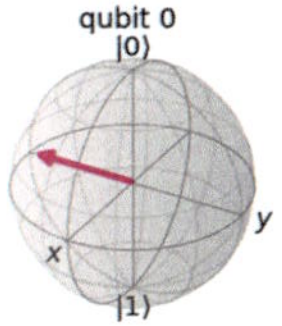	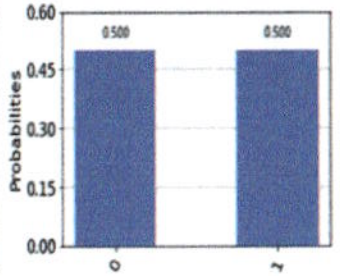

$$
\begin{array}{llll}
x|0\rangle = |1\rangle & y|0\rangle = i|1\rangle & z|0\rangle = |0\rangle & S\,|0\rangle = |0\rangle \\
x|1\rangle = |0\rangle & y|1\rangle = -i|0\rangle & z|1\rangle = -|1\rangle & S\,|1\rangle = i\,|1\rangle \\
x|+\rangle = |+\rangle & y|+\rangle = -i|-\rangle & z|+\rangle = |-\rangle & S\,|+\rangle = |i\rangle \\
x|-\rangle = -|-\rangle & y|-\rangle = i|+\rangle & z|-\rangle = |+\rangle & S\,|-\rangle = |-i\rangle \\
x|i\rangle = i|-i\rangle & y|i\rangle = |i\rangle & z|i\rangle = |-i\rangle & S\,|i\rangle = |-\rangle \\
x|-i\rangle = -i|i\rangle & y|-i\rangle = -|-i\rangle & z|-i\rangle = |i\rangle & S\,|-i\rangle = |+\rangle
\end{array}
$$

Fig. 2 Outputs of quantum gates

4 Quantum Entanglement

Quantum entanglement [12] is a property in which a pair of particles is generated in such a way that the individual quantum states of each are indefinite until measured, and the act of measuring one determines the result of measuring the other, even when at a distance from each other. Entanglement occurs when the control qubit in a controlled gate is in a superposition. Therefore, the state of one qubit depends on another qubit, which is also in a superposition. Figure 3 gives the quantum circuit for creating entanglement. Table 3 gives the entanglement truth table for various inputs.

The following sections will discuss the implementations of generating quantum entanglement and executing the code using Qiskit simulators and using quantum real devices supported by IBM.

4.1 Executing on Simulator

The code in Block 5 creates an entanglement state [13] between two qubits, and the circuits is executed using a state vector simulator, and the output in the form of quantum circuit, state vector, qsphere, and histogram for all the inputs is consolidated in Table 4.

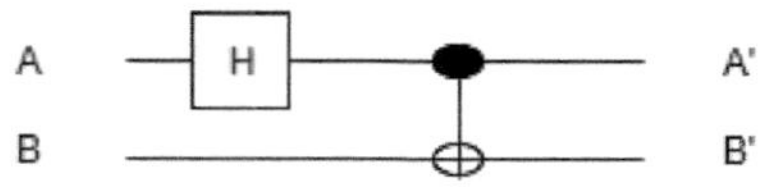

Fig. 3 Quantum entanglement circuit

Table 3 Quantum entanglement truth table

Input	Output
00	$\frac{1}{\sqrt{2}}(\|00\rangle + \|11\rangle)$
01	$\frac{1}{\sqrt{2}}(\|01\rangle + \|10\rangle)$
10	$\frac{1}{\sqrt{2}}(\|00\rangle - \|11\rangle)$
11	$\frac{1}{\sqrt{2}}(\|01\rangle - \|10\rangle)$

Table 4 Quantum entanglement-output in various forms

State	Circuit diagram	State vector	Q-sphere	Histogram
|00>	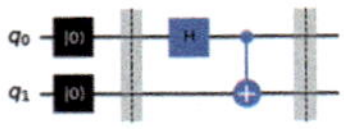	[0.707 + 0j, 0 + 0j, 0 + 0j, 0.707 + 0j]	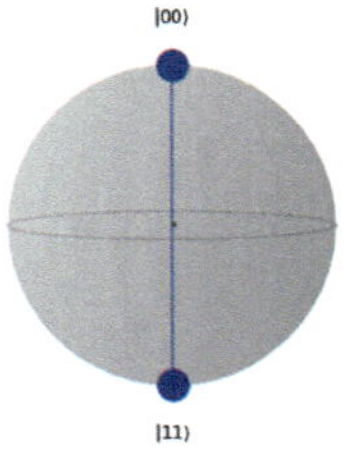	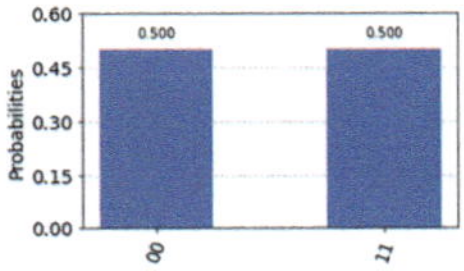
|01>	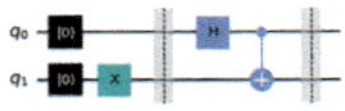	[0 + 0j, 0.707 + 0j, 0.707 + 0j, 0 + 0j]	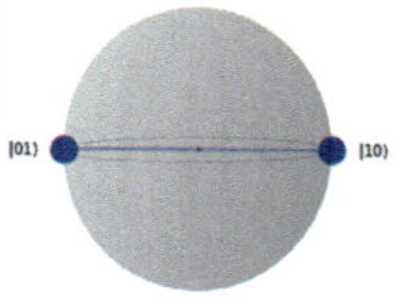	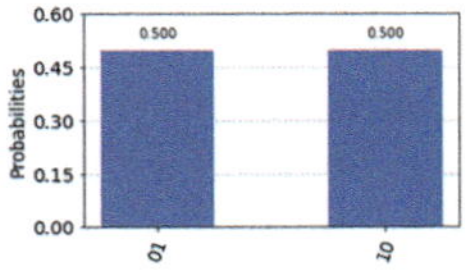
|10>	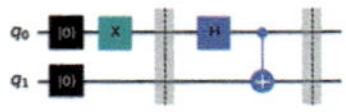	[0.707 + 0j, 0 + 0j, 0 + 0j, -0.707 + 0j]	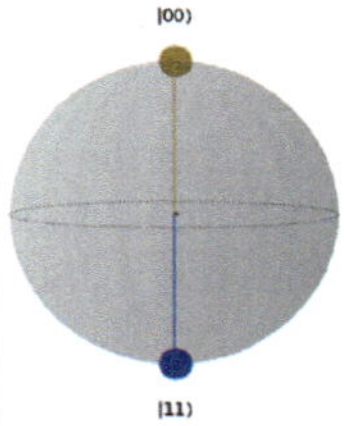	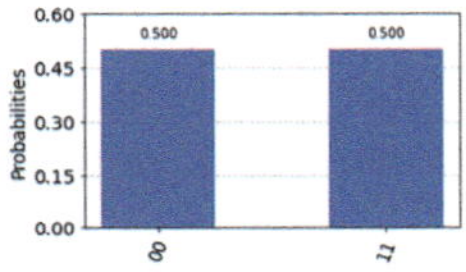
|11>	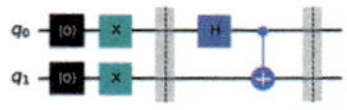	[0 + 0j,−0.707 + 0j, 0.707 + 0j, 0 + 0j]	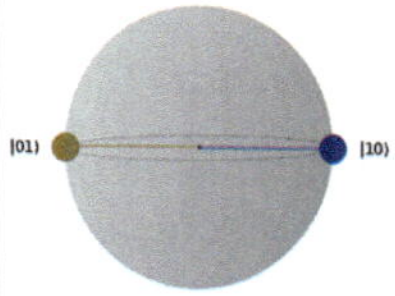	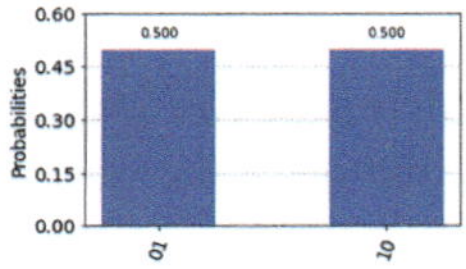

```
#Block 5
from qiskit.visualization import plot_state_qsphere, plot_bloch_multivector
from qiskit.visualization import plot_histogram, plot_state_city
from qiskit.visualization import plot_state_hinton, plot_state_paulivec
from qiskit import QuantumCircuit, Aer, execute
sim = Aer.get_backend(name = 'statevector_simulator')
import qiskit.tools.jupyter
```

```
#Block 5
def entangleCircuit(input1, input2):
    qc = QuantumCircuit(2)
    qc.reset(range(2))

    if input1==1:
        qc.x(0)
    if input2==1:
        qc.x(1)
    qc.barrier()

    qc.h(0)
    qc.cx(0,1)

    qc.barrier()
    job = execute(qc,sim)
    state = job.result()
    return qc, state
```

```
#Block 5
for inp1 in [0,1]:
    for inp2 in [0,1]:
        print(f'For input {inp1} {inp2}')
        qc, state = entangleCircuit(inp1,inp2)
        display(plot_state_qsphere(state.get_statevector()))
        display(plot_state_city(state.get_statevector()))
        display(plot_state_hinton(state.get_statevector()))
        display(plot_state_paulivec(state.get_statevector()))
```

4.2 *Executing on Real Quantum Device*

The code in Block 6 creates an entanglement state between two qubits and the state $\frac{1}{\sqrt{2}}(|01\rangle - |10\rangle)$ is generated as output. The circuit is executed using a quantum real device, ibmq_quito [14].

```
#Block 6
import numpy as np

# Importing standard Qiskit Libraries
from qiskit import QuantumCircuit, transpile, Aer, IBMQ,execute
from qiskit.tools.jupyter import *
from qiskit.visualization import *
from ibm_quantum_widgets import *
from qiskit.providers.aer import QasmSimulator
import qiskit as q

# Loading your IBM Quantum account(s)
provider = IBMQ.load_account()

qr = q.QuantumRegister(2)
c = q.ClassicalRegister(2)
qc=QuantumCircuit(qr,c)

qc.h(qr[0])
qc.cx(qr[0],qr[1])
qc.z(qr[0])
qc.x(qr[0])

qc.measure(qr,c)
display(qc.draw('mpl'))
```

The circuit for the code in Block 8 is shown in Fig. 4.

The code in block 7 will display a least busy real quantum device, which supports more than two qubits, not a simulator and whose operational status is true.

```
#Block 7
from qiskit.providers.ibmq import least_busy
backend = least_busy(provider.backends(filters=lambda x:
          x.configuration().n_qubits >= 2 and not x.configuration().simulator
          and x.status().operational==True))
backend
```

The output of Block 7 is 'ibmq_quito'.

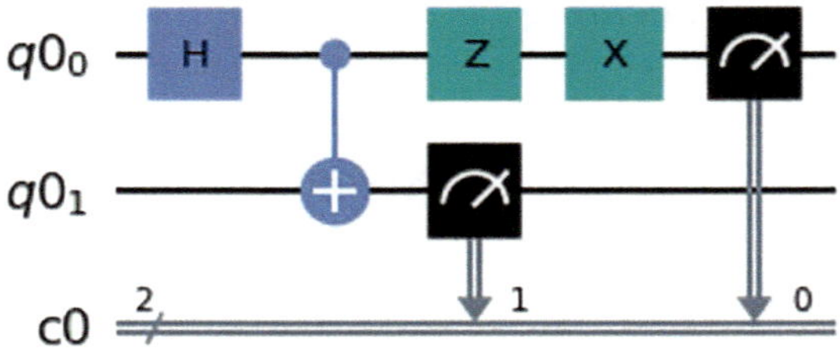

Fig. 4 Quantum Entanglement Circuit

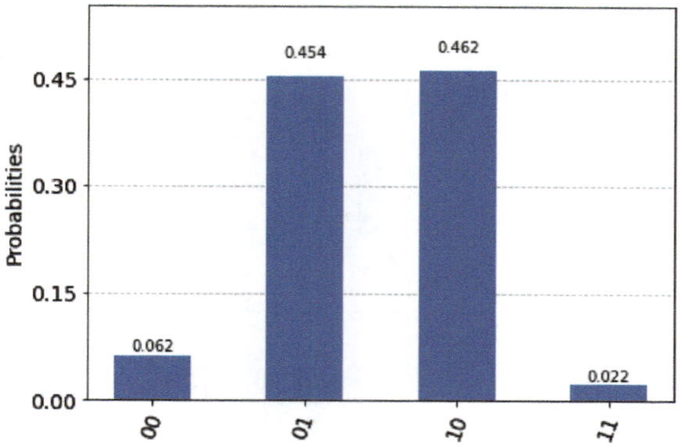

Fig. 5 Quantum entanglement output—executed on real quantum device

```
<IBMQBackend('ibmq_quito') from IBMQ(hub='ibm-q', group='open', project='main')
>
```

The code in Block 8 sets 'ibmq_quito' as the backend device and executes the circuit and displays the output in the form of a histogram.

```
#Block 8
backend=provider.get_backend('ibmq_quito')
job = q.execute(qc, backend=backend)
q_result = job.result()
q.visualization.plot_histogram(q_result.get_counts(qc))
```

The output of Block 8 is shown in Fig. 5. In addition to states $|01\rangle$ and $|10\rangle$, there are few results in which we measured the states $|00\rangle$ and $|11\rangle$. These arise due to errors in the gates and the qubits when executing using quantum real devices. In this example 'ibmq_quito' is set as a backend. The circuit is executed for a default of 1024 shots using execute function.

5 Output Visualization

Qiskit supports various output visualizations such as state vector, qsphere, histogram, bloch sphere, plot_state_city, plot_state_hinton, plot_state_paulivec [15]. Table 5 consolidates the output of quantum entanglement in plot_state_city, plot_state_hinton, plot_state_paulivec.

Table 5 Quantum Entanglement output in other forms

State	Output	Plot_state_city	Plot_state_hinton	Plot_state_paulivec
|00>	$1/\sqrt{2}$ (|00> + |11 >)	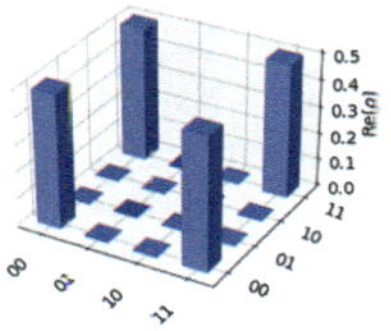	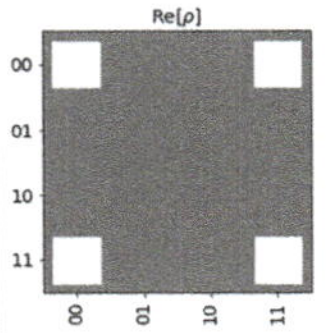	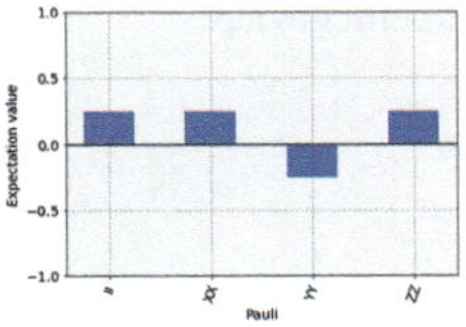
|01>	$1/\sqrt{2}$ (|01> + |10 >)	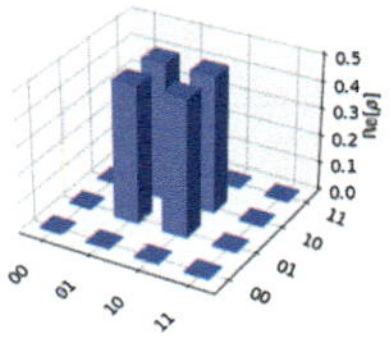	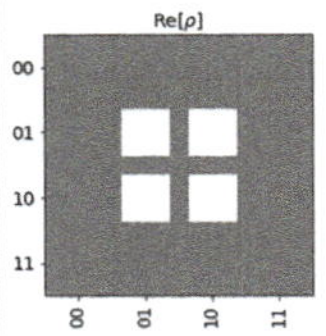	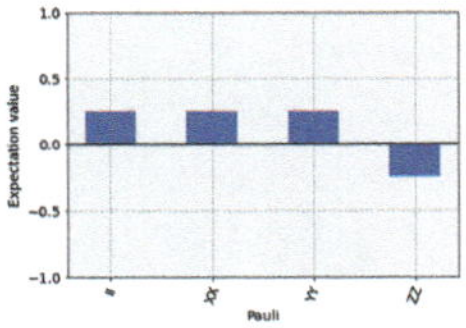
|10>	$1/\sqrt{2}$ (|00> − |11>)	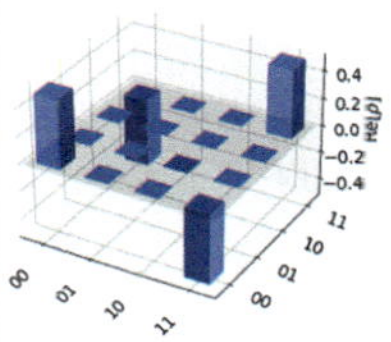	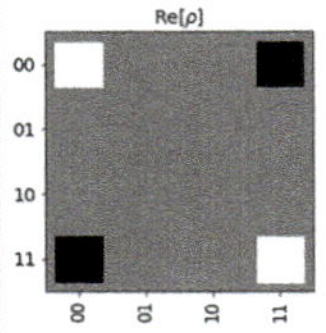	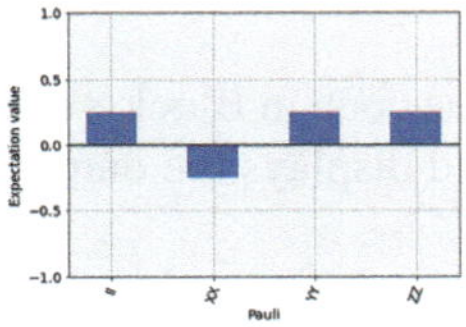
|11>	$1/\sqrt{2}$ (|01> − |10>)	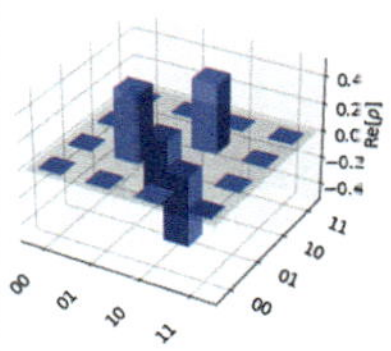	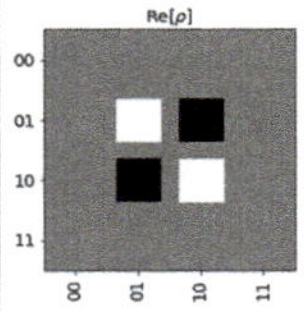	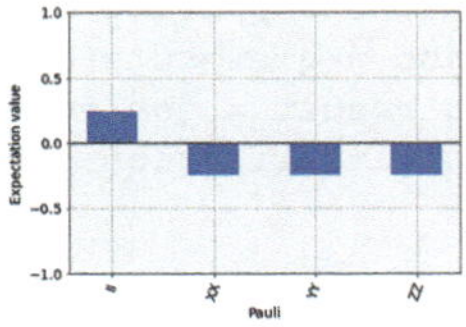

6 Conclusion

One of the open-source frameworks to write and execute quantum programs is IBM's Quantum Experience–Qiskit. In the first part of the chapter, we used Qiskit to represent various quantum states and discussed the fundamentals of building and executing quantum circuits. In the second part we discussed quantum gates, qiskit code by applying the various quantum gates and discussed the results obtained. The next part of the chapter discussed how to create quantum entanglement and represented the output of entanglement in various forms supported by qiskit such as QSphere representation, Bloch Sphere representation, Plot state city, Plot state paulivec, and Plot state hinton.

References

1. A.M. Nielsen, I.L. Chuang, *Quantum computation and quantum information*, (Cambridge University Press, 2010)
2. N.D Mermin, *Quantum computer science*, (Cambridge University Press, 2007)
3. *IBM Qiskit Textbook*. https://qiskit.org/textbook-beta/
4. IBM quantum experience is quantum on the cloud. https://quantum-computing.ibm.com/
5. O. Ezratty, *Understanding Quantum Technologies*, (2021)
6. J.L. Weaver, F.J. Harkins, *Qiskit pocket guide, quantum development with Qiskit*, (O'Reilly Media, Early Release, 2021)
7. H. Norlen, *Quantum computing in practice with Qiskit and IBM quantum experience*, (Packt Publishing, 2020)
8. R.S. Sutor, *Dancing with Qubits, how quantum computing works and how it can change the world*, (Packt Publishing Ltd., 2019)
9. N. Yanofsky, M. Mannucci, *Quantum Computing for Computer Scientists*, (Cambridge University Press, 2008)
10. E. Rieffel, W. Polak, Quantum computing: A gentle introduction, (MIT Press, 2011)
11. T.G. Wong, *Introduction to classical and quantum computing*, (Rooted Grove, 2022)
12. R. Loredo, *Learn quantum computing with Python and IBM quantum experience*, (Packt Publishing, 2020)
13. J. Gruska, *Quantum computing*, (McGraw Hill, 2011)
14. Z. Chrzastek, Assessment of IBM-Q quantum computer and its software environment, Master's Thesis, AGH University of Science and Technology, 2018
15. G. Carrascal, Alberto A del Barrio, Guillermo Botella, "First experiences of teaching quantum computing." J. Supercomput. **77**, 2770–2799 (2021)

Evolutionary Analysis: Classical Bits to Quantum Qubits

Rajiv Pandey, Pratibha Maurya, Guru Dev Singh, and Mohd. Sarfaraz Faiyaz

Abstract Quantum computing has evolved from traditional classical computing due to the integration of multiple scientific domains. Quantum physics and quantum mechanics to name a few. Classical computers worked on a 1 and 0 logic, whereby the value would be either a voltage high or voltage low respectively. Quantum computers on the contrary use Qubits to represent the states, these state vectors have the capability to represent multiple states simultaneously and thus render scalability and exponential speed to quantum computing. Quantum computers harness the unique behaviour of quantum physics by applying concepts such as superposition, entanglement, and quantum interference. This chapter is authored to take us through the transition from classical to quantum computing. The chapter shall present a comparison of features that are exhibited by classical computers and how quantum principles enable quantum to compute, The chapter is aimed to provide an introduction to the various quantum gates which are to be deliberated in the subsequent chapters.

Keywords Quantum computing · Classical computing · Classical bits · Quantum bits

1 Introduction

Quantum computing has the potential to tackle several pressing problems of our planet, including those related to the environment, agriculture, health, energy, climate, materials science, and so on [1]. The most powerful computers are not

R. Pandey · P. Maurya (✉) · G. D. Singh
Amity Institute of Information Technology, Amity University Uttar Pradesh, Lucknow Campus, Noida, India
e-mail: pmaurya@lko.amity.edu

R. Pandey
e-mail: rpandey@lko.amity.edu

Mohd. S. Faiyaz
Amity School of Engineering and Technology, Amity University Uttar Pradesh, Lucknow Campus, Noida, India

R. Pandey et al. (eds.), *Quantum Computing: A Shift from Bits to Qubits*,
Studies in Computational Intelligence 1085,
https://doi.org/10.1007/978-981-19-9530-9_7

capable to solve certain exponential problems where quantum computing comes into play. While quantum technology is simply starting to impact the computing world, it might be far-reaching and may alter the way we expect computing in the future.

In modern usage, the word quantum means the smallest possible discrete unit of any property, usually pertaining to properties of atomic or subatomic particles. Quantum computers use artificial atoms, quantum particles, or collective features of quantum particles as processing units. They are complex, large, and expensive devices.

A host of the latest computer technologies have emerged in the over a couple of years, and quantum computing is arguably the technology demonstrating the best paradigm shift on a part of developers. Quantum computers were proposed in the 1980s by Richard Feynman and Yuri Manin [2–4]. Quantum physics was developed between 1900 and 1925 and it remains the cornerstone on which chemistry, condensed matter physics and technologies ranging from computer chips to LED lighting ultimately depends. Yet despite these successes, even the variety of the sole systems seemed to be beyond the human ability to model with physics. This is since modelling even a few dozen interacting particles necessitates far more computational power than any normal computer can deliver over millions of years.

There are some ways to know why quantum physics is tough to simulate [3]. Perhaps the only is to ascertain that scientific theory are often interpreted as saying that matter, at a quantum level, is during a multitude of possible configurations (known as states). Unlike classical applied mathematics, these many configurations of the quantum state, which may be potentially observed, may interfere with one another like waves during a tide pool. This interference prevents the utilisation of statistical sampling to get the quantum state configurations. Rather, we must track every possible configuration a quantum system might be in if we would like to know the quantum evolution.

Consider an electron system in which electrons are frequently found in any of position, for assumption let's take 40 positions. The electrons therefore could also be in any of 240 configurations. This is due to the fact that each location might have or not have an electron. It would need 130 GB of memory to store the quantum state of electrons in a normal memory. This is significant, but only a few machines have access to it. If the particles may be in any of the 41 locations, there will be two times the number of configurations at 241 which successively would require quite 260 GB to save the quantum state in memory. If we want to save the state traditionally, we rapidly exhaust the memory capacity of the world's most powerful processors, thus this game of increasing the number of positions can't be played eternally [5]. With only a few hundred electrons, the amount of memory required to store the system surpasses the number of particles in the universe; hence, our current computers will never be able to mimic quantum dynamics. Yet, in nature, such systems evolve in time in accordance with quantum mechanical rules, blissfully oblivious of the absence of computational capacity to engineer and simulate their evolution.

This discovery inspired individuals with quantum computing vision to ask a simple but profound question: can we convert this difficulty into an opportunity? What would happen if we created hardware with quantum effects as fundamental operations, if quantum dynamics are difficult to simulate? Could we mimic systems of interacting particles using a system that uses a law similar to the one that governs them in nature? Could we investigate jobs that aren't found in nature but yet adhere to or enjoy quantum mechanical laws? These inquiries sparked the pursuit of quantum computing.

Quantum computing is based on the storage of information in quantum states of matter and the computation of that information using quantum gate operations, by harnessing and learning to 'program' quantum interference. Peter Shor used programming interference to unravel a haul that was supposed to be difficult on ordinary computers in 1994 for a haul referred to as factoring [6]. Solving factoring has the potential to disrupt many of the public key cryptosystems that underpin today's e-commerce security, such as RSA and Elliptic Curve Cryptography. Since then, quantum computer algorithms for a variety of challenging classical tasks have been created, including modelling physical systems in chemistry, physics, and materials science, solving systems of linear equations, searching an unordered database, and machine learning.

The chapter has been organised as follows. Section 2 deals with the concepts of classical computing, Sect. 3 elaborates on the principles of quantum computing, and Sect. 4 gives a comparative analysis of the two schools of computing. Section 5 describes the quantum gates in detail. The chapter is concluded in Sect. 5.

2 Classical Computer

Classical computers are machines that use discrete types of digital logic to solve any issue. Binary logic is employed, and inputs might be numerical, alphabetic, special characters, video, or even audio. The data processing procedure was broken down into six phases as shown in Fig. 1.

- Data collection
- Data preparation
- Data input
- Processing
- Data output/interpretation
- Data storage

Data is received in three ways by traditional computers.

- Create a file with a name and a type.
- Make a recording with a specific size in mind.

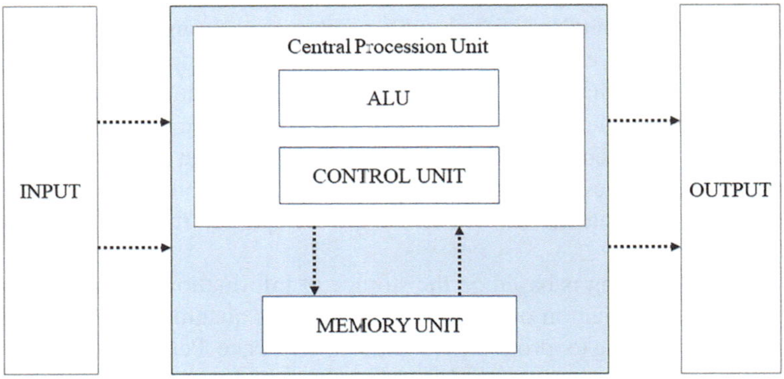

Fig. 1 Architecture of a classic computer

- A field with a storage place.

A record contains information on a certain entity and its attributes, as well as operations on those properties, the type of data format the entity uses, and so on. A customer's information from a bank or any of the organisations may be used as an example. A customer's name, customer id, client address, phone number, email id, and other characteristics are examples of its properties in a bank, and this information about the consumer will be kept in the category of a record. Similar records might be created for a large number of customers.

- Data preparation
- Programming
- Verification of the program's validity

Classic computers are classified into three categories based on their applications.

- In the realm of science and technology for data processing
- In business settings for data processing
- Computers can be used in a variety of ways.

Computers are primarily employed in the scientific and technological industries to accomplish activities that need data in various formats, such as assessing mathematical equations based on collected data, creating hypotheses, and so on. The computers employed in this sector have enormous capacity, a wide range of characteristics, and can process data in a wide range of formats. Computers used in business applications, on the other hand, are capable of collecting data (symmetric or asymmetric), statistically analysing data, drawing observations in various formats, and eventually making choices based on them. Examples include allocating a loan to a consumer in a banking environment, determining discounts for product purchases in online marketing, and so forth. The third category comprises flexible computers, which are a combination of science, technology, and business. The bulk of today's computers is adaptable.

The Versatile Computers are classified into two types:

- General-purpose computer
- Special-purpose computers

General-purpose computers can be used to perform fundamental scientific applications or mathematical evaluations/calculations, as well as to carry out corporate activities. Special-purpose computers, such as supercomputers, are used when a specialised sort of processing is required, such as.

- Numerical figures are used to control mechanical tools.
- Switching operations in electronic devices.
- Assisting satellites in determining the best path to rotate around the Earth, and so on.

2.1 Classical Bits

- In computers, a bit is the most fundamental unit of data storage. A computer is known to operate by manipulating binary digits, or 1 and 0s. These logical values are recorded in Bits, which is a unit of measurement.
- A bit is a fundamental unit in computers that transports data. Only a binary digit, 0 or 1, may be carried by these single bits. Bits are too tiny to be utilised and are frequently incapable of conveying any significant information on their own, therefore they're organised into eight-bit units called bytes. In a computer, a bit is the smallest unit of data, while a byte is a collection of bits. Bytes are used to organise all of the data that we create and require.
- The organisation of data in bytes improves data processing speed and efficiency. Because a byte is eight times larger than a bit, the number 8 and its multiples are significant quantities to remember while doing computational operations. The first four multiples of eight will appear often in various settings in computing computations. The byte, which is 8 bits long, is the foundation of everything. It is the fundamental building block of information in computers. However, it's essential to keep in mind that the entire system's sophisticated functionality is built on only two digits: 1 and 0. A bit is created in a computer by electrical current fluctuations that pass through its components. These oscillations are converted into 1 or 0s, which are utilised to send bits, do computations, and transport data over the network. When any data or information is sent between networks, the bits are encrypted to provide the highest level of data security and integrity.

3 Quantum Computer

Quantum computing is a new computing approach in which the computer is built using atomic and subatomic quantum concepts [7, 8]. On a small scale, certain industries are presently building quantum computers. The development of quantum computers is limited for microscopic sizes due to numerous variables that impact a quantum computer's functionality. IBM just finished the design of a 51-qubit quantum computer. Any technique for constructing quantum computers can correlate a high number of qubits with each other if the quantity is in the thousands or lakhs. There is no question that quantum computers will become the dominant technology.

It is difficult to properly identify an elementary particle's location and momentum at the same time. Even if the particle's position is properly measured, there is always a flaw in estimating its momentum. As a result, while measuring an elementary particle's location and momentum at the same time, these characteristics are inversely proportional to each other in terms of measurement.

3.1 Quantum Bits

A qubit (or quantum bit) is the quantum mechanical equivalent of a bit. In traditional computing, information is represented in bits, with each bit having a value of zero or one. Information in quantum computing is encoded in qubits [9]. A qubit is a two-level quantum system with the two base qubit states denoted as $|0\rangle$ and $|1\rangle$. A qubit can be in one of three states: $|0\rangle$ $|1\rangle$ or unlike a conventional bit, a linear mixture of both. This phenomenon is known as superposition.

A qubit's two orthogonal z-basis states are defined as

- $|0\rangle$
- $|1\rangle$

z-basis states are implicitly referred to as the computational basis states when we discuss qubit basis states.

The two orthogonal x-basis states are as follows:

$$|+\rangle = \frac{|0\rangle + |1\rangle}{\sqrt{2}} |-\rangle = \frac{|0\rangle - |1\rangle}{\sqrt{2}} \tag{1}$$

The two orthogonal y-basis states are as follows:

$$|R = \frac{|0\rangle + i|1\rangle}{\sqrt{2}} |L\rangle = \frac{|0\rangle - i|1\rangle}{\sqrt{2}} \tag{2}$$

3.2 Bloch Sphere

The Bloch sphere is a mathematical representation of qubit states as points on a unit sphere's surface [10, 11] and is depicted in Fig. 2.

The Bloch sphere picture may represent various operations on single qubits that are commonly used in quantum information processing in a concise manner [12]. It turns out that a single qubit state may be expressed as

$$|\psi\rangle = e^{i\gamma}\left(\cos\frac{\theta}{2}|0\rangle + e^{i\varphi}sin\frac{\theta}{2}|1\rangle\right) \tag{3}$$

where, θ, φ, and γ represent real numbers. A point on a unit three-dimensional sphere is defined by the integers $0 \leq \theta \leq \pi$ and $0 \leq \varphi \leq 2\pi$. Because the component of $e^{i\gamma}$ has no observable consequences, all qubit states with arbitrary values are represented by the same point on the Bloch sphere, and we may essentially write

$$|\psi\rangle = \cos\frac{\theta}{2}|0\rangle + e^{i\varphi}sin\frac{\theta}{2}|1\rangle \tag{4}$$

The Bloch Sphere is a refinement of the depiction of a complex number z as a point on the complex plane's unit circle with |z|2 = 1.

If z = x + iy, and x and y are both real numbers, then:

$$|z|2 = z * z = (x - iy)(x + iy) = x^2 + y^2 \tag{5}$$

where x2 + y2 = 1 is the equation of a circle centred on the origin with radius one.

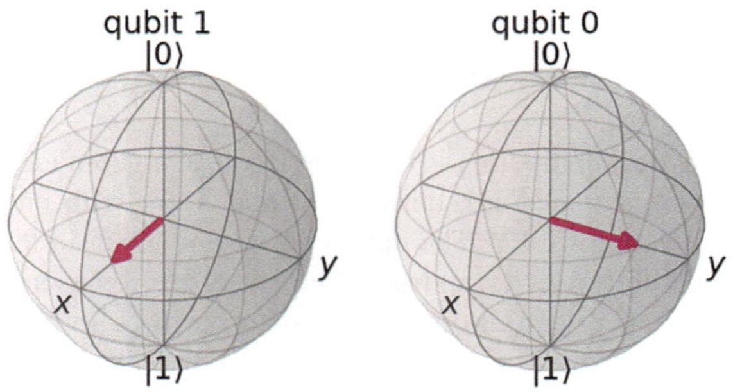

Fig. 2 Visual representation of bloch sphere

3.3 Quantum State Formats for Representing Classical Data

Consider the issue of converting a matrix of single-bit values into a quantum state (b_{ij}). Take the 2 by 2 matrix $A = \begin{bmatrix} b_{00} & b_{01} \\ b_{10} & b_{11} \end{bmatrix}$ as an example.

The i subscript for bit bij denotes the row, with i = 0 or 1, while the j subscript denotes the column, with j = 0 or 1. A is the quantum state that corresponds to the matrix A bit values. The four single-bit values bij, as well as the associated locations of the bits in the matrix, are represented by three (3) qubits in the quantum state A.

$$\begin{aligned} |\psi A\rangle &= |00\rangle \otimes |b00\rangle + |01\rangle \otimes |b01\rangle + |10\rangle \otimes |b10\rangle + |11\rangle \otimes |b11\rangle \\ &\equiv |00b00\rangle + |10b01\rangle + |10b10\rangle + |11b11\rangle \end{aligned} \tag{6}$$

The tensor product operator is the symbol ⊗ that will be used to demarcate groups of qubits within a quantum state [13, 14]. As seen in the second line of Eq. (1), quantum information notation frequently omits the tensor symbol when writing the state.

The i index corresponds to the initial, leftmost qubit of the state, which represents the row inside the matrix. The second qubit, corresponding to the index j, represents the column. The single-bit value of the associated (row, column) matrix item is represented by the third, rightmost qubit. The total quantum state normalisation constant is omitted for readability, as is customary in quantum information. It's simple to calculate and reinsert the overall state normalisation constant as needed. Regardless of the values of the bij, the normalisation constant for the state in Eq. (1) is $\frac{1}{2}$.

4 Comparative Analysis

Classical computing is based on Boolean algebra ideas and generally uses a 7-mode logic gate standard, however, it may work with only three modes, notably NOT, AND, and COPY. The data will only be given in terms of two-state logic, that is, 0 (off/false) or 1 (on/true). The states are represented by binary numbers or bits. Millions of capacitors and transistors will be in one of the computer's states. The state of capacitors or transistors (0 or 1) will be monitored in millions of seconds. Traditional computers' speeds are completely governed by how long it takes the computer to measure these states [15]. The conventional computer is developing with technological development. However, because classical computers are computationally restricted, this leads to the discovery of new computation methods [15], and this is how quantum computing began to evolve using quantum physics as a basis.

In quantum computing, qubits can be in the states 0 or 1, or in a superposition of 0 and 1. That is, a single qubit in a quantum computer can be in more than two states, whereas this is not possible with conventional computers. Atoms, electrons, neutrons, ions, and other fundamental components of qubits can be used in quantum computers. Furthermore, when these fundamental components are exposed to a force, they change their spin or rotations and acquire a state [17]. When a state changes, the most fundamental force that happens is polarisation. After deciding on the type of element for a qubit and the mechanism of changing states, these qubits should exhibit superposition and entanglement properties. Based on the study done in this paper on classical and quantum computers, the following observations are made utilising designs, methodologies, and problem types as examples. Over the span of decades, conventional computers have made significant advances in terms of calculation rates and memory sizes. As a result, conventional computers are employed in almost every industry.

However, cloud servers are still inefficient in many businesses in terms of speed, efficiency, accuracy, and so on. Another type of computer examined in this study is the quantum computer, which employs quantum physics concepts to solve the problem. It solves problems by utilising quantum characteristics like as superposition and entanglement on qubits, resulting in a novel operating principle. The examples also show that quantum computers are superior in tackling specific problems in which neural networks underperform.

Classical Computer	Quantum Computer
The conventional computer uses binary logic to process data, and data is stored in the form of bits. Almost all efficient computers today are built on the same basis and solve all problems using binary logic; nevertheless, there are some scenarios for which traditional computers require an unreasonable amount of time to solve. These challenges are solved by employing a new computing paradigm, called quantum computing/computer, which uses quantum physics rules to solve problems with qubits as the fundamental unit	Quantum computers are expected to solve problems more quicker than conventional computers. The primary objective of this study is to discover and analyse the drawbacks of traditional computing, as well as to turn these drawbacks into potential ideas. The research concentrated on discovering the limitations of traditional computers in this respect, and it was revealed that certain challenges require unique viewpoints in order to be solved quickly

5 Quantum Gates

A quantum gate, also known as a quantum logic gate, is a simple quantum circuit that operates on a limited number of qubits. They are quantum computers' equivalents to ordinary digital computers' classical logic gates. Unlike many classical logic gates, quantum logic gates are reversible [18]. Certain universal classical logic gates, such as the Toffoli gate, are reversible and may be transferred directly into quantum logic gates. Unitary matrices are used to depict quantum logic gates.

The most popular quantum gates operate on one or two-qubit spaces. This means that quantum gates may be characterised as matrices by 2×2 or 4×4 matrices with orthonormal rows.

5.1 Single Qubit Gate

Single qubit gates represent spin rotations around an axis. The simplest gates are rotations around axes in the xy-plane, which may be accomplished using resonant RF pulses [9, 19]. The angle by which the spin is rotated (The flip angle of the pulse) is determined by the duration and power of the Radiofrequency pulse, but the phase angle of the pulse (and thus the azimuthal angle produced by the rotation axis in xy-plane) can be adjusted by changing the Radiofrequency's starting phase angle. Rotations around the z-axis can be performed using precession periods under the Zeeman Hamiltonian [20, 21], whereas off-resonance Radiofrequency stimulation can be used to create rotations around slanted axes. However, avoiding these last two approaches and instead constructing all single qubit gates using rotations in the xy-plane is usually easier. The composite z-pulse, which performs a z-rotation utilising x and y-rotations, is the easiest example.

$$\Theta_z \equiv 90 -_x \theta_y 90_x \equiv 90_y \theta_x 90_{-y} \tag{7}$$

Nuclear Magnetic Resonance notation was used to write pulse sequence, with the flow of time flowing from left to right, rather than operator notation, in which operators are applied progressively from right to left. A similar technique is used to create slanted rotations in the Hadamard gate, for example,

$$H \equiv 180_z 90_y \equiv 90_y 180_x 90_{-y} 90_y \equiv 90_y 180_x \tag{8}$$

This method may be used to create any single qubit gate.

Even this approach, although, is too complex, as z-rotations may be created in a considerably more straightforward manner. Rotating the spin's reference frame is easier than rotating the spin itself. This may be done by utilising a pulse sequence to forward or reverse z-rotations.

$$\psi_z \theta_\phi \equiv \theta_{\phi-\psi} \psi_z \tag{9}$$

in addition adjusting the pulse phase to correspond to the new frame of reference this technique, also known as an abstract frame of reference, has the advantage of allowing z-rotations to be done with no time or resources! Many existing implementations of NMR quantum logic gates rely solely on xy-plane rotations and reference frame modifications to produce all single qubit gates.

5.2 Hadamard Gate

To get a particle into a superposition state, some really complex physics is necessary. To be in a superposition state, you may simply apply a certain gate to a qubit.

A Hadamard gate or H gate is a fundamental quantum gate. It enables us to travel away from the Bloch sphere's poles [14], resulting in a superposition of $|0\rangle$ and $|1\rangle$. It has the following matrix:

$$H = \frac{1}{\sqrt{2}}\begin{bmatrix} 1 & 1 \\ 1 & -1 \end{bmatrix} \tag{10}$$

What happens if we apply this gate to a qubit in the $|0\rangle$ state. By multiplying the gate matrix by the qubit vector, this can be seen.

$$H = \frac{1}{\sqrt{2}}\begin{bmatrix} 1 & 1 \\ 1 & -1 \end{bmatrix}\begin{bmatrix} 1 \\ 0 \end{bmatrix} = \frac{1}{\sqrt{2}}\begin{bmatrix} 1 \\ 1 \end{bmatrix} \tag{11}$$

When the Hadamard Gate is introduced to a qubit in the $|0\rangle$ state, the qubit reaches a new state with a probability of measuring 0.

$$\mathrm{H} = \left(\frac{1}{\sqrt{2}}\right)^2 = \frac{1}{2} \tag{12}$$

Besides that, the possibility of measuring 1 is.

$$\mathrm{H} = \left(\frac{1}{\sqrt{2}}\right)^2 = \frac{1}{2}$$

Consequently, applying the Hadamard Gate on a state $|0\rangle$ qubit places it in a superposition state where the possibility of measuring 0 is equal to the possibility of measuring 1.

What probability occurs if the Hadamard Gate is applied to a qubit in state $|0\rangle$?

In such state, the vector representation of a qubit is

$$H = \begin{bmatrix} 0 \\ \overline{1} \end{bmatrix}$$

As a result, using a Hadamard Gate on this qubit entails multiplying the Hadamard matrix by the vector above:

$$H = \frac{1}{\sqrt{2}}\begin{bmatrix} 1 & 1 \\ 1 & -1 \end{bmatrix}\begin{bmatrix} 1 \\ 0 \end{bmatrix} = \frac{1}{\sqrt{2}}\begin{bmatrix} 1 \\ -1 \end{bmatrix} \tag{13}$$

The chances of measuring 0 if we measured the qubit at this point is

$$\mathrm{H} = \left(\frac{1}{\sqrt{2}}\right)^2 = \frac{1}{2}$$

and the probability of measuring 1 is

$$\mathrm{H} = \left(\frac{-1}{\sqrt{2}}\right)^2 = \frac{1}{2}$$

As a result, in both situations (qubit $|0\rangle$ and qubit $|1\rangle$), using a Hadamard Gate[21] provides the qubit an equal probability of being 0 or 1' when measured.

5.3 *Phase-Shift Gate*

A phase-shift gate is simply made by adding m inductors solely in the lower wire compared to the higher wire [22]. The coefficients of transmission and reflection are given by

$$\mathrm{T(k)} = \mathrm{e2imk},\ \mathrm{R(k)} = 0$$

Because $|\mathrm{T\ (k)}| = 1$, the transmission flawless regardless of momentum. As a result, the φ mixing-gate $\mathrm{U}\varphi$ with $\varphi = 2\mathrm{mk}$ is obtained.

$$U_{2m} = \begin{pmatrix} 1 & 0 \\ 0 & e^{i2mk} \end{pmatrix} \tag{14}$$

Physically, it's well recognised that added inductors induce a delay in the lower wire.

In this case, the phase-shift gate's momentum should be the same as the mixing gate in one circuit. Then, for each circuit, we must pick one momentum from $\mathrm{k0} = 3\pi/8,\ 11\pi/8,\ 5\pi/8,\ 13\pi/8$. Following that, we'll use $\mathrm{k0} = 11\pi/8$ since it takes the fewest inductors to build the $\pi/4$ phase-shift gate.

The $\pi/4$ phase-shift gate is obtained by employing three inductors, $\mathrm{m} = 3$ and $\mathrm{k0} = 11\pi/8$.

$$U_{\pi/4} = \begin{pmatrix} 1 & 0 \\ 0 & e^{i\pi/4} \end{pmatrix} \tag{15}$$

The $\pi/4$ phase-shift gate yields two operations in a succession.

$$U_{\pi/2} = U_{\pi/4}^2 = \begin{pmatrix} 1 & 0 \\ 0 & i \end{pmatrix} \tag{16}$$

It is referred to as the phase gate.

5.4 Universal Quantum Gate

A series of universal quantum gates is a collection of gates which can be reduced to any operation conceivable on a quantum computer [23]. The H gate also known as Hadamard gate, a phase rotation gate R($cos^{-1}\frac{3}{5}$), and the controlled-NOT gate, a particular instance of controlled-U, are basic two-qubit universal quantum gates.

Controlled gates use more than two qubits, with more than one of them acting as a control for a specific operation. The controlled-NOT gate (or CNOT or CX) is a two-qubit gate that performs the NOT operation on the second qubit only when the first qubit is $|1\rangle$ and otherwise leaves it unchanged [23]. Referring to the basis $|00\rangle$, $|01\rangle$, $|10\rangle$, $|11\rangle$ the representation of the matrix is shown.

$$CNOT = \begin{bmatrix} 1\,0 & 0\,0 \\ 0\,1 & 0\,0 \\ 0\,0 & 0\,1 \\ 0\,0 & 1\,0 \end{bmatrix}$$

The three-qubit Deutsch gate D(θ) can also be used to create a single-gate set of universal quantum gates.

$$D(\theta) : ||i, j, k\rangle \rightarrow \begin{cases} icos(\theta)|i,\ j,\ k\rangle + \sin(\theta)|i, j, 1-k\rangle \\ \qquad\qquad ||i, j, k\rangle \end{cases} \tag{17}$$

5.5 Measurement (Z Gate)

The Z Basis or Computational basis is a single-tier gate that operates on a single qubit. It specifically maps 1 to−1 and leaves 0 alone. It accomplishes this by rotating the qubit by radians around its Z axis (180°) [25, 26]. This causes the phase of the qubit to be flipped.

The following matrix describes the operation of Z-gates:

$$Z = \begin{bmatrix} 1 & 0 \\ 0 & -1 \end{bmatrix} \tag{18}$$

By multiplying the column vector of the qubit's state by the Pauli Z matrix, we can observe how the Pauli Z gate acts on the qubit. For example, suppose the qubit is set to $|0\rangle$:

$$\begin{bmatrix} 1 & 0 \\ 0 & -1 \end{bmatrix}\begin{bmatrix} 0 \\ 1 \end{bmatrix} = \begin{bmatrix} 1(1) + 0(0) \\ 0(1) + -1(0) \end{bmatrix}\begin{bmatrix} 1 \\ 0 \end{bmatrix} \tag{19}$$

This has left |0⟩unaltered. Let's now set the qubit to |1⟩and examine how Z gate changes the qubit's state:

$$\begin{bmatrix} 1 & 0 \\ 0 & -1 \end{bmatrix}\begin{bmatrix} 0 \\ 1 \end{bmatrix} = \begin{bmatrix} 1(0) + 0(1) \\ 0(0) + -1(1) \end{bmatrix}\begin{bmatrix} 0 \\ -1 \end{bmatrix} \tag{20}$$

which is accurate since it has altered the state of the qubits from |1⟩ to -|1⟩ This is just |1⟩ with the phase reversed such that when measured, it will be |1⟩.

6 Conclusion

Quantum information processing is concerned with quantum mechanics-based information processing and computation. Quantum computers aren't restricted to two states, unlike contemporary digital computers, which encode data in binary numbers (bits). Quantum bits, or qubits, are used to encode information and can exist in superposition. Qubits can be created using atoms, ions, photons, or electrons, as well as appropriate control devices, to function as computer memory and processors. Quantum computers have intrinsic parallelism since they can hold many states at the same time.

This will allow them to complete some jobs far faster than a traditional computer utilising the most advanced techniques now available, such as integer factorization or quantum many-body system simulation. The quantum computer is still in its early stages. The basic building elements, such as quantum logic gates and memory Superposition and entanglement are examples of real quantum phenomena, which are the first steps on that path. This chapter is an attempt to provide the basics of quantum computing. It also compares the two different types of computing and emphasises why do we need quantum computing in today's time.

References

1. Y. Kanamori, S-M. Yoo, F. T. Sheldo, A short survey on quantum computers. Article in Int. J. Comput. Appl, (2006)
2. R. Feynman, Quantum mechanical computers. Found. Phys. **16**, 507–531 (1986)
3. R.P. Feynman, Simulating physics with computers. Int J Theor Phys **21**, 467–488 (1982). https://doi.org/10.1007/BF02650179
4. Y. I. Manin, *Mathematics as metaphor. Selected essays*, (American Mathematical Society, 2007), pp. 77–78
5. K. Manay, *Quantum computers could be a billion times faster than Pentium III*, (1998). USA Today, Dec, 2002 from:http://www.amd1.com/quantum_computers.html
6. P. W. Shor, Algorithms for quantum computation: discrete logarithms and factoring, in Proceedings of 35th IEEE symposium on foundations of computer science, (IEEE, 1994)
7. T. Hey, Quantum computing: an introduction. Comput. & Control. Eng. J. **10**(3), 105–112 (1999)

8. A. Ekert, P. Hayden, H. Inamori, Basic concepts in quantum computation, in *Coeherent atomic matter waves Vol. LXXII of Nato Advanced Study Institute-Les Houches*, (2001)
9. D.P. DiVincenzo, Two-bit gates are universal for quantum computation. Phys. Rev. A **51**, 1015–1022 (1995)
10. M.A. Nielson, I.L. Chuang, *Quantum computation and quantum information* (Cambridge University Press, Cambridge, 2000)
11. S. Hill, W.K. Wooters, Entanglement of a pair of quantum bits. Phys. Rev. Lett. **78**, 5022 (1997)
12. C.R. Wie, *Bloch sphere model for two-qubit pure states,* (2014). arXiv preprint arXiv:1403.8069v2
13. J. A. Cortese, T. M. Braje, Loading classical data into a quantum computer, (Lincoln laboratory massachusetts institute of technology Lexington, Massachusetts, USA, 2018) arXiv:1803.01958v1 [quant-ph]
14. J. Watrous, An introduction to quantum information and quantum circuits, Published in The Association of Computing Machinery—Special Interest Group in Algorithms and Computation Theory (ACM SIGACT). **42** (2), 52–67
15. A. Khrennikov, *Probability and randomness: Quantum versus classical* (Imperial College Press, London, 2016)
16. D. Deutsch, R. Jozsa, Rapid solution of problems by quantum computation, in *Royal Society London*, (1992)
17. A. V. Navaneeth, M. R Dileep, A study and analysis of applications of classical computing and quantum computing: A survey, (2021). https://doi.org/10.1007/978-981-15-8354-4_25
18. M.A. Nielsen, I.L. Chuang, Programmable quantum gate arrays. Phys. Rev. Lett. **79**(2), 321 (1997)
19. H. Sleator, Weinfurter, "Realizable universal quantum logic gates." Phys. Rev. Lett. **74**, 4087–4090 (1995)
20. G.E. Hale, On the probable existence of a magnetic field in sun-spots. Astrophys. J. **28**, 315 (1908)
21. R.M. Crutcher, Magnetic fields in molecular clouds. Ann. Rev. Astron. Astrophys. **50**, 29–63 (2012). https://doi.org/10.1146/annurev-astro-081811-125514
22. J. Li, X. Mao, X. Gu, S. Xie, Z. Geng, H. Chen, Phase shift induced by gate-controlled quantum capacitance in graphene FET. IEEE Electron Device Lett. **42** (4), (2021)
23. F Delgado, Universal quantum gates for quantum computation on magnetic systems ruled by Heisenberg-Ising interactions, in *Journal of Physics: Conference series,* (2017)
24. T. Bækkegaard, L. B. Kristensen, N. J. S. Loft, C. K. Andersen, D. Petrosyan, *Realization of efficient quantum gates with a superconducting qubit-qutrit circuit*, (2019)
25. M.A. Nielsen, Quantum computation by measurement and quantum memory. Phys. Lett. A **308**(2–3), 96–100 (2003)
26. S. Perdrix, State transfer instead of teleportation in measurement-based quantum computation. Int. J. Quantum Inf. **3**(01), 219–223 (2005)

Non-silicon Computing with Quantum Superposition Entanglement Using Qubits

N. Vidhya, V. Seethalakshmi, and S. Suganyadevi

Abstract The transistors, which can represent either 0 or 1 (bits) at the same time, are computed in classical computing. Even though classical computers have become smaller and faster, they are still incapable of solving issues like factoring huge integers, messages are sent in coded form using large digit prime integers. Currently, hundreds of electrons are used to turn on or off a single transistor on a chip in classical computing. To achieve great performance, we must progress past the limitations of conventional computing using electrons as well as silicon and into the non-silicon era. One solution that is gaining some traction is quantum computing. Quantum computers typically built on quantum bits (qubits) and take advantage of quantum features like quantum superposition and quantum entanglement to overcome the miniaturization issues of classical computing. Quantum superposition simply states that a quantum unit can exist in 2 different places at once. According to this hypothesis, simultaneously qubits can represent a wide range of 1 and 0 combinations. Quantum computers are acquiring processing power at a twice exponential rate, according to this observation. To gain an exponential growth in power scaling, the qubits are replaced with bits. The best part is that there are numerous choices available, from quantum computing to miraculous materials such as nanotechnology, and customized processors. Regardless of which path we choose, the future of computers promises to be exciting.

Keywords Classical computing · Moore's law · Quantum computing · Qubits · Quantum superposition · Exponential growth · Non-silicon computers · Quantum tunneling · Quantum entanglement · Quantum scaling · Bijection

---

N. Vidhya (✉) · V. Seethalakshmi · S. Suganyadevi
Department of ECE, KPR Institute of Engineering and Technology, Coimbatore, India
e-mail: vidhya.n@kpriet.ac.in

V. Seethalakshmi
e-mail: seethav@kpriet.c.in

S. Suganyadevi
e-mail: suganya3223@gmail.com

R. Pandey et al. (eds.), *Quantum Computing: A Shift from Bits to Qubits*,
Studies in Computational Intelligence 1085,
https://doi.org/10.1007/978-981-19-9530-9_8

1 Introduction

Some problems can be solved by a quantum computer by several times greater. This makes today's difficult problems simple to solve tomorrow. Quantum Superposition is the driving force behind tremendous computation power. We'll look at how this affects computing quantitatively in this post. At various scales, how well does Quantum Computing compare to Classical Computing. Excessive power dissipation affects dependability and raises the cost imposed by cooling systems and packaging in high-performance systems, and battery technology cannot keep up with rising expectations on devices featuring lightweight batteries and long-term between recharges in portable systems. The expanding market for portable devices like cell phones, gaming consoles, and battery-powered electronic systems necessitates the development of microelectronic circuits with extremely reduced power consumption. Moore's Law asserts that every two years, the number of transistors in a dense integrated circuit double. In general, having many transistors on a one computer chip and a lesser physical distance among them (nm technology) results in improved performance and power efficiency. In VLSI design, the nm technology defines the channel length of the transistor. For example, 45 nm technology represents the channel length of the transistor is 45 nm. The main goal is to achieve the high-performance and low power consumption by reducing the transistor length. This depicts the correlation between Moore's law and the energy efficiency in VLSI design.

- Moore's law has mostly resulted in energy efficiency improvements (low power dissipation).
- Every two years, the transistor count on a chip doubles thru ever-tiny circuitry.
- Power efficiency developed in parallel with CPU speed as transistors became smaller (nm technology). This depicts the correlation between Moore's law and energy efficiency.

The Intel 4004 was the initial Intel microprocessor, with 2300 transistors that were each measuring 10 μm in size. A single transistor upon this wider market is 14 nm on average as of 2019, with several 10 nm devices hitting the consumers (market) in 2018. On each square millimeter, Intel was able to fit over 0.1 billion transistors into a single chip. The tiniest transistors are 1 nm in size. It doesn't get much more compact than that (Fig. 1).

The light speed is finite and constant, imposing a general limit on the total amount of calculations that a single transistor is capable to perform. After all, data cannot travel speedier than the light's speed ($c = 3 \times 10^8$ m/s). Data are currently depicted as the movement of electrons via semiconductors, limiting computation speed to that of an electron moving thru matter. We'll almost certainly run against Heisenberg's uncertainty principle as we proceed to miniaturize chips, which restricts accuracy at the quantum level, therefore restricting potential computing capabilities. Moore's Law will then be outdated around 2036 owing to the uncertainty principle alone. However, it's possible that we've already reached. The year 2020 is used 7 nm is being used as the last processor node technology by Robert Colwell, head of

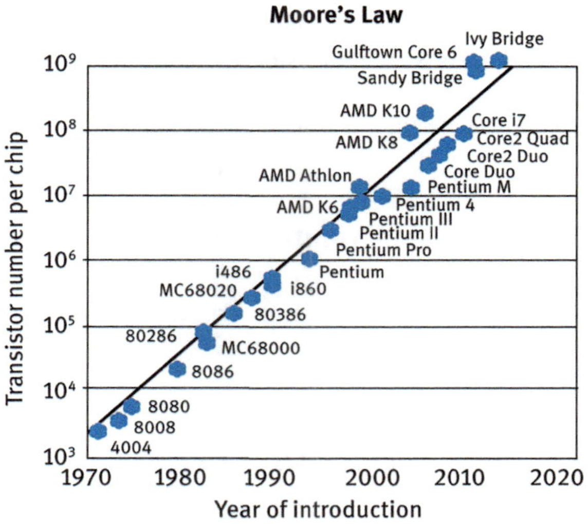

Fig. 1 Moore's law and the processors till 2020

the Microsystems Technology Office at the Defense Advanced Research Projects Agency. It's possible, however, that we've already arrived. Moore's law came to an end. Because transistors can no longer continue inside smaller circuits at greater temperatures, computers might eventually reach their limits. This is due to the fact that the transistors must be kept cool, which consumes more energy than the energy that flows through the transistor itself. Regarding all of these factors into an account, it's essential to explore for different computing methods that aren't based on electrons and silicon transistors. To move beyond [1] Moore's Law, we must abandon traditional computing with silicon and electrons and begin to the future of non-silicon computer systems. Moore's Law is being replaced by Neven's Law. Neven's law is named after the inspiration of Hartmut Neven, the director of Google's Quantum Artificial Intelligence Lab. "Neven's law" refers to the fact that quantum computers increase computational power at a double exponential pace.

2 Concept of Quantum Computing

Quantum computing is a field of computing technology that seeks to create computer technology that is built on quantum concepts (where it describes materials and energy behavior at the atomic levels). Computer systems presently can only encode information in bits with 1 or 0 values, greatly constraining their possibilities. The reversible logic processes are unable to erase data and emit zero heat. The circuit really works backwards, allowing the inputs to be reproduced from the outputs while consuming zero power. Logic gates are utilized to actualize Boolean functions as the essential elements of any logic circuit. Any paradigm of computation in which the computational process is time-reversible to some extent is referred to as reversible computing. Quantum circuits are reversible due to quantum physics' unitarity, as long as the quantum states they act on do not "collapse."

3 Reversible Evolution in Quantum Computing

Reversible computation is a primary concept to attain an effective quantum computation.

3.1 Basic Logic Gates in Classical Computing

We already knew the basic gates for classical binary computation. Some basic gates for binary calculations are NOT, AND, OR, and XOR.

NOT Gate [2]

The NOT gate flips a single bit, as we might expect. The NOT gate is reversible at all times: When we use a NOT gate to a certain signal twice, we obtain the same value as before: NOT(NOT(X)) = X (Tables 1 and 2).

AND, OR and XOR Gate [2]

These three gates are not reversible, since it is having 4 various possible inputs (00,01,10,11) and only 2 feasible output stages (0 and 1). As a result, there isn't enough data in the output to determine whatever the inputs were with certainty.

In this case, an input (11) can be determined only when the output is 1but the remaining three input conditions (00,01,10,11) cannot be determined (Tables 3 and 4).

Table 1 Classical truth table for NOT gate

Input	Output
0	1
1	0

Table 2 Classical truth table for an AND gate

Input (A)	Input (B)	Output (C)
0	0	0
0	1	0
1	0	0
1	1	1

Table 3 Classical truth table for an OR gate

Input (A)	Input (B)	Output (C)
0	0	0
0	1	1
1	0	1
1	1	1

Table 4 Classical truth table for an XOR gate

Input (A)	Input (B)	Output (C)
0	0	0
0	1	1
1	0	1
1	1	0

The behavior of the OR gate is similar to that of the AND gate in many aspects. In this case, the input (0,0) can be determined only when the output is 0 but the remaining three input conditions (01,10,11) cannot be determined.

The XOR gate is virtually reversible, with 2 input bits and 1 output bit but still there isn't sufficient data in the output to determine with certainty whatever the inputs have been. However, if one of the inputs is given, undoubtedly the other input can be determined. Hence, the XOR gate is not fully reversible.

3.2 Information and Entropy

IBM's Charles Bennett developed the foundations for binary reversible logic functions in the early 1970s, later joined by Caltech's Richard Feynman and MIT's Tommaso Toffoli and Ed Fredkin [2]. Bennett was a follower of Rolf Landauer, who realized that destroying information increases the Universe's entropy and must produce excess heat. The amount of disorder in something is measured in entropy. When learning about thermodynamics in a physics or chemistry class, we will be introduced to entropy. We can employ a lot of energy that is confined in one location to conduct work, such as operating a motor, which disperses the energy. And the entropy is reached highest after the energy is distributed uniformly, and it is no longer possible to accomplish effective work with this. The entropy of information is also discussed in computer science. If you've a specific number of information, it may be entirely zeros or entirely ones, in this scenario telling a friend how to replicate the same state is as simple as saying, "I have 700 ones." However, as your data becomes increasingly complicated and it becomes more difficult to describe; it has a high entropy.

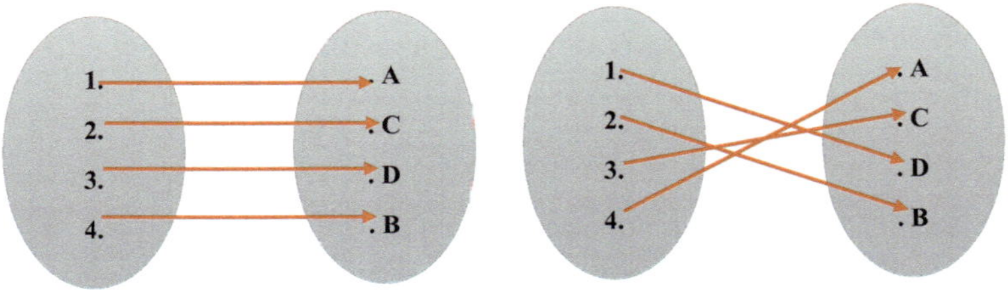

Fig. 2 Bijection (One-to-one and onto)

3.3 Reversible Computing

Whenever we lost data, or even when the data is deleted, entropy increases. Because the AND and OR gates start using 2 input bits and end using only one, they ultimately lose information, making it difficult to carry all of the information across. If we need the whole of our gates to have same quantity of input and output signal, we may be capable to undo the computations. To make the gate reversible, a second additional criterion is required: every possible output state originates from exactly 1 input signal. The function is thus said to be "one to one," or bijective Fig. 2.

Reversible circuits are usually referred to as lossless circuits since they do not lose energy or information. Digital communications, Very Large-Scale Integration (VLSI) technologies with low power consumption, electro-optic computing technologies, and nanotechnology all benefit from any such circuits. We'll look at several logic functions that 1, 2, and 3-bit reversible gates can compute.

4 Quantum Gates and Circuits

Identity Gate [2]

The identity gate has no effect on the position because its input and output are identical. It is evidently reversible, as it has no effect, all we have to do now is do nothing once again to get back to where we started! (Table 5).

NOT Gate

The NOT gate has previously been seen. Performing NOT twice in a succession returns one to the beginning: **NOT(NOT(X)) = X.** Hence this NOT gate is always reversible.

Table 5 Reversible truth table for Identity gate

Input	Output
0	0
1	1

C-NOT (Complementary NOT) Gate [2]

The C-NOT (controlled NOT gate) receives 2 input bits and produces 2 output bits. Another bit (named as target bit) is inverted if one of the bits (called the control bit) is one. The target bit is transmitted via unaltered if the control bit is zero Fig. 3 and (Tables 6 and 7).

Each of the B inputs is used as a control signal in this case: If A = 0, the output A' is just a copy of the input A.; if A = 1, the output B' = B ⊕ 1, which is the inverse of the input A, is produced (Refer Table 3.5) Because of its widespread use in the field of quantum computer technology, the Controlled-NOT (C-NOT) gate is also referred as Feynman Gate (FG).

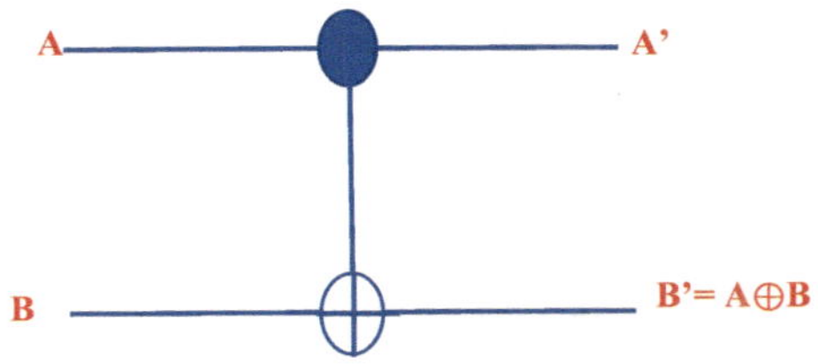

Fig. 3 C-NOT gate

Table 6 Reversible truth table for C-NOT gate

Input		Output	
A	B	A'	B'
0	0	0	0
0	1	0	1
1	0	1	1
1	1	1	0

Table 7 Reversible truth table for C^2NOT gate

Input			Output		
A	B	C	A'	B'	C'
0	0	0	0	0	0
0	0	1	0	0	1
0	1	0	0	1	0
0	1	1	0	1	1
1	0	0	1	0	0
1	0	1	1	0	1
1	1	0	1	1	1
1	1	1	1	1	0

C^2NOT (Toffoli Gate) [2]

The control-control-NOT gate, or CCNOT gate, is also called the Toffoli gate, named after Tommaso Toffoli. Instead of one control bit, it employs two. The number of outputs and inputs are equal in a reversible multi-output Boolean function. Toffoli Gate (TG) is an exemplification of a (3,3) reversible gate. TG also is a universal (worldwide) reversible logic circuits, meaning it may be utilized to make any conventional reversible circuit. Because of how it operates, it's also referred as the "controlled-controlled-not (C^2NOT)" gate. It has 3 inputs and outputs; the third bit is reversed if the first two bits are both 1, otherwise all bits are intact Fig. 4.

CSWAP (Fredkin Gate) [2]

The Fredkin gate, also known as the CSWAP or control-SWAP gate, is named after Ed Fredkin. If the control bit is 0, another 2 bits are discarded. If the control bit is 1, another 2 bits are switched. The following truth table explains the operation of CSWAP gate.

In Table 8, input A is a control bit, when A is 1, the remaining bits B and C are swapped otherwise bits B and C remain same, which resembles output bits B' and C'. Figure 5 shows the Fredkin or CSWAP gate.

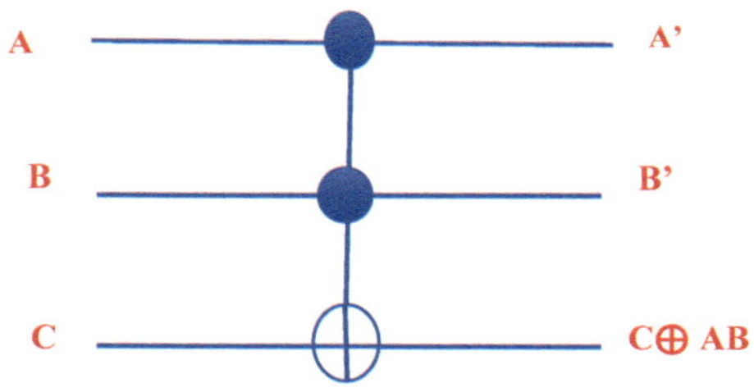

Fig. 4 C^2NOT gate (or) Toffoli gate

Table 8 Reversible truth table for CSWAP gate

Input			Output		
A	B	C	A'	B'	C'
0	0	0	0	0	0
0	0	1	0	0	1
0	1	0	0	1	0
0	1	1	0	1	1
1	0	0	1	0	0
1	0	1	1	1	0
1	1	0	1	0	1
1	1	1	1	1	1

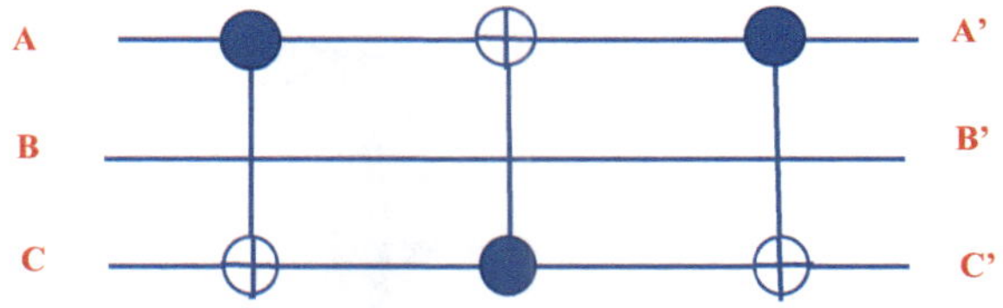

Fig. 5 CSWAP or Fredkin gate

Either C^2NOT or the CSWAP gate is capable of constructing any conventional logic circuit design.

5 Quantum Superposition and Quantum Entanglement

The quantum theory concepts of quantum entanglement and quantum superposition are the focus of these quantum computers. This allows quantum computers to do operations at speeds factors of magnitude quicker than traditional computers while consuming lesser energy.

5.1 Qubits

We have to think beyond binary bits, which is known as Qubits. One of the features that distinguishes a qubit from a classical bit is its ability to be in superposition. One of the basic principles of quantum theory is superposition. Quantum computing, and from the contrary, makes utilization of quantum bits, generally referred as qubits [8]. It involves the advantage of the subatomic particles' one-of-a-kind ability to live in a variety of states (i.e., a one and a zero simultaneously).

5.2 Quantum Superposition

Quantum superposition simply asserts that the quantum particle can exist in 2 distinguish places at once [7]. Unless the measurement procedure is performed, a quantum particle can exist in numerous states at the same time, according to this theory. Entanglement is another counter-intuitive phenomenon in quantum physics. When each particle's quantum state cannot be characterized independently of the quantum state of the other, a pair or group of particles is said to be entangled (s). Although the parts of the system are not in a definite state, the quantum state of the system as a whole can be stated.

Consider the game of coin tossing. You can say that the coin is both heads and tails while it is whirling in the air. We have two alternative coexisting states of being in

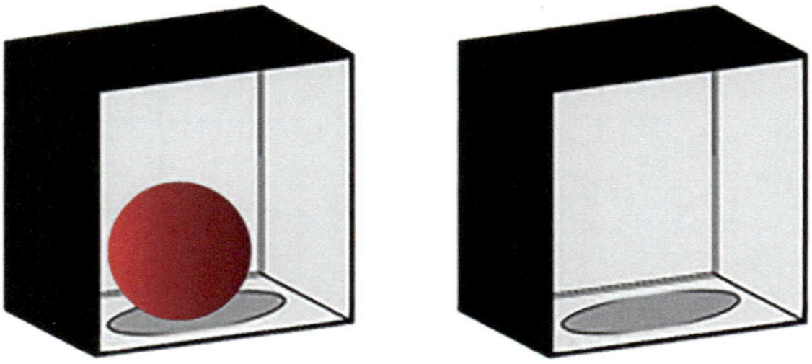

Fig. 6 Example of quantum superposition and entanglement

superposition before the coin lands (When the coin lands, the likelihood of one result is reduced to 0%, while the probability of the other result is increased to 100%.). This spinning coin, or "qubit," can be in one of three states: one, zero, or both one and zero.

Qubits are in the "|0> " state (also known as a zero-ket), the "|1 > " state (also known as a one-ket), or a linear combination of the two states (also known superposition). Qubits, as opposed to ordinary bits, are usually denoted by the half-angle bracket notation "|>". Consider an experiment involving ping-pong balls and boxes. (This is preferable to Schrödinger's famous thought experiment, which involved killing cats).

Imagine you have a device that, when you press a button, drops a ball into one of two boxes at random. You only have one ball and no way of knowing which box it will fall into. Three things are true after pressing the button and before opening the box [7].

- The boxes are in a state of superposition, which means that they both have or don't have balls, as shown in Fig. 6.
- There's a probability that the ball will end up in one of the boxes.
- When you open the box and see that it contains a ball, you know that the other box does not contain a ball.

This last point demonstrates entanglement [7]. Those boxes are quantumly entangled, regardless of where they are in the universe. You can know what happened in the other box quicker than the speed of the light (3×10^8 m/s) by looking at the contents of an either box. We could say that each of those boxes had information, a zero or a one, a single bit of data in classical computing, but we'd have to open the box to retrieve that information. Before we open the box in quantum computing, we have a qubit, or a quantum bit—that is, a zero, one, or both—and we don't really have to open the box (A qubyte is made up of 8 qubits).

5.3 Quantum Entanglement

The term quantum entanglement is another one that underpins quantum mechanics and gives quantum computers their incredible strength. If you're having trouble in digesting the next bit [4], don't worry: even Albert Einstein, who co-authored the first educational article proposes the idea in 1935 with physicists Podolsky and Rosen, explained this as "spooky action at a distance" [14] and decided that "no reasonable notion of reality will be possible to permit this." Quantum entanglement appeared to Einstein to violate physical rules because it proposed that one electron sent information to the other instantly, meaning that the information actually passing greater than the speed of light. That's why it seemed "spooky" and impossible to explain. Nowadays, quantum entanglement is universally perceived, and scientists have demonstrated that it is possible, and let's see this impact on a quantum computer.

We already knew that the equivalence of zero and one occur simultaneously due to the quantum theory of superposition." Superposition allows particles to exist in several states at the same time, analogous to the famed, in the thought experiment of [12] Schrödinger's cat, a cat can also be alive and dead at the same time. When a system has more than one qubit, those qubits are likely to be entangled due to quantum entanglement, which means that the state of one qubit affects the state of the other. Even at great distances, the measurements of the qubits will be completely coupled in this entangled condition. This is why Einstein believed the information was being conveyed at a faster rate than the speed of light. The quantum entanglement principle allows us to create data correlations that are impossible to obtain with a traditional computer. Whenever a qubit has entangled with the other, the number of qubits that may be computed at any given time increases dramatically. Parallelism, which permits a quantum system to work on several calculations at the same time, is enabled via superposition. And that's why a processing power of quantum computer doubles with every single-qubit speed improvement. To put it more simply, when the ideas of superposition and entanglement are combined, a quantum computer can achieve potentially unexpected speeds.

5.4 Gates with Single-Qubit

Every above-mentioned reversible gate has quantum counterparts, which we can employ to build quantum algorithms. Qubits, on the other hand, are more complicated objects than plain binary (classical) bits, enabling for a broader range of unitary procedures. We'll examine other single-qubit gates that aren't as simple as the NOT and Identity gates.

5.5 The Bloch Sphere's Rotations

While we first discussed qubits, we explained the Bloch sphere and how can the state of a single-qubit be considered as a single point on a sphere. A single qubit gates can all be thought of as movements around the X, Y, or Z axis of a Bloch sphere (Fig. 7).

One (Single) qubit states that have not been entangled and do not have a global phase can really be described as dots on the Bloch sphere's surface, written as follows.

$$|\varphi\rangle = \cos(\theta/2)|0\rangle + e^{i\varphi}\sin(\theta/2)|1\rangle \quad (1)$$

The rotation operator gates represent rotations around the x, y, and z axes of the Bloch sphere. Rotation is a generally reversible process: simply rotate the same amount in the opposite way around. Each and every point mostly on the sphere has a beginning and finishing point, and thus no 2 points ever overlap. There are no two beginning positions that lead to the same finishing position. It's a crucial aspect of reversibility. If 2 points collided, we'd have no other path of identifying them and bringing the same to their respective originating places if we tried to reverse the process. We could say that each of those boxes had information, a zero or a one, a single bit of data in classical computing, but we'd have to open the box to retrieve that information. Before we open the box in quantum computing, we have a qubit, or a quantum bit that is, a zero, one, or both and we don't really have to open the box. (A qubyte is made up of 8 qubits).

Now let us consider the classical computing, which is a binary system—a complex system of 1's and 0's, "on" and "off" resulting through logic gates (currently mostly about 14 nm in diameter). A "bit" is a binary value of 1 or 0, and a traditional byte comprises 8 bits. We can build whatever character, number, letter, or symbol we need

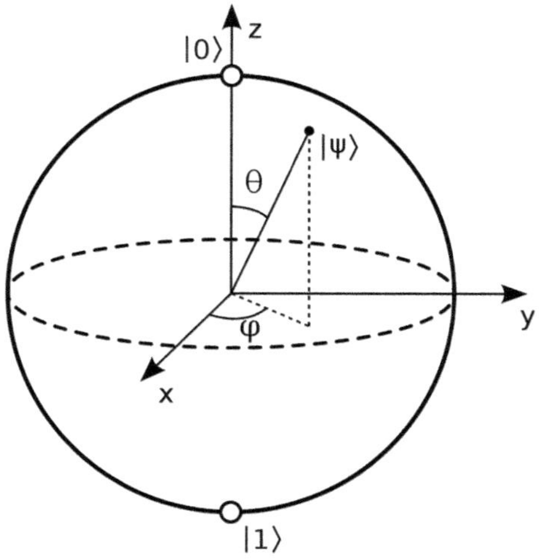

Fig. 7 Bloch sphere of single qubits

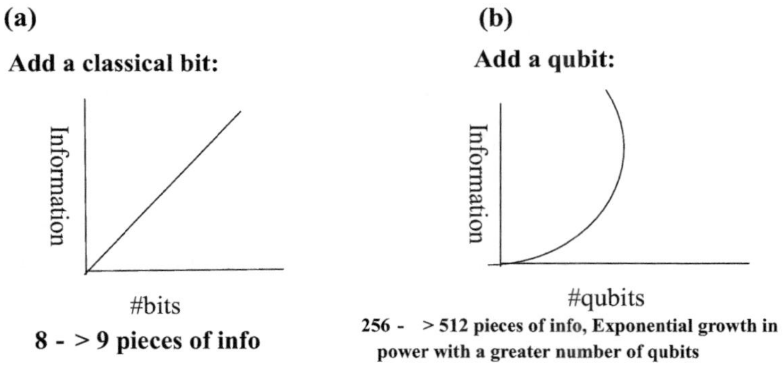

Fig. 8 **a, b** The relation between classical bits and qubits

using 2^8 or 256 possible permutations of 0 and 1 s to execute all of the individual functions we required from a computer.

Quantum computers process the data in a unique way. Transistors, which are either one or zero, are used in conventional computers. Qubits, which can be 1 and 0 at the very same moment, are used in quantum computers. Quantum computing power grows exponentially (shown in Fig. 8b) as the number of qubits coupled together grows. Meanwhile, connecting further more transistors increases only power in a linear fashion.

However, on a qubyte, we could theoretically store all 256 possibilities at once; one qubit contains all possible combinations of 0 and 1 s. And this grows exponentially—if you have 20 qubits, or 2^{24} permutations of 0 and 1, we can store over a million information simultaneously.

Figure 9 shows the Bloch sphere of qubits with the multiple bunch of possibilities [7].

5.6 *Why Quantum Computing? (Keys to Take Away)*

- Quantum computing is a technique for generating innovative computer methods by utilizing quantum physics phenomena [3].
- Qubits seem to be the basic foundation block of quantum computing.
- A qubit is not like a standard binary bit, would be either 0 or 1, or a superposition of both 0 and 1.
- Quantum computers' power increases exponentially with the greater number of qubits.
- This is in contrast to conventional computers, where increasing further more transistors increases power in a linear fashion as shown in Fig. 8a.

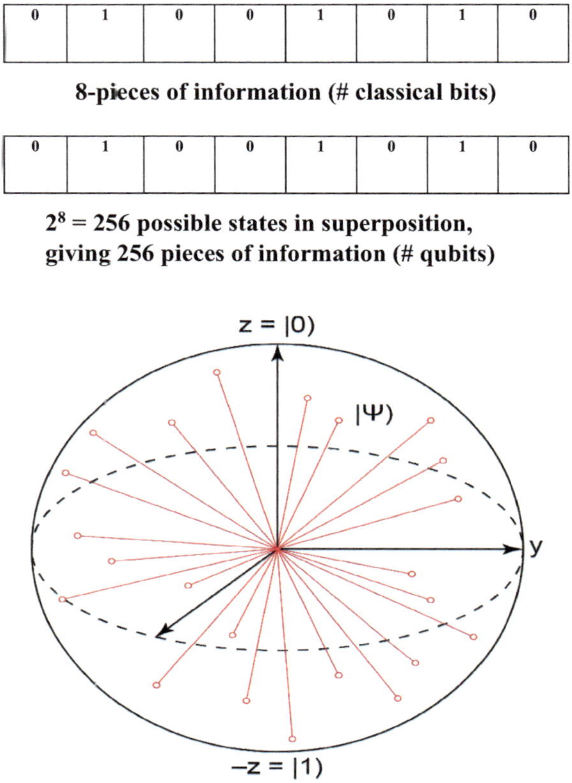

Fig. 9 Bloch sphere of qubits

5.7 *Scaling of Qubits*

For a better understanding, take a look into the scaling of qubits with the memory. Here's a table that shows how classical and quantum bits compare at various scales with memory. We could carry out all traditional state-of-the-art operations [4]. This would necessitate more memory. We might employ one classical state at a time instead. The number of iterations and thus the compute time would increase as a result of this.

From Table 9, we can understand that the memory utilizes only 13 qubits to store a kB (kilobyte). By comparison, 13 bits are only one byte and a half. For the same 1 kB memory, it takes $2^{13} = 8192$ classical bits to store. Clearly, it's not just a speedup for the sake of convenience. Quantum computing opens a new realm of possibilities. The train does not halt at 1000 qubits. It goes on from there, with the ability of problem solving for which we currently lack the words to articulate!

A conventional bit can be 0 or 1. A quantum bit, also known as a qubit, is a superposition of the bits 0 and 1. As a result, a single-qubit can hold two classical values at the same time. Every action on the qubit affects both values at the same

Table 9 Scaling of qubits with memory

No. of qubits	No. of classical bits (Binary)	RAM
1	$2^1 = 2$	2 bits
2	$2^2 = 4$	4 bits
3	$2^3 = 8$	1 byte
4	$2^4 = 16$	2 bytes
5	$2^5 = 32$	4 bytes
6	$2^6 = 64$	8 bytes
7	$2^7 = 128$	16 bytes
8	$2^8 = 256$	32 bytes
9	$2^9 = 512$	64 bytes
10	$2^{10} = 1024$	128 bytes
11	$2^{11} = 2048$	256 bytes
12	$2^{12} = 4096$	512 bytes
13	$2^{13} = 8192$	1 kB

time. That is why we frequently see that such a qubit can store more data than a conventional bit. When looking at the two bits, we can see that they will have the preceding values:

- **0, 0**
- **0, 1**
- **1, 0**
- **1, 1**

Every one of those values is taken by two qubits at the same time. The trend is obvious. Two bits can be represented by 1 qubit. Four bits can be represented by 2 qubits. In general, the values of n qubits can be 2^n, which is shown in Table 9.

A standard logic gate (like AND, OR, and NOT) classical computing accepts a fixed number of inputs and generates a single consistent output. Due to the obvious way, the logical gates are established and the inputs generate one answer, that the answer is assured (Refer Sect. 3.1).

A quantum gate rotates probabilities and manages superpositions as an input before producing other superpositions as an output. Specific algorithms can be used to determine the likelihood of the right answer based on this output, collapsing the superpositions into an actual sequencing of 0 and 1s. This means you get a whole bunch of the calculations possible with the setup accomplished simultaneously. Actually, we only can measure one of the outcomes, and it will very certainly be the one we want, so we may have to double-check and try once again. However, by intelligently utilizing superposition and entanglement, it can be enormously more effective than on a traditional computer, which must complete all the calculations in order to reach the final result.

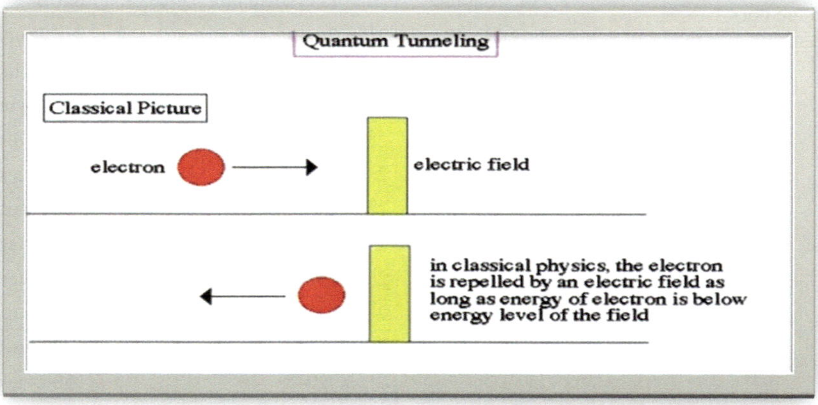

Fig. 10 Quantum tunneling in classical computing

6 Quantum Tunneling

Here, let us discuss the impact of quantum tunneling in both the classical computing and quantum computing.

6.1 *Impact of Quantum Tunneling on Classical Computing*

In classical mechanics, if a barrier's gravitational potential is too large—that is, more than the initial kinetic energy of a free particle attempting to pass through the barrier—the particle literally cannot pass through the barrier. In this case, the particle just bounces against the barrier finally coming to a halt. The particle bounces back and forth between the walls in classical physics, finally becoming stuck.

The particle has enough energy to escape the confines of the walls, but not enough to get there halt. Figure 10 [6] shows the quantum tunneling in classical computing.

6.2 *Impact of Quantum Tunneling on Quantum Computing*

The particle in quantum mechanics acts like a wave. Because the wave is strongest here between walls, the particle is most likely present. The quantum wave at the walls dims but does not go to zero; it penetrates slightly into walls. Outside the walls, a low-intensity wave extends. As a result, there's a slight possibility that particles will

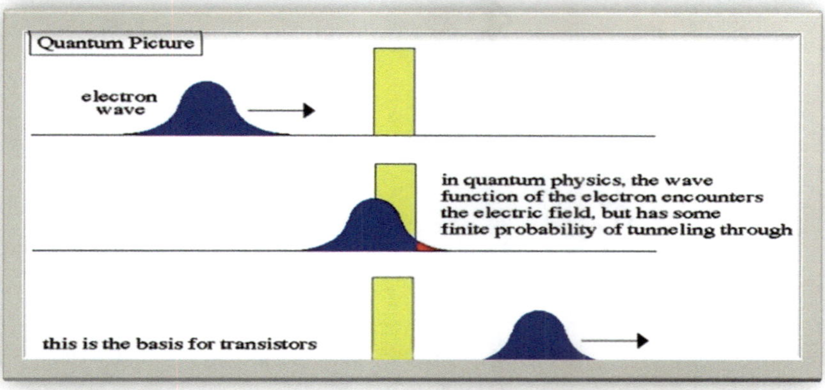

Fig. 11 Quantum tunneling in quantum computing

now be discovered outside the walls. Figure 11 [6] shows the quantum tunneling in quantum computing.

The macroscopic example of a rock rolling up a steep mountain can be used to describe the phenomena of quantum tunneling [5]. In this case, we know the ball can only go over the mountain if its kinetic energy is greater than the mountain's gravitational potential. The ball would try to roll down the mountain, if this is not done. To travel over the mountain, the ball must either utilize more energy to overcome the mountain's gravitational potential (e.g., a person moving the ball up the mountain) or employ quantum tunneling to keep passing it through the barrier as a wave. Quantum tunneling is what allows some quantum computers to not only accomplish tasks quickly, but also to complete activities that a conventional computer would be unable to complete within the limits of classical physics. Figure 12 shows the quantum tunneling principle in both the classical and quantum physics. Even though its energy appears to imprison it on one side, a quantum particle sometimes can sneak thru a barrier. This phenomenon, known as quantum tunneling, explains how confined nuclear particles can occasionally escape their nuclei, causing radioactive decay. Tunneling also permits visible light photons to escape the sun's interior and electric currents to run. The tunneling probability is the chance of being detected beyond the barrier.

The possibility of tunneling for a tiny particle and an atomic-sized barrier can be high; for a heavy object and a powerful thick barrier, the possibility of tunneling is extremely low and can be statistically impossible.

Why the Quantum Computing is so Excited?

While classical computers are enormously powerful, capable of doing an unimaginable number of tasks in a short period of time, they have limitations. Adding exponentially increasing variables is an issue that current computers can't solve with

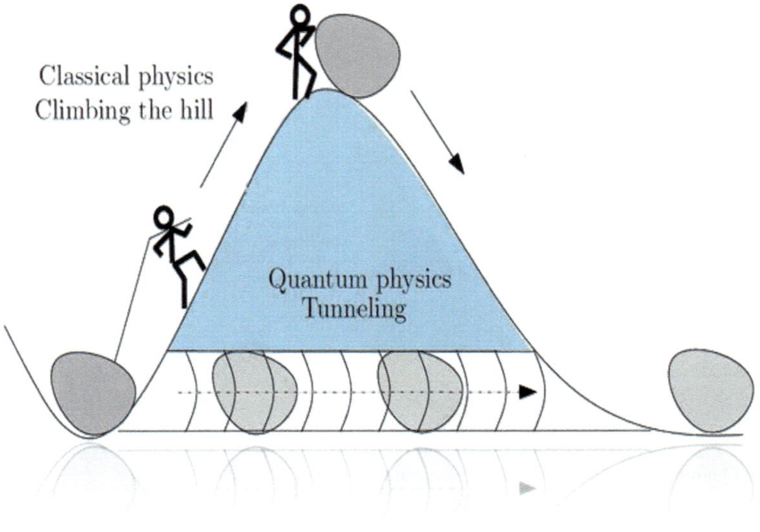

Fig. 12 Quantum tunneling principle in both classical and quantum computing

logic gates and a binary manner of managing data. Security algorithms, efficiency applications, AI applications, and, most importantly for Cadence, simulation applications, are the most common practical uses for quantum computing that I've read about. Any system with variables can be simulated to determine all of the different outcomes that could occur. Changing just one logic gate, as any complicated system designer knows, adds another level of complexity. Before developing any system—chip, board, or system designers must simulate, imitate, and verify whether their system is most efficient, strong, and consumes the minimum amount of possible power. Currently, organizations must utilize hundreds of servers to execute some of these simulations, and offering verification via the cloud adds even more computers to the mix. This is still a lot of processing power, but even with the world's most powerful supercomputers thrown in, no classical computer can handle all those computations. However, using quantum computers, all of those calculations might theoretically be completed at the same time, using a fraction of the computation power. Verification, simulation, and emulation will not be as complex as previously thought.

7 Conclusion

However, Moore's law's energy-related advantages are slowing, potentially jeopardizing future computing advances. We've reached a point where transistor miniaturization is hitting physical limits. This is because relatively small transistors possess extremely thin insulating layers, resulting in more heat dissipation and electric current leakage, necessitating increasing energy to compensate for the dropped current. Leakage current has become an increasingly greater technological challenge as transistors become smaller. This has begun to challenge whether Moore's law will be going to survive in its present form. "Even though Moore's law fails, alternative forms of computing may once again drive the industry on dizzying development speeds." That alternative one is the quantum computing. Quantum computers will enable data storage and processing in ways that we can't even imagine right now. They can also perform more sophisticated computations than classical computers, allowing them to solve issues that would otherwise take many years to solve on a classical computer. Nowadays, researchers can have plenty of options regarding quantum computing with the numerous applications from graphene-like materials to optical computing technology and customized chips Whatever path the future of computing takes, the benefits of excellent parallelism, quick information speed of processing, enormous amounts of data storage, and performance make it very intriguing (includes speed and low power dissipation).

Future of Quantum Computing in the Market

Quantum computing appears to be the next big thing in IT, the number of firms in the space and the millions of dollars in funding. Expert predictions back this up: the quantum computing business is expected to reach $770 million by 2025.

References

1. P. Chojecki. https://builtin.com/hardware/moores-law. (2021). Updated 9 Feb 2022
2. https://www.futurelearn.com/info/courses/intro-to-quantum-computing/0/steps/31561
3. https://www.investopedia.com/terms/q/quantum-computing.asp#:~:text=Quantum%20computing%20is%20the%20study,of%20both%200%20and%201
4. G. Popkin, China's quantum satellite achieves spooky action at record distance. (2017). http://www.sciencemag.org/news/2017/06/china-s-quantum-satellite-achieves-spooky-action-record-distance. Science Magazine, Online. Accessed 8 July 2017
5. https://www.linkedin.com/pulse/physics-behind-quantum-computers-entanglement-sahar-shoja/
6. The curious quantum world: Part 6—quantum tunneling. (2016). https://steemit.com/science/@pjheinz/the-curious-quantum-world-part-6. Quantum Tunneling Image, Online, Accessed 15 July 2017
7. C. Meera. https://community.cadence.com/cadence_blogs_8/b/on-the-beat/posts/quantum-computing-101. published on 31 July 2018
8. 2 Quantum computing: A new paradigm. National academies of sciences, engineering, and medicine. Quantum computing: progress and prospects, (The National Academies Press, Washington, DC, 2019). https://doi.org/10.17226/25196
9. M. Nielsen, I. Chuang, *Quantum computation and quantum information* (Cambridge Univ, Press, 2000)
10. J. P. Dowling, G. J. Milburn, Quantum technology: the second quantum revolution. Phil. Trans. R. Soc. A. **361,** 1655–1674 (2003) .https://doi.org/10.1098/rsta.2003.1227
11. A. Einstein, B. Podolsky, N. Rosen, Can quantum-mechanical description of physical reality be considered complete? Phys. Rev. **47**, 777–780 (1935). https://doi.org/10.1103/PhysRev.47.777

12. P. Mosley, Physicists prove quantum spookiness and start chasing schrodinger's cat. (2015). http://theconversation.com/physicists-prove-quantum-spookiness-and-start-chasing-schrodingers-cat-48190. The conversation—Independent academic news website. Accessed 18 July 2017
13. C. Weedbrook, *How to build a universal photonic quantum computer* (Unpublished white paper by CEO of quantum rm, Xanadu, 2016)
14. How does quantum computing work? https://plus.maths.org/content/how-does-quantum-com muting-work, 2015, Plus Magazine, Online. Accessed 6 July 2017
15. M. Joudi, *A look into quantum computing*. (2021). https://doi.org/10.13140/RG.2.2.11839.74409.
16. M. Rahaman, M. M. Islam, A review on progress and problems of quantum computing as a service (Qcaas) in the perspective of cloud computing. J. Comput. Sci. Technol. **15**(4),15-18 (2015)
17. https://www.idquantique.com/quantum-computing-review-q3-2021/. 27 Oct 2021
18. C. Dilmegani, Future of quantum computing in 2022: In-depth guide. Published on june 20. (2020). https://research.aimultiple.com/future-of-quantum-computing/
19. L. Gopal et al., Design and synthesis of reversible arithmetic and logic unit (ALU). Conference Paper, (2014). https://doi.org/10.1109/I4CT.2014.6914191
20. R. Hongal, Design of reversible logic based basic combinational circuits. Commun. Appl. Electron. **5**(9), 38–43 (2016). https://doi.org/10.5120/cae2016652372

Quantum Algorithms–Theory and Applications

A Reversible Hybrid Architecture for Multilayer Memory Cell in Quantum-Dot Cellular Automata with Minimized Area and Less Delay

Suparba Tapna, Debarka Mukhopadhyay, and Kisalaya Chakrabarti

Abstract CMOS innovation shows limited features when diminishing the size and region of a circuit. The burden of such a technology incorporates higher force utilization and also shows some temperature issues. Quantum-Dot Cell Automata is another innovation which is useful to defeat any of its weaknesses. The reversible rationale is innovation used to diminish the force misfortune in QCA. QCAs are utilized to plan recollections requiring a high working rate. In the following research, construction of reversible memory is proposed in QCA. It is designed by using a 3-layer innovation that altogether has an effect on the decreased size of the circuit. The reversible memory proposed here has 61% increase in cell number, with a 74% enhancement in the territory inhabitance, and 59% decrease in delay contrasted with any previous optimal designs.

Keywords Multilayer memory cell · QCA · Reversible · Hybrid architecture · Three-layer structure

1 Introduction

QCA elaborated as the Quantum-Dot Cell Automata are among the latest advancements in the plan of computerized nanocircuits [1]. Some fundamental highlights of the innovation incorporate minuscule measurements, rapid, and low force consumption [1]. This innovation has pulled in the consideration of scientists to plan diverse computerized designs, for example, adder [2–11] shift register [12] counter [13] as well as the comparator circuits [14].

S. Tapna (✉)
Brainware University, Barasat, India
e-mail: suparba7@gmail.com

D. Mukhopadhyay
Christ (Deemed to be University), Bengaluru, India

K. Chakrabarti
Haldia Institute of Technology, Haldia, India

R. Pandey et al. (eds.), *Quantum Computing: A Shift from Bits to Qubits*,
Studies in Computational Intelligence 1085,
https://doi.org/10.1007/978-981-19-9530-9_9

For storing any type of data in a technical computing system, memory plays an important role [15–18]. Hence, planning a memory with high processing speed remains the biggest requirement, especially in QCA. Reversibility in the nanocircuits assumes a significant part in blunder identification because of no missing information and furthermore receiving correspondence without loss [19].

Circuits with reversible constructions are the ones with higher number of yields equivalent to that of the number of data sources, with a balanced correspondence between the data sources and output vector. It not just assists us with deciding the output from the data source and yet additionally assists us with exceptionally recuperating the results from outputs [20].

The principle motivation behind the research is to present a proficient memory based on QCA memory that beats different circuits introduced already as far as cell check, territory, and inactivity. The following are the objectives of this study: • A three-level design for lessening the utilization of energy is proposed which also diminishes the number of cells as well as consumption territory in a QCA-based memory; • The plan is compared to other best-in-class plans as far as cell tallies and region.

2 Background and Related Work

The following part gives foundation material as well as the previous work conducted to comprehend the QCA.

2.1 QCA Cell

A cell is comprised of four quantum dabs with two electrons. Columbic repugnance among two electrons compels them to go into inverse cell corners. As demonstrated in Fig. 1, The elements give two stable states with rep-hating double data P = 1 (for "0") and P = +1 (for "1") [1, 21].

QCA wires are developed by setting QCA cells one next to the other. The data is communicated from cell to cell in a wire. A QCA wire structure is demonstrated in Fig. 2 [1, 3, 15, 19, 22–25].

2.2 Basic Gates

Gates are among the primary design feature in a QCA, where the rationale capacity of a larger part door is presented in Eq. (1). When the polarization of one of the three contributions is fixed to a rationale "1" or rationale "0," AND or OR door are easily developed. Various designs are outlined in Fig. 3 [26].

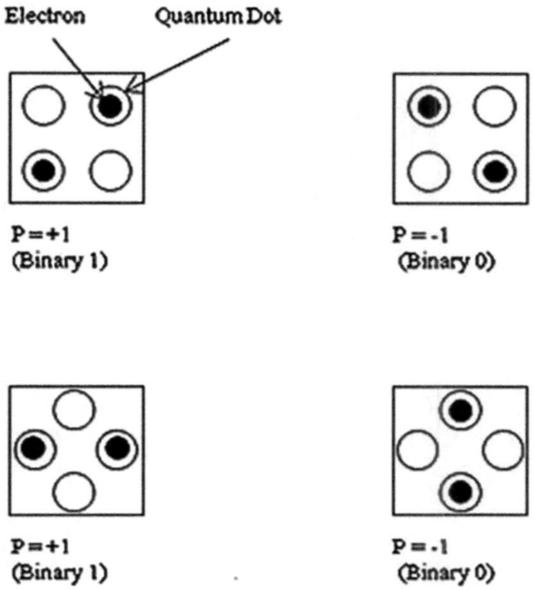

Fig. 1 QCA cell and two potential stable states [21]

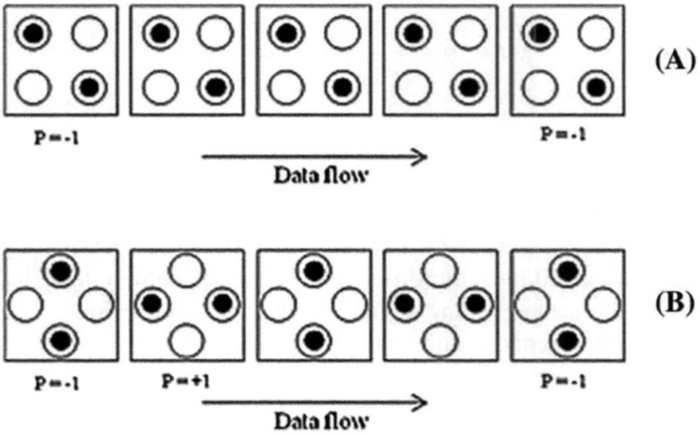

Fig. 2 QCA types of wires [21]

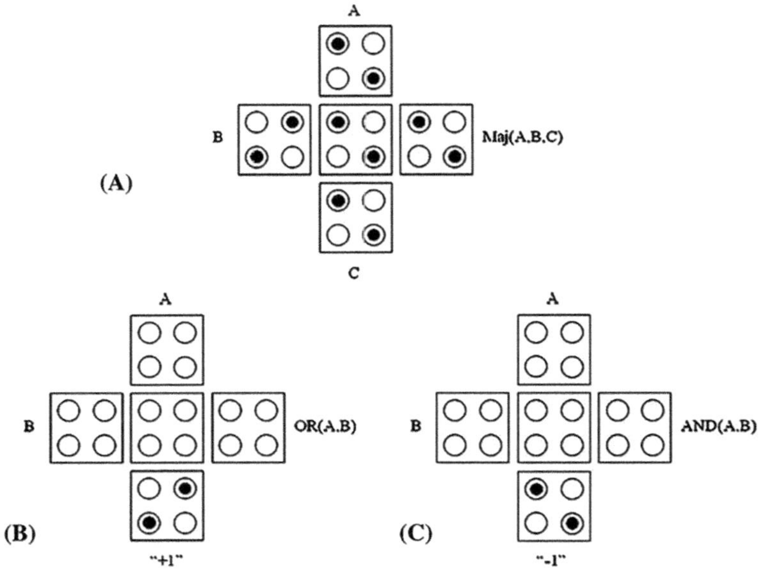

Fig. 3 **a** 3-input MG, **b** 2-info OR gate, **c** 2-input AND gates [27]

$$M(A, B, C) = AB + BC + CA \tag{1}$$

2.3 Clock Mechanism

Timing component is presented with four extraordinary and progressive stages, where a circuit can be arranged into four different clock zones with systems in a specific region that gets the same clicking signal. As demonstrated below, the four clock stages: hold, switch, discharge, and unwind.

2.4 Related Work

Rezai and Rashidi [28] have proposed another construction for a QCA 2:1 Multiplexer. This construction in a 0.01-μm^2 region has 15 cells and the deferral of two clock zones. They at that point utilized the construction to plan 4:1 and 8:1 multiplexer (Fig. 4).

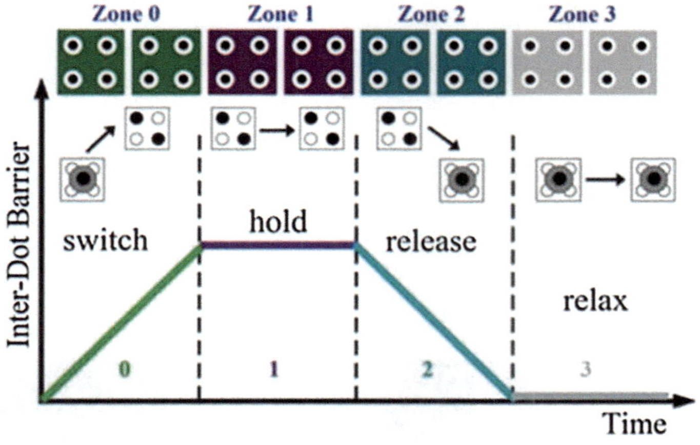

Fig. 4 QCA clock mechanism [27]

Rezai and Rashidi [30] have proposed another construction for a QCA 2:1 Multiplexer. This construction in a 0.02-μm^2 region has 17 cells and the deferral of two clock zones. At that point it utilizes the construction to plan 4:1 and 8:1 multiplexers.

Asfestani et al. [29] in their research worked on a special construction for 2:1 QCA multiplexer as a MUX door, shown in Fig. 5. In this case with S = 0 the yield is equivalent to "I0" and with S = 1 yield is equivalent to "I1."

Angizi et al. [30] also came up with another arbitrary structure of access memory with reset and set capacity. Such plan included 3-input dominant part doors with 1 five-input lion's share entryway as demonstrated in Fig. 6. Such a system was made with two isolated reset and set signs.

Mubarakali et al. [31] presented another irregular access memory structure having a select capacity. Their system was constructed with five three-layer part entryways and one MUX door as demonstrated in Fig. 7.

3 Proposed Structure

An important factor to include in a plan of reversible memory is to initially allude to the reality table in QCA memory. Table 1 demonstrates the truth table of proposed memory. The construction of the proposed configuration includes three data sources with three yields. R W, I, and PO are lines of input, and O, O1, are the output lines. Here the past state is PO, demonstrated in Table 2, with fundamental yields of reversible memory given by O. Both the Yields O1 and O2 are trash yields which are also added to the list of output for the reversible feature. Yield O1 is equivalent to include R|W. Information R|W is the selector. On account of R|W = 1, the yield O2 = I and on account of R|W = 0, the yield O2 = PO.

Fig. 5 **a** MUX gate Logic outline, **b** Layout of QCA [24]

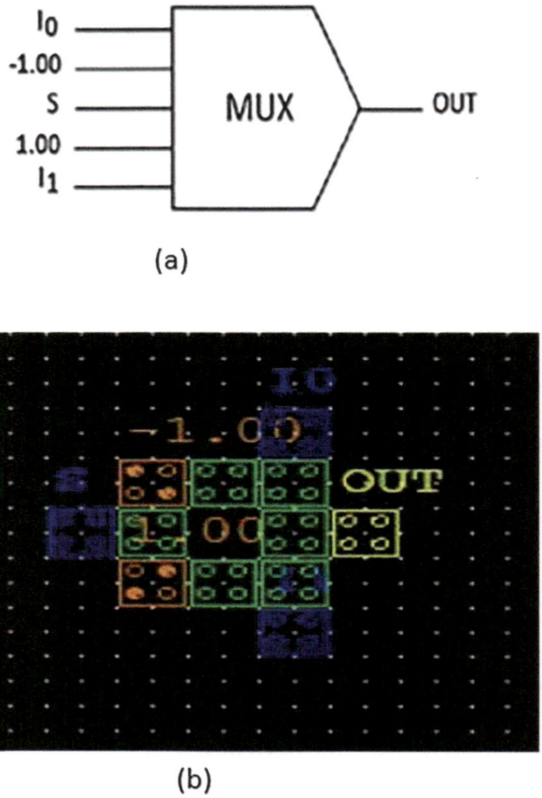

The conditions of yields are gotten as Eqs. (2)–(4).

$$O1 = R \mid W \tag{2}$$

$$O2 = R \mid W.PO + R \mid W.I \tag{3}$$

$$O = R \mid W.I + R \mid W.PO \tag{4}$$

Figure 8 demonstrating the structural arrangement for reversible memory proposed in this research. Two multiplex doors are utilized to execute the structure proposed in this research. The constructions of the multiplex door were the same as that given by Heikalabad and Asfestani [24]. Figure 9 depicts QCA design with 3-layer

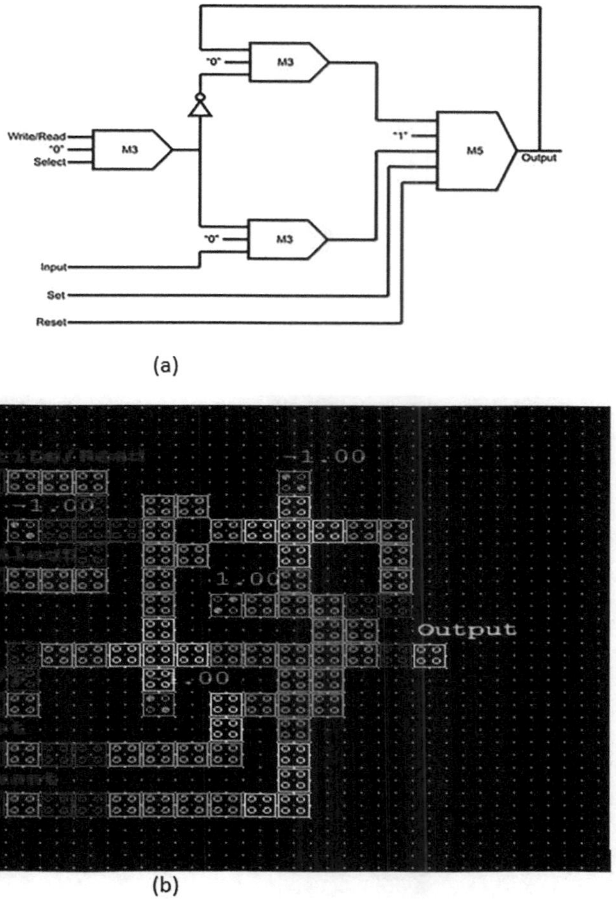

Fig. 6 **a** Logic chart of introduced memory, **b** QCA layout [25]

reversible memory proposed here. QCA format of the 4-digit reversible memory proposed is given in Fig. 10. The plan was overseen in 3 layers to have an improvement in equipment prerequisites, cell tally, and territory.

4 Result and Discussion

The following part includes the structure proposed for reversible memory which is reproduced using a QCA Designer variant 2.0.3 which works in the same way as the other reenactment instrument for QCA circuits. Recreation boundaries are expected to be as demonstrated in Table 5 [32].

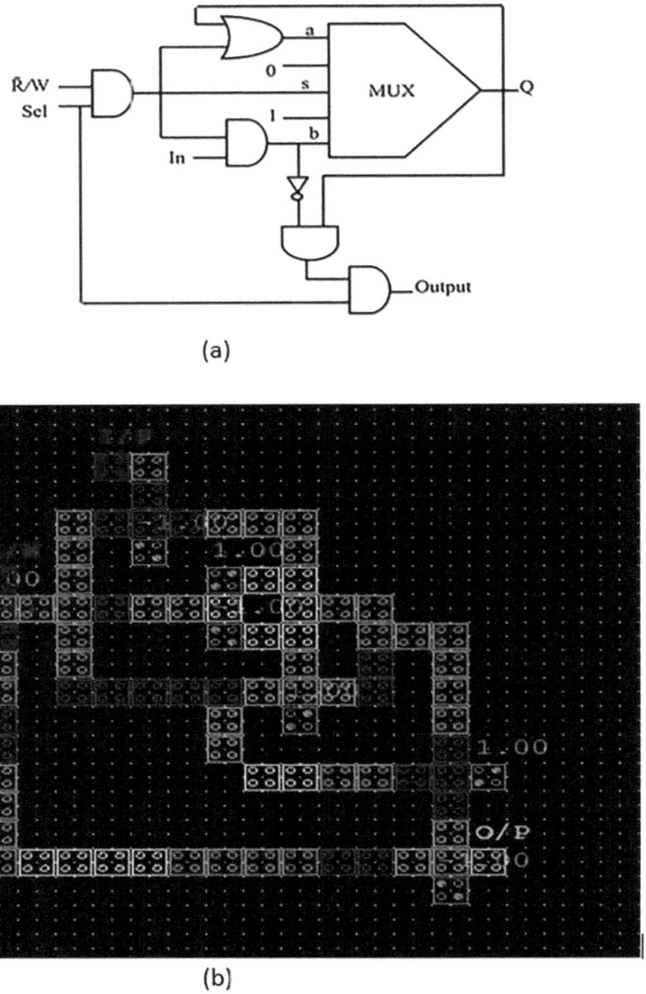

Fig. 7 **a** Logic outline of introduced memory, **b** Format of QCA [31]

Table 1 Truth table for the operation of a basic memory

R/W	I	O
0	0/1	0/1
1	×	Po

Table 2 Proposed reversible memory truth table

R/W	I	PO	O	$O1$	$O2$
0	0	0	0	0	0
0	0	1	0	0	1
0	1	0	1	0	0
0	1	1	1	0	1
1	0	0	0	1	0
1	0	1	1	1	0
1	1	0	0	1	1
1	1	1	1	1	1

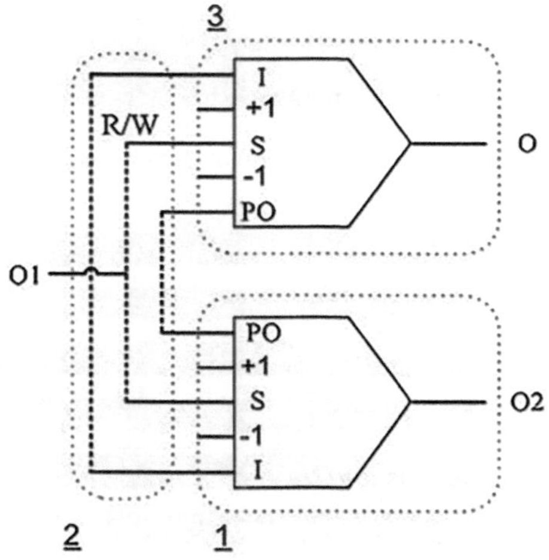

Fig. 8 Proposed arrangement for the reversible memory

Figures 9 and 10 depict the simulated output waveform of basic memory cell operation. Figure 13 shows the reenactment aftereffect of the proposed reversible memory. The outcome affirms the right activity in the proposed system. The execution examination and comparison between both previous and present systems are listed in Table 6. It shows that the proposed system shows improvements in the number of cells, as well as the deferral contrasted with the other previously conducted systems (Fig. 11).

Energy and Power Analysis

Let the superposition condition of an electron burrowing between the spots the x way be ψ (x). Fourier change communicates the superposition [23] an electron's state (x),

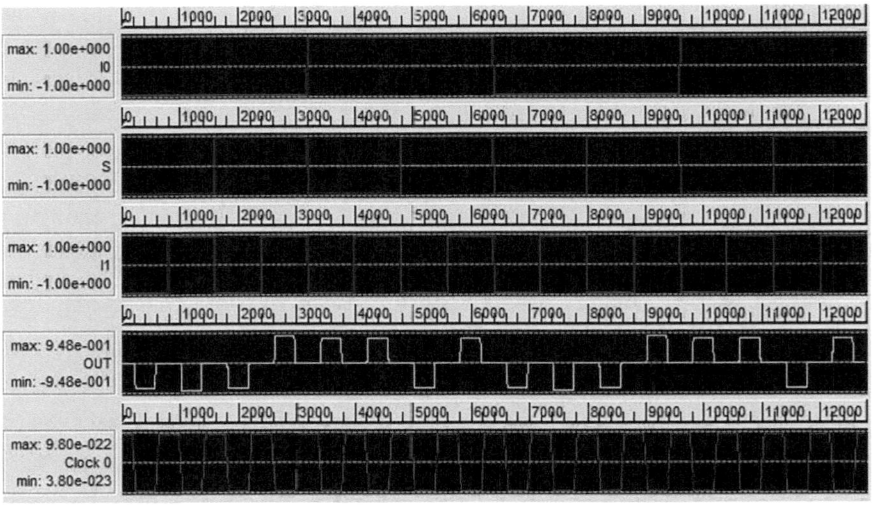

Fig. 9 Simulation waveform of QCA memory cell with Mux gate

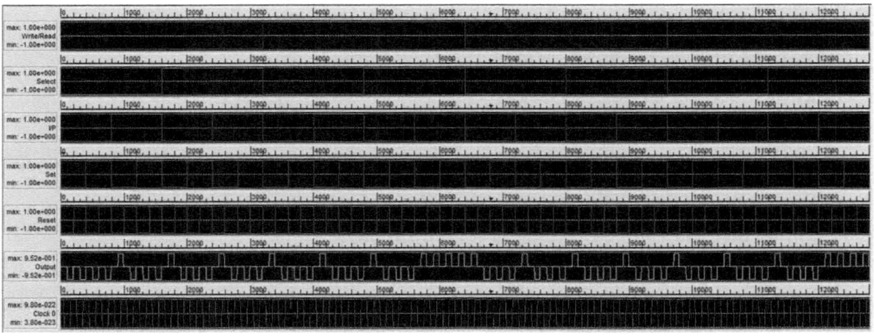

Fig. 10 Simulation waveform o: QCA basic memory cell operation

Fig. 11 QCA cell in 2D representation [33]

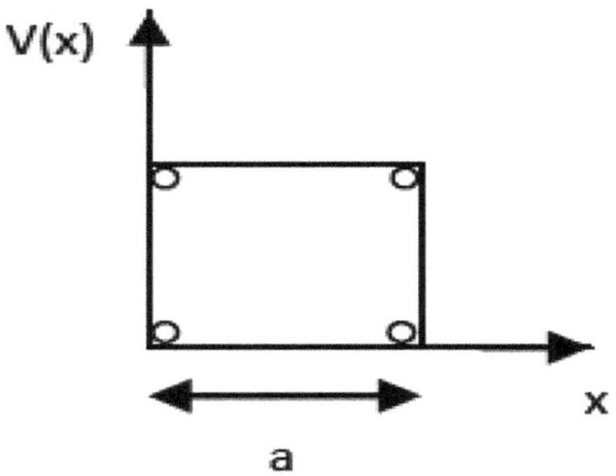

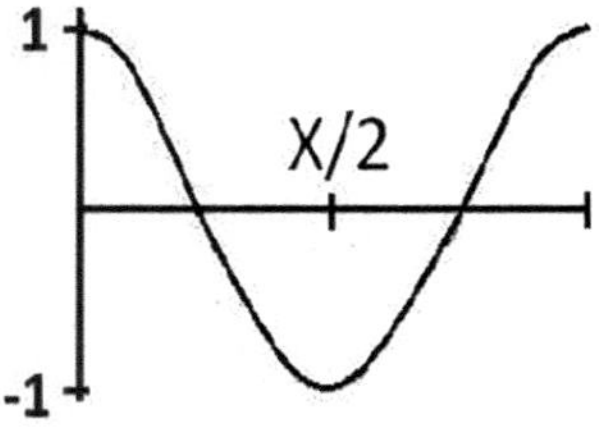

Fig. 12 Characteristic curve [33]

$$\psi(x) = \frac{1}{2\pi}\int_{-\infty}^{\infty} \phi(k)e^{i(kx)}\,dk \tag{5}$$

where the adequacy of the superposition wave is represented by ψ (k). k = $\frac{2\pi}{\lambda}$ is the wave spread speed, the wavelength is denoted by λ and $i = \sqrt{-1}$. The trademark bend of an electron wave is illustrated by Fig. 12 while moving through channels [16]. At whatever point the electron is confined, it is situating at x = 0 and $e^{i(kx)} = 1$ achieved this value. It implies electron influxes with varied frequencies meddle valuably and no motions are announced. Henceforth ψ(x) achieves top at x = 0 with varying upsides of x, the segments of $e^{i(kx)}$ are inserted in Eq. (9). In this way bringing about motions and the worth of ψ(x) is acquired. At $\frac{x}{2}$, $e^{i(kx)}$ accomplishes the least worth and trademark bend accomplishes the negative pinnacle. At the point when x develops starting here, the worth of $e^{i(kx)}$ likewise develops bringing about the development of ψ(x).

As expressed before, a clock signal is the energy provider to the electrons for changing their state. We accept that between spot channels are encountering infinite V(x) potential energy in the positive x direction. At that point time, autonomous Schrodinger wave condition is,

$$\frac{d^2\psi(x)}{dx^2} + \frac{2m}{\hbar^2}(E_n - V(x))\psi(x) = 0 \tag{6}$$

where electron mass is denoted by m, decreased plank constant by $\hbar$ [34, 35]. Infinite arrangement of discrete energy levels is expressed by E_n relating to all conceivable non-negative basic upsides of n. whereas the quantum number is given by n. Condition 7 can be diminished to

$$E_n = \frac{n^2\pi^2\hbar^2}{2md^2} + V(x) \tag{7}$$

where d is the atomic cell measurement [16]. Whenever voltage is given to the atom as V volt with C intersection capacitance at that point, articulation will be produced as [16],

$$\frac{n^2\pi^2\hbar^2}{2md^2} + V(x) = \frac{1}{2}CV^2 \tag{8}$$

This articulation will help to compute the working RMS voltage of the framework.

We need to additionally investigate energy and power as far as Schrodinger Equation [19].

On the off chance that expects the Quantum number to be meant by n, decreased Plank's constant be signified by $\hbar$, electron mass be m, cell region be a^2, number of cells in the design be N, number of clock stages utilized by k. Here, n = 10 and $n_2 = 3$.

The energy provided to the whole circuit = Energy provided by the clock signal

$$E_{lat} = \frac{n^2\pi^2\hbar^2 N}{2ma^2} \tag{9}$$

For Energy dispersal

$$E_{disp} = \frac{\pi^2\hbar^2(n^2 - 1)N}{2ma^2} \tag{10}$$

Occurrence energy for frequency

$$F_{occ} = \frac{\pi\hbar((n^2) - (n_2^2))}{2ma^2} \tag{11}$$

Recurrence of dissemination energy

$$F_{rec} = \frac{\pi\hbar((n^2) - (n_2^2))}{2ma^2} \tag{12}$$

Difference in the frequency level

$$F_{rec} - F_{occ} = \frac{\pi\hbar(n_2^2 - 1)N}{2ma^2} \tag{13}$$

Time expected to arrive at the Quantum level

$$T_1 = \frac{1}{F_{occ}} \tag{14}$$

Disperse time to loosen up state

$$T_2 = \frac{1}{F_{rec}} \tag{15}$$

The time prerequisite that cells in a clock zone go to the succeeding polarization

$$T = T_1 + T_2 \tag{16}$$

At conclusive time needed to spread through the whole design

Table 3 Different parameter level optimization for power and energy analysis of implemented design in TQCA

Theoretical aspect	Expression
The energy provided to the whole circuit = Energy provided by the clock signal	$E_{lat} = \frac{n^2\pi^2\hbar^2 N}{2ma^2}$
Energy dispersal	$E_{disp} = \frac{\pi^2\hbar^2(n^2-1)N}{2ma^2}$
Occurrence energy for frequency	$F_{occ} = \frac{\pi\hbar((n^2)-(n_2^2))}{2ma^2}$
Recurrence of dissemination energy	$F_{rec} = \frac{\pi\hbar((n^2)-(n_2^2))}{2ma^2}$
Difference of the frequency level	$F_{rec} - F_{occ} = \frac{\pi\hbar(n_2^2-1)N}{2ma^2}$
Time expected to arrive at the Quantum level	$T_1 = \frac{1}{F_{occ}}$
Disperse time to loosened up state	$T_2 = \frac{1}{F_{rec}}$
The time prerequisite that cells in a clock zone to go to the succeeding polarization	$T = T_1 + T_2$
At conclusive time needed to spread through the whole design	$T_p = T + (k-1)T_2 N$

Table 4 Power and energy analysis of implemented design in TQCA

Design	Energy (J)	Power (W)
Multilayer memory	4.86×10^{-22}	1.5×10^{-13}

$$T_p = T + (k-1)T_2 N \tag{17}$$

Here we have to obtained the value from Eq. (9), E_{lat} is 4.90×10^{-22} J and for Eq. (10), the resultant value for, E_{disp} is 4.86×10^{-22} J. we have also find the requirement incident energy for frequency F_{occ} that is get the value form Eq. (11), $F_{occ} = 1.16 \times 10^9$ Hz. It is extracted the value for finding from Eq. (12), $F_{rec} = 2.33 \times 10^{11}$ Hz. We have also shown that, $T_1 = \frac{1}{F_{occ}}$, so It is being find $T_1 = 8.62 \times 10^{-10}$ s and $T_2 = 4.29 \times 10^{-12}$ s from Eqs. (14) and (15). The time required to go the polarization T is equal to 8.66×10^{-10} s. The total time required to propagate through the entire design of the proposed phenomenon is T_p is equivalent to 3.24×10^{-9} s or 3.24 ns. The ultimate analysis for power we have to write, $P = \frac{E_{disp}}{T_p}$ is find to consider the power 1.5×10^{-13} W or 0.15 pW.

The power and energy for obtaining the requirement are briefly discussed in Tables 3 and 4.

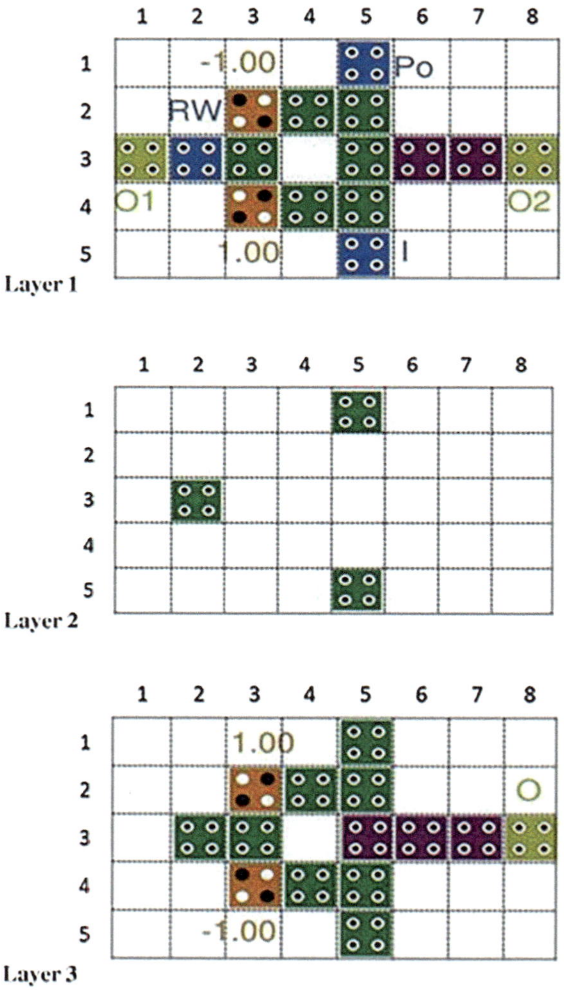

Fig. 13 Proposed system with three unique layers in QCA-based reversible memory

This relative logic phenomenon is developed at the gate level behind the proposed circuit and the circuit layout is then created. In QCA, these structures are then simulated in the aforementioned sections, as a consequence of ongoing research work done by the Walus Group of the University of British Columbia for developing a plan as well as a simulation tool. The system is converted into a QCA design using main gates and inverters. This tool allows the designer for developing and simulating QCA plan quickly. The QCA designer has provided the engineers with a new stage; various international organizations have released the results of simulations that use this tool [15, 16]. The results obtained from this tool then contrast with hypothetical attributes to check that the circuit is correct (Figs. 14, 15 and Tables 5, 6).

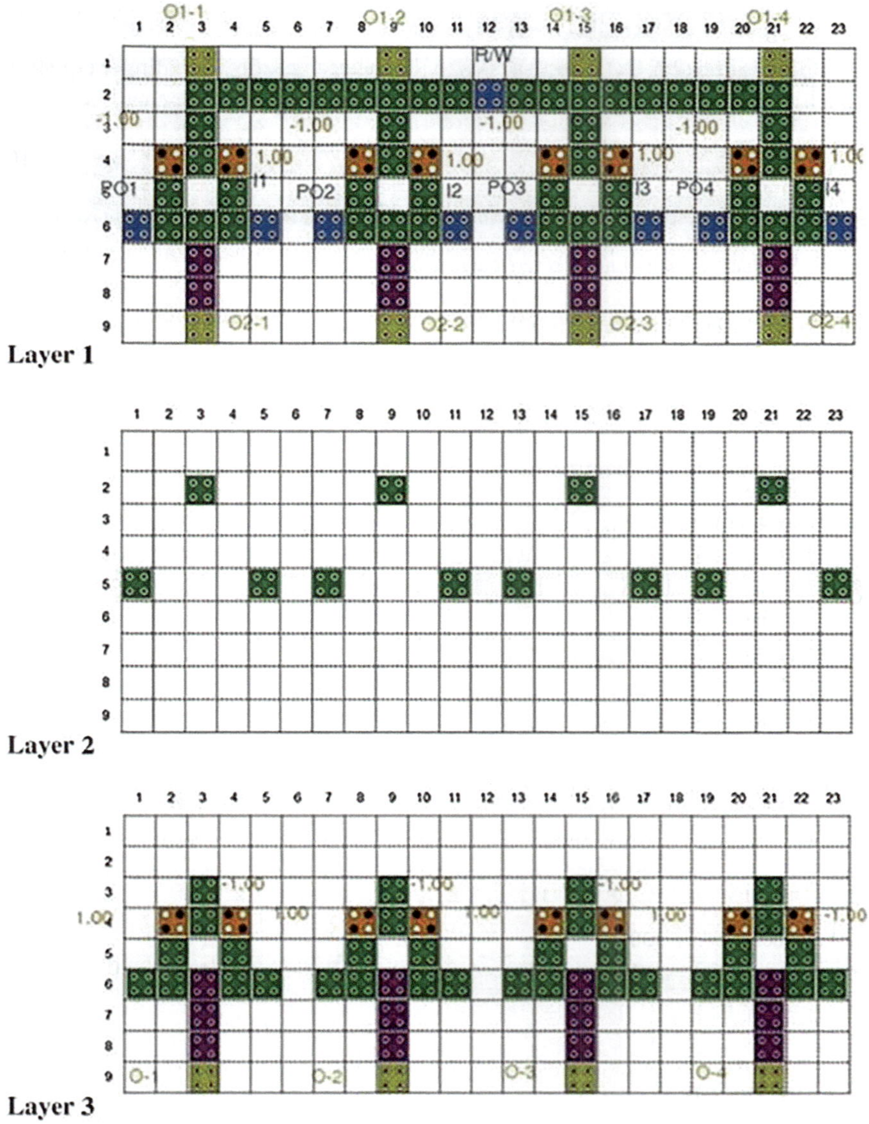

Fig. 14 Proposed system with three unique layers in QCA-4-bit reversible memory

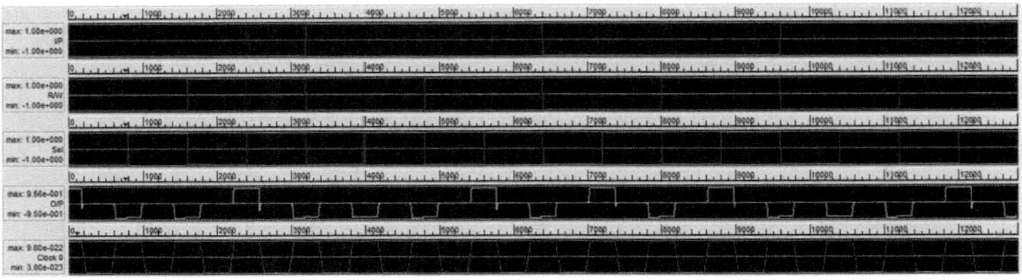

Fig. 15 Simulation result of novel reversible memory

Table 5 Parameters for the simulation

Parameter value	Value
Cell size	18 nm * 18 nm
Relaxation time	1.0e-015
Time step	1.0e-016
Radius of effect	80.0
Relative permittivity	12.90
Clock high	9.80e-022
Clock low	3.80e-023
Clock amplitude factor	2.0
Layer separation	11.5
Clock shift	0.0e+000
Total simulation time	7.000000e-011

Table 6 Execution result

Structure	No. of cells	Area (μm^2)	Delay
[26]	88	0.08	5
[27]	87	0.12	6
Proposed design	30	0.01	1
Comparison	61%	74%	59%

5 Conclusion

The research details a novel design for reversible slightest-bit memory along with the 4-bit memory used with QCA. The structure of the memory proposed here makes use of a 3-layer arrangement that altogether affects to decrease the circuit size. The support for the presented structure of a reversible memory has 61% increase in cell number, 74% enhancement in the territory inhabitance, and with a 59% decrease in the delay contrasted to the previously conducted optimal designs. In the future this construction could be upgraded by using additional control inputs.

References

1. F. Salimzadeh, S. Heikalabad, Design of a novel reversible structure for full adder/subtractor in quantum-dot cellular automata. Phys. B: Condens. Mat. **556**, 163–169 (2019)
2. H. Roshany, A. Rezai, Novel efficient circuit design for multilayer qca rca. Internet J. Theor. Phys. **58**, 1745–1757 (2019)
3. S.-S. Ahmadpour, M. Mosleh, S. Heikalabad, A revolution in nanostructure designs by proposing a novel qca full-adder based on opti- mized 3-input xor. Phys. B: Condens. Mat. **550**, 383–392 (2018)
4. S. Ahmadpour, M. Mosleh, S.R. Heikalabad, Robust qca full-adders using an efficient fault-tolerant five-input majority gate (2019)
5. M. Hayati, A. Rezaei, Design of novel efficient adder and subtractor for quantum-dot cellular automata. Internet J. Circuits Theory Appl. **43**, 1446–1454 (2015)
6. S. Babaie, A. Sadoghifar, A. Bahar, Design of an efficient multilayer arithmetic logic unit in quantum-dot cellular automata (qca). IEEE Trans. Circuits Syst. II: Express Brie **66**, 963–967 (2018)
7. S. Heikalabad, M. Asfestani, M. Hosseinzadeh, A full adder structure without cross-wiring in quantum-dot cellular automata with energy dissipation analysis. J. Supercomput. **74**, 1994–2005 (2018). https://doi.org/10.1007/s11227-017-2206-4
8. M. Norouzi, S. Heikalabad, F. Salimzadeh, A reversible alu using hng and ferdkin gates in qca nanotechnology. Int. J. Circuit Theory Appl. https://doi.org/10.1002/cta.2799
9. S.-S. Ahmadpour, M. Mosleh, S. R. Heikalabad, The design and implementation of a robust single-layer qca alu using a novel fault-tolerant three-input majority gate, The Journal of Supercomputing: https://doi.org/10.1007/s11227-020-03249-3
10. S.-S. Ahmadpour, M. Mosleh, S. Heikalabad, An efficient fault-tolerant arithmetic logic unit using a novel fault-tolerant 5-input majority gate in quantum-dot cellular automata. Comput. Electr. Eng. https://doi.org/10.1016/j.compeleceng
11. S. Heikalabad, M. Gadim, Design of improved arithmetic logic unit in quantum-dot cellular automata. Int. Theor. Phys. **57**, 1733–1747 (2018). https://doi.org/10.1007/s10773-018-3699-1
12. M. Divshali, A. Rezai, A. Karimi, Towards multilayer qca siso shift register based on efficient d-ff circuits. Internet J. Theor. Phys. **57**, 3326–3339 (2018)
13. M. Divshali, A. Rezai, S. Hamidpour, Design of novel coplanar counter circuit in quantum dot cellular automata technology. Internet J. Theor. Phys. **58**, 2677–2691 (2019)
14. A. Shiri, A. Rezai, H. Mahmoodian, Design of efficient coplanar comprator circuit in qca technology. Facta Universitatis, Ser.: Electr. Energy **32**, 119–128 (2019)
15. A. Sadoghifar, S. Heikalabad, A content-addressable memory structure using quantum cells in nanotechnology with energy dissipation analysis. Phys. B: Condens. Mat. **537**, 202–206 (2018)
16. X. Yang, L. Cai, H. Huang, X. Zhao, A comparative analysis and design of quantum-dot cellular automata memory cell architecture. Internet J. Circuits Theory Appl. **40**, 93–103 (2012)
17. S. Heikalabad, A. Navin, M. Hosseinzadeh, T. Oladghaffari, Midpoint memory: a special memory structure for data-oriented models implementation. J. Circuits, Syst. Comput. **24**, 1550063-1–1550063-14 (2015)
18. S. Heikalabad, A. Navin, M. Hosseinzadeh, Content addressable memory cell in quantum-dot cellular automata. Microelectr. Eng. **163**, 140–150 (2016)
19. A.N.S. Heikalabad, Design of reversible parity generator and checker for the implementation of nano-communication systems in quantum-dot cellular automata. Photon. Netw. Commun. 1–13 (2019)
20. M. Ali, M. Hossin, M. Ullah, Design of reversible sequential circuit using reversible logic synthesis. Internet J. VLSI Design Commun. Syst. **2**, 37–45 (2011)
21. E. Karkaj, S. Heikalabad, Binary to gray and gray to binary converter in quantum-dot cellular automata. Optik. **130**, 981–989 (2017). https://doi.org/10.1016/j.ijleo.2016.11.087
22. C. Lent, P. Tougaw, W. Porod, G. Bernstein, Quantum cellular automata. Nanotechnology **4**, 49–57 (1993)

23. Y. Barughi, S. Heikalabad, A three-layer full adder/subtractor structure in quantum-dot cellular automata. Internet J. Theor. Phys. **56**, 2848–2858 (2017)
24. E. Karkaj, S. Heikalabad, A testable parity conservative gate in quantumdot cellular automata. Superlatt. Microstruct. **101**, 625–632 (2017)
25. S. Rad, S. Heikalabad, Reversible flip-flops in quantum-dot cellular automata. Internet J. Theor. Phys. **56**, 29990–30004 (2017)
26. H. Hosseinzadeh, S. Heikalabad, A novel fault tolerant majority gate in quantum-dot cellular automata to create a revolution in design of fault tolerant nanostructures, with physical verification. Microelectr. Eng. **192**, 52–60 (2018)
27. C. Lent, P. Tougaw, A device architecture for computing with quantum dots (1997)
28. S. Ali, H.S. Rasouli, A content-addressable memory structure using quantum cells in nanotechnology with energy dissipation analysis. Phys. B: Condens. Matter **537**, 202–206 (2018). https://doi.org/10.1016/j.physb.2018.02.024
29. H. Rashidi, A. Rezai, Design of novel efficient multiplexer architecture for quantum-dot cellular automata. J. Nano- Electron. Phys. **9**. https://doi.org/10.21272/jnep.9(1).01012
30. H. Rashidi, A. Rezai, S. Soltany, High-performance multiplexer architecture for quantum-dot cellular automata. J. Comput. Electron. **15**, 968–981 (2016)
31. M. Asfestani, S. Heikalabad, A unique structure for the multiplexer in quantum-dot cellular automata to create a revolution in design of nanostructures. Phys. B: Condens. Mat. **512**, 91–99 (2017)
32. S. Angizi, S. Sarmadi, S. Sayedsalehiand, K. Navi, Design and evaluation of new majority gate-based ram cell in quantum-dot cellular automata. Microelect. J. **46**, 43–51 (2015)
33. D. Mukhopadhyay, P. Dutta, A study on energy optimized 4 dot 2 electron two dimensional quantum dot cellular automata logical reversible flip-flops. Microelectron. J. **46**, 519–530 (2015)
34. K. Chakrabarti, Realization of original quantum entanglement state from mixing of four entangled quantum states, in O. Castillo, D. Jana, D. Giri, A. Ahmed, (eds.), *Recent Advances in Intelligent Information Systems and Applied Mathematics. ICITAM 2019*. Studies in Computational Intelligence, vol. 863 (Springer, Cham, 2020). https://doi.org/10.1007/978-3-030-34152-712
35. K. Chakrabarti, Is there any spooky action at a distance?, in A.K. Maji, G. Saha, S. Das , S. Basu, J.M.R.S. Tavares, (eds.), *Proceedings of the International Conference on Computing and Communication Systems*. Lecture Notes in Networks and Systems, vol. 170 (Springer, Singapore, 2021), https://doi.org/10.1007/978-981-33-4084-865

Quantum Neural Network for Image Classification Using TensorFlow Quantum

J. Arun Pandian and K. Kanchanadevi

Abstract Quantum neural networks are artificial neural networks that are designed using the principles of classical artificial neural networks and quantum information. Developing a quantum neural network requires knowledge of both classical neural networks and quantum computing techniques. TensorFlow Quantum is a library for developing quantum and hybrid neural network models. The MNIST handwritten digits dataset was used as a sample dataset for quantum neural network development. This chapter introduces the design and development process of a simple quantum neural network for image classification tasks using the TensorFlow quantum library. Also, it compared the classification performances of the quantum neural network and classical neural network on the MNIST handwritten digits classification. This chapter identified that the quantum neural network performed better than the classical neural network on digit classification. Also, the chapter discussed the advantages and limitations of quantum neural networks in image classification. The first section of the chapter introduced the quantum neural network models and the TensorFlow quantum library. Afterward, the data preparation steps such as data loading, downscaling, contradictory removal, and Tensorflow Quantum circuit conversions were discussed. Furthermore, the building process of quantum neural networks for image classification, such as quantum neural network designing, the model circuit to the Tensorflow Quantum model binding, and model training, was discussed in the third section of the chapter. Subsequently, the testing performance of the simple quantum neural network was discussed in section four of this chapter. Finally, the conclusions of the chapter were discussed in the fifth section.

Keywords Classical neural networks · Quantum neural networks · Quantum dataset · TensorFlow quantum · MNIST handwritten digits dataset

---

J. Arun Pandian
School of Information Technology and Engineering, Vellore Institute of Technology, Vellore, India

K. Kanchanadevi (✉)
Department of Computer Science and Engineering, Vel Tech Rangarajan Dr. Sagunthala R&D Institute of Science and Technology, Chennai, India
e-mail: kanchanadevi95@gmail.com

R. Pandey et al. (eds.), *Quantum Computing: A Shift from Bits to Qubits*,
Studies in Computational Intelligence 1085,
https://doi.org/10.1007/978-981-19-9530-9_10

1 Introduction

Quantum computing and artificial intelligence are emerging cutting-edge technologies in computer science [1]. Artificial intelligence techniques are used to solve complex decision-making and optimization problems in real-world applications. In general, artificial neural networks, genetic algorithms, and fuzzy logic approaches are widely used to solve decision-making problems in artificial intelligence [2]. Artificial neural networks extend the limitations of artificial intelligence in decision-making applications [3]. The performance and reliability of recent artificial neural network techniques such as machine learning and deep learning in decision-making applications are equivalent to human accuracy [4]. These techniques perform better than human experts in computer vision and natural language processing applications. The availability of a large volume of data and improved computation capability are the most important reasons for the performance of artificial intelligence techniques [5]. Machine learning and deep learning techniques require large amounts of data samples for their learning processes [6]. Also, more computing power is needed to process the large volume of data and train the learning techniques. Graphics Processing Unit (GPU) enabled computers are trying to address the processing power requirement challenge of artificial intelligence techniques [7]. But the GPU-enabled computers also take more time to complete the training process of machine learning and deep learning applications [8]. So, the computational requirement challenge is not completely addressed by existing techniques.

Quantum computing is an alternative technique to traditional computing approaches. It is motivated by the principles and characteristics of quantum physics [9]. Quantum computing uses quantum state properties such as superposition, interference, and entanglement to perform computation [10]. It proposed the information be used by quantum bits or qubits. A single qubit can represent an exponential number of states, which means that quantum computing can process more information in a minimum amount of time [11]. Recently, quantum computing algorithms have been used in numerous fields for advantages in computation capabilities [12]. The artificial neural network is one of the most focused application areas of quantum computing. It is used in artificial neural network design to satisfy computational requirements. TensorFlow is one of the most widely used packages for developing neural network applications [13]. TensorFlow Quantum is a specific package for implementing quantum and hybrid classical and quantum networks [14]. The TensorFlow quantum is an open-source package; it uses the Google CIRQ framework in the backend for handling quantum circuits and states. The CIRQ framework was used to construct the quantum circuits and schedules for quantum neural network implementation.

The TensorFlow quantum simulator simulates the quantum circuits on traditional computers and interfaces the traditional TensorFlow with the quantum environment [15]. TensorFlow version 2.7.0 is the backend of the TensorFlow quantum package. The TensorFlow quantum will be installed using the "pip install tensorflow-quantum"

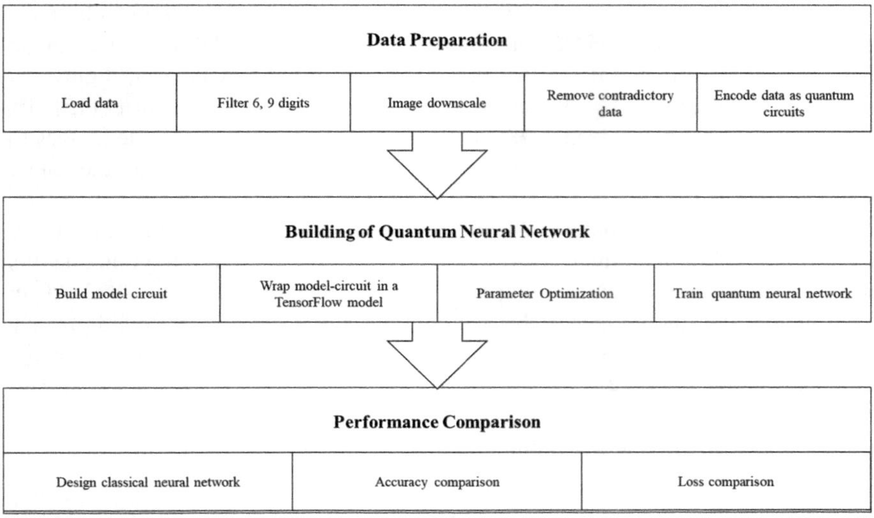

Fig. 1 Implementation phases of quantum neural network

command. The complete experiment in the chapter uses the Google Colab environment. The TensorFlow quantum was installed on Google Colab using the first code block of the experiment page [8]. A Graphics Processing Unit (GPU) resource was used in the Google Colab to speed up the implementation process. This chapter introduced a binary image classifier for image data using a Parametrized Quantum Circuit (PQC) layer network [9, 10]. Also, it compared the performance of the quantum model with that of the classical neural network using the MNIST handwritten digits dataset. The quantum neural network implementation process in this chapter consists of three phases. The implementation process started with the data preparation phase. The data preparation phase prepares the dataset for quantum neural network training. After the data preparation, the building of the quantum neural network phase was introduced. The performance comparison phase compared the performance of the quantum neural network with classical neural networks using accuracy and loss values. Each phase of the quantum neural network implementation was discussed in the subsequent sections. The complete flow of the quantum neural network implementation is shown in Fig. 1.

2 Data Preparation

Data preparation is the first phase of the quantum neural network implementation. The MNIST handwritten digits dataset was downloaded from the TensorFlow data repository [16]. The original data contains 600,000 samples for training and another 10,000 samples for testing. The original data consists of ten classes of images. Each class represents one digit between 0 and 9. This chapter focused on binary classification. So, only the digits 6 and 9 were filtered from the original dataset. A new binary class dataset was created using the digits 6 and 9. In the new dataset, digit 6 was labeled as 0 (false) and digit 9 was labeled as 1 (true) in the new dataset [15]. The images in the dataset are sized at 28*28 pixels. The dataset has grayscale images for the training and testing of the model. A random sample of images from the dataset is shown in Fig. 2.

The filtered dataset contains 11,867 images for training and 1967 images for testing. The 28*28 pixel images are very heavier for the simulator to build quantum circuits. The current quantum computers are not capable to process 28*28 pixel-sized images. So, the images in the dataset were downscaled from 28*28 pixels to 4*4 pixels. Figure 3 shows the downscaled images of the randomly selected images which are shown in Fig. 2.

After the downscaling, some of the data in both classes are looking similar. The models may not easily classify the contradicting data. There are 58 contradicting data available in the filtered dataset. The flattening and mapping function was used

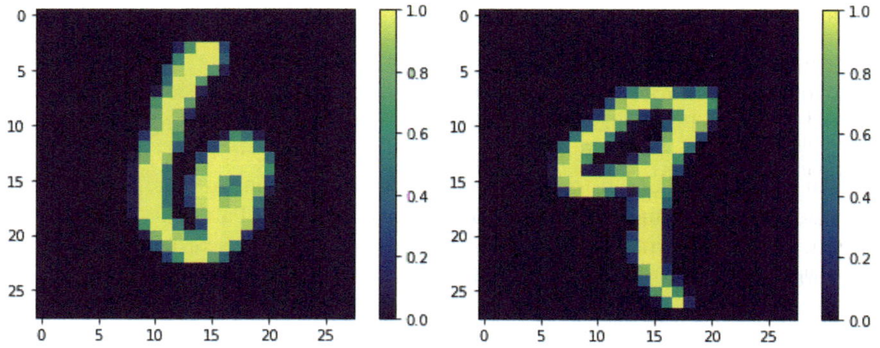

Fig. 2 Random samples from filtered data

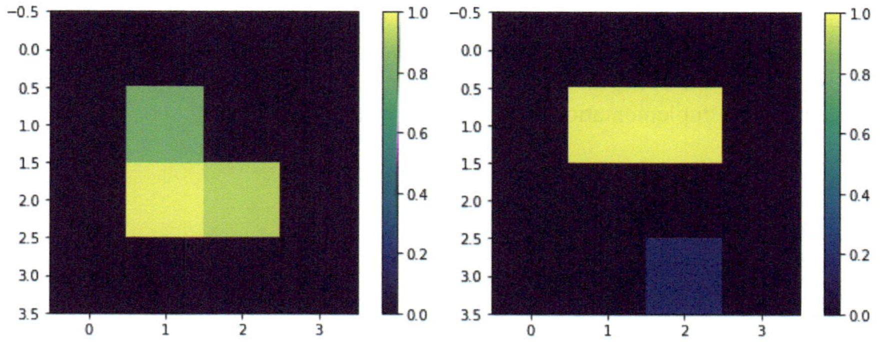

Fig. 3 Downscaled Random samples from filtered data

for removing the contradicting data. After the removal of contradicting data, the number of training images in digits 3 and 6 are 5417 and 5601, respectively. In the non-contradictory training dataset, there are 11,018 images are available. The quantum circuit design for each data was processed using binary encoding techniques with a threshold value of 0.5. The pixels which are less than the threshold limit are considered as 0 (False).

The quantum circuit conversion process used the google Cirq framework. Figure 4 illustrates the quantum circuit of the randomly selected image data. The left column of the figure shows the pixel representation of downscaled images. The right column of the figure shows the respective quantum circuit of the digit image. The quantum circuit considered the pixel positions of the images as an input which are higher than the threshold value. After the conversion of all training images to the quantum circuits, the quantum neural network building process began. The quantum neural network building process was discussed in the subsequent section.

3 Building of Quantum Neural Network

The quantum neural network (QuantumNN) design process uses five qubits, including one ancillary qubit. The quantum circuit of the QuantumNN was designed using the GridQubit function in the Google Cirq framework. The quantum circuit for QuantumNN was illustrated in Fig. 5.

The QuantumNN model was developed using the data qubits and the model qubits for classifying the digit data. The QuantumNN model was constructed using two layers for balancing the data circuit size and performing preparation and readout processes. The readout circuit was constructed using the GridQubit function on the Google Cirq framework with arguments of -1 and -1 values. The model was constructed using the Parametrized Quantum Circuit layer (PQC) from the TensorFlow quantum library. The QuantumNN model used the Binary Cross entropy as a loss function. Because the model was trained to classify binary classes of data. Also, the QuantumNN model used the Adam optimizer to optimize the learning process with a variable learning rate. The layered architecture of the QuantumNN model was shown in Fig. 6.

The QuantumNN model has been trained on quantum data circuits for ten training epochs. The dataset was split for training and validation with a ratio of 90 and 10%, respectively. The training process of the QuantumNN was performed on Google Colab with GPU resources. A custom training metric called "hind accuracy" was used for measuring the training performance of the QuantumNN model. The hinge accuracy ranged between -1 and 1. Figure 7 illustrates the training progress of the QuantumNN on the digit data circuit.

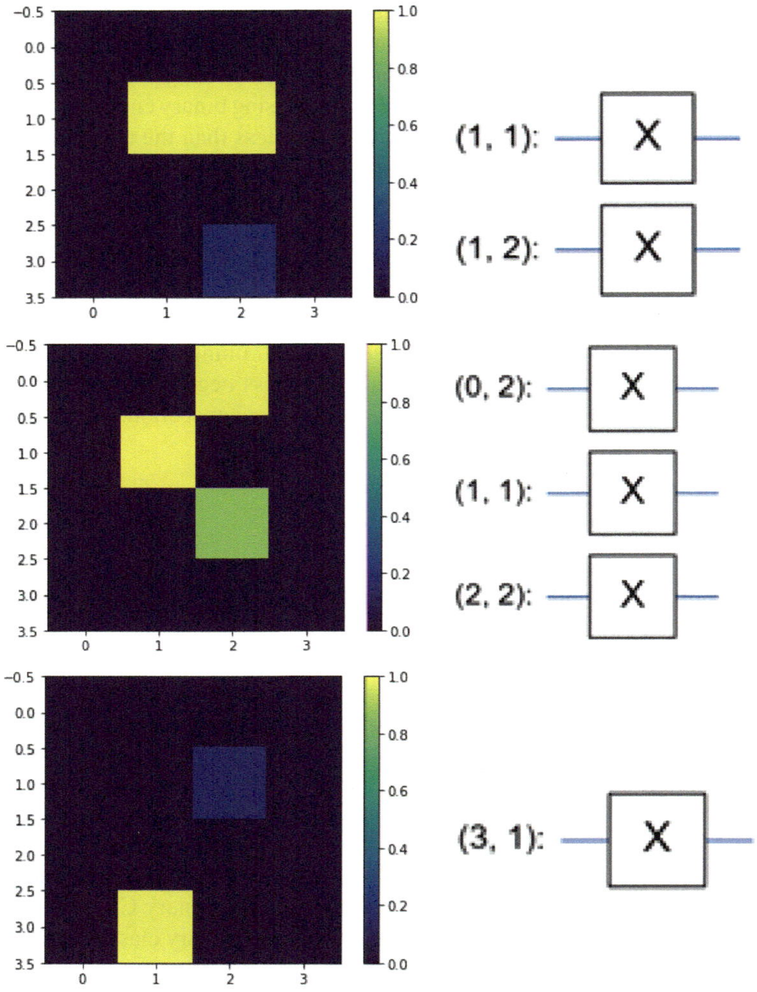

Fig. 4 Quantum circuits of random sample images

There are no remarkable improvements in the training and validation performance of the QuantumNN model after ten training epochs. So, the training process was stopped with ten training epochs. The QuantumNN model achieved the training accuracy and loss after the completion of ten training epochs of 0.8564 and 0.4244, respectively. The performance of the QuantumNN model was compared with that of a simple classical neural network in the next section of this chapter.

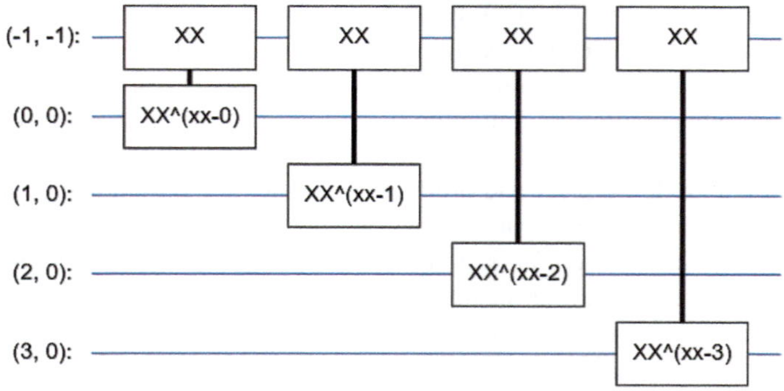

Fig. 5 Quantum circuit for QuantumNN

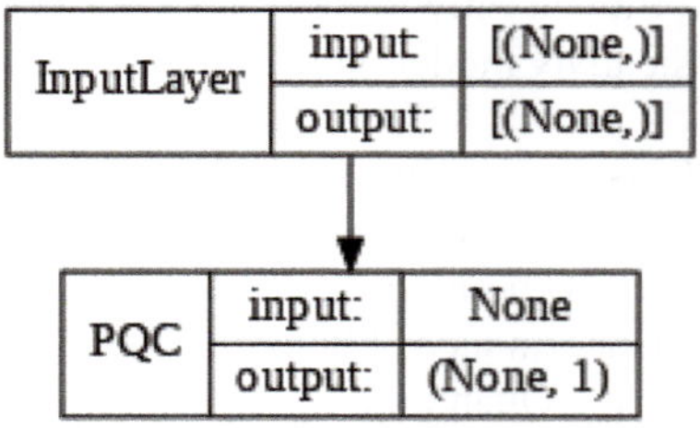

Fig. 6 The layered architecture of QuantumNN

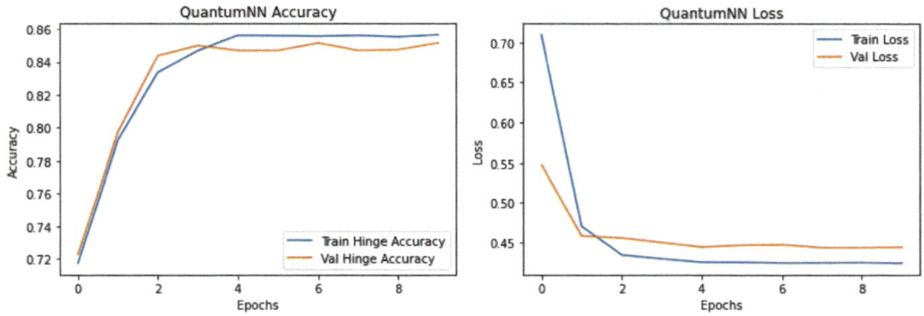

Fig. 7 Accuracy and loss of QuantumNN in training and validation data

4 Results and Discussions

The trained QuantumNN model was compared with a simple classical fully connected neural network. The developed QuantumNN model uses a single PQC layer. So, the classical neural network was also developed with two dense layers for digit classification. The simple classical neural network (ClassicalNN) was developed with an input size of 4*4. The ClassicalNN was trained on the downscaled dataset. The collected input was converted into single-dimensional data using a flattening layer. The output of the flattening layer was sized with 16 values. The first dense layer was introduced after the flattening layer with the size of 16 neurons. The first dense layer produced two values as an output. The second dense layer received the outputs of the first dense layer and produced classification output using the sigmoid activation function. The same loss function and optimizer that were used in the QuantumNN model are used for training the ClassicalNN model. Figure 8 shows the layered architecture of the ClassicalNN model.

In the training dataset, 90% of the data was used for training and the remaining 10% was used for the validation process. The ClassicalNN model was also trained on the Google Colab with the GPU resource for 10 training epochs. The ClassicalNN model achieved training accuracy and validation accuracy of 0.8500 and 0.3030, respectively. The QuantumNN was only trained for 100 epochs. Therefore, the ClassicalNN model was also trained only until 10 epochs. The training and validation performance of the ClassicalNN model on the downscaled digit dataset is shown in Fig. 9.

The test data of the dataset was preprocessed for testing the performance of the trained QuantumNN and ClassicalNN models. There are 100 random samples were used for the testing process of the models. The test accuracy and test loss metrics were used to compare the performance of both models. The hinge accuracy was used to measure the accuracy of the QuantumNN model on test data quantum circuits. A similar data quantum circuits conversion was used in the test dataset. The model was tested using the predict measure function in the TensorFlow Keras library.

The QuantumNN model achieved a test accuracy of 85.16% on test data circuits. And, the ClassicalNN achieved an accuracy of 84.49% on downscaled test data. The

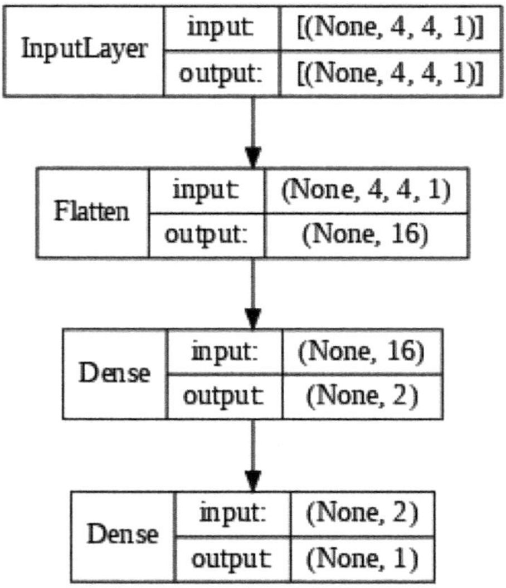

Fig. 8 The layered architecture of ClassicalNN

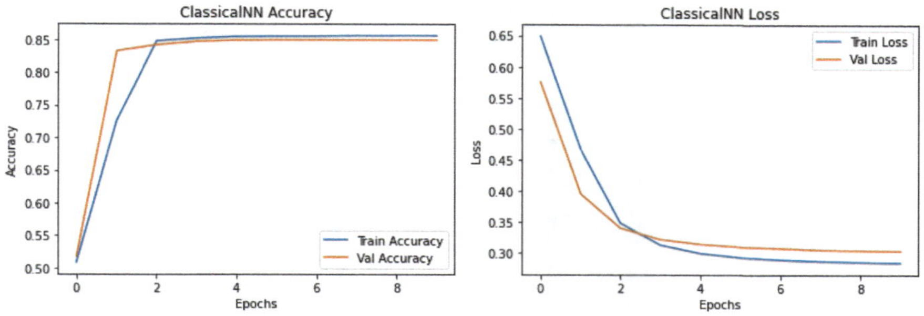

Fig. 9 Accuracy and loss of ClassicalNN in training and validation data

QuantumNN achieved 0.67% higher classification accuracy than the ClassicalNN on binary digit classification tasks. Figure 10 illustrates the test accuracy comparison between the QuantumNN and ClassicalNN models on the test dataset.

Similarly, the test loss of QuantumNN and ClassicalNN was compared in Fig. 11. The QuantumNN achieved and ClassicalNN achieved the test loss of 0.3307 and 0.4443, respectively. The test loss of the QuantumNN was 0.1136 better than the ClassicalNN on digits image classification.

The test accuracy and test loss comparison between QuantumNN and ClassicalNN shows that the performance of QuantumNN was superior to ClassicalNN on MNIST handwritten digits classification. However, the experimentation with digit classification using QuantumNN revealed three major challenges. At first, the QuantumNN took more training time than the ClassicalNN on the image dataset. It restricts training the QuantumNN to a greater number of training epochs. Second, QuantumNN requires more qubits to handle the higher resolution images. It may slow down the training process and increase the computational requirements. Finally,

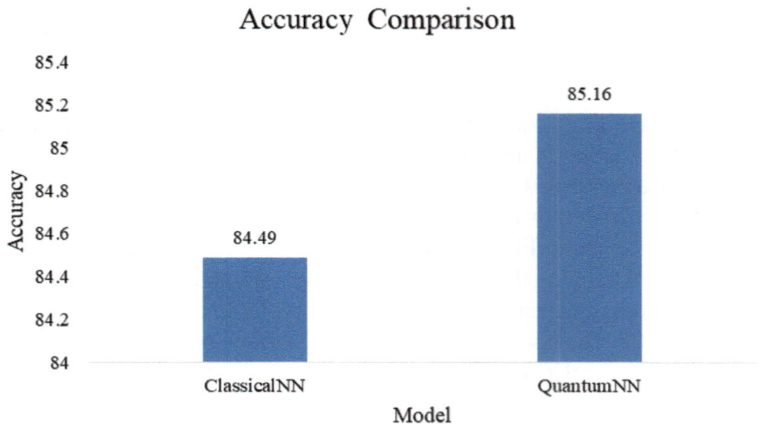

Fig. 10 Accuracy comparison between QuantumNN and ClassicalNN

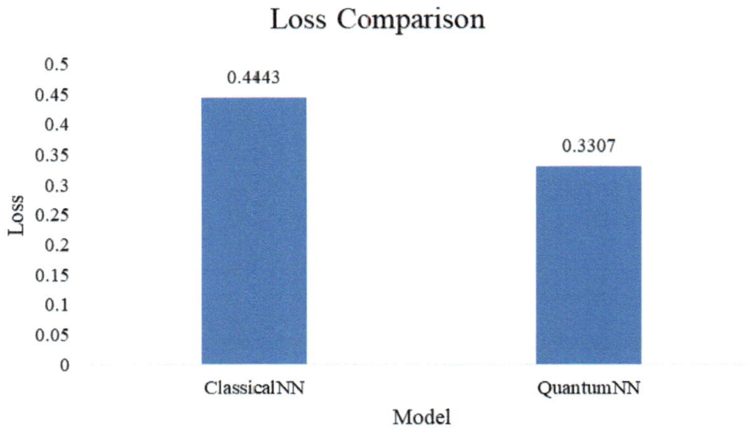

Fig. 11 Loss comparison between QuantumNN and ClassicalNN

more quantum data circuits need to be created when handling the large volume of training data. The above limitations have reduced the significance of quantum neural network development for high-resolution image classification tasks. This chapter's summary was discussed in the following section. Testing accuracy and loss values of the quantum neural network and classical neural networks were compared in the performance comparison section. The MNIST handwritten digit dataset was used to train and test the quantum neural network.

5 Conclusions

Quantum computing is a key tool to overcome the limitations of traditional computing techniques. A quantum neural network is a neural network that uses quantum principles for computation. This chapter discussed the building process of the quantum neural network using the TensorFlow quantum library. The complete building process of the quantum neural network was split into three stages, such as data preparation, the building of the quantum neural network, and performance comparison. The data preparation phase comprises data loading, filtering of 6 and 9 digits, image downscaling, contradictory removal, and data encoding. Building model circuits, warping model circuits in the TensorFlow model, parameter optimization, and model training are the processes of building the quantum neural network phase. On digit 6 and digit 9 image classifications, the quantum neural network achieved a test accuracy of 85.16% and a loss of 0.3307. The test performance of the quantum neural network was better than the classical neural network. Also, the chapter discusses the challenges and limitations of quantum neural network implementation in image classification.

Acknowledgements The authors are grateful for the infrastructure support of Vel Tech Technology Business Incubator, Chennai.

References

1. L. Yang, Z. Leng, G. Yu, A. Patel, W.-J. Hu, H. Pu, Deep learning-enhanced variational Monte Carlo method for quantum many-body physics. Phys. Rev. Res. **2** (2020). https://doi.org/10.1103/PhysRevResearch.2.012039
2. Y. Li, R.-G. Zhou, R. Xu, J. Luo, W. Hu, A quantum deep convolutional neural network for image recognition. Quantum Sci. Technol. **5** (2020). https://doi.org/10.1088/2058-9565/ab9f93
3. M. Henderson, S. Shakya, S. Pradhan, T. Cook, Quanvolutional neural networks: powering image recognition with quantum circuits. Quantum Mach. Intell. **2** (2020). https://doi.org/10.1007/s42484-020-00012-y
4. Y. Kumar, S.K. Verma, S. Sharma, Quantum-inspired binary gravitational search algorithm to recognize the facial expressions. Int. J. Mod. Phys. C. **31** (2020). https://doi.org/10.1142/S0129183120501387
5. S. Oh, J. Choi, J. Kim, A tutorial on quantum convolutional neural networks (QCNN), in *International Conference on ICT Convergence* (2020), pp. 236–239. https://doi.org/10.1109/ICTC49870.2020.9289439
6. S. Lu, L.-M. Duan, D.-L. Deng, Quantum adversarial machine learning. Phys. Rev. Res. **2** (2020). https://doi.org/10.1103/PhysRevResearch.2.033212
7. M. Ahmadi, A. Sharifi, S. Hassantabar, S. Enayati, QAIS-DSNN: tumor area segmentation of MRI image with optimized quantum matched-filter technique and deep spiking neural network. Biomed Res. Int. (2021). https://doi.org/10.1155/2021/6653879
8. J. Liu, K.H. Lim, K.L. Wood, W. Huang, C. Guo, H.-L. Huang, Hybrid quantum-classical convolutional neural networks. Sci. China Phys. Mech. Astron. **64** (2021). https://doi.org/10.1007/s11433-021-1734-3
9. H.-L. Huang, Y. Du, M. Gong, Y. Zhao, Y. Wu, C. Wang, S. Li, F. Liang, J. Lin, Y. Xu, X. Zhu, J.-W. Pan, Experimental quantum generative adversarial networks for image generation. Phys. Rev. Appl. **16** (2021). https://doi.org/10.1103/PhysRevApplied.16.024051
10. Y. Lu, Q. Gao, J. Lu, M. Ogorzalek, J. Zheng, A quantum convolutional neural network for image classification, in *Chinese Control Conference, CCC* (2021), pp. 6329–6334. https://doi.org/10.23919/CCC52363.2021.9550027
11. G. Hellstem, Hybrid quantum network for classification of finance and MNIST data, in *Proceedings-2021 IEEE 18th International Conference on Software Architecture Companion, ICSA-C 2021* (2021), pp. 106–109. https://doi.org/10.1109/ICSA-C52384.2021.00027
12. Y. Jing, X. Li, Y. Yang, C. Wu, W. Fu, W. Hu, Y. Li, H. Xu, RGB image classification with quantum convolutional ansatz. Quantum Inf. Process. **21** (2022). https://doi.org/10.1007/s11128-022-03442-8
13. T. Hur, L. Kim, D.K. Park, Quantum convolutional neural network for classical data classification. Quantum Mach. Intell. **4** (2022). https://doi.org/10.1007/s42484-021-00061-x
14. T. Nguyen, I. Paik, Y. Watanobe, T.C. Thang, An evaluation of hardware-efficient quantum neural networks for image data classification. Electronics **11** (2022). https://doi.org/10.3390/electronics11030437
15. E. Farhi, H. Neven, Classification with Quantum Neural Networks on Near Term Processors (2018). https://doi.org/10.48550/ARXIV.1802.06002
16. Y. LeCun, C. Cortes, C.J. Burges, MNIST handwritten digit database. ATT Labs 2 (2010). http://yann.lecun.com/exdb/mnist

Quantum Network Architecture and Its Topology

Supriyo Banerjee, Biswajit Maiti, and Banaini Saha

Abstract In this chapter, the idea of the deployment of quantum key distribution (QKD) techniques within a more realistic insecure network has been discussed for achieving unconditional secure communication between two or more legitimate users in presence of eavesdroppers. The architecture of the network can play a crucial role in protocol design with trusted or un-trusted repeaters/relays in between these legitimate users. Point-to-point link (P2P) communication for long distance-based communication is not desirable as per the security concern. The present QKD techniques can achieve low-distance group communications where one trusted server can authenticate the users within the group. Stand-alone QKD communication architecture can be extended to create a wide network using the relays at regular intervals depending on the network topology. Efficient QKD protocols with proper network topologies have been discussed for achieving better security in different eavesdropping scenarios.

Keywords QKD · Quantum network · P2P communication · Topology · Secure group communication

1 Introduction

In 1984 [1], after conceiving the idea of secure communication using qubit, quantum secure communication achieves better security than the classical way by the inherent law of quantum mechanics. Quantum key distribution (QKD) emerges as a useful technique to create a secure communication network between two or more legitimate users even in the presence of the intruder, Eve. The most required part was to make

S. Banerjee (✉)
Kalyani Government Engineering College, Kalyani, West Bengal, India
e-mail: supriyocal@gmail.com

B. Maiti
Maulana Azad College, Kolkata, West Bengal, India

B. Saha
University of Calcutta, Kolkata, West Bengal, India

R. Pandey et al. (eds.), *Quantum Computing: A Shift from Bits to Qubits*,
Studies in Computational Intelligence 1085,
https://doi.org/10.1007/978-981-19-9530-9_11

it more useful in long-distance communication. In case of long-distance communication using fiber optic cable or free space, a repeater may be trusted or un-trusted. Using fiber optic communication up to 100 km secure QKD-based communication can be achieved [2–5]. But due to in-fiber attenuations and atmospheric losses, inappropriate quantum detectors, and the inadequate development of quantum repeaters [6–10], building a network is the next required step to emerge the idea of QKD from point-to-point communication to multipoint communication.

In 2003, a group of researchers from BBN technologies, Boston University and Harvard University engaged in creating a secure network under DARPA [11–13]. Their goal was to create an end to end secure communication between groups of legitimate users using a high-speed QKD network. This was the first step in creating such network using QKD techniques.

In the recent development of satellite-based QKD communication, the requirements of such high-speed secure network emerge. This emerging technique opens a new door for researches from political as well as military point of view.

A group of researchers Sheng-Kai Liao et al. developed QKD-enabled satellite Micius which provides unconditional security using entangle-based QKD techniques from space to ground satellite communication [14]. This development initiates a requirement of a secure network from any countries' internal security point of view.

At present, QKD based network scheme uses virtual private network (VPN) to distribute the secret key between groups of users. Due to the geographic limitation of VPN technology, QKD-based communication suffers a lot. Regular interval-based relays or repeaters are required to install to make this practicable for a wide range of secure communication. But these relays can be trusted or un-trusted. In the practical scenario, a fault-tolerant protocol design is the most required design nowadays to overcome this problem. A good protocol design is extensively dependent on network attributes such as network model and topology [15].

In the case of end-to-end secure communication, the link must be alive throughout the session. But in the practical scenario, due to inefficient implementation of router/relays, increasing link error on communicating path, not proper tamper-resistant repeater, etc., reasons the path has to change. But the frequency of changing this path must be within some limit otherwise it will raise the issues of inefficient design of protocol. Dynamic changing or updating path creates emerging issues in implementing efficient quantum network. Proper network topology adaption can be feasible solutions to overcome this problem.

In case of random link, error occurs more frequently in a network then static topology will not be the better solutions. A dynamic topology can handle this random link error more efficiently. In the quantum-based network, most of the links are entangled which are more susceptible in case of collective, side-channel attacks. In these types of attacks, Eve collects the information by attaching a probe with the entangled photons and in the later part measures collectively. The dynamic topology will provide better security in case of the ability to change the link after detecting the increasing link error in the specific path between two or more nodes.

The commercial-based QKD network requires the interpretability of the devices and software/process. This uniformity can only be achieved by the proper adaption of network models in which every layer's activities will be properly predefined. The layers are responsible for establishing physical connectivity using P2P communication, intra-network secure and error-free communication, defining inter-network routing process creation, and reliable end-to-end connection suppressing all layers activities from users.

In this chapter, the three components of network attributes and their recent developments have been discussed in Section 1. In Sect. 2, the recent applications of these three components in creating an efficient quantum network have been discussed.

2 Network Attributes

It defines the properties of data communication between groups of nodes/users using wired or unwired (free space) link. The major components are.

2.1 Types of Connection

Two or more nodes or users can use a link, the common pathway to transfer data. The two communicating devices can use the same link known as Point-to-Point (P2P) communication. More than two users can use the same link known as multipoint communications.

2.1.1 Point-to-Point Communication

A dedicated link has been reserved for this type of communication. Entire capacity of this dedicated link is reserved for communications.

In case of point-to-point QKD link, two legitimate users exchange the secure data using this link which is completely dedicated to that user. This kind of communication can be implemented using Virtual Private Network (VPN) but distance limitation is the disadvantage of this technique. Two distinct paths are used for this communication.

The block diagram of the point-to-point communication is described in Fig. 1 where QKD Endpoints act as the gateways for both encryption and decryption for key exchange process which needs to be secure.

Using Quantum Teleportation Procedure in the multi-port network two distant parties can exchange information where the in-between nodes acted as trusted relays. Two distant parties A and B use the Bell state $|\beta_{xy}\rangle$ for exchanging information. Both the sending and receiving nodes share an entangled pair

$$|\beta_{00}\rangle = \frac{|0_{N_A}\rangle|0_{N_B}\rangle + |1_{N_A}\rangle|1_{N_B}\rangle}{\sqrt{2}} = \frac{|00\rangle + |11\rangle}{\sqrt{2}},$$

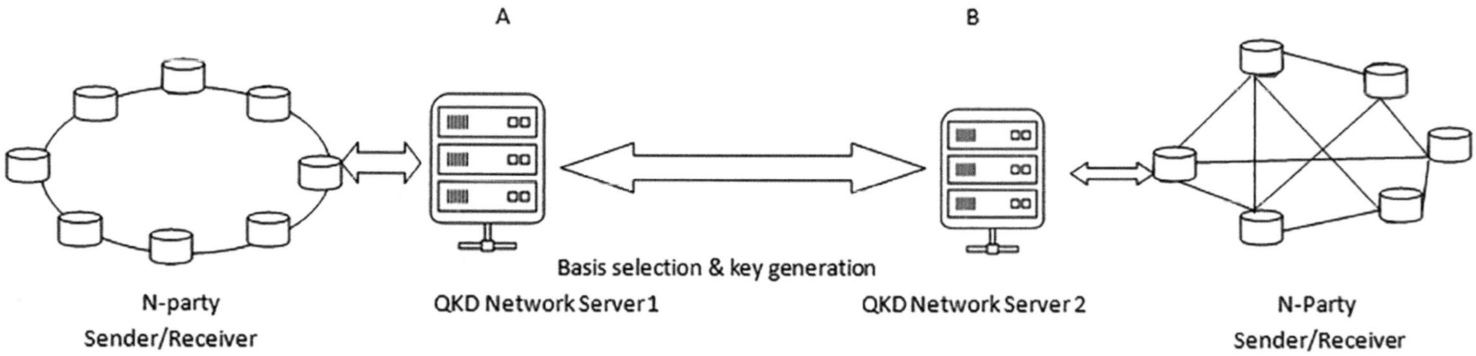

Fig. 1 Basic block of P2P communication in QKD network Server [15]

Table 1 Receiver's unitary operation according to Sender's measurement

Sender's measurement	Receiver's measurement	Unitary operation
$\|00\rangle$	$\frac{1}{2}[(\alpha\|0\rangle+\gamma\|1\rangle)]$	I
$\|01\rangle$	$\frac{1}{2}[(\alpha\|1\rangle+\gamma\|0\rangle)]$	σ_x
$\|10\rangle$	$\frac{1}{2}[(\alpha\|0\rangle-\gamma\|1\rangle)]$	σ_z
$\|11\rangle$	$\frac{1}{2}[(\alpha\|1\rangle-\gamma\|0\rangle)]$	$\sigma_z\sigma_x$

where the first particle belongs to node A and the second belongs to node B. The node A prepares a state

$$\begin{aligned}|\psi\rangle &= (\alpha|0\rangle+\gamma|1\rangle)\otimes|\beta_{00}\rangle = (\alpha|0\rangle+\gamma|1\rangle)\otimes\frac{|00\rangle+|11\rangle}{\sqrt{2}}\\ &= \frac{\alpha(|000\rangle+|011\rangle)+\gamma(|100\rangle+|111\rangle)}{\sqrt{2}}\end{aligned}$$

to send it using an insecure communication fiber optic channel node B. The first two particles belong to the sender node A, whereas the remains last one belongs node B. The sender can interact with the first two qubits using CNOT followed by the Hadamard operation. The resultant state will be

$$\begin{aligned}|\psi'\rangle = H(U_{CNOT}|\psi\rangle) &= H\left(\frac{\alpha(|000\rangle+|011\rangle)+\gamma(|110\rangle+|101\rangle)}{\sqrt{2}}\right)\\ &= H\left(\frac{\alpha|0\rangle(|00\rangle+|11\rangle)}{\sqrt{2}}+\frac{\gamma|1\rangle(|10\rangle+|01\rangle)}{\sqrt{2}}\right)\\ &= \alpha\left(\frac{|0\rangle+|1\rangle}{\sqrt{2}}\right)\frac{(|00\rangle+|11\rangle)}{\sqrt{2}}+\gamma\left(\frac{|0\rangle-|1\rangle}{\sqrt{2}}\right)\frac{(|10\rangle+|01\rangle)}{\sqrt{2}}\\ &= \frac{1}{2}\begin{bmatrix}|00\rangle(\alpha|0\rangle+\gamma|1\rangle)+|01\rangle(\alpha|1\rangle+\gamma|0\rangle)\\ +|10\rangle(\alpha|0\rangle-\gamma|1\rangle)+11(\alpha|1\rangle-\gamma|0\rangle)\end{bmatrix}\end{aligned}$$

In case of $|00\rangle$ measurement at the sender's end, the receiver will use an identity operator to extract the unknown state $|\psi\rangle=\alpha|0\rangle+\gamma|1\rangle$. Table 1 shows the results of sender and receiver's proper measurement. The receiver confirms the correct state with the sender using classical confirmation like public announcement, phone calls, etc.

2.1.2 Multipoint Communications

In this type of communication, multiple nodes share same link. In these types of communication, the link capacity has been shared between the communicating users. The single link shares by the users is more practical and cost-effective but it can breach the data security. In Fig. 2, the initial block diagram of architecture of multipoint network has been shown.

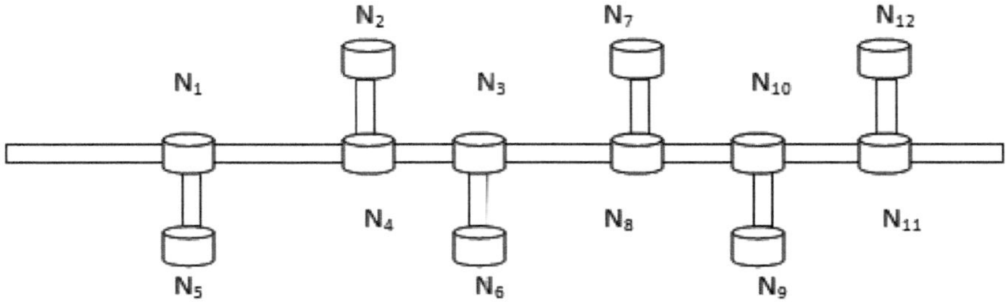

Fig. 2 Basic Block of Multipoint communication in QKD-based network

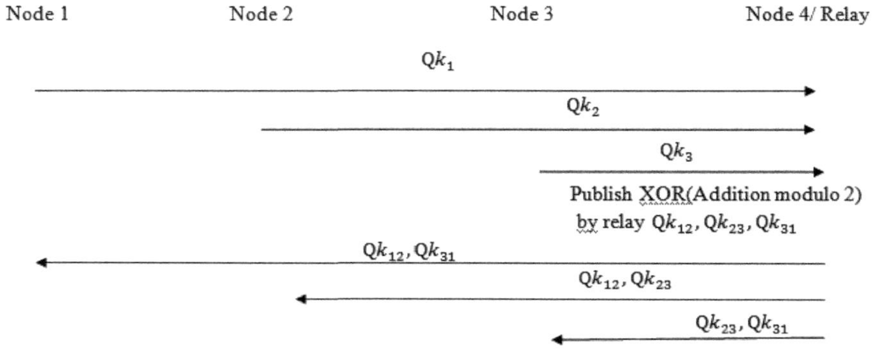

Fig. 3 Multipoint-based trusted relay with public XOR scheme

One or more multipoint trusted relays can be used to exchange the information between two distant parties. In Fig. 2, Nodes N_1, N_2, N_3 are the distant parties where N_4 acts as multipoint trusted relay. Nodes N_1, N_2, N_3 send the prepared key using QKD protocol at trusted relay N_4 which performs XOR (Addition modulo 2) operation $Qk_{ij} = Qk_i \oplus Qk_j$ and publish it. Then it sends the results to the intended receiver where it again performs the reverse XOR operation to extract the keys, i.e., $Qk_i = Qk_{ij} \oplus Qk_j$ and $Qk_j = Qk_{ij} \oplus Qk_i$ (see Fig. 3).

2.2 *Network Topology*

It defines the physical or logical layout or geometrical arrangement of the nodes in a network. Two or more users can form a topology. It can be mixed or hybrid topology to enhance the throughput of the network. Physical topology refers to the actual physical layout of the nodes in a network, whereas logical topology shows flow of data in some instances. Logical topology is dynamic in nature but physical topology is static.

2.2.1 Mesh Topology Network

In mesh topology, one-to-one mapping is done between two nodes. Every node can communicate with other nodes using a dedicated link. In this mesh topology-based network, two types of nodes are used—one is capable of sending and receiving classical encoded data, while other qubits are referred to as classical code and quantum nodes.

In 2018, Xiao-Qin Gao et al. [16] used the mesh topology for creating a quantum teleportation-based network. In this paper, they have shown fault-tolerant one to multi-hop-based communication systems where if one node's link is not working another node can act as a relay to transmit data from one to another. Thus, they solve one of the major problems of mesh-based network. Using some chosen basis, a quantum state is prepared $|\psi\rangle = \alpha|ii\rangle + \beta|ij\rangle + \gamma|ji\rangle + \delta|jj\rangle$ where $|\alpha^2| + |\beta^2| + |\gamma^2| + |\delta^2| = 1$.

In Fig. 4, the quantum teleportation-based network has been shown where the undirected link is referred to as classical link and the directed line are fiber optic communication link for qubit transmission.

Two EPR pairs $|\phi^{\pm}\rangle = \frac{1}{\sqrt{2}}(|ii\rangle \pm |jj\rangle)$, $|\varphi^{\pm}\rangle = \frac{1}{\sqrt{2}}(|ij\rangle \pm |ji\rangle)$ are used for transmitting data for P2P communication. The Bell measurement is performed at the receiver's end to decode the data. A classical channel is used to convey the acknowledgment of the correct basis selection at both ends. Upon receiving the right selection, they create a session for further data transmission using the link. In case of single-hop communication, adjacent nodes share two EPR pairs. Alice sends $|\psi\rangle$ using QKD channel. Alice performs two Bell measurements and sends the results to its adjacent node. Out of 16 results of Bell measurements, Bob has to pick the correct results with proper unitary operation. Table 2 gives the details of the process.

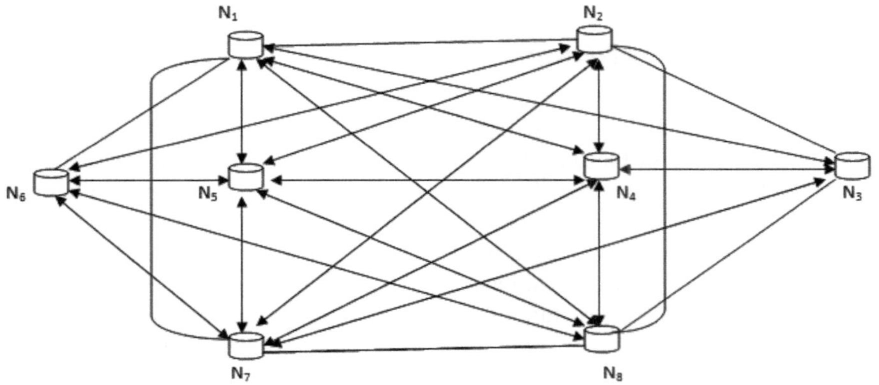

Fig. 4 Mesh Network with 8 nodes [16]

Table 2 Relations between quantum state, bell measurement, and unitary operation

Alice	Bell measurement	Bob	Unitary operation
$\alpha\|00\rangle + \beta\|01\rangle + \gamma\|10\rangle + \delta\|11\rangle$	$\|\phi^{\pm}\rangle_{13}\|\phi^{\pm}\rangle_{25}$	$(\alpha\|00\rangle \pm \beta\|01\rangle \pm \gamma\|10\rangle \pm \delta\|11\rangle)_{46}$	I
	$\|\phi^{\pm}\rangle_{13}\|\varphi^{\pm}\rangle_{25}$	$(\alpha\|01\rangle \pm \beta\|00\rangle \pm \gamma\|11\rangle \pm \delta\|10\rangle)_{46}$	$(\sigma_x)_6$
	$\|\varphi^{\pm}\rangle_{13}\|\phi^{\pm}\rangle_{25}$	$(\alpha\|10\rangle \pm \beta\|11\rangle \pm \gamma\|00\rangle \pm \delta\|01\rangle)_{46}$	$(\sigma_x)_4$
	$\|\varphi^{\pm}\rangle_{13}\|\varphi^{\pm}\rangle_{25}$	$(\alpha\|11\rangle \pm \beta\|10\rangle \pm \gamma\|01\rangle \pm \delta\|00\rangle)_{46}$	$(\sigma_x)_4(\sigma_x)_6$

The corresponding unitary operations are $I = \begin{bmatrix} 1\,0\,0\,0 \\ 0\,1\,0\,0 \\ 0\,0\,1\,0 \\ 0\,0\,0\,1 \end{bmatrix}$, $(\sigma_x)_4 = \begin{bmatrix} 0\,0\,1\,0 \\ 0\,0\,0\,1 \\ 1\,0\,0\,0 \\ 0\,1\,0\,0 \end{bmatrix}$,

$(\sigma_x)_6 = \begin{bmatrix} 0\,1\,0\,0 \\ 1\,0\,0\,0 \\ 0\,0\,0\,1 \\ 0\,0\,1\,0 \end{bmatrix}$, $(\sigma_x)_4(\sigma_x)_6 = \begin{bmatrix} 0\,0\,0\,1 \\ 0\,1\,0\,0 \\ 1\,0\,0\,0 \\ 0\,0\,0\,1 \end{bmatrix}$.

In case of any link failure, one node acts as a relay such as if n_1 to n_3 link is not working then $n_1 \rightarrow n_2 \rightarrow n_3$ will be a new route for transmitting the data. EPR pairs are used to create a session between two nodes using a dedicated link. The multi-hop procedure is also discussed in [16].

The mesh topology has inherent advantages over other topology that it can handle huge load during high traffic as multiple channels can be used to reach from one end to another. Addition or deletion of nodes affects minimum in this topology which makes it more dynamic in nature. However, $\frac{n(n+1)}{2}$ links are required to create n nodes-based mesh topology, which incurs huge cost.

2.3 Star Topology Network

In 2020, Poderini et al. [17] used the directed acyclic graph (DAG) approach to analyze the activity of the star topology-based network. In this chapter, DAG was applied to visualize the graphical and mathematical structure of the complex network in a casual modeling design approach. Figure 5 shows the EPR pair sharing scenario where single source λ_1 is sharing the EPR pair with stations A and B. The Bell measurement-based choice of stations A and B is X, Y, respectively.

In case, the extended structure of three or more independent single sources can form the star topology-based network where the central station or Hub has measured the outcome depending on the Bell basis choice. In Fig. 6, four single sources $\lambda_1, ..., \lambda_4$ form the star network with central station B acting as a Hub gets its measurement outcome using Y chosen basis.

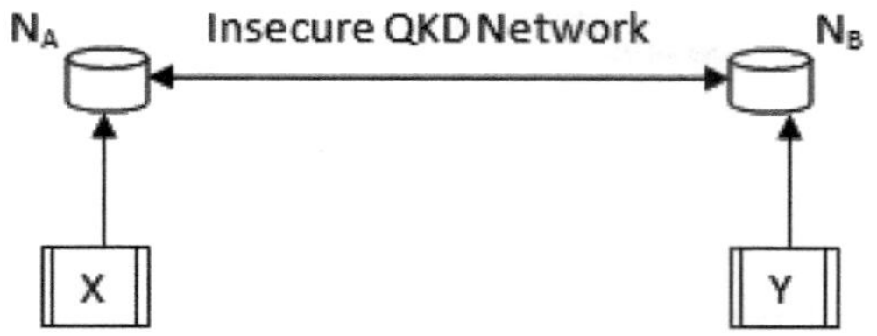

Fig. 5 Directed acyclic graph approach to visualize the network structure

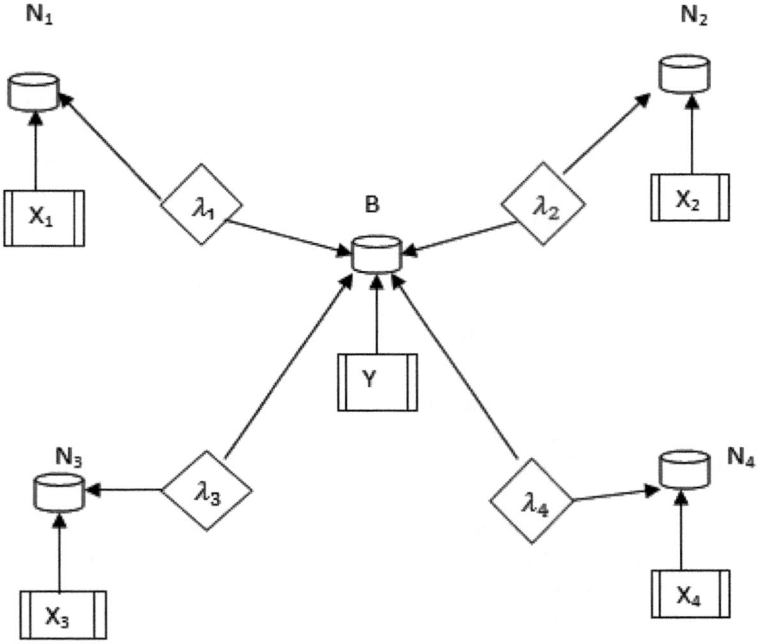

Fig. 6 Star topology-based network with central station B act as a Hub [17]

B^y acts as observable of central Node and $N_i^{x_i}$ in case of external node. For all the four sources, λ_i generates singlet entangled Bell State $|\varphi^-\rangle = (|01\rangle - |10\rangle)/\sqrt{2}$. This will achieve nonlinear Bell inequality $k\cos(\pi/2k)$.

$$S_n^k = \sum_{i=1}^{k} |I_i|^{1/n} \leq k - 1$$

$$\text{where } I_i = \frac{1}{2^n} \sum_{x_1,\ldots,x_n=i-1}^{i} \langle N_1^{x_1} \ldots N_n^{x_n} B^{i-1} \rangle \text{ with } N_i^k = -N_i^0$$

All the external node N_i perform the projective measurement on their own subsystem:

$$\left|\psi_{k,x}^{0}\right\rangle = \cos(x\pi/2k)|0\rangle + \sin(x\pi/2k)|1\rangle, \left|\psi_{k,x}^{1}\right\rangle = \cos(x\pi/2k)|1\rangle - \sin(x\pi/2k)|0\rangle$$

In case of Central Node B, the measurement will be

$$\begin{aligned}\left|\Phi_{k,y}^{0}\right\rangle &= \cos\frac{(2y+1)\pi}{4k}|0\rangle + \sin\frac{(2y+1)\pi}{4k}|1\rangle, \left|\Phi_{k,y}^{1}\right\rangle \\ &= \cos\frac{(2y+1)\pi}{4k}|1\rangle + \sin\frac{(2y+1)\pi}{4k}|0\rangle.\end{aligned}$$

The main idea of this network is to certify the presence of non-classical correlation among the nodes in the network. The n-locality chain inequality will violate the implementation of this correlation. This will strengthen the possibility to increase the number of nodes in a network. A multi-party secretes sharing scheme can be implemented using this strategy.

2.4 Bus Topology Network

In this topology, various nodes share a common channel which is known as backbone cable. When a node or user sends a message, it uses the common link to reach every connected node and only intending nodes or users will receive this and others will reject.

In the bus topology-based quantum network, some nodes create a passive network with fixed energy and are permanently coupled. Other than the nodes which are engaging in the formation of passive network can be used as a backbone cable or common bus to provide data communication between two or more legitimate users or nodes. In the same classical way of bus communication, the excitation or echo is shared among all the nodes using the common bus. After a certain time limit, it is only localized by the destination user or node with the highest probability [18] (see Fig. 7).

In this structure, they use $S_d = \{1, ..., d-1\}$ of d numbers of physical nodes which creates the physical layout where $L \subseteq S_d$ is a subset of physical layout used as a logical layout. In the initial phase all the nodes in ground state $|\phi\rangle^{\otimes d}$, $|l_s\rangle$ and $|l_d\rangle$ denote the source state and destination state, respectively. PST or flow of information can be used in this topology by concatenation of multiple numbers of point-to-point PST links. The single excitation is observed by preserving the total excitation of the system during data transfer from source to destination. $|K\rangle = |\phi\rangle^{\otimes(K-1)}|\psi\rangle|\phi\rangle^{\otimes(d-K)}$ denotes the excitation at the site K and $|\psi\rangle$ is the prepared data for transmission. At time $\mathrm{T} = \varsigma$, the data transfer is performed between two

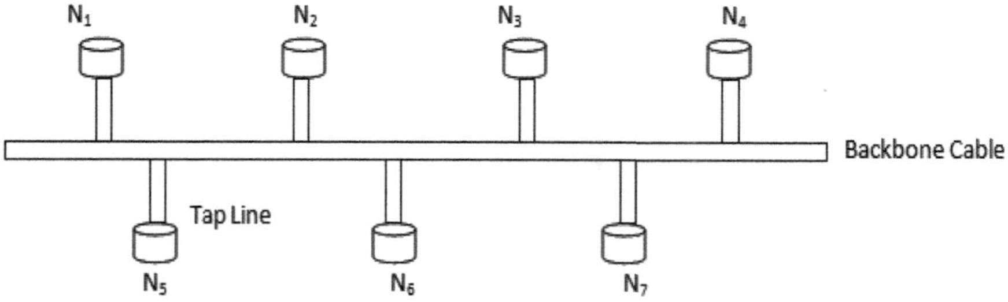

Fig. 7 Common Bus structure with four connecting nodes

nodes or users, i.e., $|l_s\rangle \rightarrow |l_d\rangle$. The order of excitation transferred is generally 0.....d − 1. But same sites are not allowed to get a second time within the time $T = \varsigma$.

In this topology, the inclusion or deletion of users or nodes is easier than in other topologies. But if the backbone cable is much longer then regenerators or amplifiers are required to strengthen the data.

2.5 *Tree Topology Network*

It is an extended form of Bus Topology. In the tree structure, higher level nodes get higher priority and leaf nodes get the least priority. The root node point is generally known as the head node. Two or more major branches emerged out from the head nodes. These branches can create other sub-branches. In Fig. 8, the basic structure of tree topology is shown.

In 2020, Dong et al. [19] published a quantum key distribution process using a complete binary tree topology with both trusted and un-trusted relays. In this paper, two types of channels are used—quantum channel and physical channel for carrying the qubit and classical bit, respectively. The quantum key generation rate between two nodes depends on the distance, environment, and physical properties of the quantum channel. It is purely a key transport protocol with multi-relay-based trusted network. In the physical design of this network, they have proposed five modules—transceiver for transmitting and receiving qubit signals, post-process module for the post-processing, Key Store (KS) module for storing and managing the keys, beam splitter used for splitting the signal, and encryption and decryption module used for encrypting and decrypting the keys. N_0 node used top-down approach to distribute the unique global key gk for secure communication. N_5 uses the beam splitter to split the signals sk_1, $sk_{2,}$ and sk_3 and store those signals in the KS module. Then the encrypted key $gk + sk_1$ is again encrypted by $sk_1 + sk_2$ and $sk_1 + sk_3$. The N_5 transmits the encrypted data to N_{10}, $N_{11,}$ *and* N_{14}. The intended user sends the acknowledgment using classical channel. In this process, N_0 can send the global key gk to N_{10}, $N_{11,}$ and N_{14} but the consumption of key will be reduced to one between N_0 and N_5 which indicates the higher efficiency of the protocol.

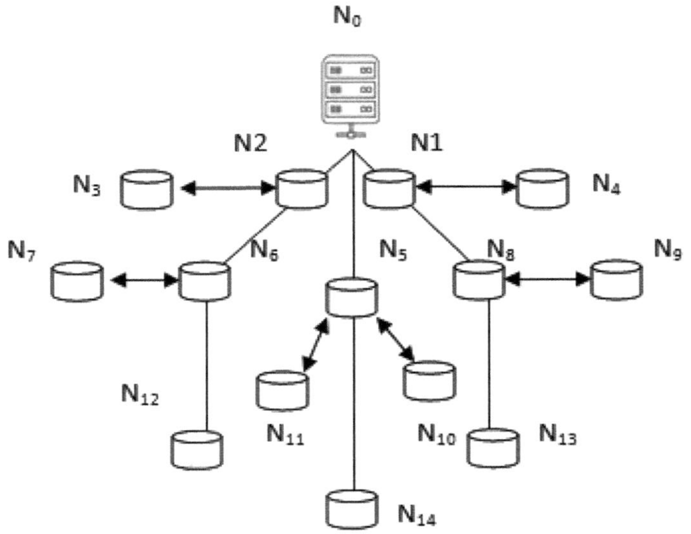

Fig. 8 Tree Topology with degree ≤ 3

2.6 Ring Topology Network

In this network, some nodes create a closed loop where each node has a dedicated point-to-point connection with its neighbor. Basic ring topology data movement is unidirectional which is susceptible to link failure problem. Bi-directional data movement provides fault-tolerant computation in this type of network.

In 2002, Hutton et al. [20] published a paper based on the performance comparison between star and ring topology-based quantum networks.

In this article, the ring network comprises six users or nodes. The distance between two parties and the radius of the network is d and R, respectively.

$E(d, n)$ is a function for measuring the entanglement between two parties of n wire length and each has length $d = 2R\sin\left(\frac{\pi}{6}\right)$.

In case of ring network, $E_{\text{avg}}^{\text{ring}}$ defines the averaged distributed entanglement between two communicating parties $E_{\text{avg}}^{\text{ring}} = \frac{1}{N-1}\left[2\sum_{n=1}^{U} E\left(2R\sin\left(\frac{\pi}{N}\right), n\right) + \mu E\left(2R\sin\left(\frac{\pi}{N}\right), N/2\right)\right]$, where $U = (N-1)/2$ or $N/2 - 1$, $\mu = 0$ or 1 if N is odd or even, respectively.

In this network, the longest communication will occur when the two communicating parties are in half of the network apart from themselves. The channel acts on state as $|\psi^+\rangle\langle\psi^+| \to \Lambda|\psi^+\rangle\langle\psi^+| + (1-\Lambda)|\phi^+\rangle\langle\phi^+|$, where $\Lambda = \frac{1+e^{-d}}{2}$, the related channel length. Figure 9 describes the general structure of ring topology. Undirected path shows bidirectional movement of data from source node to destination node.

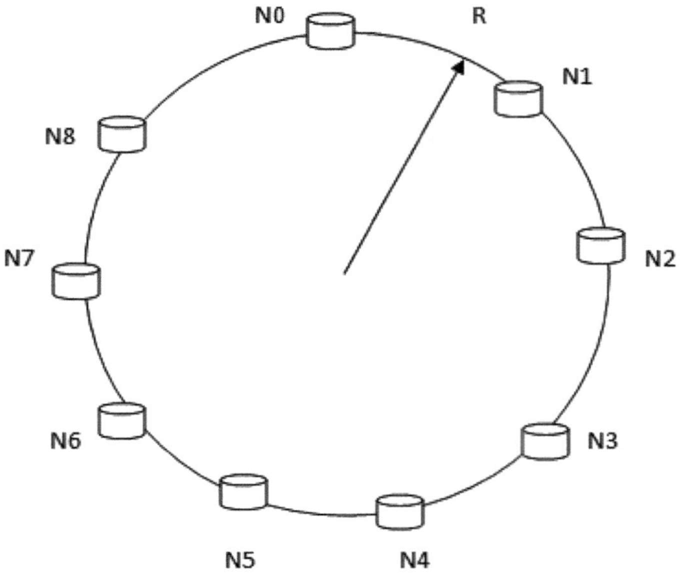

Fig. 9 Bidirectional quantum ring topology network

2.7 *Hybrid Topology*

It is an integrated formation of two or more different kinds of topologies with mixed advantages and disadvantages of the specific topologies. One of the hybrid topologies is lattice-based network.

In this figure, lattice structure is used for the formation of quantum network. Using entanglement swapping at intermediate nodes (Fig. 10), the legitimate users N_1 and N_{12} share the data. Initially, they use flooding techniques to send the initial echo state to all the nodes. After getting the acknowledgment from proper receiver, the path is usually created between sender and receiver. The swapping occurs between two adjacent nodes using EPR pairs. Individual EPR pair can be labeled $|\beta_{00}\rangle_{12} = \frac{|00\rangle_{12}+|11\rangle_{12}}{\sqrt{2}}$ where qubits 1 and 2 are entangled each other.

Another EPR pair $|\beta_{00}\rangle_{34} = \frac{|00\rangle_{34}+|11\rangle_{34}}{\sqrt{2}}$ where qubits 3 and 4 are entangled.

The product state can be expressed as

$$|\beta_{00}\rangle_{12}|\beta_{00}\rangle_{34} = \left(\frac{|00\rangle_{12}+|11\rangle_{12}}{\sqrt{2}}\right)\left(\frac{|00\rangle_{34}+|11\rangle_{34}}{\sqrt{2}}\right)$$
$$= \frac{1}{2}(|00\rangle_{12}|00\rangle_{34} + |00\rangle_{12}|11\rangle_{34} + |11\rangle_{12}|00\rangle_{34} + |11\rangle_{12}|11\rangle_{34})$$

Using some arrangements, this can be expressed as

$|\beta_{00}\rangle_{12}|\beta_{00}\rangle_{34} = \frac{1}{2}(|00\rangle_{14}|00\rangle_{23} + |01\rangle_{14}|01\rangle_{23} + |10\rangle_{14}|10\rangle_{23} + |11\rangle_{14}|11\rangle_{23})$.

But,

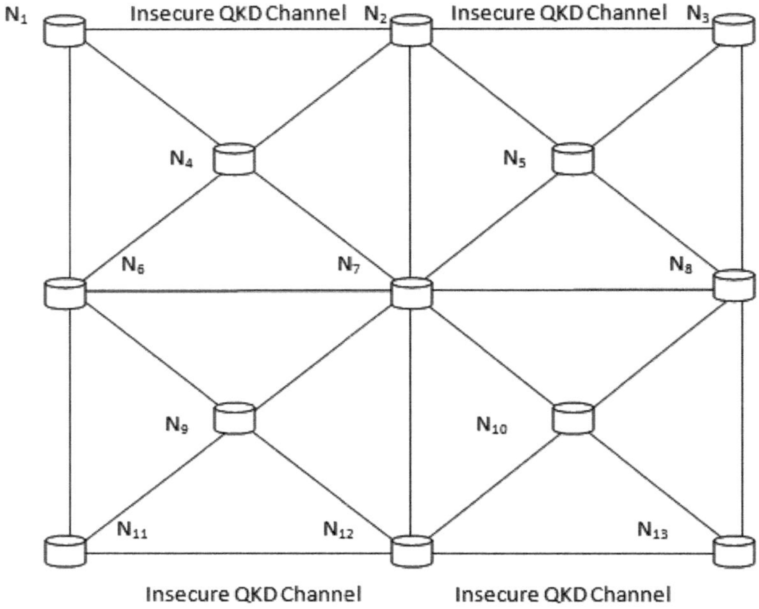

Fig. 10 Lattice (Hybrid) topology for connecting node N_1 to Node N_{12} [21]

$$|\beta_{00}\rangle_{14}|\beta_{00}\rangle_{23} = \left(\frac{|00\rangle_{14}+|11\rangle_{14}}{\sqrt{2}}\right)\left(\frac{|00\rangle_{23}+|11\rangle_{23}}{\sqrt{2}}\right) = \frac{1}{2}\begin{pmatrix}|00\rangle_{14}|00\rangle_{23}+|00\rangle_{14}|11\rangle_{23}\\ +|11\rangle_{14}|00\rangle_{23}+|11\rangle_{14}|11\rangle_{23}\end{pmatrix}$$

$$|\beta_{00}\rangle_{12}|\beta_{00}\rangle_{34} = \frac{1}{2}\begin{pmatrix}|\beta_{00}\rangle_{14}|\beta_{00}\rangle_{23}+|01\rangle_{14}|01\rangle_{23}+|10\rangle_{14}|10\rangle_{23}\\ -|00\rangle_{14}|11\rangle_{23}-|11\rangle_{14}|00\rangle_{23}\end{pmatrix}$$
$$= \frac{1}{2}\left(|\beta_{00}\rangle_{14}|\beta_{00}\rangle_{23}+|\beta_{10}\rangle_{14}|\beta_{10}\rangle_{23}+|\beta_{01}\rangle_{14}|\beta_{01}\rangle_{23}+|\beta_{11}\rangle_{14}|\beta_{11}\rangle_{23}\right)$$

Sender makes Bell measurement on particles 1 and 4 with possible outcomes $|\beta_{00}\rangle_{14}, |\beta_{10}\rangle_{14}, |\beta_{01}\rangle_{14}, |\beta_{11}\rangle_{14}$ each with uniform probability is $\frac{1}{4}$. The corresponding receiver's state will collapse into one of the $|\beta_{00}\rangle_{23}, |\beta_{10}\rangle_{23}, |\beta_{01}\rangle_{23}, |\beta_{11}\rangle_{23}$ according to Alice measurement. Figure 11 describes the entanglement swapping procedure for lattice-based network topology.

In this lattice-based structure, multiple paths may be possible to reach from sender to receiver but less number of intermediate nodes is the only criterion to choose the next path. As the number of intermediate nodes increase, the secrecy rate decreases.

But, using the hybrid topology, it is easier to increase the network and it is more flexible according to the requirement of the computational laboratories and its position. Detection upon fault or eavesdropping, the affected part can be easily isolated from the rest of the network. But, it will be more complex than others and costly.

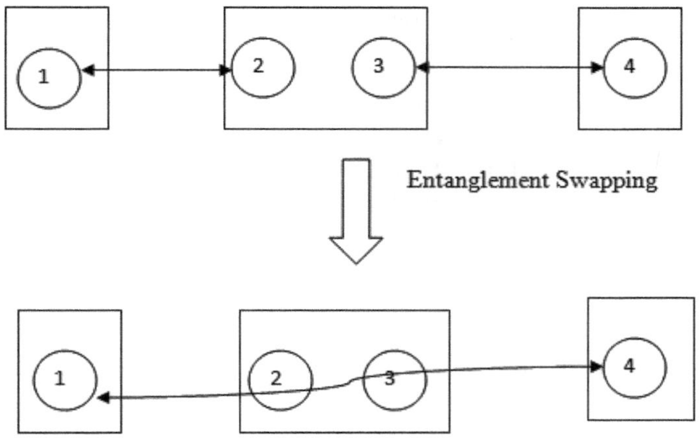

Fig. 11 Entanglement swapping can connect to distant parties using some intermediate nodes [22]

3 Current Status of the Quantum Network and Its Development

After the remarkable progress in secure multi-party communication, the idea of quantum network-based communication has evolved. In October 2003, BBN technologies first implemented a fully operational six nodes quantum network popularly known as DARPA Network. In the year 2004, using fiber optic communication they created a network between Harvard University, Boston University, and BBN successfully. In the later year, Secure Communication on Quantum Cryptography (SECOQC) has been commissioned by European Commission including 11 European Countries with Russia. AVPN tunnel-based communication has been set up to communicate within the nodes which provides added security. After these remarkable achievements, Japan introduced Japan Giga Bit Network 2 plus which is BB84 protocol-based secure fiber optic network. In the later part, most of the networks developed by the Chinese researcher using Decoy state-based BB84 which are more realistic as single-photon generation is not proper due to imperfection of the photon devices. In 2017, the Japanese Physicists Hideki Takenaka et al. from the National Institute of Information and Communications Technology successfully implement the microsatellite-based secure communication using QKD which created a new era in the area of secure long-distance communication. After these remarkable achievements, microsatellite CubeSat-based quantum network has been set up using point-to-point communication. This creates a global network using satellite-based communication.

Year	Name of the network	Channel type	QKD network type	QKD protocol	Topology
2002–06	DARPA network [22–24]	Fiber optic channel	Trusted node	BB84 protocol [1]	Bus
2006	Butterfly network [25]	Fiber optic channel	Trusted node		Hybrid
2004–08	SECOQC network [26]	Fiber optic channel	Trusted node	Decoy state BB84 [39], SARG04 [40], COW [41], BBM92 [42], CV-QKD [43]	Mesh + Bus
2009	Beijing network [27]	Fiber optic channel	Trusted node	BB84 [1]	Star
2010	TokyoUQCC network [28]	Fiber optic channel	Trusted node	BB84 [1] + BBM92 [42]	Bus + Ring
2011	Geneva QKD network [29]	Fiber optic channel		BB84 [1] and SARG protocol [40]	Ring
2014	Hefei-Chaohu-Wuhu QKD network [30]	Fiber optic channel	Trusted node	Decoy state BB84 [39]	Bus
2016	High-data bandwidth CV QKD link [31]	Fiber optic channel	Trusted + un-trusted	MDI QKD [45]	Star
2017	Beijing-Shanghai [32]	Fiber optic channel	Trusted repeater	Decoy state BB84 [39]	Bus
2017	Jinan private network [33]	Fiber optic channel	Trusted repeater	Decoy state BB84 [39]	Tree
2017	Wuhan Metropolitan network [34]	Fiber optic channel	Trusted repeater	Decoy state BB84 [39]	Tree
2018	Zhucheng–Huangshang QKD link [35]	Fiber optic channel	Trusted repeater	Decoy state BB84 [39]	Bus
2019	Cambridge quantum network [36]	Fiber optic channel	Trusted node	DV-QKD [44]	Ring
2020	CubeSat-based global QKD network [37]	Fiber optic channel + free space	Trusted node	2-decoy state protocol [39]	Point-to-point
2021	Satellite based QKD network [38]	Free space	Trusted repeater		P2P + dynamic topology

References

1. C.H. Bennett, G. Brassard, Quantum cryptography: public key distribution and coin tossing, in *Proceedings of the IEEE International Conferences on Computers, Systems and Signal Processing* (Dec. 1984), pp. 175–179
2. M. Campagna, L. Chen, Ö. Dagdelen, J. Ding, J.K. Fernick, N. Gisin, D. Hayford, T. Jennewein, N. Lütkenhaus, M. Mosca, B. Neill, M. Pecen, R. Perlner, G. Ribordy, J.M. Schanck, D. Stebila, N. Walenta, W. Whyte, Z. Zhang, *Quantum Safe Cryptography and Security*, vol. 8 (2015), pp. 1–63
3. N. Gisin, G. Ribordy, W. Tittel, H. Zbinden, Quantum cryptography. Rev. Mod. Phys. **74**, 145 (2002)
4. W.K. Wootters, W.H. Zurek, A single quantum cannot be cloned. Nature **299**, 802–3 (1982)
5. H. Takesue, et al., Quantum key distribution over a 40-dB channel loss using superconducting single-photon detectors. Nat Photonics **1**, 343–348 (2007)
6. D. Stucki et al., High rate, long-distance quantum key distribution over 250 km of ultra low loss fibres. New J. Physics **11**, 075003 (2009)
7. Y. Liu, et al., Decoy-state quantum key distribution with polarized photons over 200 km. Opt. Express **18**, 8587–8594 (2009)
8. S. Aleksic, D. Winkler, G. Franzl, A. Poppe, B. Schrenk, F. Hipp, Quantum key distribution over optical access networks, in *Proceedings of 18th European Conference on Network and Optical Communications & 8th Conference on Optical Cabling and Infrastructure (NOC-OC&I)* (2013)
9. V. Scarani, et al., The security of practical quantum key distribution, Rev. Modern Phys. **81**, 1301–1350 (2009)
10. E. Kerstel, A. Gardelein, M. Barthelemy, M. Fink, S.K. Joshi, R. Ursin, Nanobob: a cubesat mission concept for quantum communication experiments in an uplink configuration. EPJ Quantum Technol. **5** (2017)
11. C. Elliott, Building the quantum network. New J. Phys. **4**, 46 (2002)
12. C. Elliott, D. Pearson, G. Troxel, Quantum cryptography in practice, in *Proceedings of the ACM SIGCOMM2003*
13. C. Elliott, The DARPA quantum network," forthcoming book chapter. Springer Verlag
14. Tiangong2,*China Space Report* (2017)
15. C. Elliott, Building the quantum network. New J. Phys. **4** (Jan. 2002)
16. X.-Q. Gao, Z.-C. Zhangy, B. Sheng, Multi-hop teleportation in a quantum network based on mesh topology. Front. Phys. **13**(5), 130314 (2018)
17. D. Poderini et al., Experimental violation of n-locality in a star quantum network. Nat. Commun. **11**, 2467 (2020)
18. T. Brougham et al., Communication in quantum networks of logical bus topology. Phys. Rev. A **80**, 052325 (2009)
19. K. Dong, et.al., Tree-topology-based quantum-key-relay strategy for secure multicast services. J. Opt. Commun. Netw. **12**(5) (May 2020)
20. A. Hutton et al., Comparison of star and ring topologies for entanglement distribution. Phys. Rev. A **66**, 032320 (2002)
21. S. Das et al., Robust quantum network architectures and topologies for entanglement distribution. Phys. Rev. A **97**, 012335 (2018)
22. W. Kozlowski, et.al., Designing a quantum network protocol, in *CoNEXT '20: Proceedings of the 16th International Conference on emerging Networking Experiments and Technologies* (Nov. 2020), pp. 1–16
23. A. Pirker, et.al., A quantum network stack and protocols for reliable entanglement based networks. New J. Phys. **21**, 033003 (2019)
24. C. Elliott, et al., Current status of the DARPA quantum network, in *Proceedings of SPIE-The International Society for Optical Engineering* (2005)
25. M. Hayashi, et al., Quantum network coding, in *Annual Symposium on Theoretical Aspects of Computer Science STACS 2007* (2007), pp. 610–621

26. M. Peev, C. Pacher, et al., The SECOQC quantum key distribution network in Vienna. New J. Phys. **11**, 75001 (Jul. 2009)
27. T.-Y. Chen, et.al., Field test of a practical secure communication network with decoy-state quantum cryptography. Opt. Expr. **17**(8), 6540 (Apr. 2009)
28. M. Sasaki, et al., Field test of quantum key distribution in the Tokyo QKD network. Opt. Expr. **19**(11), 10387 (May 2011)
29. D. Stucki, et al., Long term performance of the SwissQuantum quantum key distribution network in a field environment. New J. Phys. **13**(12) (Mar. 2012)
30. S. Wang, et al., Field and long-term demonstration of a wide area quantum key distribution network. Opt. Express **22**(18) (Sep. 2014)
31. a Z. Qu, et al., High-speed free-space optical continuous-variable quantum key distribution enabled by three-dimensional multiplexing. Opt. Express **25**(7), 7919–7928 (2017). b L.C. Comandar, et al., Quantum key distribution without detector vulnerabilitiesusing optically seeded lasers. Nat. Photon **10**, 312–315 (2016)
32. Q. Zhang, et al., Large scale quantum key distribution: challenges and solutions. Opt. Expr. **26**(18), 24260 (Sep. 2018)
33. ChinaDaily, Quantum tech to link Jinan governments (2017). http://www.chinadaily.com.cn/china/2017-07/11/content_30065215.htm
34. Y. Zhao, The integration of QKD and security services. In *Proceedings of the ITU QIT4N Workshop Shanghai* (2019). https://www.itu.int/en/ITU-T/Workshops-and-Seminars/2019060507/Documents/Yong
35. Y. Mao et al., Integrating quantum key distribution with classical communications in backbone fiber network. Opt. Express **26**, 6010 (2018)
36. a C.-W. Tsai, et al., Quantum key distribution networks: challenges and future research issues in security. Appl. Sci. **11**, 3767 (2021). b J.F. Dynes, et al., Cambridge quantum network. npj Quantum Inf. (2019)
37. E. Aykut, et al., Feasibility Study for CubeSat Based Trusted Node Configuration Global QKD Network. arXiv:2102.13536 [quant-ph]
38. Y.-A. Chen et al., An integrated space-to-ground quantum communication network over 4,600 kilometres. Nature **589**, 214–219 (2021)
39. H.-K. Lo, et al., Decoy state quantum key distribution. Phys. Rev. Lett. **94**, 230504 (2005)
40. V. Scarani, A. Acin, G. Ribordy, N. Gisin, Quantum cryptography protocols robust against photon number splitting attacks for weak laser pulse implementations. Phys. Rev. Lett. 92057901 (2004)
41. C. Kollmitzer, M. Pivk, *Applied Quantum Cryptography*, vol. 797. Springer Science Quantum Key Distribution: A Networking Perspective (2010)
42. C.H. Bennett, Quantum cryptography using any two nonorthogonal states. Phys. Rev. Lett. **68**, 3121 (25 May 1992)
43. F. Laudenbach, et al., Continuous-variable quantum key distribution with gaussian modulation–the theory of practical implementations. https://doi.org/10.1002/qute.201800011, Corpus ID: 51419358
44. I.B. Djordjevic, Discrete variable (DV) QKD, in *Physical-Layer Security and Quantum Key Distribution* (Springer, Cham, 2019). https://doi.org/10.1007/978-3-030-27565-5_7
45. W. Wang, et al., Measurement-device-independent quantum key distribution with leaky sources. Sci. Rep. **11** (2021). Article number: 1678

Quantum Computing-Enabled Machine Learning for an Enhanced Model Training Approach

Jayesh Soni, Nagarajan Prabakar, and Himanshu Upadhyay

Abstract Machine learning is an exciting area where ever-growing problems such as anomaly detection using sensor data, natural language processing, image processing, etc., are solved using complex yet fascinating algorithms. Such algorithms learn the function that maps input to output from the training examples. The algorithms are evaluated with the validation data and then used to predict the output for an unknown dataset. For the past few years, scholars have been researching to improve such classical machine learning algorithms using quantum computing. Some of the latest research is the optimization of computationally expensive algorithms with quantum computing and transforming stochastic procedures into the semantics of quantum theory. The quest for the learning-based algorithm is aspiring: the discipline seeks to comprehend what learning is and studies how algorithms approximate learning. Quantum machine learning takes these aspirations further by looking at the subatomic level to aid learning. Machine learning-based algorithms minimize a constrained multivariate function. Different algorithms have different hyper-parameters that need to be tuned for the trained model to generalize well. Such optimization has high time and space complexity, which is central to learning theory. This contribution gives an organized overview of the evolving arena of quantum machine learning. It presents the methods as well as practical details. We start this chapter by introducing the major components of Quantum Computers, where we provide an overview of quantum computing with an in-depth explanation of the superposition of state, which will be crucial for all quantum algorithms. Next, we exploit a fascinating phenomenon called entanglement in quantum computations. Parallelism is the key to speeding up the training process of learning algorithms. One of the significant advantages of quantum

J. Soni (✉) · N. Prabakar
Knight Foundation School of Computing and Information Sciences, Florida International University, Miami, FL, USA
e-mail: jsoni@fiu.edu

N. Prabakar
e-mail: prabakar@cis.fiu.edu

H. Upadhyay
Applied Research Center, Florida International University, Miami, FL, USA
e-mail: upadhyay@fiu.edu

R. Pandey et al. (eds.), *Quantum Computing: A Shift from Bits to Qubits*,
Studies in Computational Intelligence 1085,
https://doi.org/10.1007/978-981-19-9530-9_12

computing is quantum parallelism which we will explore through Grover's search algorithm. Next, we go over the learning mechanism of a traditional machine learning algorithm, namely a Support Vector Machine (SVM) trained on classical computers. Furthermore, we exploit the Quantum SVM (QSVM) trained on quantum computers. Finally, we solve the famous malignant breast cancer classification problem using the quantum SVM algorithm. We study various simulations in Python language on the mentioned dataset to analyze their time complexity and performances on standard evaluation metrics, namely accuracy, precision, recall, and F1-score. Two simulations are conducted using classical machine learning algorithms using the Python library Scikit-learn. Finally, the last simulation is based on IBM's real quantum computer using its quantum machine learning library called Qiskit.

Keywords QSVM · Breast cancer classification · Quantum computing

1 Introduction

In the modern world, machine learning algorithms made a big impact in almost all industries due to their predictive capability. Such algorithms need big data with high computational power for training purposes. There are multiple hyper-parameters needed to be tuned to get the optimized model. Training the model with multiple hyper-parameters requires high computational power and is highly time-consuming. This leads to the notion of utilizing quantum mechanics for computation purposes which can reduce the search time for optimal hyper-parameters. Richard Feynman in 1982 researched that with the increase in system size, the simulation of quantum theory on traditional computers becomes highly infeasible, but the quantum particles would not suffer from the same limitations [1]. The idea is generalized by David Deutsch, who represents quantum computers as universal Turing machines with an experiment on quantum parallelism. He noted that some particular probabilistic-based tasks could be implemented quicker than by any traditional means of computation [2]. There are three main branches of quantum information currently. They are quantum cryptography, quantum computing, and quantum information theory. Quantum cryptography is the study of secure communication for information exchange. Quantum information theory deals with storing and transmitting data coded in quantum states. Quantum Computing is the study of the operation of quantum phenomena on the data denoted in quantum states. Superposition, Entanglement, and Interference are different quantum phenomena. Many quantum-based algorithms were created decades later; the first suggested theory of quantum computing. Shor created an algorithm that runs exponentially faster in factorizing integers [3], and Grover developed an algorithm that runs quadratically faster than the traditional computation in finding an item from an unordered list [4]. Even with such advancements in the quantum algorithms from the pioneers, the development is still complex [5]. Since quantum computing is still unable to solve the problems of the class NP in sub-exponential time [6], it is feasible to experiment at a small scale with few qubits and restrictions to particular domain problems. For example, the experiment optimizing the combinatorial problem with 500 binary variables shows a higher speedup on a quantum computer than a traditional one [7]. However, their experimental result is somewhat controversial [8]. Thus, in this chapter, we study the effect of quantum algorithms over the classical algorithms using an open-source dataset. The rest of the chapter is summarized as follows. Section 2 discusses quantum computers. Section 3 highlights the computation difference between the traditional computer and the quantum computer. Section 4 explains the quantum phenomenon. Section 5 discusses the quantum algorithms. Section 6 discusses SVM for traditional and quantum computers. Section 7 highlights various libraries. Section 8 explains various evaluation metrics. The proposed framework is discussed in Sect. 9 and, finally, we conclude in Sect. 10.

2 Why Quantum Computers

Small Particles behave differently than the objects we encounter in our day-to-day life. Quantum theory studies the behavior of matter and light at atomic and subatomic levels. One of the earliest discoveries that hold the argument for quantum computing is the calculations for integer factoring and element search, which can be done exponentially and quadratically faster, respectively, with quantum computation than classical computation. Such astonishing theoretical results lead to an increase in quantum-based algorithms' research. However, it is very challenging to detect particular applications for quantum computing. Furthermore, the design of an optimal quantum-based algorithm is an additional challenge. The main problem is not on the side of quantum mechanics but instead on the ideal thought that the specific problem can be solved exponentially faster using a quantum-based algorithm than traditional classical computers. According to our recent research into this field, machine learning seems to be one of the applications for quantum computing. Machine learning is nothing but the optimization of the multivariate function and thus can relate to quantum computation. A quantum device has to be used to perform such quantum computation. Such a quantum device triggers the qubit states in a controlled fashion to perform computation. Even with the countless errors in scaling the approach, this particular form of learning has been demonstrated with good results. Quantum neural networks allow a high level of abstraction with qubit implementation. Quantum-like learning is the study of quantum mechanics in designing and analyzing algorithms suited for classical computers. Superposition and entanglement are some of the behavior of such quantum mechanics. There are many practical applications for such type of learning in the systems that show contextual behavior. Failure of the traditional models in the correlation findings leads to the successful exploitation of the entanglement phenomenon [9], whereas the phenomenon of obtaining unusual outcomes when multiple attributes are combined is the quantum superposition [10]. Particle swarm optimization has its quantum version [11]. In this algorithm, the particles act like an agent that can show multiple

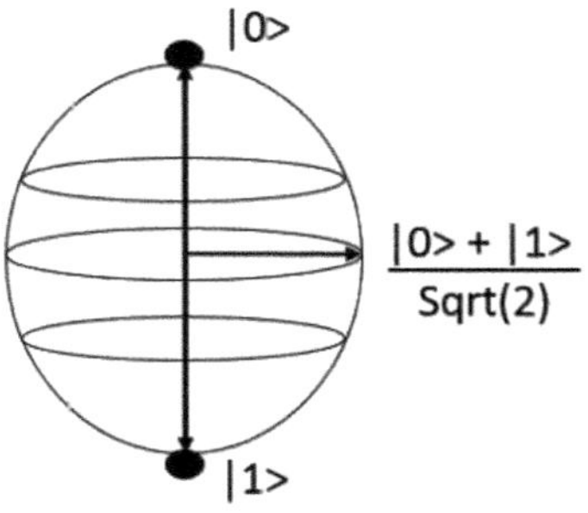

Fig. 1 Qubit representation

patterns. Each such patterns exhibit a particular solution. One of the main future goals of high-performance computing is to design an algorithm that can run efficiently on parallel computing systems quickly. Quantum-like learning brings this ease of algorithmic design to machine learning.

3 Classical Bits and Qubits

In a traditional computer, information is encoded in a classical bit, which can be either 0 or 1 at a time, whereas, in quantum computing, a quantum bit encodes the information and can be 0 or 1 at the same time. A single classical bit takes one value at a time, whereas a single qubit can take two values simultaneously due to the superposition phenomenon. The qubit states are represented as |0> and |1> as shown in Fig. 1.

Every operation is performed only for the current one value of the classical bit, whereas in qubit, it can be done on all values of the qubit at the same time. Thus, a single qubit can store more information than a single classical bit.

Let us understand this with a simple example. Suppose we have two classical bits, then a total of four operations can be performed with one value for each operation at a time (00, 01, 10, 11), whereas all the four operations can be performed simultaneously using two quantum bits.

4 Quantum Phenomenon

Superposition and Entanglement are two of the exciting quantum phenomena.

4.1 Superposition

The quantum bit is different from the classical bit due to its superposition phenomenon. It is one of the natural principles of quantum mechanics. In traditional physics, the tone of the music is described in a wave format where multiple distinct waves are joined together linearly. In the same way, a quantum bit in its superposition state can be thought of as the linear combination of various unique quantum states. The double-slit is the one famous experiment to visualize the superposition state. At a particular time, a single qubit can be in superposition with both the states |0> and |1>. Upon observing the qubit for further calculation, it will fall into one of the values reflecting the states at the period of time. For instance, during its superposition with equal weights for both states, a measurement taken from a qubit can have an equal chance of being |0> and |1>. |0> state if collapsed upon measurement gives the result 0, and |1> state gives the result 1. A quantum computer with n qubits can have 2^n values and thus it can be in superposition that can represent all 2^n superposition states simultaneously.

4.2 Entanglement

One of the other exciting phenomena of quantum physics is entanglement. A group of elements or particles is considered to be in an entangled state when the quantum state of one particle in a group can be used to detect or identify the quantum state of another particle in the same group. There exists a unique connection when two particles are entangled together. As we know that the qubit outcome can be either 0 or 1 upon the measurement. However, when the measurement of the outcome of a particular element is taken, the measurement of the other element in the same group is always correlated. Fundamentally, this case is universal and occurs every single time, even if there is a significant distance gap between such particles in the group. Let us look at this phenomenon in the spin direction. Let us assume that the measurement of the spin of a particle on a particular axis results in counterclockwise, and then it is always true that the spin measurement of another particle on the same axis will also be in a counterclockwise direction if those two particles are in entanglement. It seems strange since it appears like there is an exchange of information, but this is not the case. This phenomenon is still valid even if the particles are billions of miles away from each other. A common misconception is that this particular quantum physics phenomenon can be used to transfer information spontaneously. However, this is not possible because though it is possible to know the state of the particular particle upon measurement, such measurement results are always random. There is no algorithmically predetermined way to get the correct discrete result.

5 Quantum Algorithms

Algorithms are a series of instructions executed step by step to accomplish a particular task. All classical algorithms can be executed on the quantum computer. Quantum algorithms are where at least one of the instructions is specifically "quantum," meaning it has to run either in the superposition or entanglement state of quantum physics. Such quantum algorithms cannot be executed on classical computers; instead, it needs their circuit, a.k.a. quantum circuits.

A quantum circuit is a quantum computational model where quantum gates are utilized to execute the quantum steps of an algorithm. The quantum steps are performed on qubits using quantum gates. The quantum state of the qubit changes upon the application of the quantum gate. These quantum gates can be divided into single qubits or two qubits, depending on the number of qubits on which they were executed. The quantum circuit is said to be concluded once the measurement is taken from one or more qubits.

In theoretical computer science, there are many problems known universally and cannot be solved by traditional computation methods. Such problems are called undecidable problems, which even quantum computers cannot solve. The significant advantage of quantum algorithms over the traditional algorithm is that they can solve some problems numerically faster. Shor's and Grover's algorithms are among those quantum algorithms.

Shor algorithm performs integer factorization that runs exponentially faster than the classical known algorithm. Grover algorithm can search quadratically faster in an unstructured database than the best-known algorithm.

5.1 *Grover Algorithm*

Grover algorithm solves the problem of searching for unstructured data mathematically. It is a quantum algorithm that finds the input value X0 for the function f(x), where f (X0) = 1 and f (X) = 0 for all remaining X. The significant advantage of this algorithm is that it runs in O(sqrt(N)) time complexity for the function with N unordered values where the time complexity for the same operation reaches O(N) when executed on a classical computer.

The search for a specific card in the unordered deck of N cards is one such application of this algorithm. This can be done using the classical way where we search the card in the deck one by one. This operation will take N steps (For N cards in the deck). The same task can be performed using Grover's algorithm in only (O(sqrt(N)) steps. However, the caveat is that the probability of finding the correct result is not 100%.

6 Machine Learning

Machine learning algorithms learn the patterns from the dataset for prediction. There are three types of machine learning algorithms, as shown in Fig. 2.

Supervised Learning: Input data with a label is used for training purposes. Based on the target label value, such a learning algorithm is divided into classification and regression. Classification is when the output label is categorical, and the problem is a regression when the output label is a real value. Decision trees, random forest, K-nearest neighbor, support vector machine, and linear regression are supervised learning algorithms.

Unsupervised Learning: Input data without the target label is used for training. Such algorithms learn the hidden patterns from the unlabeled dataset. Clustering is one such type. It clusters the data into different groups. K-means, agglomerative, and divisive are some of the unsupervised algorithms.

Semi-Supervised Learning: It contains a small part of the labeled dataset, with most data being unlabeled. The algorithms learn from the labeled data to predict unknown samples.

We restrict our discussion to the support vector machine algorithm in this chapter.

Support Vector Machine

Support vector machine is a supervised machine learning algorithm that can be used for classification and regression purposes. However, it is widely used to solve classification types of algorithms.

The main idea behind SVM is to find the hyperplane that can divide the dataset into two different classes. Figure 3 shows the high-level view of the algorithm.

Support Vectors: Points nearest to the hyperplane are known as support vectors. Suppose those points are removed from the dataset, then there will be a change in the position of the hyperplane. Thus, such points are critical in deciding the hyperplane.

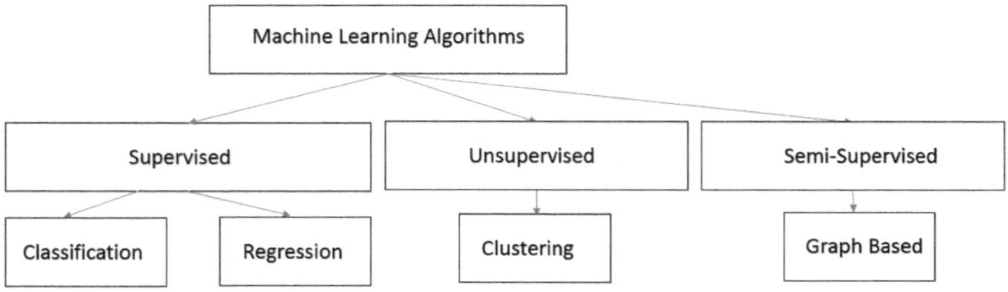

Fig. 2 Supervised ML algorithms

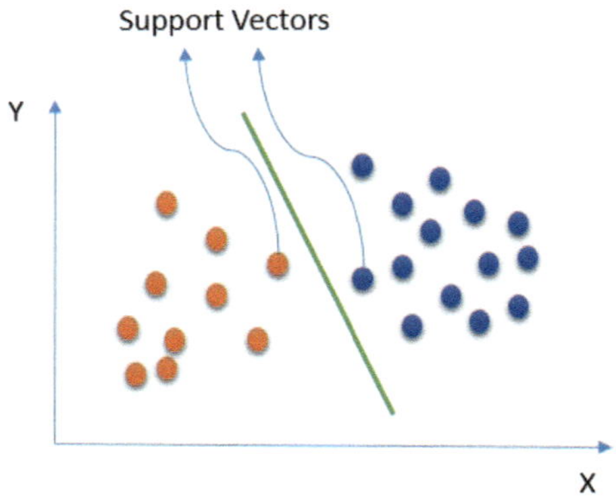

Fig. 3 SVM

Hyperplane: In a dataset with two features, a hyperplane is a line that divides the dataset into two different classes. The further the data points from the hyperplane, the more accurate the model is.

The distance between the nearest data points from each class and the hyperplane is called the margin as shown in Fig. 4. The margin is one of the hyper-parameters that needs to be tuned. The primary objective is to select the hyperplane that has a considerable margin distance so that there is less chance of misclassification.

The dataset shown in the above figure was simple, where finding a hyperplane is easy, but the real-world dataset is messy, where it may not be possible to divide data into multiple classes with just a linear hyperplane. We have to perceive the 2D data into a 3D axis in such cases as depicted in Fig. 5.

The transformations of the data points to the higher dimension are called kernel tricks. So, the basic idea is to transform the data into higher and higher dimensions until a hyperplane is formed. There are various kernel tricks. Radial bias function and polynomial kernel are the most popular ones.

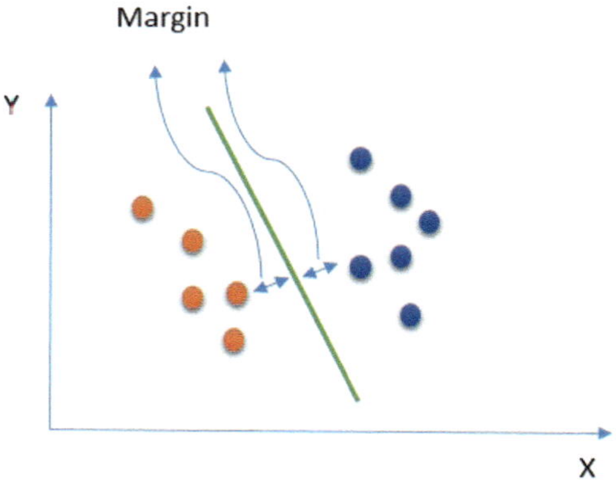

Fig. 4 SVM margin

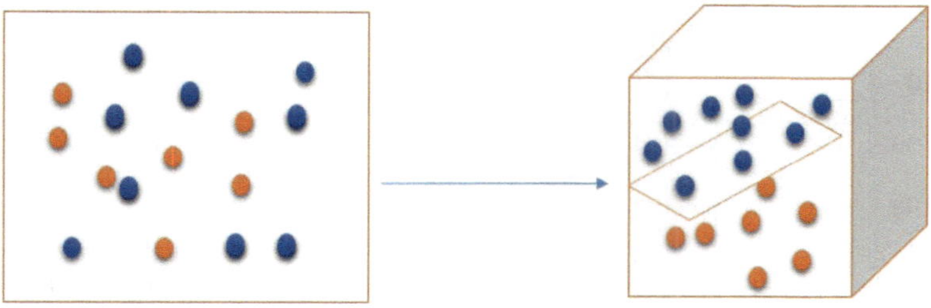

Fig. 5 SVM transformation

Advantages of SVM

- It provides high accuracy.
- Memory is efficient since it needs a subset of training data to find the hyperplane.

Disadvantages of SVM

- Less effective for data with lots of noise.

Quantum SVM

Let us see the steps to translate the supervised classification problem onto a quantum computer problem.

- The classical data point X must be translated into a quantum datapoint $|\Phi(\vec{x}) >$. To perform this operation, we need circuit $C(\Phi(\vec{x}))$.
- To process the data, we need parameterized quantum circuit $P(\Theta)$ with parameters Θ.
- Finally, we can apply the measurement that can return the output from the quantum circuit in a classical format (-1 or 1) for each input. This will identify the label for the input.

The quantum kernel is created by taking the inner products of the quantum feature maps $C(\Phi(\vec{x}))$. Since simulating such quantum feature maps is difficult on traditional computer systems, the quantum advantage is obtained by quantum simulation using the following equation:

$$C(\Phi(\vec{x})) = M(\Phi(\vec{x})) . Hn \quad (1)$$

where Hn is the Hadamard [12] gate and is applied to all the qubits. A total number of qubits is denoted by n. In this case, we use two qubits to have interaction among themselves, and that leads to ZZ interaction. Furthermore, the inner product has to be repeated based on the number of qubits. So, the above equation now becomes

$$C(\Phi(\vec{x})) = M(\Phi(\vec{x})) . Hn . C(\Phi(\vec{x})) = M(\Phi(\vec{x})) . Hn \quad (2)$$

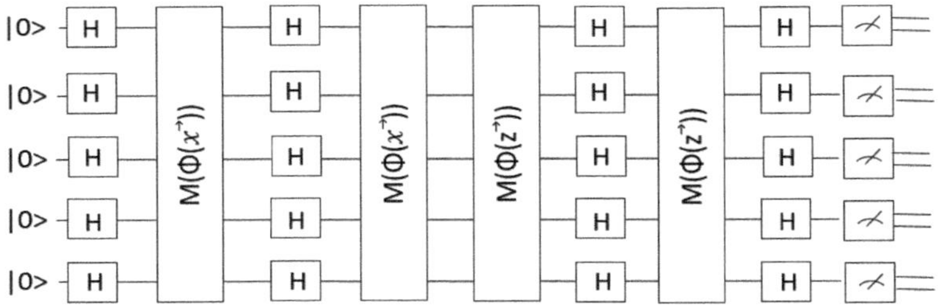

Fig. 6 Q circuit

Finally, to extract the quantum kernel information from the quantum circuit, the measurement of the two overlap states is needed, which is nontrivial. The author in [13] used the circuit below to calculate the estimated overlap (see Fig. 6).

They use the simple fact that $C(\Phi(\vec{x})) = C(\Phi(\vec{z}))$.

7 Libraries

(1) *For Algorithms on Traditional Computer*

The experimentation of the proposed research framework is performed with the following three libraries.

TensorFlow: TensorFlow is an open-source library from google for deep learning and machine learning.

Keras: Keras is a wrapper on TensorFlow and is used by many societies worldwide. The code written in Keras is internally converted to TensorFlow for further execution. It has functional API (Application Programming Interface) and Sequential API.

Scikit-learn: Scikit-learn deals with various learning-based algorithms (both supervised and unsupervised).

(2) *For Algorithms on Quantum Computer*

Qiskit: It is an open-source library for developing quantum programs that can run on quantum computers at the level of circuits. It is founded by IBM Research.

Qiskit has the following components:

Qiskit Terra: It provides libraries to model the quantum circuits, which can be close to quantum machine code.

Qiskit Aer: It provides a high-performance simulation of quantum computing.

Qiskit Ignis: It provides different libraries and modules for hardware verification and error correction at the quantum level.

Qiskit Machine Learning: It provides some of the quantum classification algorithms such as Quantum Support Vector Machine (QSVM) and Variational Quantum Classifier (VQC) and some other types of an algorithm like QGAN (Quantum Generative Adversarial Network).

8 Libraries

We used the following metrics to evaluate the model.

$$\text{Recall} = \text{TP}/(\text{TP} + \text{FN}) \tag{3}$$

$$\text{Precision} = \text{TP}/(\text{TP} + \text{FP}) \tag{4}$$

$$\text{Specificity} = \text{TN}/(\text{TN} + \text{FP}) \tag{5}$$

$$\text{Accuracy} = (\text{TP} + \text{TN})/(\text{TP} + \text{FP} + \text{FN} + \text{TN}) \tag{6}$$

$$\text{False Negative Rate (FNR)} = \text{FN}/(\text{TP} + \text{FN}) \tag{7}$$

$$\text{False Positive Rate (FPR)} = \text{FP}/(\text{FP} + \text{TN}) \tag{8}$$

where TP: True positive, FN: False Negative, TN: True Negative, and FP: False Positive.

9 Proposed Framework

See Fig. 7.

Stage 1: Data Collection

We use the Breast Cancer benchmark dataset collected from [14]. It has 30 features with two classes (Benign and malignant) in the target label. The following features are computed for each cell nucleus:

(1) Radius
(2) Perimeter
(3) Texture
(4) Area
(5) Compactness
(6) Smoothness

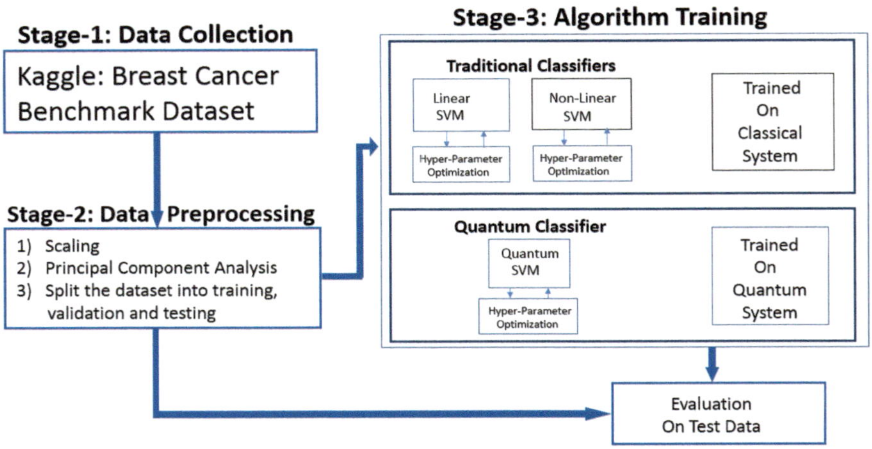

Fig. 7 Proposed framework

(7) Symmetry
(8) Concave points
(9) Concavity
(10) Fractal Dimension.

For each image, the average, standard error, and largest (mean of the three most significant values) of the above features are calculated, which results in a total of 30 features. Table 1 depicts the distribution of the dataset.

Stage 2: Data Preprocessing

In this stage, we perform the following data preprocessing task:

- Scale the data using MinMaxScalar.
- Perform PCA Analysis [15] to reduce the features set to the two-dimensional dataset.
- Split the dataset for training, validation, and testing using stratified sampling.

Stage 3: Algorithm Training

For the experiment purpose, we are using 2-qubits using a quantum system. Thus, it was essential to reduce the feature set to a two-dimensional vector. Therefore, after scaling the data, PCA analysis was performed. It solves the problem of limited qubit quantity and allows us to visualize the data.

Table 1 Dataset

Benign	Malignant
357	212

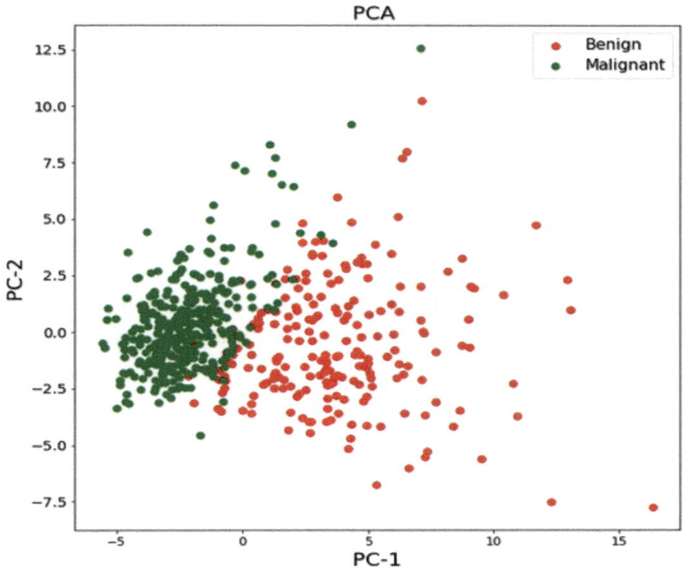

Fig. 8 PCA

Figure 8 shows two principal components having a combined explained variance of 70%.

The following algorithms are trained to learn the features of the data:

(1) Traditional System: Linear SVM and Nonlinear SVM.
(2) Quantum System: QSVM.

For any machine learning and deep learning algorithms, multiple hyper-parameters need to be tuned. This is called hyper-parameter optimization [16, 17]. There are various techniques such as follows:

(1) GridSearchCV: This technique performs an exhaustive search on all the parameters of the algorithms. Its time complexity is high but gives the optimal set of hyper-parameters.
(2) RandomSearchCV: This technique searches for the random value of parameters rather than an exhaustive search. It takes less time than the GridSearchCV and gives a near-optimal set of hyper-parameters.

Here, CV is K-Fold cross-validation with K being 10.

Experimental Results

This subsection discusses the experimental results from the algorithm trained on the traditional and quantum computer systems.

Traditional System

The following hyper-parameters are tuned:

- *C parameter*: It adjusts the trade-offs between the classification correctness of the training points and the decision boundary.
  - *Small C*: Cost of misclassification is low.
  - *Large C*: The cost of misclassification is high.
- *Gamma Parameter*: It adjusts the influence of boundary training points.
  - *Large Gamma*: High Weightage given to closer data points.
  - *Small Gamma*: Weightage goes beyond the nearby points.
- Kernel: Type of function (Linear or Nonlinear).

  We trained the SVM algorithm with the following parameters:

C: {0.01, 0.1, 0.5, 1, 10, 100}

Gamma: {1, 0.75, 0.5, 0.25, 0.1, 0.01, 0.001}

Kernel: {Nonlinear, Linear}

The table shows the best hyper-parameter for each kernel function (see Table 2).

Table 3 depicts the evaluated metrics. We can say that the nonlinear kernel function achieves higher accuracy than the linear function. We also noticed that QSVM performed better than the Linear SVM. Figures 9, 10 and 11 show the confusion matrix for linear SVM, nonlinear SVM, and QSVM. Nonlinear SVM has the highest accuracy, but at the same time, the time complexity increases with the increase in nonlinearity. Linear SVM has the least time complexity with the lowest accuracy due to the absence of the kernel trick operation. QSVM has a trade-off between time complexity and accuracy.

Table 2 Best hyper-parameter value

Kernel	Gamma	C
Linear	0.1	1
Nonlinear	1	100

Table 3 Evaluated metrics

Type	Precision	Recall	F1-Score	Accuracy
Linear	0.84	0.85	0.85	0.89
Nonlinear	0.93	0.94	0.93	0.95
QSVM	0.85	0.87	0.85	0.90

Fig. 9 Linear SVM

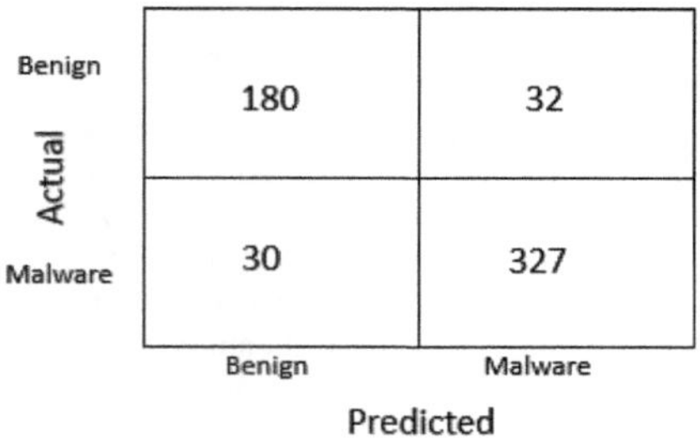

Fig. 10 Nonlinear SVM

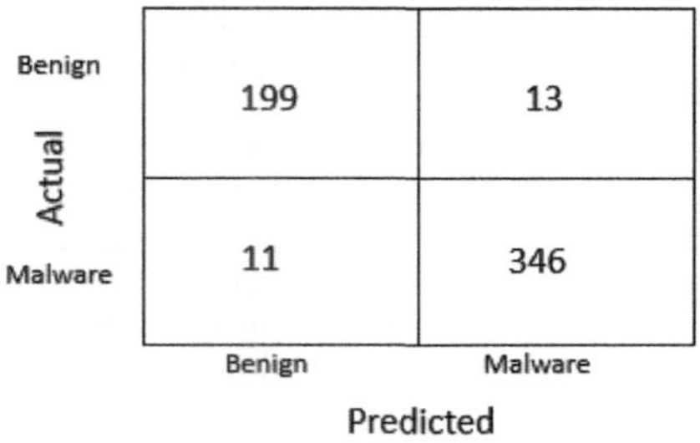

Fig. 11 QSVM

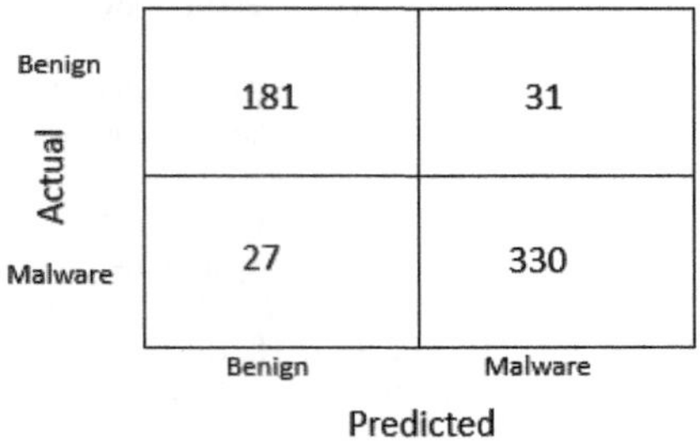

10 Conclusion

With the advent of the noteworthy progress of quantum computing in many applications such as cryptography and optimization, a new field called quantum machine learning has emerged. It is the application of quantum information processing with quantum machine learning. Therefore, in this chapter, we discuss the fundamentals of quantum algorithms. Furthermore, we presented a practical implementation of solving the well-known breast cancer classification problem using the traditional machine learning algorithms, namely linear and nonlinear SVM trained on traditional computers and using QSVM trained on IBM Quantum Computer. This work can be extended by employing more than two qubits to train QSVM.

References

1. R.P. Feynman, Simulating physics with computers. Int. J. Theor. Phys. **21**(6), 467–488 (1982)
2. D. Deutsch, Quantum theory, the Church-Turing principle and the universal quantum computer. Proc. R. Soc. A **400**(1818), 97–117 (1985)
3. P. Shor, Polynomial-time algorithms for prime factorization and discrete logarithms on a quantum computer. SIAM J. Comput. **26**, 1484 (1997)
4. L.K. Grover, A fast quantum mechanical algorithm for database search, in *Proceedings of STOC0-96, 28th Annual ACM Symposium on Theory of Computing* (1996), pp. 212–219
5. D. Bacon, W. van Dam, Recent progress in quantum algorithms. Commun. ACM **53**(2), 84–93 (2010)
6. C. Bennett, E. Bernstein, G. Brassard, U. Vazirani, Strengths and weaknesses of quantum computing. SIAM J. Comput. **26**(5), 1510–1523 (1997)
7. C.C. McGeoch, C. Wang, Experimental evaluation of anadiabiatic quantum system for combinatorial optimization, in *Proceedings of CF-13, ACM International Conference on Computing Frontiers* (2013), pp. 23:1–23:11.
8. T.F. Rønnow, Z. Wang, J. Job.S. Boixo, S.V. Isakov, D. Wecker, J.M. Mar-tinis, D.A. Lidar, M. Troyer, Defining and detecting quantum speedup (2014). arXiv:1401.2910
9. P. Bruza, R. Cole, Quantum logic of semantic space: An ex-ploratory investigation of context effects in practical reasoning, in *We Will Show Them: Essays in Honour of Dov Gabbay*, ed. by S. Arte-mov, H. Barringer, A.S. d'Avila Garcez, L. Lamb, J. Woods (College Publications, 2005)
10. D. Aerts, M. Czachor, Quantum aspects of semantic analysis and symbolic artificial intelligence. J. Phys. A Math. Gen. **37**, L123–L132 (2004)
11. J. Sun, B. Feng, W. Xu, Particle swarm optimization with particles having quantum behavior, in *Proceedings of CEC-04, Congress on Evolutionary Computation*, vol. 1 (2004), pp. 325–331
12. A. Tipsmark, R. Dong, A. Laghaout, P. Marek, M. Ježek, U.L. Andersen, Experimental demonstration of a Hadamard gate for coherent state qubits. Phys. Rev. A **84**(5), 050301 (2011)
13. V. Havlíček, A.D. Córcoles, K. Temme, A.W. Harrow, A. Kandala, J.M. Chow, J.M. Gambetta, Supervised learning with quantum-enhanced feature spaces. Nature **567**(7747), 209–212 (2019)
14. https://archive.ics.uci.edu/ml/datasets/breast+cancer+wisconsin+(diagnostic)
15. J. Soni, N. Prabakar, H. Upadhyay, Visualizing high-dimensional data using t-distributed stochastic neighbor embedding algorithm, in *Principles of Data Science* (Springer, Cham, 2020), pp. 189–206
16. J. Soni, N. Prabakar,KeyNet: enhancing cybersecurity with deep learning-based LSTM on keystroke dynamics for authentication, in *Intelligent Human Computer Interaction. IHCI 2021*. Lecture Notes in Computer Science, vol. 13184, ed. by J.H. Kim, M. Singh, J. Khan, U.S. Tiwary, M. Sur, D. Singh (Springer, Cham, 2022). https://doi.org/10.1007/978-3-030-98404-5_67
17. J. Soni, N. Prabakar, H. Upadhyay. Behavioral analysis of system call sequences using LSTM Seq-Seq, cosine similarity and jaccard similarity for real-time anomaly detection, in *2019 International Conference on Computational Science and Computational Intelligence (CSCI)* (IEEE, Dec. 2019), pp. 214–219

Numerical Modeling of the Major Temporal Arcade Using a Quantum Genetic Algorithm

Jose A. Soto-Alvarez, Ivan Cruz-Aceves, Arturo Hernandez-Aguirre, Martha A. Hernandez-Gonzalez, and Luis M. Lopez-Montero

Abstract The Major Temporal Arcade (MTA) is the thickest vessel in the retina, which can be useful to analyze different pathologies related to the retina such as diabetic retinopathy. Consequently, its numerical modeling plays a vital role in systems that perform computer aided-diagnosis in Ophthalmology. In the present chapter, a novel method for the automatic modeling of the MTA is introduced. The method consists of the steps of automatic MTA segmentation and numerical modeling based on spline curves and the use of the Quantum genetic algorithm (QGA). In this step, the QGA is analyzed and implemented in order to determine the optimal control points on a set of previously segmented vessel pixels of the MTA in retinal fundus images. These control points are used to generate the best curve to fit the MTA through spline curves. In the experimental results, the proposed method was evaluated in terms of the Mean distance to the closest point and Hausdorff distance obtaining the average values of 9.91 and 53.32, respectively, using a test set of images. Finally, in terms of computational time, the proposed method achieved an average of 7.51 s per image, which makes it suitable for computer-aided diagnosis in ophthalmology.

J. A. Soto-Alvarez · A. Hernandez-Aguirre
Centro de Investigación en Matemáticas (CIMAT), A.C., Jalisco S/N, Col. Valenciana, C.P. 36000 Guanajuato, Gto, México

I. Cruz-Aceves (✉)
CONACYT - Centro de Investigación en Matemáticas (CIMAT), A.C., Jalisco S/N, Col. Valenciana, C.P. 36000 Guanajuato, Gto, México
e-mail: ivan.cruz@cimat.mx

M. A. Hernandez-Gonzalez · L. M. Lopez-Montero
Unidad Médica de Alta Especialidad (UMAE) - Hospital de Especialidades No.1. Centro Médico Nacional del Bajio, IMSS, León, Gto, Mexico

R. Pandey et al. (eds.), *Quantum Computing: A Shift from Bits to Qubits*,
Studies in Computational Intelligence 1085,
https://doi.org/10.1007/978-981-19-9530-9_13

1 Introduction

The automatic modeling of vessel-like structures in the retina can help in the diagnosis and monitoring of different pathologies related to the retina. The Major Temporal Arcade (MTA) can be detected as the thickest vascular structure along the retinal fundus image. The MTA is composed by the Superior Temporal Arcade (STA) and Inferior Temporal Arcade (ITA) [1]. In clinical practice, the opening between the ITA and STA is analyzed by visual inspection and it is used as indicator of the severity of diabetic retinopathy. Consequently, the importance of the MTA numerical modeling plays a vital role in systems that perform computer-aided diagnosis in ophthalmology. Generally, the problem of automatic modeling of the MTA has been addressed by the steps of automatic vessel segmentation and then the numerical modeling is performed using different parametric object detection methods. In Fig. 1, two images of the MTA outlined by specialist are presented.

In general, the problem of automatic blood vessel segmentation has been carried out by using the independent steps of vessel detection and vessel segmentation or classification because of the disadvantage of non-uniform illumination. The step of vessel detection also known as vessel enhancement consists on applying a spatial or frequency filter in order to enhance vessel-like structures from the background image. Moreover, the second step of segmentation is commonly performed by applying a thresholding strategy or binary classification method on the filter response obtained from the previous step. In literature, methods based on mathematical morphology [2–4], Hessian matrix [5–7], Gabor filters [8–10], and Gaussian matched filters [11, 12] have been mainly applied.

The main idea behind the Gaussian matched filters (GMF) is to approximate the shape of blood vessels in the spatial image domain by using a Gaussian curve as matching template. The GMF is governed by four parameters, where the parameter σ controls the amplitude of the Gaussian profile, which is of interest to detect the thickest vessels, specifically the Major Temporal Arcade. The parameter σ has been tuned in different works according to the type of blood vessel to be detected. Kang et al. [13, 14] modified ($\sigma = 1.5$) the Gaussian profile to detect the coronary artery

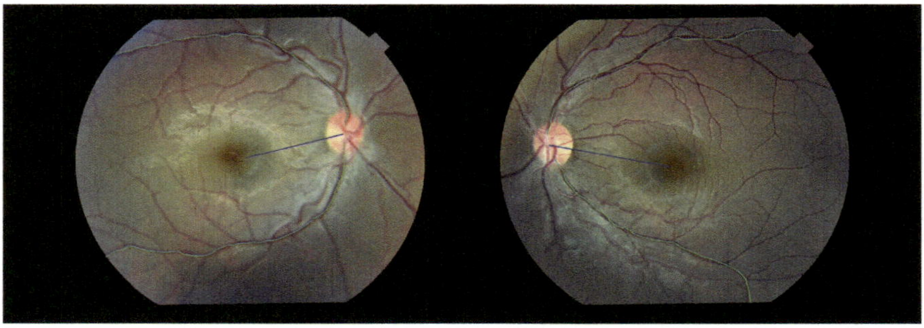

Fig. 1 Retinal fundus images with manual delineation of the major temporal arcade

tree. Al-Rawi et al. [15] introduces a range to determine the optimal value through an exhaustive search in retinal fundus images. Later, Al-Rawi and Karajeh [16] uses a Genetic algorithm to perform the search instead of the exhaustive strategy. Finally, Cruz et al. [17] perform a comparative analysis between four different evolutionary algorithms to select the optimal parameters of the GMF for the segmentation of the coronary artery tree.

On the other hand, to perform the numerical modeling on the segmented MTA, different strategies have been proposed. Oloumi et al. [18–20] introduced a method based on Gabor filters and the Hough transform for parabolic modeling. Later, Oloumi et al. [21] presented an improved method based on a dual-parabolic modeling of the MTA using two parabolas obtained from the Hough transform. The first parabola was used for the STA, and the second one for the ITA. Moreover, Guerrero et al. [22] proposed a method based on the evolutionary algorithm called estimation of distribution algorithm to reduce the computational time of the Hough transform by solving the parabolic equation using three pixels from the previously segmented MTA. Following this idea, Valdez et al. [23] proposed a hybrid evolutionary algorithm based on simulated annealing and the univariate marginal distribution algorithm obtaining a more robust search strategy.

The use of evolutionary algorithms (EAs) for optimal parameter selection and discrete optimization problems is a well-known practice. Recently, the Quantum Genetic Algorithm has begun to attract more attention for solving discrete optimization problems such as automatic clustering and feature selection in stenosis detection [24]. In the present chapter, a novel method for the automatic numerical modeling of the Major Temporal Arcade in retinal fundus images is introduced. The proposed method consists of the steps of segmentation and numerical modeling of the MTA. The first step is performed by using the Gaussian matched filter for the automatic detection of the MTA, which is the thickest vessel structure in the image. In the second step, the MTA is modeled by using Quantum Genetic Algorithms for the automatic location of the control points that best fit the spline curve on the segmented MTA. Finally, the proposed method is evaluated in terms of the Hausdorff distance and mean distance to the closest point (MDCP) with respect different methods of the state of the art based on the Hough transform.

The remainder of this chapter is organized as follows. In Sect. 2, the fundamentals of the Gaussian matched filters for automatic vessel segmentation and Quantum genetic algorithms for discrete optimization are presented. In Sect. 2.6, the proposed method for modeling the major temporal arcade is described in detail. The experimental results are discussed in Sect. 3, and finally, Sect. 4 presents the conclusions of the present work.

2 Background

Due to the uneven illumination of the vessel-like structures in retinal fundus images, the Gaussian matched filters are used for the automatic detection of the MTA, and the stage of numerical modeling of the MTA is performed using Quantum genetic algorithms for working with spline curves. Since these methods are of interest in the proposed method, they are described in detail in the present section.

2.1 *Database of Major Temporal Arcade Images*

The publicly available DRIVE database consisting of 40 retinal fundus images has been used to perform the present research. Each image is of size 565×584 pixels in RGB format. To assess the performance of the proposed and the state-of-the-art methods, the database has been divided into the training and testing subsets using 20 images each.

Since the present work is focused on the numerical modeling of the MTA and the DRIVE database has been used for the segmentation of vessel-like structures of different amplitudes, the ground-truth images of the MTA have been outlined by an specialist of the Ophthalmology Department of the Mexican Social Security Institute, UMAE T1-León.

2.2 *Gaussian Matched Filters*

The Gaussian matched filters (GMF) method was initially proposed by Chaudhuri et al. [11] for detecting blood vessels in retinal fundus images. The main idea of the GMF is that the shape of vessel-like structures can be detected in the spatial image domain by using a Gaussian curve as matching template. The Gaussian profile, can be defined as following:

$$G(x, y) = -exp\left(-\frac{x^2 + y^2}{2\sigma^2}\right), |y| \leq L/2, \quad (1)$$

where L is the length in pixels of the vessel segment to be enhanced, and σ is the average width of the blood vessels to be detected. In order to form the Gaussian template, two additional parameters have to be determined; width template and number of oriented filters. The width (in pixels) of the template represents a discrete parameter (T), which is used to define the position where the Gaussian curve trails will cut, since the Gaussian curve has infinitely long double sided trails. Moreover, to enhance vessel-like structures at different orientations, the Gaussian kernel $G(x, y)$

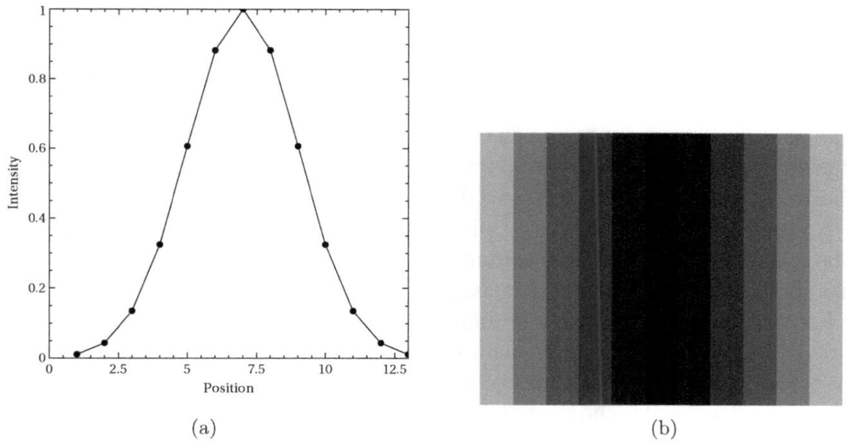

Fig. 2 **a** Gaussian profile. **b** Gaussian template for vessel enhancement

can be rotated at different angles (θ) using $\kappa = 180/\theta$ oriented filters. The rotation matrix is defined as following:

$$\kappa = \begin{bmatrix} \cos\theta_i & -\sin\theta_i \\ \sin\theta_i & \cos\theta_i \end{bmatrix}, \tag{2}$$

where κ represents the number of evenly spaced filters in the range $[-\pi/2, \pi/2]$. Since the parameter κ is also the number of oriented filters, κ filter responses will be obtained, where pixels with the maximum response over the filter bank are conserved to generate the final filtered image.

The correct selection of the optimal set of GMF parameters represents an essential task. Since the GMF is governed by the four parameters of L, T, κ, σ, the σ parameter is the most relevant for detecting the Major Temporal Arcade. In Fig. 2, a Gaussian profile and the respective Gaussian template with parameters $L = 9$, $T = 13$, and $\sigma = 2.0$ is illustrated.

2.3 Spline Curves

Let us start by assuming that the point values of a function $f(x)$ are known, whose functional form is not know. Furthermore, the data set (points) $x_1, x_2, \ldots, x_n$ was not necessarily chosen, however it is desired to estimate the values $f(x)$ for each point in the interval (x_1, x_n). To solve this problem an interpolation methods are necessary, which depending on the distribution of the data set they can be classified as: power series, Fourier series, least squares and spline fit [25]. However, it is not possible to say that one is superior to the other, this will be totally determined by the distribution of the data.

Let us remember that the Major Temporal Arcade (MTA) vein has a parabolic shape, however, it is not symmetrical since it can be present tortuosity related to damage to the vascular structure due to some disease. That is why choosing a type of adjustment based on a piecewise parametric approach can be a good option. Spline adjustment fitting does just this, so a deeper explanation it is necessary.

A spline curve is defined as a piecewise polynomial curve passing through a dataset of points and satisfying certain continuity condition. When the spline curves pass approximately through the dataset, they are smoothing splines [26].

Let us consider the dataset of ordered pairs $\{(x_1, y_1), (x_2, y_2), \ldots, (x_n, y_n)\}$ and $y_i = f(x_i)$, where f is an unknown function. Approximating f through polynomials f_i restricted to each interval $[x_i, x_{i-1}], i = 1, 2, \ldots, n-1$ where generally the degree of polynomial is chosen as 3, however it is possible to assign other degree. The set of points in each interval are known as knots and when the knots coincide with the x-data the a classical spline appears. Keeping this in mind, $y_i = f(x_i) = s(x_i)$, which guarantees that the approximating spline goes exactly through the data.

Piecewise polynomials are defined as:

$$
\begin{aligned}
s(x) &= f_i(x), x \in [x_i, x_i + 1], i = 1, 2, \ldots, n-1 \\
s(x) &= f_1(x), x \leq x_1 \\
s(x) &= f_{n-1}(x), x \geq x_n
\end{aligned}
$$

A problem with cubic splines is that unwanted bends can be generated between the data intervals, fortunately there are alternatives to overcome this issue, come of them are smoothing splines or B-splines. However, the cost to pay is a minor fitting to the data set. An important thing to note is that any type of spline can be expressed as a linear combination o basis functions called B-spline basis functions. These bases functions are non-zero piecewise polynomials on the small intervals, and this intervals are defined by a sequence of numbers or knots. It can be understood like an orthogonal basis of functions in the B-splines, something similar as in Fourier series. The degree of the base will be the same as the resulting spline.

For a d-degree base functions set:

$$B_{i,d}(x), i = 1, 2, \ldots, n \tag{3}$$

the B-spline can be expressed as

$$s(x) = \sum_{i=1}^{n} c_i B_{i,d}(x) \tag{4}$$

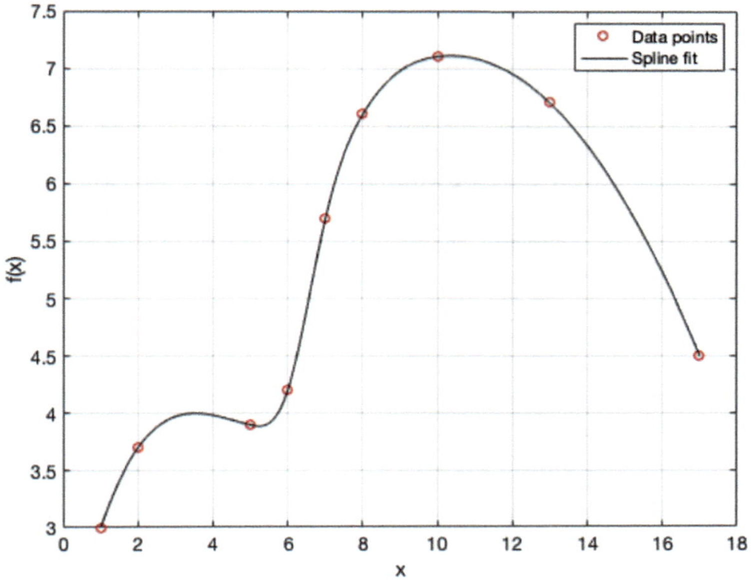

Fig. 3 Spline fitting for a data set

being able to determinate the coefficients c_i when applying the last equation to the data set $\{(x_i, y_i)\}i = 1, 2, \cdots, n$ previously defined above of this section, which results in a linear system of equations to be solved,

$$y_i = \sum_{i=1}^{n} c_i B_{i,d}(x_i) \tag{5}$$

for example, using least squares, $\mathbf{B}c = \mathbf{y} \Longrightarrow \mathbf{c} = \left(B^T B\right)^{-1} B^T \mathbf{y}$. In Fig. 3 an example of data fitting by a spline curve is shown.

In the proposed method, the knots of required for the construction of the spline curve are obtained by using the Quantum Genetic Algorithm and the skeleton of the previously segmented MTA.

2.4 *Genetic Algorithms*

Genetic Algorithm (GA) is an stochastic optimization strategy which uses a number of potential solutions to address a complex problem that would be very difficult to solve using classical (deterministic) methodologies. The GAs are based on the principles proposed by Charles Darwin, the "survival of the fittest". In 1975 Holland visualized the concept, which was able to publish [27].

In general, a GA starts with a random population of individuals (solutions), which is usually encoded in a binary string. The population is evaluated using a fitness value which is directly related to the function of the optimization problem. The initial population is evolved through the natural genetic operators which are crossover, mutation and selection. This is repeated iteratively until a stopping criteria is reached.

As mentioned above, GAs work with encoded variables. This is a great advantage since it allows working with a discretized search space even when the function is continuous. Also, in contrast to traditional optimization methods which work with one solution, GA uses a population of potential solutions at a time. Namely, GAs, being an iterative optimization technique, work with a variety of solutions called population in each generation.

The genetic operators are applied in each iteration, and the number of iterations in the algorithm indicates the number of generations that will be performed on the population. Below the basic vocabulary used in the GAs is explained:

- **Population:** at the beginning of the algorithm, a search space (potential solutions) is initialized randomly, where each element of the population is known as individual. From this, new populations will be generated from which the best solution will be extracted when the stop criteria is reached.
- **Chromosome:** the elements that make up a population are called individuals or chromosomes.
- **Gene:** each element that make up a chromosome is known as gene. In a binary coding string, genes can take values of 0 or 1.
- **Parent:** within the entire population, in each iteration or generation, candidates will be chosen to generate the offspring for the next generation.
- **Child:** new candidates generated from the parents of the previous generation.

Figure 4 shows a graphic representation of the elements involved in the GAs.

On the other hand, the genetic operators are described below:

1. **Selection operator:** depending on the objective of the problem, the most common strategy consists on ordering the individuals according to the fitness value, such that the most suitable are chosen to continue in the next generation.
2. **Crossover operator:** it consists of selecting a section of genes from two parents which will be exchanged, resulting in the creation of a new individual (offspring). Cross of one or several points can be performed. Generally, it is established a percentage of the population (60%) will be selected to be part of the next generation (Figs. 5 and 6).
3. **Mutation operator:** this operator makes a change from 0 to 1 and vice versa in some individual, a small mutation probability p_m must be considered.

Next, the main idea behind the procedure performed by the Genetic Algorithm is described below:

First, a population of n individuals, each one with a desired length is initialized. This set of n individuals will form the initial population of possible solutions to the problem to be solved. Each individual is made up of a string of randomly chosen

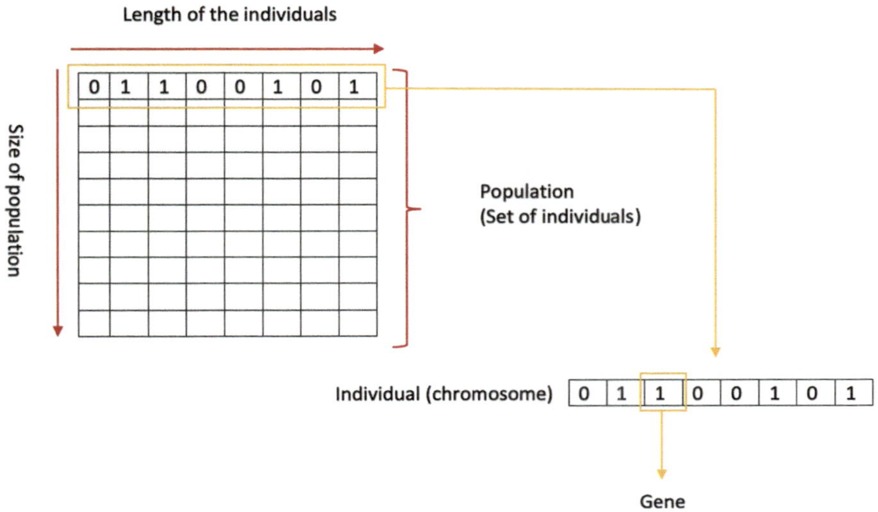

Fig. 4 Schematization of the elements that make up a population of solutions

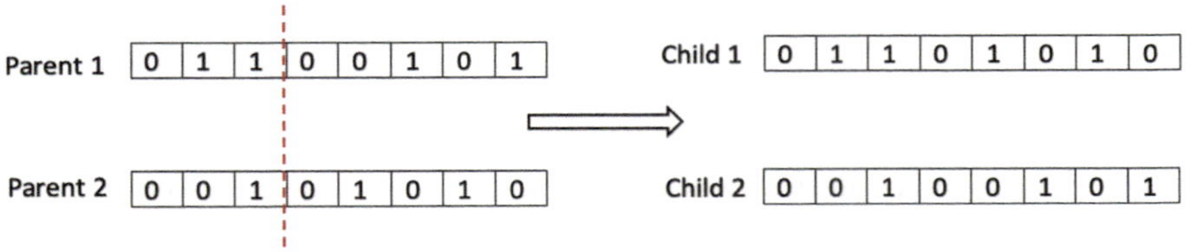

Fig. 5 Crossover operator using the one point strategy

Fig. 6 Mutation of a gene in an individual

zeros and ones, i.e., binary string. After the population is created, it is evaluated by means of a fitness function, in order to assign a fitness value for each individual. The aim is now to improve the population, this is achieved by discarding the least fit individuals, now it is necessary to perform a ranking and make a selection of a part of the population, generally between 60–70% of the individuals that will serve for the next generation are chosen.

Must now proceed to create offspring from the parents in the population, to do this traits (genes) will be chosen that will be exchanged between them to form a new child individual, this is achieved by applying the crossover operator. In order to maintain a diversity of the population, a quantity of genes of certain individuals are changed from 0 to 1 or vice versa, this is called mutation and must be kept at a low percentage ($p_m = 5\%$). Mutation is useful, for local improvement of a solution. After

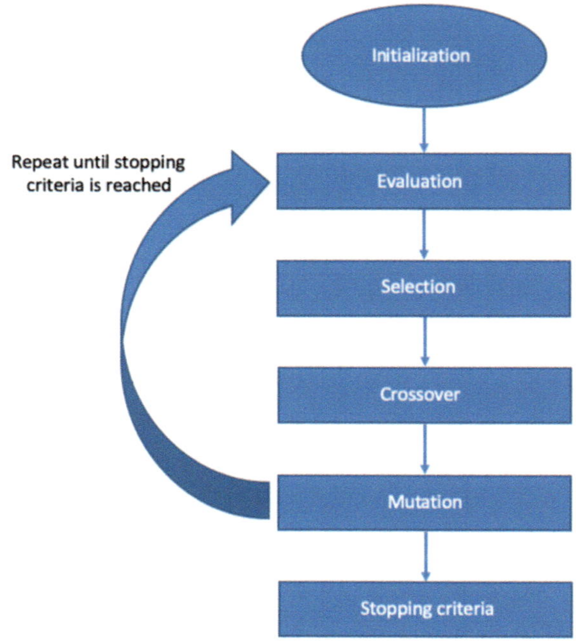

Fig. 7 Block diagram of the genetic algorithm

the previous steps are finished, a generation of the GA is completed. The process is repeated until a stopping condition is reached, for example, number of generations (iterations). Figure 7 shows a block diagram of the process to follow the algorithm.

2.5 *Quantum Genetic Algorithms*

In the above section, a explanation of what a genetic algorithm is was addressed, however, in the last decades the union of quantum computation and genetic algorithms has been studied, resulting in Quantum Genetic Algorithms (QGAs). In 1990s *Narayanan and Moore* proposed such an algorithm and it was successfully used in the TSP problem [28]. The main advantage of the QGA compared to a conventional GA is that its efficiency is significantly better, it also has a fast convergence speed and robustness [29].

QGAs employ concepts from quantum mechanics; a classical computer deals with "bits", but a quantum computer uses quantum bits or "qubits", a really important concept since it is the smallest unit of information in a quantum-inspired evolutionary algorithm [30]. Now working with quantum states, namely, the states "0" and "1", but here is the important thing, a qubit will also be the superposition of these two states. Let the new state be $|\psi\rangle$, such that $|\psi\rangle = \alpha|0\rangle + \beta|1\rangle$, where α y β are the

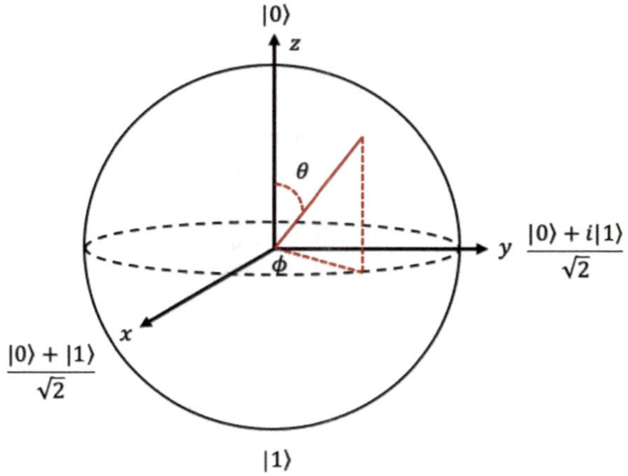

Fig. 8 Graphical representation of a qubit. States $|0\rangle$ and $|1\rangle$ are represented at the top and the bottom and any point on the sphere will be a linear combination of both as shown in the two examples over the x and y axes

probability amplitudes of the states, which also may satisfy the condition $|\alpha|^2 + |\beta|^2 = 1$, $(\alpha, \beta \in \mathbb{C})$. Using the Dirac notation:

$$|\psi\rangle = \begin{pmatrix} \alpha \\ \beta \end{pmatrix}, \tag{6}$$

and the representation of the states "0" and "1" in qubits will be as:

$$|0\rangle = \begin{pmatrix} 1 \\ 0 \end{pmatrix}, \quad |1\rangle = \begin{pmatrix} 0 \\ 1 \end{pmatrix}. \tag{7}$$

It is also possible to have geometrical representation of the qubits, this is done by means of a Bloch sphere. Figure 8 shows a better seen.

For a set of m qubits, the probability amplitude that in turn will define a quantum population will be expressed as:

$$Q(t) = \begin{bmatrix} \alpha_1^t & \alpha_2^t & \cdots & \alpha_m^t \\ \beta_1^t & \beta_2^t & \cdots & \beta_m^t \end{bmatrix}. \tag{8}$$

When starting a population, the probability amplitudes at $t = 0$ will be $\frac{1}{\sqrt{2}}$, in order to all possible states have the same probability.

Let's analyze some concepts of quantum mechanics before continuing. When dealing with an isolated quantum system, if an evolution from an initial state to a final state is required, the Schrödinger equation comes into play, since it is the one that governs these processes:

$$i\hbar \frac{\partial}{\partial t} |\psi\rangle = H(t)|\psi\rangle, \tag{9}$$

being $i = \sqrt{-1}$ and $\hbar = h/2\pi$ the reduced Planck constant. The above equation is a partial differential equation whose analysis variable is the vector $|\psi\rangle$, which describes the state of a quantum system at time t. $H(t)$ is the Hamiltonian of the system and it is an operator that is applied on the state vector (wave function), when the Hamiltonian is known, then the Schrödinger equation can be solved. This operator can also be understood as the sequence of operations executed on an initial state in a quantum computer.

Let us assume that $U(t)$ corresponds to the solution of the Schrödinger equation, in such a way that given an initial state $|\psi(0)\rangle$, the solution at time t will be:

$$|\psi\rangle = U(t)|\psi(0)\rangle. \tag{10}$$

Considering that $|\psi(0)\rangle$ corresponds to the input of a computer, then $|\psi(t)\rangle$ will be output as the result of the evolution of the Schrödinger equation and a subsequent measurement [31]. A measurement is when the states of the qubits are observed, leading to the collapse to the wave function $|\psi\rangle$. This means that at the moment of observing the qubit leaves its superposition form, i.e., it takes one and only one state, $|0\rangle$ or $|1\rangle$. U (solution of the Schrödinger equation) is called unitary evolution operator and in QGAs is know as a quantum gate (Q-gate).

Returning to the description of a quantum genetic algorithm, having understood the basic concepts of quantum mechanics, then:

- A QGA will adopt a new representation for the population of individuals, the qubit representation, which will probabilistically provide a linear superposition of multiple states.
- After the population of possible solution has been generated, it must be evolved in such a way to obtain better solutions, this is achieved by means of the Q-gates. With this, the individuals of the next generation will be produced.

There are several types of quantum genetic operators (Q-gates) to evolve the population, let's talk about the most characteristic:

Rotation gate: this is the most characteristic operator for updating a population. It operates as a process of unitary transformations and it is defined as:

$$U(t) = \begin{pmatrix} \cos\left(\delta\theta_j\right) & -\sin\left(\delta\theta_j\right) \\ \sin\left(\delta\theta_j\right) & \cos\left(\delta\theta_j\right) \end{pmatrix}. \tag{11}$$

Then the process of the evolution of the quantum state will be given by:

$$\begin{pmatrix} \alpha_j^{t+1} \\ \beta_j^{t+1} \end{pmatrix} = \begin{pmatrix} \cos\left(\delta\theta_j\right) & -\sin\left(\delta\theta_j\right) \\ \sin\left(\delta\theta_j\right) & \cos\left(\delta\theta_j\right) \end{pmatrix} \begin{pmatrix} \alpha_j^{t} \\ \beta_j^{t} \end{pmatrix} \tag{12}$$

where $\delta\theta_j$ is a rotation angle, which one is looked up in a table that ensures the convergence. This quantum genetic operator seeks to amplify or decrease the probability amplitudes of the qubits in correlation with the chromosome of the highest fitness value (Fig. 9).

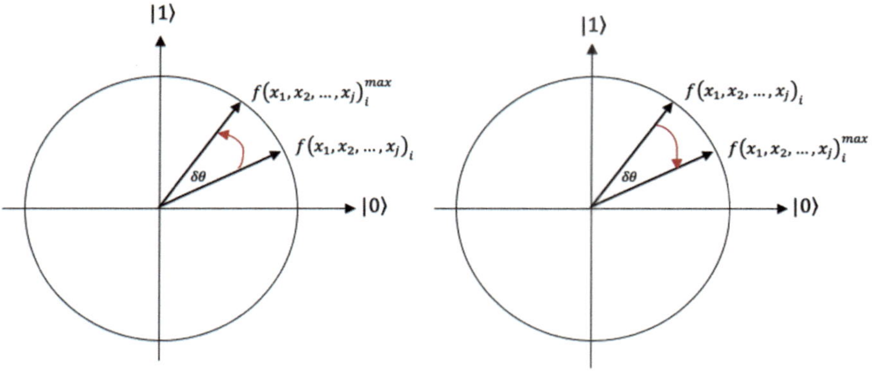

Fig. 9 Rotation gate

Quantum Mutation Gate: just as the classical version of the GAs for the mutation operator, there is a similar version for the QGAs. Essentially the process consists of swapping the amplitudes of the *j-th* qubit using the Pauli matrix X:

$$X = \begin{pmatrix} 0 & 1 \\ 1 & 0 \end{pmatrix}, \tag{13}$$

which results in:

$$\begin{pmatrix} \beta_j^{t+1} \\ \alpha_j^{t+1} \end{pmatrix} = \begin{pmatrix} 0 & 1 \\ 1 & 0 \end{pmatrix} \begin{pmatrix} \alpha_j^t \\ \beta_j^t \end{pmatrix}. \tag{14}$$

Quantum Crossover Gate: the quantum analogous to the classical crossover operator is practically the same, just now it will be operating on the amplitudes. For example, for one point of crossover, and assuming that the point was randomly chosen between the first and the second position:

$$\begin{pmatrix} \alpha_1 & \alpha_2 & \alpha_3 & \cdots & \alpha_j \\ \beta_1 & \beta_2 & \beta_3 & \cdots & \beta_j \end{pmatrix}, \quad \begin{pmatrix} \alpha_1' & \alpha_2' & \alpha_3' & \cdots & \alpha_j' \\ \beta_1' & \beta_2' & \beta_3' & \cdots & \beta_j' \end{pmatrix}$$

which results in:

$$\begin{pmatrix} \alpha_1 & \alpha_2' & \alpha_3' & \cdots & \alpha_j' \\ \beta_1 & \beta_2' & \beta_3' & \cdots & \beta_j' \end{pmatrix}, \quad \begin{pmatrix} \alpha_1' & \alpha_2 & \alpha_3 & \cdots & \alpha_j \\ \beta_1' & \beta_2 & \beta_3 & \cdots & \beta_j \end{pmatrix}$$

Below is a block diagram that follows a quantum genetic algorithm (Fig. 10):

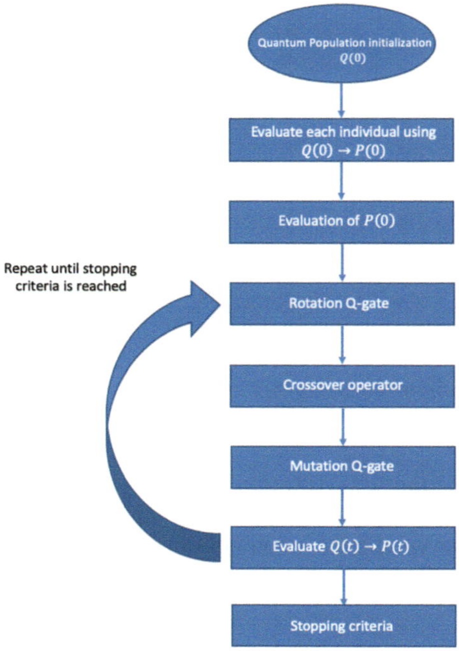

Fig. 10 Block diagram of quantum genetic algorithm

2.5.1 Implementation Details

In order to exemplify the potential of Quantum Genetic Algorithms, its operation was adapted with the intention of being able to determine the optimal solution to a set of five points (control points) that will be used to make the best adjustment of the vein Major Temporal Arcade. From a randomly generated population $Q(0)$ of possible solutions, the QGA evolves it and the most optimal solutions that guarantee a god fit are obtained.

The Algorithm 1 shows a description of the operation to be followed by the QGA:

As mentioned in the last section, the population of individuals must be evolved in order to achieve diversity and avoid to stuck in some local minima of the search space, for this by means of quantum genetic operators namely quantum gates (Q-gates) the process is performed. Let's analyze a little bit more the one of them, the quantum mutation operator, which purpose is to slightly disturb the evolution states of some individuals, thus seeking not to fall into a local minimum during the iterative process of search.

The process followed for quantum mutation is described by the Algorithm 2.

It is easy to visualize that the mechanism carried out would be in a compact form as $\left(\beta_j^{t+1}, \alpha_j^{t+1}\right)^T = U \cdot \left(\alpha_j^t, \beta_j^t\right)^T$, where for this case U represents the quantum mutation operator. Another clarification would be that the mutation process describe

Algorithm 1 Procedure Quantum Genetic Algorithm

Begin
$t \leftarrow 0$
initialize $Q(0)$
make $P0$ by measuring $Q(0)$
evaluate $P(0)$
store the best solution among $P(0)$
while (not stopping criteria is reached) **do**
  $t \leftarrow t + 1$
  make $P(t)$ by observing $Q(t-1)$ population
  evaluate $P(t)$
  update $Q(t)$ using quantum genetic operator (Q-gates)
  store the best solution among $P(t)$
end while
end

Algorithm 2 Procedure Quantum Mutation Operator

Begin
Given some probability of mutation Pm, select a set of chromosomes
Perform a randomly selection of qubit for each chromosome
Exchange the information of probability amplitude, $\alpha_j^t \rightarrow \beta_j^t$
end

above acts on a single qubit, however, it works on situations where there are multiple qubits.

2.6 *Proposed Method for Modeling the Major Temporal Arcade*

The modeling of the MTA is a problem that has not been explored in depth. In literature, different approaches of its anatomical structure through parabolas using methods based on the Hough transform are presented. However, the MTA is not a symmetrical parabola even when the image comes from a healthy patient, since for patients who present some ocular disease their deformity will be greater and therefore their deviation from the parabolic shape. In order to perform a comparison of fundus images for patients who present some disease with respect to an image of a healthy patient, a modeling of the MTA is necessary, however, taking into account the problems mentioned above, a new methodology is addressed. Instead of considering a single function that fits the entire vein, a piecewise parametric curve is considered.

It is known that the spline curves has the peculiarity of adjusting complex shapes through low-order piecewise polynomials functions, as long as the set of ordered pairs (coordinates) that represent the shape to be adjusted. Besides to a number of control points that help the construction of the functions at the intervals specified by them. In the present work, an adjustment to the MTA by means of splines curves is made, where a Quantum Genetic Algorithm is used to obtain the five best (optimal) control points that are necessary to draw the best curve.

From a set of fundus images form the DRIVE database, a vein segmentation process is started, from which the MTA is extracted using the Gaussian Matched filter and a binary thresholding strategy. Subsequently, a skeletonization of major vein is made and the ordered pairs (pixel coordinates) are extracted. Subsequently, the Quantum Genetic Algorithm is initialized, generating an initial population of individuals which will be evolve through the steps described in the Algorithm 1.

Once the number of generations established as the stopping criteria is reached, the five best coordinates are extracted from the population, and they will be used as control points for adjust the pixel coordinates by means of a spline curve. The QGA configuration to work was a population of 100 individuals, 50 generations (stopping criteria), and crossover and mutation rates of 0.7 and 0.03, respectively.

In order to measure the closeness to the spline-fit curve data to the original data, Mean Distance to the Closest Point (MDCP) and Hausdorff distance were used, since both distance metrics were employed by Oloumi et al. [18]. For understanding MDCP, let be two sets of points A and B, in A there are the points obtained by the approximation and in B there are the points of the original coordinates, namely, the ground-truth delineation. The MDCP can be defined as:

$$\mathrm{MDCP}(A, B) = \frac{1}{N}\sum_{i=1}^{N} \mathrm{DCP}(a_i, B), \tag{15}$$

where N is the cardinality of A, a_i is its *i-th* element and DCP is the distance to the closest point, which is calculated as follows:

$$\mathrm{DCP}(a_i, B) = \min \parallel a_i - b_j \parallel, \tag{16}$$

with $j = 1, 2, \ldots, M$, being M the cardinality of B, its *j-th* element b_j and $\parallel \cdot \parallel$ a norm operator, which is usually the Euclidean distance. Then, it is conclude that the MDCP measures an average of the distance for each point of one set to the closest point of the other set.

On the other hand, the Hausdorff distance uses a similar calculation to the previous one, it also makes use of the DCP, thus finding the distance form each point of set A to the elements belonging to set B; however instead of calculating an average, it takes the maximum value:

$$\mathrm{H}(A, B) = \max \mathrm{DCP}(a_i, B). \tag{17}$$

As long as small values are obtained at both distances, it is inferred that the model is a good fit for the original data set.

3 Computational Experiments

The computational experiments introduced in this section, were performed on a computer with an Intel Core i3, 2.13 GHz processor, and 4 GB of RAM using the Matlab version 2019a. Since the proposed method consists on different steps, they are presented from the segmentation stage to numerical modeling using the Quantum genetic algorithm. The stage of automatic segmentation of the Major Temporal Arcade, has been divided into enhancement, binary segmentation, and length filtering steps. This stage plays an essential role, since the numerical modeling is performed on the segmented MTA.

In the MTA enhancement, the GMF was used since the parameter σ of vessel width is very useful to discriminate thin vessels. The set of GMF parameters was experimentally determined as $\kappa = 12$, $L = 15$, $T = 15$, and $\sigma = 6.0$ using the training set of retinal fundus images. In Fig. 11, the enhancement or vessel detection results are presented on a subset of retinal images of the test set.

To classify vessel and non vessel pixels from the Gaussian filter response, the Otsu's thresholding method [32] was applied. In Fig. 12, the binary segmentation results on a subset of retinal images is presented. The segmentation results illustrates the Major Temporal Arcade as the main vessel-like structure in the images. Finally, in order to remove isolated pixels from the segmented image, a length filtering was used to erase structures smaller than 1500 pixels. In Fig. 13, the final segmentation result, which is the input for the numerical modeling stage is illustrated.

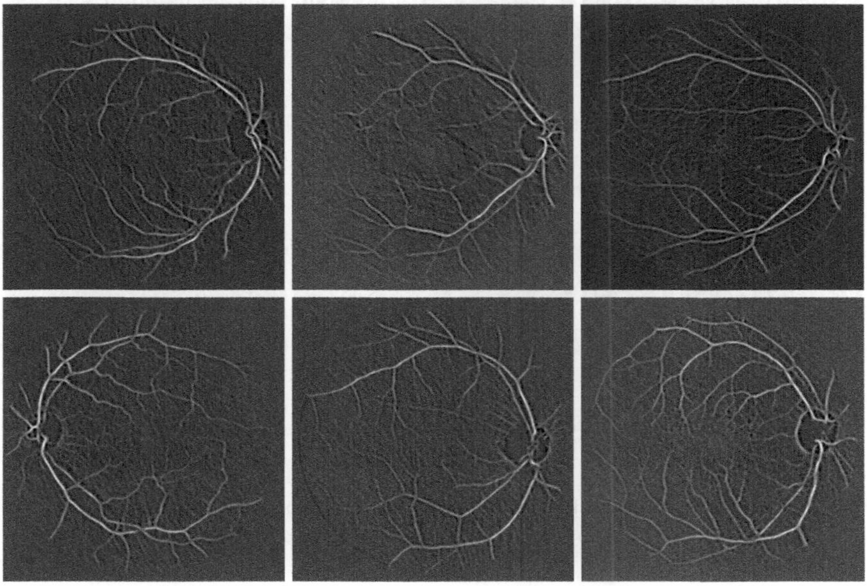

Fig. 11 Filter response of the Gaussian matched filters using the test set of retinal images

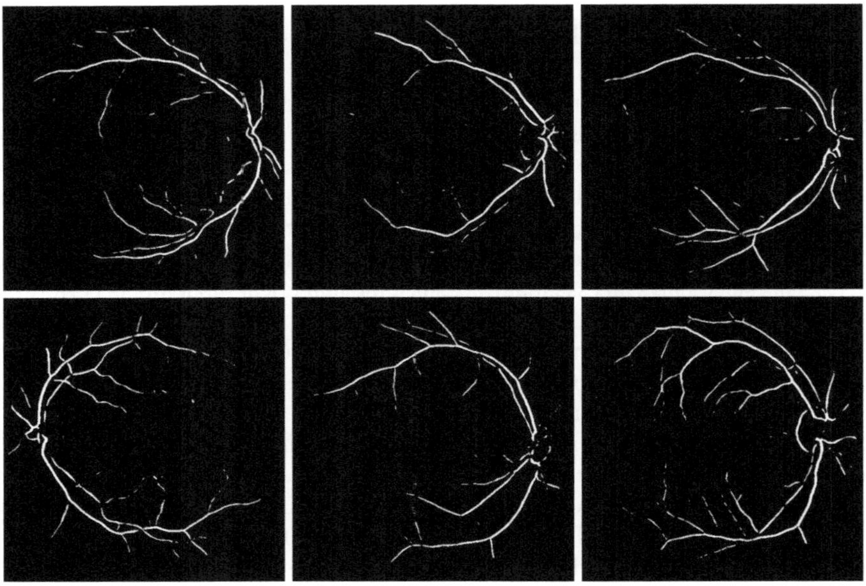

Fig. 12 Segmentation of the Gaussian filter response in order to obtain the MTA

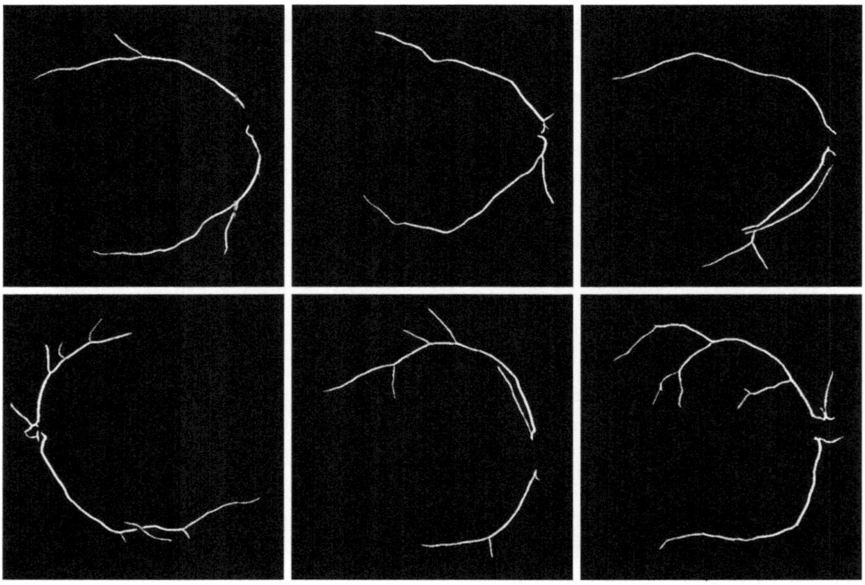

Fig. 13 Final segmentation result of the MTA after applying a length filtering

Table 1 Results of 30 independent runs of the QGA for determining the spline curves using the test set of images

Iteration	MDCP (px.)	Hausdorff (px.)	Time (s)
1	11.72	52.35	7.25
2	9.92	53.15	7.54
3	7.82	53.15	7.52
4	6.79	52.88	7.93
5	8.00	53.85	7.43
6	9.47	52.63	7.24
7	9.21	52.83	7.61
8	8.23	52.47	7.31
9	11.14	52.88	7.24
10	9.56	52.88	7.40
11	9.43	53.14	7.27
12	10.53	52.80	7.22
13	10.20	52.80	7.21
14	9.85	52.80	7.53
15	8.08	52.63	7.30
16	9.19	52.95	7.36
17	13.59	52.83	7.78
18	7.97	53.23	7.60
19	9.00	52.77	7.64
20	17.12	63.12	7.49
21	9.76	52.88	7.72
22	8.72	53.16	7.46
23	10.19	52.47	7.56
24	9.85	52.63	7.34
25	10.05	52.47	7.50
26	7.58	53.16	7.52
27	11.42	53.00	7.96
28	14.47	56.08	7.98
29	9.34	52.80	7.97
30	8.88	52.80	7.31

After the MTA segmentation step, the numerical modeling stage is performed using the skeleton of the MTA. Since the proposed strategy is determine the optimal control points position by using the Quantum genetic algorithm, which is a stochastic optimization method, a set of 30 independent runs is presented in Table 1.

It is important to point out that according to Table 1, the use of the QGA obtained an average of 9.91 and 53.32 pixels, and a standard deviation of 2.13 and 1.96, in terms of the MDCP and Hausdorff distance, respectively.

Table 2 Different methods comparison about the closeness between modeling approximation and ground-truth MTA delineation. Average of 30 runs of the MDCP and Hausdorff distances

Method	MDCP (px.)	Hausdorff(px.)
	Mean ± Std.	Mean ± Std.
General hough transform	31.28 ± 0.00	64.49 ± 0.00
MIPAV-software	25.69 ± 0.00	59.91 ± 0.00
UMDA+SA	30.45 ± 12.94	105.8 ± 27.54
Proposed method	**9.91 ± 2.13**	**53.32 ± 1.96**

Table 3 Average execution time comparison of MTA for modeling using the test set of images

Method	Execution time (s)
General Hough	4.7641 (per pixel)
MIPAV	230
UMDA+SA	**1.68**
Proposed method	7.51

In order to evaluate the numerical modeling results of the proposed methods with different methods of the state of the art, in Table 2 a comparative analysis is presented. The comparison was performed in terms of the closeness between the numerical approximation model and the original image data (ground-truth MTA delineation). In this analysis, the results of the proposed method represent the average values obtained from the previously presented 30 runs. Moreover, the comparative methods are based on the generalized Hough transform, the publicly available MIPAV software using a variant of the Hough transform, and a method based on a hybrid evolutionary algorithm.

On the other hand, in Table 3 a comparative analysis in terms of computational time is shown. In general, the proposed method is faster than the Hough-based methods (They use an exhaustive search), and slower than the evolutionary method since the QGA requires a $2n$ population to perform the Quantum search.

To illustrate the spline curves modeling the Major temporal arcade, a subset of retinal fundus images is presented in Fig. 14. The modeling of the MTA consists of the construction of cubic spline curves of 5 control points located by the QGA. The number of control points was determined experimentally by using the training set of images.

Obtaining the polynomial functions of the numerical modeling performed through spline curves can be very useful for monitoring the changes in the MTA structure and to simulate different application scenarios. The polynomial functions of the spline curves are shown in Table 4, where column image is the number in the test set of the DRIVE database, and four polynomial functions are presented since the proposed method works with 5 control points.

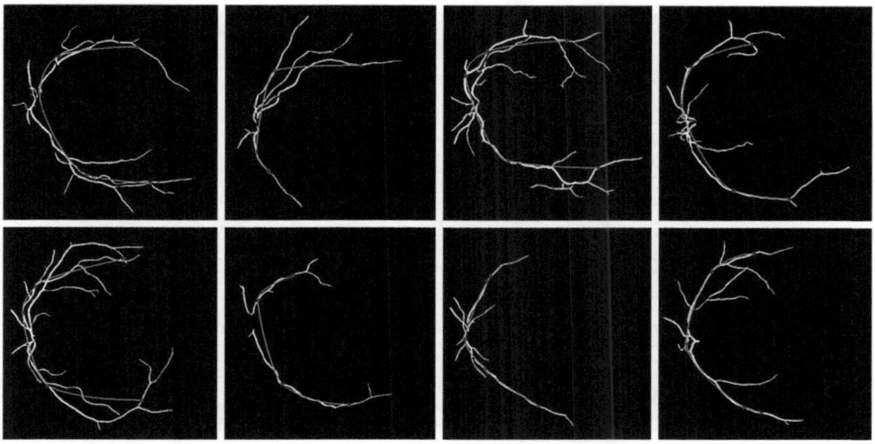

Fig. 14 Modeling of the MTA using Spline curves on a subset of retinal images of the test set

4 Conclusions

In this chapter, a novel method for the automatic numerical modeling of the Major Temporal Arcade (MTA) was introduced. The method consists of two different steps; vessel segmentation and numerical modeling. In the segmentation step, Gaussian matched filters were tuned to detect the main vessel-like structure (MTA) in retinal fundus images. In the second stage of the proposed method, the numerical modeling was performed using spline curves where the control points position was determined using a Quantum genetic algorithm (QGA). In order to determine the optimal set of parameters for the proposed method, a training set of 20 retinal fundus images was used. In the experimental results, the proposed method was evaluated in terms of distance modeling and computational time with respect different state-of-the-art methods. Using a test set of 20 retinal fundus images, the proposed method obtained a Mean distance to the closest point of 9.91, Hausdorff distance of 53.32 and an average computational time of 7.51 s. According to the experimental results, the use of the QGA for the curve fitting problem in the modeling of the MTA, is appropriate for computer-aided diagnosis in ophthalmology.

Table 4 Polynomial functions of the cubic spline curves modeled on the images presented in Fig. 14

Image	Spline $S(y)$	Function
	Interval (y-axis)	
01_test	[132,150)	$-0.001y^3 + 0.439y^2 - 67.726$
	(150,272)	$-0.001y^3 + 0.576y^2 - 105.21y + 6353.25$
	(272,460)	$0.0002y^3 - 0.21y^2 + 69.849y - 7027.18$
	(460,472]	$-41.914y^3 + 58010.333y^2 - 26762393y + 4115474996$
02_test	[160,244)	$17.474y^3 - 8420.148y^2 + 1352327.231y - 72391643$
	(244,394)	$-0.013y^3 + 1.025y^2 - 2653.2y + 228260$
	(394,443)	$-0.587y^3 + 694.178y^2 - 273570y + 35938000$
	(443,444]	$-0.00001y^3 + 0.022y^2 - 9.481y + 1790$
03_test	[95,105)	$-0.0003y^3 + 0.1547y^2 - 26.305y + 1595.6$
	(105,233)	$-0.008y^3 + 3.204y^2 - 392.5y + 15805$
	(233,419)	$0.0002y^3 - 0.191y^2 + 55.161y - 4690.4$
	(419,431]	$-0.0002y^3 + 0.434y^2 - 212.929y + 34186$
06_test	[85,152)	$-0.0003y^3 + 0.177y^2 - 31.315y + 1909.7$
	(152,340)	$-0.003y^3 + 1.848y^2 - 325.239y + 18878$
	(340,468)	$0.003y^3 - 3.751y^2 + 1380y - 167870$
	(468,506]	$-0.715y^3 + 1016.9y^2 - 481690y + 76041000$
07_test	[141,157)	$-0.0007y^3 + 0.455y^2 - 90.181y + 6109$
	(157,299)	$-476y^3 + 224919y^2 - 35422968y + 1859668611$
	(299,395)	$0.696y^3 - 634.477y^2 + 192520y - 58404000$
	(395,488]	$0.0001y^3 - 0.226y^2 + 89.446y - 116360$
09_test	[136,140)	$0.001y^3 - 0.812y^2 + 160.248y - 10090$
	(140,193)	$0.015y^3 - 7.312y^2 + 1178.1y - 62282$
	(193,926)	$0.008y^3 - 7.631y^2 + 2330.6y - 237020$
	(926,486]	$0.00002y^3 - 0.016y^2 + 1.258y + 626.163$
14_test	[97,157)	$-0.001y^3 + 0.508y^2 - 67.054y + 3107.4$
	(157,212)	$-0.023y^3 + 12.108y^2 - 2061.8y + 116470$
	(212,444)	$0.0001y^3 - 0.094y^2 + 26.216y - 1845.2$
	(444,474]	$-0.002y^3 + 4.236y^2 - 20203y + 319720$
17_test	[91,171)	$0.015y^3 - 4.362y^2 + 405.638y - 12073$
	(171,271)	$-0.0008y^3 + 0.525y^2 - 108.81y + 7458$
	(271,485)	$0.007y^3 - 6.705y^2 + 1962.2y - 189870$
	(485,492]	$0.0001y^3 - 0.16y^2 + 85.544y - 13739$

Acknowledgements This research has been supported by the National Council of Science and Technology of México under project Cátedras-CONACYT No. 3150-3097, and posdoctoral studies 290388-2159529.

Appendix A

In this appendix, the main structure of the Quantum Genetic Algorithm in Matlab code is provided.

```
clear variables;
close all;
clc;

% QUANTUM GENETIC ALGORITHM (MATLAB CODE)

I=imclearborder(imread('Input_image.png'));
I=imrotate(I,180); %Only in case need to rotate image
popSize = 100;
maxIters = 50;
howManyPolygonPoints = 5; %Number of control points
searchedValue = 1;

[BestSol, BestSols, WorstSol, searchSpace] = qga_max(...
I, ...
popSize, maxIters, ...
howManyPolygonPoints, ...
searchedValue);

cells = searchSpace(BestSol.Chromosome);
cps = [];
for i = 1 : size(cells, 1)
[row, col] = ind2sub(size(I), cells(i));
cps = [cps; row, col];
end
[puntosx, puntosy] = interpolar(cps, 1, size(I, 2), 1, size(I, 1));
points_approx = sp2p(puntosy, puntosx);
IO2 = drawShapeRGB(I, points_approx, 0, 1, 0, 4);
IO2 = drawSplines(IO2, cps, 1, 0, 0, 64);
figure(); imshow(IO2);

%MDCP and Hausdorff distance
mdcp=MDCP(XY,newXY);
H=hausdorff(XY,newXY);

%FUNCTIONS TO RUN CODE

function [BestSol, BestSols, WorstSol, searchSpace] = qga_max(...
I, ...
popSize, ...
maxIters, ...
howManyPolygonPoints, ...
```

```
searchedValue, ...
fig)
% GA Parameters
MaxIt = maxIters;
nPop = popSize;
pc = 0.7;
nc = 2 * round(pc * nPop/2);
pm = 0.3;
nm = round((pm * nPop) / 2.0);
mu = 0.02;
TournamentSize = 3;

% GA Initialization
% Determine the Search Space:
I2=bwmorph(bwmorph(I,'skel',inf),'spur',10);
searchSpace = find(I2 == searchedValue);
indexes = 1:size(searchSpace, 1);

% Generate First Population:
struct_individual = struct_ga();
pop = repmat(struct_individual, nPop, 1);
disp('Generating Initial population...');
for k = 1 : nPop
pop(k).Chromosome = getRandomPoints(searchSpace, 0, size(I),
     howManyPolygonPoints, 2);
end

% Add Quantum Elements to First Population:
pop = qpop(pop, indexes, searchSpace, size(I));

% CROSSOVER
popc = repmat(struct_individual, nc / 2, 2);
for k = 1 : nc/2
% Select Parents Indices
i1 = TournamentSelection(pop, TournamentSize);
i2 = TournamentSelection(pop, TournamentSize);

% Select Parents
p1=pop(i1);
p2=pop(i2);
mp = round(numel(p1.Chromosome) / 2.0);
% Perform Crossover
%[popc(k,1).Chromosome, popc(k,2).Chromosome]= Crossover(p1.Chromosome, p2.
     Chromosome);
[c1, c2] = SinglePointCrossover(p1.Chromosome, p2.Chromosome);
if numel(p1.Chromosome) < 1 || mp < 1
p1.Chromosome = getRandomPoints(searchSpace, 0, size(I), howManyPolygonPoints
     , 2);
elseif p1.Chromosome(mp) ~= c1
p1.Chromosome = getRandomPoints(searchSpace, c1, size(I),
     howManyPolygonPoints, 2);
end

```

```
if numel(p2.Chromosome) < 1 || mp < 1
p2.Chromosome = getRandomPoints(searchSpace, 0, size(I), howManyPolygonPoints, 2);
elseif p2.Chromosome(mp) ~= c2
p2.Chromosome = getRandomPoints(searchSpace, c2, size(I), howManyPolygonPoints, 2);
end

popc(k, 1) = p1;
popc(k, 2) = p2;

% MUTATION
popm = repmat(struct_individual, nm, 1);
for k = 1 : nm
% Select Parent
i = randi([1 nPop]);
p = pop(randi([1 nPop]));

% Perform Mutation
popm(k).Chromosome = MutateContinuos(p.Chromosome, mu, 1, size(searchSpace, 1));
%Adjust Individual:
%Adjust Individuals before evaluation_
popm(k).Chromosome = adjustChromosome(popm(k).Chromosome, searchSpace, size(I));
popm(k).Chromosome = getRandomPoints(searchSpace, 0, size(I), howManyPolygonPoints, 2);
end
```

References

1. C. Wilson, M. Theodorou, K.D. Cocker, A.R. Fielder, The temporal retinal vessel angle and infants born preterm. Br. J. Ophthalmol. **90**(6), 702–704 (2006)
2. S. Eiho, Y. Qian, Detection of coronary artery tree using morphological operator. Comput. Cardiol. **24**, 525–528 (1997)
3. Y. Qian, S. Eiho, N. Sugimoto, M. Fujita, Automatic extraction of coronary artery tree on coronary angiograms by morphological operators. Comput. Cardiol. **25**, 765–768 (1998)
4. K. Sun, N. Sang, Morphological enhancement of vascular angiogram with multiscale detected by gabor filters. Electron. Lett. **44** (2008)
5. A. Frangi, W. Niessen, K. Vincken, M. Viergever, Multiscale vessel enhancement filtering, in Medical Image Computing and Computer-Assisted Intervention (MICCAI'98). Springer LNCS **1496**, 130–137 (1998)
6. N. Salem, A. Nandi, Unsupervised segmentation of retinal blood vessels using a single parameter vesselness measure, in *Sixth Indian Conference on Computer Vision, Graphics and Image Processing*, vol. 34 (IEEE, 2008), pp. 528–534
7. T. Tsai, H. Lee, M. Chen, Adaptive segmentation of vessels from coronary angiograms using multi-scale filtering, in *International Conference on Signal-Image Technology and Internet-Based Systems* (2013), pp. 143–147
8. J. Soares, J. Leandro, R. César, H. Jelinek, M. Cree, Retinal vessel segmentation using the 2-D Gabor wavelet and supervised classification. IEEE Trans. Med. Imaging **25**, 1214–1222 (2006)

9. F.J. Ayres, R.M. Rangayyan, Design and performance analysis of oriented feature detectors. J. Electron. Imag. **16**(2), 023007 1–12 (2007)
10. I. Cruz-Aceves, F. Oloumi, R.M. Rangayyan, J.G. Aviña-Cervantes, A. Hernandez-Aguirre, Automatic segmentation of coronary arteries using gabor filters and thresholding based on multiobjective optimization. Biomed. Signal Process. Control **25**, 76–85 (2016)
11. S. Chaudhuri, S. Chatterjee, N. Katz, M. Nelson, M. Goldbaum, Detection of blood vessels in retinal images using two-dimensional matched filters. IEEE Trans. Med. Imaging **8**, 263–269 (1989)
12. T. Chanwimaluang, G. Fan, S. Fransen, Hybrid retinal image registration. IEEE Trans. Inf Technol. Biomed. **10**, 129–142 (2006). (Jan)
13. W. Kang, K. Wang, W. Chen, W. Kang, Segmentation method based on fusion algorithm for coronary angiograms, in *2nd International Congress on Image and Signal Processing (CISP)* (2009), pp. 1–4
14. W. Kang, W. Kang, Y. Li, Q. Wang, The segmentation method of degree-based fusion algorithm for coronary angiograms, in *2nd International Conference on Measurement, Information and Control* (2013), pp. 696–699
15. M. Al-Rawi, M. Qutaishat, M. Arrar, An improved matched filter for blood vessel detection of digital retinal images. Comput. Biol. Med. **37**(2), 262–267 (2007)
16. M. Al-Rawi, H. Karajeh, Genetic algorithm matched filter optimization for automated detection of blood vessels from digital retinal images. Comput. Methods Programs Biomed. **87**(3), 248–253 (2007)
17. I. Cruz-Aceves, A. Hernandez-Aguirre, S.I. Valdez, On the performance of nature inspired algorithms for the automatic segmentation of coronary arteries using gaussian matched filters. Appl. Soft Comput. **46**, 665–676 (2016)
18. F. Oloumi, R.M. Rangayyan, A.L. Ells, Parabolic modeling of the major temporal arcade in retinal fundus images. IEEE Trans. Instrum. Meas. **61**(7), 1825–1838 (2012)
19. F. Oloumi, R.M. Rangayyan, P. Casti, A.L. Ells, Computer-aided diagnosis of plus disease via measurement of vessel thickness in retinal fundus images of preterm infants. Comput. Biol. Med. **66**(1), 316–329 (2015)
20. F. Oloumi, R.M. Rangayyan, A.L. Ells, Computer-aided diagnosis of retinopathy of prematurity in retinal fundus images, in *Medical Image Analysis and Informatics* (CRC Press, 2017), pp. 57–83
21. F. Oloumi, R. Rangayyan, A L. Ells, Dual-parabolic modeling of the superior and the inferior temporal arcades in fundus images of the retina, in *2011 IEEE International Symposium on Medical Measurements and Applications* (2011)
22. J.d.J. Guerrero-Turrubiates, I. Cruz-Aceves, S. Ledesma, J.M. Sierra-Hernandez, J. Velasco, J.G. Avina-Cervantes, M.S. Avila-Garcia, H. Rostro-Gonzalez, R. Rojas-Laguna, Fast parabola detection using estimation of distribution algorithms, in *Computational and Mathematical Methods in Medicine*, vol. 2017 (2017)
23. S.I. Valdez, S. Espinoza-Perez, F. Cervantes-Sanchez, I. Cruz-Aceves, Hybridization of the univariate marginal distribution algorithm with simulated annealing for parametric parabola detection, in *Hybrid Metaheuristics for Image Analysis* (Springer, 2018), pp. 163–186
24. S. Bhattacharyya, M. Köppen, E. Behrman, I. Cruz-Aceves, Automatic feature selection for coronary stenosis detection in x-ray angiograms, in *Hybrid Quantum Metaheuristics Theory and Applications*, vol. 1 (CRC Press, 2022), pp. 117–140
25. N.S.-N. Lam, Spatial interpolation methods: a review (1983)
26. V.-M. Taavitsainen, Denoising and Signal-to-Noise Ratio Enhancement: Splines **12**, 67–83 (2009)
27. J.R. Sampson, Adaptation in natural and artificial systems (John H. Holland). SIAM Rev. **18**(3), 529–530 (1976)
28. A. Narayanan, M. Moore, Quantum-inspired genetic algorithms, in *Proceedings of IEEE International Conference on Evolutionary Computation* (1996), pp. 61–66
29. H. Wang, J. Liu, J. Zhi, C. Fu, The improvement of quantum genetic algorithm ans its application od function optimization, in *Mathematical Problems in Engineering*, vol. 10 (2013), p. 30749

30. T. Hey, Quantum computing: an introduction. Comput. Control Eng. J. **10**(3), 105–112 (1999)
31. R. Lahoz-Beltra, Quantum genetic algorithms for computer science. Computers **5** (2016)
32. N. Otsu, A threshold selection from gray-level histograms. IEEE Trans Syst., Man Cybern. **9**(1), 62–66 (1979)

Entangled Quantum Neural Network

Qinxue Meng, Jiarun Zhang, Zhao Li, Ming Li, and Lin Cui

Abstract Quantum entanglement (QE) is the phenomenon that when several particles interact, the properties of each particle will be integrated into the properties of the overall system, and the properties of each particle cannot be described independently from others. QE can be proved by violating Bell Inequality, that is, it can describe strong statistical correlation (i.e., quantum correlation). By introducing QE into machine learning, the adjusted framework would have advantages such as faster execution time of the learning algorithms and stronger capacity. Therefore, here we introduce our novel framework called Entangled Quantum Neural Network, EQNN. Using quantum entanglement to development on neuron networks can be described in three different ways: By replacing the hidden layer nodes of the Multi-Layer Perception (MLP) with a measurement process of entangled states (QECA, QCCA); by replacing the output layer of MLP with a quantum measurement operation (ECA); or by reconstructing the neurons in NN using regularizer to constrain state vectors to entangled states (QNN). With extensive experiments on the three most frequently used machine learning datasets from UCI, Abalone, Wine Quality (Red), and Wine Quality (White), we demonstrate that all QCCA, QECA, ECA, and QNN outperform the baseline algorithms. Under the same parameter settings, which are: learning rate

---

Q. Meng
College of Information Engineering, Suzhou University, Suzhou, China

J. Zhang (✉)
University of California San Diego, San Diego, USA
e-mail: jiz727@ucsd.edu

Z. Li
Zhejiang University, Link2Do Technology Ltd., Hangzhou, Zhejiang, China
e-mail: zhao_li@zju.edu.cn

M. Li
The Key Laboratory of Intelligent Education Technology and Application of Zhejiang Province, Zhejiang Normal University, Jinhua, China
e-mail: mingli@zjnu.edu.cn

L. Cui (✉)
College of Information Engineering, Suzhou University, Suzhou 234000, China
e-mail: cl@ahszu.edu.cn

R. Pandey et al. (eds.), *Quantum Computing: A Shift from Bits to Qubits*,
Studies in Computational Intelligence 1085,
https://doi.org/10.1007/978-981-19-9530-9_14

is 0.001, mini-batch is 1, training epochs is 500 and initial weight is 0.01, the performance among themselves in descending order are QNN > ECA > QECA > QCCA.

Keywords Quantum entanglement · Quantum probability · Multi-layer perceptron

1 Introduction

After 30 years of development since Rosenblatt F. proposed the Perceptron in 1958 laying the ground for Neural Networks [1], in 1986 Rumelhart et al. [2] invented the Backpropagation algorithm and solved the complex computational problem required by the two-layer neural network. Since then, neural network, NN, becomes one of the most popular and dominant frameworks in the field of machine learning. In 2006 Hinton et al. further proposed the concept of deep learning [3]. Different from traditional training methods, "Deep Learning Network" has a "pre-training" process, which can easily allow the weights in the neural network to find a value close to the optimal solution, and then use Fine-tuning techniques to optimize the training of the entire network. The use of these two techniques greatly reduces the time to train multi-layer neural networks. Compared to traditional machine learning algorithms, Multi-layer Neural Networks, also known as Multi-Layer Perceptron (MLP), can infinitely approximate continuous functions. This means that it has better performance on complex nonlinear classification tasks and thus has its decisive advantage on tasks such as speech recognition and image recognition.

To improve the learning ability of deep learning models to solve more complicated tasks, research work based on classic paradigms imitates the human brain nervous system to build larger and more complex model structures. However, the largest NN we know still has less than two percent of the perceptron's compared to the neurons of a human brain. Larger models require more computing power and face constraints [4]. The non-classical paradigm, such as quantum mechanics theory, has attracted the attention of researchers due to its unique advantages. Quantum mechanics theory has some characteristics that are not possessed by classical theory. For example, observational behavior affects the observed system [5, 6] and the observations of the system are affected by the observed context [7]. These differences (or advantages) may help us to improve and perfect the current classical models and algorithms.

Among all the quantum mechanics theories, Entanglement, or called Quantum Entanglement (QE), is an important quantum resource for quantum computing and quantum information processing. It is a phenomenon that when several particles interact, the properties of each particle will be integrated into the properties of the overall system, and the properties of each particle cannot be described independently from others. QE can provide a statistical correlation between subsystems (or attributes) that is stronger than what classical systems are able to produce [8]. To further develop the performance of neural network, researchers tried to introduce the idea of QE into NN. It can be described in three different ways: by replacing the hidden layer nodes of the MLP with a measurement process of entangled states (QECA, QCCA); by replacing the output layer of MLP with a quantum measurement operation (ECA); or by reconstructing the neurons in NN using regularizer to constrain state vectors to entangled states (QNN). In this paper, we conclude them as a novel framework called Entangled Quantum Neural Network, or EQNN. We first briefly introduce the basic concepts of the quantum entanglement and the multi-layer perceptron and then look deeper into each of the four proposed designs of EQNN models and discuss their differences. With extensive experiment on the three most frequently used machine learning datasets, we demonstrate that all four designs outperforms the baseline algorithms and the descending order of performance among themselves are QNN > ECA > QECA > QCCA.

2 Related Background

Our framework combines the idea of quantum entanglement with multi-layer perceptrons, by using a measurement process of the entangled states to replace the certain layer of the MLP. Here we briefly introduce their background knowledge.

2.1 Quantum Entanglement

In quantum theory, when several particles interact, the properties of each particle will be integrated into the properties of the overall system, and the quantum state of each particle of the group cannot be described independently of the state of the others. This phenomenon is called Quantum Entanglement (QE). This phenomenon shows a strongly correlated system and it was first discussed by Einstein et al. [9]. They formulated the Einstein-Podolsky-Rosen paradox (EPR paradox) in which they claimed that "the quantum-mechanical description of physical reality given by wave functions is not complete" [9]. Schrödinger later, first used the term "entanglement" to refer that, and shortly after published a paper defining the notion of "entanglement" in detail [10]. The EPR paper aroused the interest in the basic theory of quantum mechanics and in 1964, John Bell proved part of the EPR wrong and proposed Bell's inequality. Roughly speaking, Bell's inequality qualitatively gives the difference that if the spins of two particles are measured along different axes, the statistical correlation result obtained by quantum mechanics is much stronger than that of the localized hidden variable theory [11].

The basic definition of an entangled system is.

Definition 1 Suppose an arbitrary composite system is composed of two quantum subsystems A and B, and the Hilbert spaces of these two subsystems are H_A and

H_B, then The Hilbert space of the composite system is the tensor product $H_{AB} = H_A \otimes H_B$. If the two subsystems A, B are in quantum state $|\alpha\rangle_A$, $|\beta\rangle_B$ and the quantum state of the composite system $|\psi\rangle_{AB}$ cannot be written as a tensor product $|\alpha\rangle_A \otimes |\beta\rangle_B$, Then this composite system is called an entangled system of subsystems A and B, and the two subsystems A and B are entangled with each other.

We begin with an arbitrary bipartite entangled state in the bases $\sigma_3|\pm\rangle = \pm|\pm\rangle$[1] that

$$|\psi\rangle = \alpha|+-\rangle + \beta|-+\rangle \tag{1}$$

where α and β are the normalization condition with $|\alpha|^2 + |\beta|^2 = 1$ but $\alpha, \beta \neq 0$. Without losing generality, α and β can be parameterized as $\alpha = e^{i\eta}\sin(\xi)$, $\beta = e^{-i\eta}\cos(\xi)$, where i is the imaginary number with $i^2 = -1$ and η, ξ are two real parameters but $\sin(\xi), \cos(\xi) \neq 0$. The density matrix of the entangled state, $\rho = |\psi\rangle\langle\psi|$, can be separated to the local and non-local parts [12], $\rho = \rho_{lc} + \rho_{nlc}$. The local part

$$\rho_{lc} = \sin^2(\xi)|+-\rangle\langle+-| + \cos^2(\xi)|-+\rangle\langle-+|, \tag{2}$$

describes the classic statistical correlation between subsystems (or properties), which belongs to the classical statistics. The non-local part

$$\rho_{nlc} = \sin(\xi)\cos(\xi)\left(e^{i2\eta}|+-\rangle\langle-+| + e^{-i2\eta}|-+\rangle\langle+-|\right) \tag{3}$$

describes the phenomenon of interference between subsystems (or properties), which belongs to the quantum statistics.

2.2 The Measurement on Density Matrix

Here we give the measurement process of the quantum probability that will be used for QE. The observable has a spectral decomposition, $M_r = \sum_m m P_r^m$, where P_r^m is the projector onto the eigenspace of M_r with eigenvalue m. The possible outcomes of the measurement correspond to the eigenvalues, m, of the observable. Upon measuring the state $|\varphi\rangle$, the probability of getting result m is given by

$$p(m_r) = Tr[P_r^m(|\varphi\rangle\langle\varphi|)] = \langle\varphi|P_r^m|\varphi\rangle \tag{4}$$

where Tr denotes the trace of the matrix.

By definition, the average value of the measurement is

[1] $\{|+\rangle, |-\rangle\}$ denotes an arbitrary orthonormal basis of the 1-qubit Hilbert space $\mathbb{C}^2$. $\sigma_3 = \sigma_z$ denotes Pauli matrix, and Pauli matrix refers to four common matrices, which are 2×2 matrix, each with its own mark, namely $\sigma_x \equiv \sigma_1 \equiv X$, $\sigma_y \equiv \sigma_2 \equiv Y$, $\sigma_z \equiv \sigma_3 \equiv Z$ and $\sigma_0 \equiv I$.

$$E(M) = \sum_m mp(m) = \sum_m m\langle\varphi|P_m|\varphi\rangle = \langle\varphi|M|\varphi\rangle. \tag{5}$$

The average value of the observable M is often written $\langle M\rangle \equiv \langle\varphi|M|\varphi\rangle$.

Therefore, the joint probability derived from QE is obtained as

$$p(+_a, +_b) = Tr[(P_a^+ \otimes P_b^+)\rho]. \tag{6}$$

It can be also divided into the local (classical probability) and non-local (quantum probability) parts

$$p_{all}(+_a, +_b) = Tr[(P_a^+ \otimes P_b^+)(\rho_{lc} + \rho_{nlc})] \tag{7}$$

$$= Tr[(P_a^+ \otimes P_b^+)\rho_{lc}] + Tr[(P_a^+ \otimes P_b^+)\rho_{nlc}] \tag{8}$$

$$= p_{lc}(+_a, +_b) + p_{nlc}(+_a, +_b). \tag{9}$$

Accordingly, the probability of other combinations, i.e., $p_{lc}(\pm_a, \pm_b)$, $p_{lc}(\mp_a, \pm_b)$, $p_{nlc}(\pm_a, \pm_b)$ and $p_{nlc}(\mp_a, \pm_b)$, can also be obtained. Moreover, the average values of a and b in the classical and quantum cases are

$$\langle ab\rangle_{lc} = -\cos(\theta_a)\cos(\theta_b) \tag{10}$$

and

$$\langle ab\rangle_{nlc} = \sin(\theta_a)\sin(\theta_b)\sin(2\xi)\cos(\phi_a - \phi_b + 2\eta), \tag{11}$$

respectively.

2.3 Multi-layer Perceptron

Multi-layer perceptron, or MLP, was designed by Rumelhart et al. [2]. MLP's the most basic unit is called perceptron, a kind of artificial neuron proposed by Rosenblatt 1958 [1] based on the earlier work from Hebb [13]. Perceptron takes several inputs $a_1, a_2, ..., a_m$, product with their weights $w_1, w_2, ..., w_m$, respectively, and sum up to get one single output $z = g(\sum_{t=1}^{m} a_t * w_x)$. On top of this, by adding a hidden layer, $a_n^{(2)}$, we get multi-layer perceptron, in which $a_1^{(2)} = g(\sum_{t=1}^{m} a_t^{(1)} * w_{1,t}^{(1)})$ and $z = g(\sum_{t=1}^{n} a_t^{(2)} * w_{2,t}^{(2)})$.

Having[2] the MLP model, we now train the model to make the parameters as close as possible to the real model. We use randomly generated parameter values to predict samples in the training data. The predicted target of the sample is y_p, the real target is y, and the value loss is $loss = (y_p - y)^2$. The gradient descent

[2] Figures 1, 2 and 3 are referenced from https://www.cnblogs.com/subconscious/p/5058741.html.

Fig. 1 Schematic diagram of a single layer of perceptron

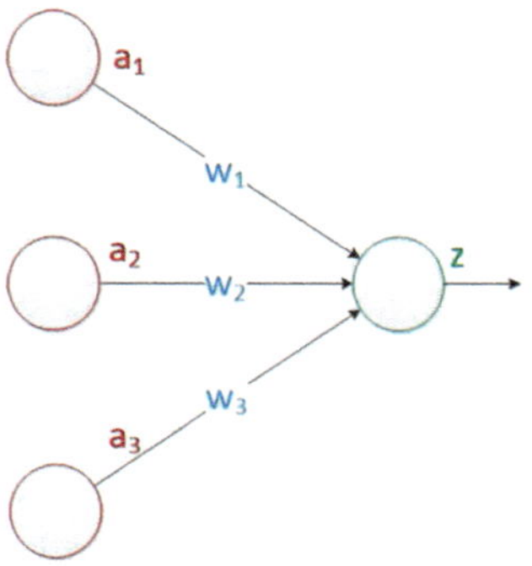

Fig. 2 Schematic diagram of a basic multi-layer perceptron

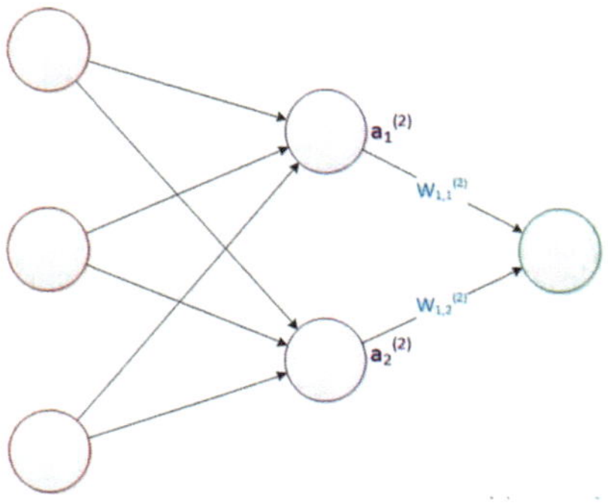

Fig. 3 Schematic diagram of the backpropagation algorithm

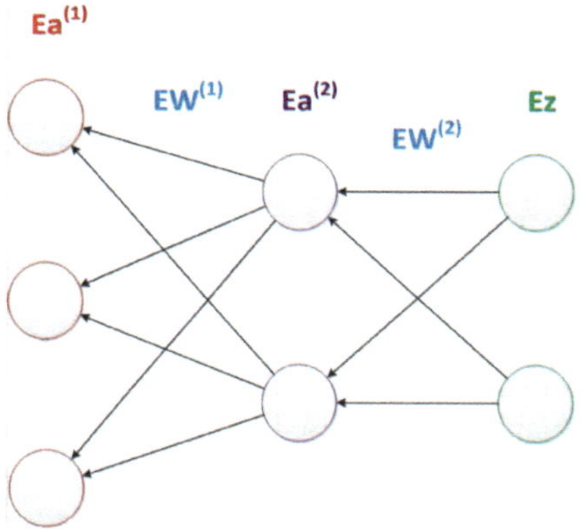

algorithm is used to optimize parameters to minimize the value of the loss function. Due to the high computational cast, a backpropagation algorithm is also required. The backpropagation algorithm can be interpreted as Fig. 3, where the gradient is calculated from the back to the front, and the layers are back-propagated. E in the figure stands for relative derivative, for a more detailed explanation, see Nielsen [14].

The unique property of quantum mechanics has attracted lots of interests of researchers to study the Interdisciplinary of quantum mechanics and machine learning. Considering the advantages of quantum computer processing high-dimensional vectors in large tensor product space, Lloyd et al. [15] proposed a supervised and unsupervised quantum machine learning algorithm for cluster assignment and cluster finding. This algorithm reduces the time complexity of the classical algorithm from the polynomial-time to the logarithmic-time. Levine et al. [16] established a contemporary deep learning architecture that effectively represents highly quantum entangled systems in the form of deep convolutional and recursive networks. By constructing tensor network equivalents of these architectures, they identify the inherent reuse of information in the network operation as a key trait that distinguishes them from standard tensor network-based representations and enhances their entanglement capacity. Schuld et al. [17] interpreted the process of encoding inputs in a quantum state as a nonlinear feature map, mapping the data into a quantum Hilbert space. According to the theory, two quantum classification models are established, which use the variational quantum circuit as a linear model to explicitly classify the data in Hilbert space. The quantum device estimates inner products of quantum states to compute a classically intractable kernel. See also Dunjko and Briegel [18], Adcock et al. [19]. There are also works that focus on redesigning the neural network. Rebentrost et al. [20] coded the network into the probability amplitudes of quantum states to store an exponentially large network in a polynomial number of qubits. Chen et al. [21] proposed a quantum probabilistic network model, which leverages quantum parallelism to track all possible network states to improve performance. In addition, Verdon et al. [22] proposed a quantum graph neural network and Cong et al. [23] proposed a quantum convolutional neural network.

Among all the works in quantum mechanics and neural networks, we here focus on one specific domain of the field, which is combining quantum entanglement into the multi-layer perceptron to optimize its performance on binary classification tasks on a traditional computer. More specifically, we focus on the researches which use a measurement process of the entangled states to replace the certain layer of the MLP. Four models, QECA, QCCA, ECA, and QNN qualify our constraints. Here we will compare the differences between these four models and explore how the QE's incorporation into MLP at different layers impacts MLP's performance differently.

3 The EQNN Framework

Again, there are hour models under EQNN that use measurement processes of the entangled states to replace certain layers of the MLP. Here we look into the details of each model and discuss their differences.

3.1 *Quantum Entanglement Inspired Correlation Learning for Classification (QECA)*

Based on the Multi-layer Perceptron, Zhang et al. [24] develop a novel QE-induced classification framework called Quantum Entanglement Inspired Correlation Learning for Classification (QECA). QECA decomposes the quantum joint probability derived from entangled systems into the classical probability and the quantum interference term uses to the model correlation between features and categories. QECA chooses the quantum system with the maximum entanglement under two qubits as the entangled system, i.e., Bell states, which form is

$$|\Psi\rangle = \frac{1}{\sqrt{2}}(|0\rangle \otimes |0\rangle + |1\rangle \otimes |1\rangle) = \frac{1}{\sqrt{2}}(|00\rangle + |11\rangle)\,. \tag{12}$$

and define the observable subsystem of the entangled system as

$$M_r = \begin{bmatrix} \cos(\theta_r) & e^{-i\phi_r}\sin(\theta_r) \\ e^{i\phi_r}\sin(\theta_r) & -\cos(\theta_r) \end{bmatrix} \tag{13}$$

also written in

$$M_r(\theta_r, \phi_r) = P_r^+(\theta_r, \phi_r) - P_r^-(\theta_r, \phi_r) = |+_r\rangle\langle +_r| - |-_r\rangle\langle -_r| \tag{14}$$

with

$$|+_r\rangle = \cos\left(\frac{\theta_r}{2}\right)|0\rangle + e^{i\phi_r}\sin\left(\frac{\theta_r}{2}\right)|1\rangle \tag{15}$$

$$|-_r\rangle = \sin\left(\frac{\theta_r}{2}\right)|0\rangle - e^{i\phi_r}\cos\left(\frac{\theta_r}{2}\right)|1\rangle \tag{16}$$

The measurement operator of the label, P_{lab} is

$$P_{lab}^+ = \frac{1}{2}(\sigma_1 + \sigma_0), \quad P_{lab}^- = \frac{1}{2}(\sigma_1 - \sigma_0). \tag{17}$$

Assuming that each instance (sample) has N attributes and one label, the positive and the negative measurement operators for the entangled system consists of the n-th attribute and the label are

$$\mathcal{P}_n^{\pm}(\theta_n, \phi_n) = P_n^{+}(\theta_n, \phi_n) \otimes P_{lab}^{\pm}. \tag{18}$$

P_n^+ can also be replaced by P_n^- and the effect is the same. By applying $\mathcal{P}_n^+$ and $\mathcal{P}_n^-$ separately to each entangled system, The measurement operator of the entangled system is defined as the probability values of both positive and negative examples:

$$p_n^{\pm}(\theta_n, \phi_n) = Tr[\mathcal{P}_n^{\pm}(\theta_n, \phi_n)(|\Psi\rangle\langle\Psi|)] = \langle\Psi|\mathcal{P}_n^{\pm}(\theta_n, \phi_n)|\Psi\rangle. \tag{19}$$

The entangled states and the measurement process is defined as above. Based on this framework, QECA uses a fully connected layer to learn the parameters of observations of the subsystems, see Fig. 4. QECA performs weighting summation on the attributes of each instance, $x \in \mathbb{R}^d$, to get the input of the hidden layer neurons. The parameters of the measurement operator of the entangled system are

$$\phi_h = ReLU(\alpha_h + b_h), \tag{20}$$

where $ReLU$ is the activation function Rectified Linear Unit [25], $\alpha_h = \sum_{i=1}^{d} v_{ih}x_i$ ($v_h \in \mathbb{R}^d$) represents the weight, and $b_h \in \mathbb{R}$ is the bias.

Finally, By performing weighting summation on $p_h^{\pm}(\theta_h, \phi_h)$, QECA gets the final output value

$$y^{\pm} = \beta^{\pm} = \sum_{h=1}^{q} w_h^{\pm} p_h^{\pm}(\theta_h, \phi_h) \tag{21}$$

where $w^{\pm} \in \mathbb{R}^q$ represents the weight. $\beta^{\pm}$ represents the input value of the output layer neurons, as shown in Fig. 4.

QECA uses the classical cross-entropy loss function to act on its loss function and Adaptive Moment Estimation (Adam) as the optimizer. The loss function is specific is

$$H(p, q) = -\sum_{j=1}^{N} p(x_j)log(q(x_j)). \tag{22}$$

3.2 *Quantum Correlation Revealed by Bell State for Classification Tasks (QCCA)*

Very likely to QECA, in Quantum Correlation Revealed by Bell State for Classification Tasks (QCCA) [26], the hidden layer nodes of the MLP are replaced with a measurement process. The difference is that in QCCA, the bell inequality is different, and different measurement operators are used, and the upper limit of the calculated correlation will be different. In specific, the two differences of QCCA comparing to QECA are: 1. entangled state in QCCA is

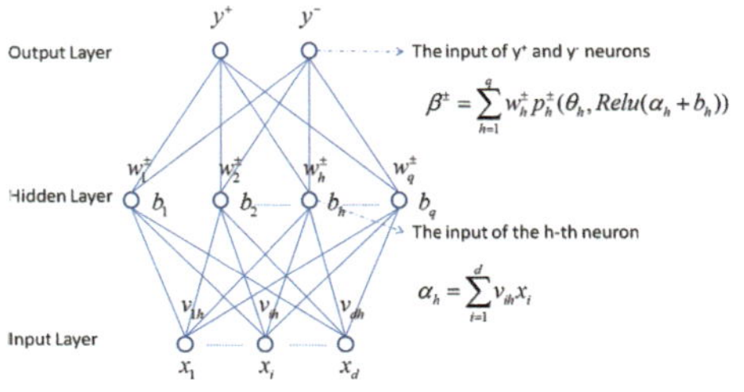

Fig. 4 Schematic diagram of QECA

$$|\Psi\rangle = \alpha|00\rangle + \beta|11\rangle, \tag{23}$$

2. In QCCA, the observations of the subsystems of the entangled system are defined as

$$M_r = \sigma_x \cos(\phi_r) + \sigma_y \sin(\phi_r) \tag{24}$$

$$= \begin{bmatrix} 0 & e^{-i\phi_r} \\ e^{i\phi_r} & 0 \end{bmatrix} \tag{25}$$

3.3 *Strong Statistical Correlation Revealed by Quantum Entanglement for Supervised Learning (ECA)*

Strong Statistical Correlation Revealed by Quantum Entanglement for Supervised Learning (ECA) [27] also shares the same overall ideas with QECA but defines the entanglement states differently. in ECA the entangled states are defined as The Greenberger-Horne-Zeilinger state (GHZ state) [28] and W states [29] (Figs. 5, 6 and 7):

$$|GHZ\rangle = \frac{1}{\sqrt{2}}\left(|0\rangle^{\otimes N} + |1\rangle^{\otimes N}\right) \tag{26}$$

$$|W\rangle = \frac{1}{\sqrt{3}}\left(|001\rangle + |010\rangle + |100\rangle\right). \tag{27}$$

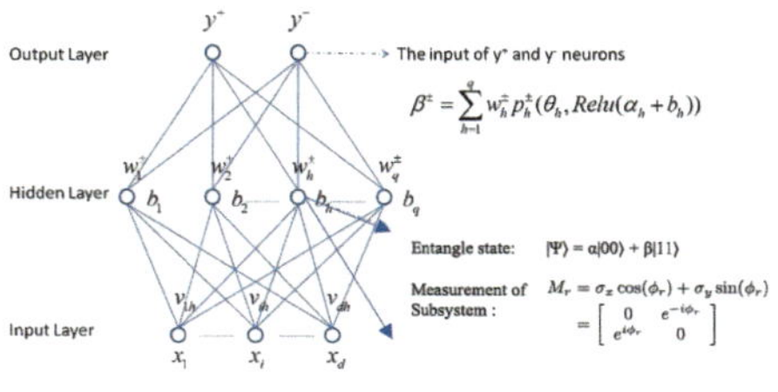

Fig. 5 Schematic diagram of QCCA

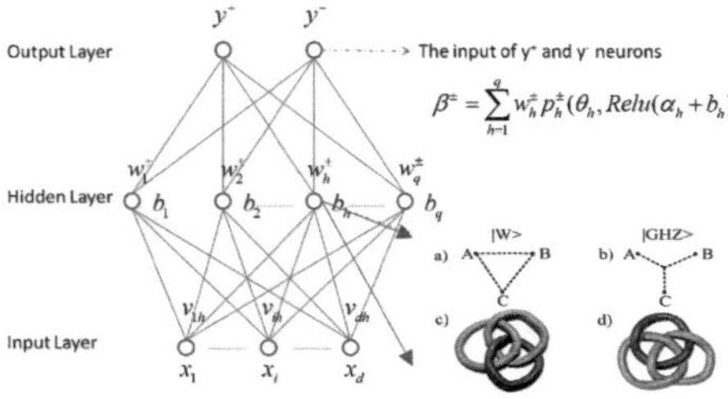

Fig. 6 Schematic diagram of ECA in which the graphical description of the entanglement relationship of the W and GHZ states is referenced from Ref. [30]

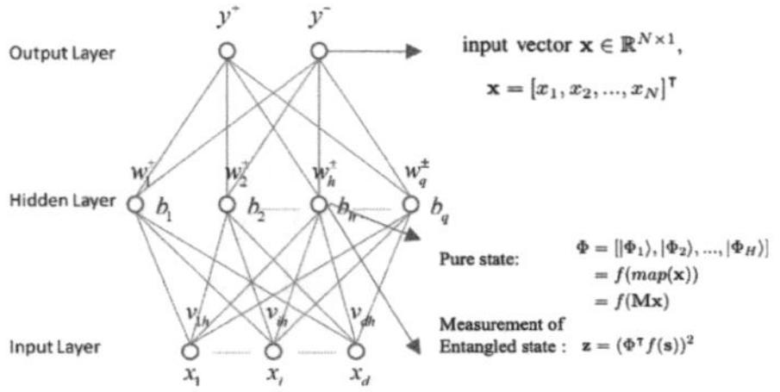

Fig. 7 Schematic diagram of QNN where in QNN, the input vector is first changed into pure states and then entangled by regularizer

As a result, the measurement process also differs where

$$P^{\pm} = \langle GHZ|p^{\pm}|GHZ\rangle . P^{\pm} = \langle W|p^{\pm}|W\rangle . \tag{28}$$

3.4 *Neural Network Model Reconstructed from Entangled Quantum States (QNN)*

Not like the previous three models, Neural Network Model Reconstructed from Entangled Quantum States (QNN) attempts to use the entangled state to reconstruct all neurons of NNs so that the network model can characterize the strong statistical relationship between the features. Specifically, based on concurrence, QNN proposes a regularizer that can constrain a state vector to an entangled state, and apply it to the optimization process to ensure that the vector passed to the neuron is a legally entangled state, then, define the measurement process of the entangled state as a neuron, and construct a network structure. To be specific, QNN defines the following operation

$$\mathbf{v} = map(\mathbf{x}) = \mathbf{M}\mathbf{x} \tag{29}$$

where $\mathbf{M} \in \mathbb{R}^{4\times n}$, and

$$|\Phi\rangle = f(\mathbf{v}) = \frac{1}{\mathcal{N}}\mathbf{v} \tag{30}$$

where $\mathcal{N} = \sqrt{\sum_{i=1}^{4} \mathbf{v}_i^2}$. and gets

$$|\Phi\rangle = f(\mathbf{Mx}) \tag{31}$$

where **M** is the parameter that need to be learned, and **x** is the input vector of the neuron. QNN uses (31) to map the n-dimensional input vector **x** to a 4-dimensional vector $\mathbf{v} \in \mathbb{R}^{4\times 1}$ and makes it a legal pure state $|\Phi\rangle$. QNN then defines a regularizer

$$\mathcal{R}(|\Phi\rangle) = 1 - |\langle\Phi|\tilde{\Phi}\rangle| \tag{32}$$

where $|\tilde{\Phi}\rangle = (\sigma_y \otimes \sigma_y)|\Phi\rangle$, $\langle\Phi| = (|\Phi\rangle)^{\dagger}$, σ_y is the Pauli operator, $\sigma_y = \begin{bmatrix} 0 & -i \\ i & 0 \end{bmatrix}$. then add the regularizer to the objective function constraining the quantum state to become an entangled state. Like the previous three models, QNN also need a measurement operator and is defined as

$$\Pi = |\mathbf{s}\rangle\langle\mathbf{s}| \tag{33}$$

where $|\mathbf{s}\rangle = f(\mathbf{s}) = |\mathbf{s}_1\rangle \otimes |\mathbf{s}_2\rangle = f(\mathbf{s}_1) \otimes f(\mathbf{s}_2)$, $\mathbf{s} = \mathbf{s}_1 \otimes \mathbf{s}_2$ and $\mathbf{s}_i$, $i = \{1, 2\}$, are a 2-dimensional vector that needs to be learned. The output of the neuron can get through measurement, and the final form of the neuron is

$$y = Tr(\Pi|\Phi\rangle\langle\Phi|) = \langle\Phi|\Pi|\Phi\rangle \tag{34}$$

$$= (\langle\Phi|\mathbf{s}\rangle)^2 = (f(\mathbf{s}^{\mathsf{T}})f(\mathbf{Mx}))^2 \tag{35}$$

3.5 Discussion

To sum up, QECA, ECA, and QCCA are similar in their network design that they all use a measurement process of the entangled states to replace a layer in the MLP. QECA and ECA share the same definition of observations of the subsystems of the entangled system but their entangled states are defined differently. QECA defines its entangled states as Bell state while ECA defines them as W state and QHZ state. QECA and QCCA have more differences as they not only define different entangled states but also have different observations of the subsystems. See from Eqs. (13) and (24), QCCA only focus on the phase but QECA adds the probability amplitude. The latter experiment data will show its impacts on the performance and will be explained in more detail. QNN, on the other hand, start from a completely different angle by using a regularizer that can constrain a state vector to an entangled state to reconstruct the neuron. Four frameworks have their different perspectives toward EQNN and their performances vary, but they all outcompetes the baseline modules such as SVM and MLP which will be shown latter in the experimental part.

4 Experiment Results

4.1 Baselines

To evaluate the effect of EQNN and comparing the difference between QECA, QCCA, ECA, and QNN, we conduct comparative experiments which are compared with standard MLP and compared with most representative classification algorithms. Among the statistical models, KNN, SVM, and LDA will be the most representative models with outstanding performance. These will also be the baseline algorithms we comparing to in our experiment. T Cover et al. proposed k-NN classification which assigns to an unclassified sample point the classification of the nearest of a set of previously classified points [31]. Cortes et al. completed the model of Support-vector machine (SVM) which maps the input vectors into some high-dimensional feature space Z through some nonlinear mapping chosen a priori [32]. In 2003, Blei et al. proposed latent Dirichlet allocation (LDA). LDA is a three-level hierarchical Bayesian model, in which each item of a collection is modeled as a finite mixture over an underlying set of topics. Each topic is, in turn, modeled as an infinite mixture over an underlying set of topic probabilities [34]. The MLP has an input layer, a hidden layer, and an output layer. The input layer has N input neurons (nodes), where N is the number of features (attributes) of the instance; the hidden layer also has N neurons and its activation function is *ReLU* (Rectified Linear Unit); the output layer has one neuron and its activation function is *Sigmoid*. Like ECA, the optimizer *Adam* is used to learn the parameters of the MLP.

4.2 Datasets

The experiments are conducted on three most frequently used and most popular machine learning datasets from UCI [25], i.e., Abalone,[3] Wine Quality[4] (Red) and Wine Quality[4] (White). The statistics of each dataset are given in Table 1. Since the performance of ECA will be verified under the two-class classification task, the above multi-class datasets need to be adjusted to meet the requirements of the task.
Abalone is a dataset that predicts the age of abalone. We divide the age <10 into one class, and the others into another. The reason for dividing the datasets in this way is that the number of instances (samples) in two categories can be as close as possible, making the data unbiased.
Wine Quality (WQ) is a dataset that scores on wine quality. The score is between 0 and 10. We divide the scores less than or equal to 5 into one class, and the others into another.

[3] http://archive.ics.uci.edu/ml/datasets/Abalone.

[4] http://archive.ics.uci.edu/ml/datasets/Wine+Quality.

Table 1 Dataset statistics: 'Instances' refers to the number of samples contained in the dataset, and the number of positive and negative samples is indicated in parentheses; 'Attributes' refers to the number of attributes of the sample, which does not contain the label of the sample; 'Species' refers to the number of sample category, such as positive and negative in the two-class classification task

Dataset	Instances	Attributes
Abalone	4177 (2096 + 2081)	8
Wine quality (red)	1599 (855 + 744)	11
Wine quality (white)	4898 (3258 + 1640)	11

All experiments use the 5-fold cross-validation method to divide the training set and test set. The experimental evaluation metrics, *F1-score*, *Accuracy*, and *AUC* (Area Under Curve), are taken as the average of 5 results 0.

4.3 Hyper-Parameter Setting

The hyper-parameters in the baselines are set to: In LR, *penalty* is *L2*. In KNN, *n-neighbors* is 5; In SVM, *C* is 1.0 and *kernel* is *rbf*; In LDA, *solver* is *svd*; In MLP, *activation* is *relu* , *solver* is *adam* and *alpha* is 0.0001.

Other parameters not listed use the default value of the framework Scikit-Learn.[5]

QECA has three hyper-parameters, which are *learning rate*, *mini-batch* and *training epoch*, respectively, and uses the same settings on all datasets: the *learning rate* is 0.0001, the *mini-batch* is 1 and the *training epoch* is 500. Their *weights* are initialized to a truncated positive distribution, and the *biases* to 0.01. The permutation and combination method are used to select the hyper-parameters.

QCCA has four hyper-parameters, which are *learning rate*, *mini-batch*, *training epochs*, and *initial weight*. The same parameter settings are used on all datasets, which are: *learning rate* is 0 001, *mini-batch* is 1, *training epochs* is 500 and *initial weight* is 0.01, except for the different *learning rate* under Abalone dataset, which is: *learning rate* is 0.0001. We compare the experimental results to determine the hyper-parameters of the model. Moreover, in order to evaluate the model as objectively as possible, we choose a fixed *initial weight*, which is set to 0.01; we also uniformly use the minimum batch (i.e., *mini-batch*) setting, which is set to 1.

ECA has three hyper-parameters, which are *learning rate*, *mini-batch*, and *training epoch*, respectively. ECA uses the same settings on three datasets, i.e., the *learning rate* is 0.0001, the *mini-batch* is 1 and the *training epoch* is 500. The entangled system uses the GHZ state.

[5] https://scikit-learn.org/stable/index.html.

4.3.1 Experiment Results

Table 2 exhibits the experiment results on datasets Abalone, WQ (Red), and WQ (White) respectively, where bold values are the best performances out of all algorithms. From the experimental results, three metrics of ECA on three datasets are significantly better than the majority of other algorithms. It verifies the effectiveness of ECA from a broader perspective. The experimental results in this section can basically show that the joint probability derived from quantum entanglement has certain effectiveness and superiority in classical classification tasks, and also show that the strong statistical correlation revealed by quantum entanglement has certain utility and contribution in specific tasks.

All QCCA, QECA, ECA, and QNN outperform the baseline algorithms and the performance among themselves in descending order are: QNN > ECA > QECA > QCCA. the reasons are: again, see from Eqs. (13) and (24), QCCA only has phase but QECA adds on the probability amplitude. Understating that from the perspective of space, QCCA looks for the optimal solution from one surface, while QECA is to find the optimal solution from the whole space. If the optimal solution can't be found on the certain surface then the entangled system is in a general superposition state, that is, there is no entanglement, QCCA will not have the ability to learn while QECA may find the optimal solution on the other surface. ECA shares the same network design but it defines its entangled states to GHZ state rather than Bell state to improve

Table 2 Experiment results on abalone, wine quality (red), and wine quality (white) dataset

Dataset	Abalone			WQ (red)			WQ (white)		
Algorithm	F1-score	ACC	AUC	F1-score	ACC	AUC	F1-score	ACC	AUC
LR	0.7708	0.7711	0.7711	0.7560	0.7404	0.7395	0.8230	0.7472	0.6799
NBM	0.7384	0.7340	0.7340	0.7396	0.7248	0.7242	0.7795	0.7051	0.6666
KNN	0.7699	0.7699	0.7699	0.6757	0.6566	0.6556	0.7871	0.7031	0.6428
SVM	0.7669	0.7536	0.7538	0.7298	0.7098	0.7080	0.8266	0.7486	0.6737
LDA	0.7742	0.7771	0.7770	0.7504	0.7379	0.7381	0.8225	0.7482	0.6840
QDA	0.7427	0.7593	0.7591	0.7561	0.7266	0.7217	0.8166	0.7445	0.6900
MLP	0.7861	0.7816	0.7818	0.7262	0.6816	0.6730	0.7933	0.7000	0.6149
QNN	**0.8736**	**N.A.**	**N.A.**	**0.8108**	**N.A.**	**N.A.**	**0.8087**	**N.A.**	**N.A.**
Over MLP	11.13%↑	N.A.%↑	N.A.%↑	11.66%↑	N.A.%↑	N.A.%↑	1.93%↑	N.A.%↑	N.A.%↑
ECA	**0.8226**	**0.8084**	**0.8203**	**0.8276**	**0.7454**	**0.7074.**	**0.8242**	**0.7610**	**0.6462**
Over MLP	4.64%↑	3.43%↑	4.93%↑	13.96%↑	9.37%↑	5.11%↑	3.90%↑	8.71%↑	5.10%↑
QECA	**0.8018**	**0.8027**	**0.8027**	**0.7633**	**0.7536**	**0.7544**	**0.7990**	**0.7034**	**0.6186**
Over MLP	1.54%↑	2.07%↑	2.05%↑	3.69%↑	2.21%↑	1.78%↑	7.23%↑	4.90%↑	5.92%↑
QCCA	**0.7883**	**0.7853**	**0.7855**	**0.7259**	**0.6813**	**0.6749**	**0.7940**	**0.7017**	**0.6185**
Over MLP	2.85%↑	4.84%↑	4.84%↑	0.04%↓	2.23%↑	2.89%↑	0.91%↑	2.44%↑	5.91%↑

its performance. ECA only uses the measurement process of the entangled state in the output layer, where the entangled state is given and only the measurement operator is learned. Our model extends the measurement process of the entangled state to all neurons in the hidden layer, and both the entangled state and the measurement operator are learned, thus the performance will be better.

5 Conclusions

The interdisciplinary of quantum mechanics combining with neural network is promising in which significant amount of work has been done. In this paper, we specifically discuss a special domain in the field which is combining quantum entanglement into multi-layer perceptron and we give it the name Entangled Quantum Neuron Network (QENN). There are four different models, QECA, QCCA, ECA, and QNN under our framework and they differ in model designs. QECA and ECA share the same definition of observations of the subsystems of the entangled system but their entangled states are defined differently. QECA defines its entangled states as Bell state while ECA defines them as W state and QHZ state. QECA and QCCA have more differences as they define both entangled states and observations of the subsystems differently. QNN uses a regularizer that can constrain a state vector to an entangled state to reconstruct the neuron. As we discussed above, because of their different ways of combining QE into NN, their performances vary but all four models' performance outcompetes the baseline algorithms. This paper aims to revel and introduce the EQNN and has achieved some meaningful results, while there are also some shortcomings: all the optimization algorithms and optimizers in the complex domain can be improved. More researches could be done in the future.

References

1. F. Rosenblatt, The perceptron: a probabilistic model for information storage and organization in the brain. Psychol. Rev. **65**(6), 386 (1958)
2. D.E. Rumelhart, G.E. Hinton, R.J. Williams, Learning representations by back-propagating errors. Nature **323**(6088), 533–536 (1986)
3. G.E. Hinton, R.R. Salakhutdinov, Reducing the dimensionality of data with neural networks. Science **313**(5786), 504–507 (2006)
4. R.S. Michalski, J.G. Carbonell, T.M. Mitchell, *Machine Learning: An Artificial Intelligence Approach* (Springer Science & Business Media, 2013)
5. N. Bohr, et al., *The Quantum Postulate and the Recent Development of Atomic Theory*, vol. 3 (Printed in Great Britain by R. & R. Clarke, Limited, 1928)
6. B. Rosenblum, F. Kuttner, The observer in the quantum experiment. Found. Phys. **32**(8), 1273–1293 (2002)
7. R. Horodecki, P. Horodecki, M. Horodecki, K. Horodecki, Quantum entanglement. Rev. Mod. Phys. **81**(2), 865 (2009)
8. T. Yu, J. Eberly, Qubit disentanglement and decoherence via dephasing. Phys. Rev. B **68**(16), 165322 (2003)

9. A. Einstein, B. Podolsky, N. Rosen, Can quantum-mechanical description of physical reality be considered complete? Phys. Rev. **47**(10), 777 (1935)
10. E. Schrödinger, Discussion of probability relations between separated systems, in *Mathematical Proceedings of the Cambridge Philosophical Society*, vol. 31, no. 4 (Cambridge University Press, 1935), pp. 555–563
11. J.S. Bell, On the einstein podolsky rosen paradox. Phys. Physique Fizika **1**(3), 195 (1964)
12. H. Zhang, J. Wang, Z. Song, J.-Q. Liang, L.-F. Wei, Spin-parity effect in violation of bell's inequalities for entangled states of parallel polarization. Mod. Phys. Lett. B **31**(04), 1750032 (2017)
13. D.O. Hebb, *The Organization of Behavior: A Neuropsychological Theory* (Psychology Press, 2005)
14. M.A. Nielsen, *Neural Networks and Deep Learning*, vol. 25. (Determination Press San Francisco, CA, USA, 2015)
15. S. Lloyd, M. Mohseni, P. Rebentrost, Quantum algorithms for supervised and unsupervised machine learning (2013). arXiv preprint arXiv:1307.0411
16. Y. Levine, O. Sharir, N. Cohen, A. Shashua, Quantum entanglement in deep learning architectures. Phys. Rev. Lett. **122**(6), 065301 (2019)
17. M. Schuld, N. Killoran, Quantum machine learning in feature hilbert spaces. Phys. Rev. Lett. **122**(4), 040504 (2019)
18. V. Dunjko, H.J. Briegel, Machine learning & artificial intelligence in the quantum domain: a review of recent progress. Rep. Prog. Phys. **81**(7), 074001 (2018)
19. J. Adcock, E. Allen, M. Day, S. Frick, J. Hinchliff, M. Johnson, S. Morley-Short, S. Pallister, A. Price, S. Stanisic, Advances in quantum machine learning (2015). arXiv preprint arXiv:1512.02900
20. P. Rebentrost, T.R. Bromley, C. Weedbrook, S. Lloyd, Quantum hopfield neural network. Phys. Rev. A **98**(4), 042308 (2018)
21. J. Chen, L. Wang, E. Charbon, A quantum-implementable neural network model. Quant. Inf. Process. **16**(10), 1–24 (2017)
22. G. Verdon, T. McCourt, E. Luzhnica, V. Singh, S. Leichenauer, J. Hidary, Quantum graph neural networks (2019). arXiv preprint arXiv:1909.12264
23. I. Cong, S. Choi, M.D. Lukin, Quantum convolutional neural networks. Nat. Phys. **15**(12), 1273–1278 (2019)
24. J. Zhang, Z. Li, J. Wang, Y. Wang, S. Hu, J. Xiao, Z. Li, Quantum entanglement inspired correlation learning for classification, in *Pacific-Asia Conference on Knowledge Discovery and Data Mining* (Springer, 2022), pp. 58–70
25. Nair V, Hinton GE (2010) Rectified linear units improve restricted boltzmann machines. In: Proceedings of the 27th international conference on machine learning (ICML-10), pp 807–814
26. J. Zhang, R. He, Z. Li, J. Zhang, B. Wang, Z. Li, T. Niu, Quantum correlation revealed by bell state for classification tasks, in *2021 International Joint Conference on Neural Networks (IJCNN)* (IEEE, 2021), pp. 1–8
27. J. Zhang, Y. Hou, Z. Li, L. Zhang, X. Chen, Strong statistical correlation revealed by quantum entanglement for supervised learning, in *ECAI* (IOS Press, 2020), pp. 1650–1657
28. D. M. Greenberger, M. A. Horne, A. Zeilinger, Going beyond bell's theorem, in *Bell's Theorem, Quantum Theory and Conceptions of the Universe* (Springer, 1989), pp. 69–72
29. W. Dür, G. Vidal, J.I. Cirac, Three qubits can be entangled in two inequivalent ways. Phys. Rev. A **62**, 062314 (2000). https://link.aps.org/doi/10.1103/PhysRevA.62.062314
30. I. Bengtsson, K. Życzkowski, *Geometry of Quantum States: An Introduction to Quantum Entanglement* (Cambridge University Press, 2017)
31. T. Cover, P. Hart, Nearest neighbor pattern classification. IEEE Trans. Inf. Theory **13**(1), 21–27 (1967)
32. C. Cortes, V. Vapnik, Support-vector networks. Mach. Learn. **20**(3), 273–297 (1995)

33. Dua, D, Graff C (2017) Machine Learning Repository. University of California, Irvine, School of Information and Computer Sciences. http://archive.ics.uci.edu/ml
34. D.M. Blei, A.Y. Ng, M.I. Jordan, Latent dirichlet allocation. J. Mach. Learn. Res. **3**, 993–1022 (2003)

Quantum Simulation Tools and Demonstrations

Exploring IBM Quantum Experience

S. Gayathri Devi, S. Manjula Gandhi, S. Chandia, and P. Boobalaragavan

Abstract Nowadays, quantum computing is a very promising technology. Quantum computers, rather than using merely 1 s or 0 s, execute calculations based on the probability of an object's condition before it is measured, allowing them to process exponentially more data than traditional computers. A bit is a single state, such as on or off, up or down, 1 or 0. Instead, operations in quantum computing utilise the quantum state of an item to produce a qubit. These are the undefined qualities of an object before they are discovered, such as an electron's spin or a photon's polarisation. Several firms are releasing tools for quantum computing practice. One of the tools we'll use to implement our quantum programming is IBM's experience. It also aids in the execution of our programs on real quantum devices. We'll look at the IBM Q experience, circuit composer and building circuits with quantum lab notebooks in this chapter. The dashboard containing the fundamental information is introduced in the IBM Q experience. The user interface for learning quantum circuits, qubits and the gates that are used to perform operations on each qubit is outlined in circuit composer. Creating quantum circuits using quantum lab notebooks shows how to build circuits with the Notebook with Qiskit that comes pre-installed on IBM Quantum Experience.

Keywords IBM Q Experience · Circuit composer · Quantum lab notebook · Qubit · Quantum state

1 Introduction

Quantum computing has gotten a lot of attention in recent years, especially when IBM released the IBM Quantum Experience (IQX) in May 2016. This was the first of its kind, with the data being stored on the cloud and made accessible to the rest of the world. The ability is to use a quantum computer for no charge and several

S. Gayathri Devi (✉) · S. Manjula Gandhi · S. Chandia · P. Boobalaragavan
Department of Computing, Coimbatore Institute of Technology, Coimbatore, Tamil Nadu 641014, India
e-mail: sgayathridevi@cit.edu.in

R. Pandey et al. (eds.), *Quantum Computing: A Shift from Bits to Qubits*,
Studies in Computational Intelligence 1085,
https://doi.org/10.1007/978-981-19-9530-9_15

characteristics exist on the IQX. A user interface that enables anyone to do experiments on both a simulator and a real computer with quantum features. It contains information about profession, accessible backends, and pending experiment results, as well as the ability to create an experiment and run it on a simulator or real quantum device. This chapter will explain how to use each view's actions and information. The following topics will be covered in this chapter:

- Navigating the IBM Quantum Experience.
- IBM Quantum composer.
- IBM Quantum Lab.
- Running Jobs on IBM real quantum machines.
- GHZ Circuit.
- Applications of Teleportation.
- Applications of quantum algorithms.
- IBM Q Experience in the field of research.

2 Navigating IBM Q Experience

IBM Q Experience offers a number of real quantum computers and simulations. Now we need to get an API token from IBM Q Experience and utilise it with Qiskit for simulations [1]. Follow the steps below to sign up for the IBM Quantum Experience:

1. Go to the IBM Quantum Experience website at https://quantum-computing.ibm.com/ and fill out the form.
2. From the login screen, as shown in Fig. 1, log in to your account. You can skip this part and go straight to the next one if you already have an account or are signed in. If you haven't registered yet, you can choose your preferred login method from the sign-in screen. As you can see, you have several options for registering, including using your IBM ID, Google, GitHub, Twitter, LinkedIn or email. If you don't have any of the above account types, you may simply create an IBMid account and use that to log in.

Fig. 1 Sign-in page of IBM Quantum Experience

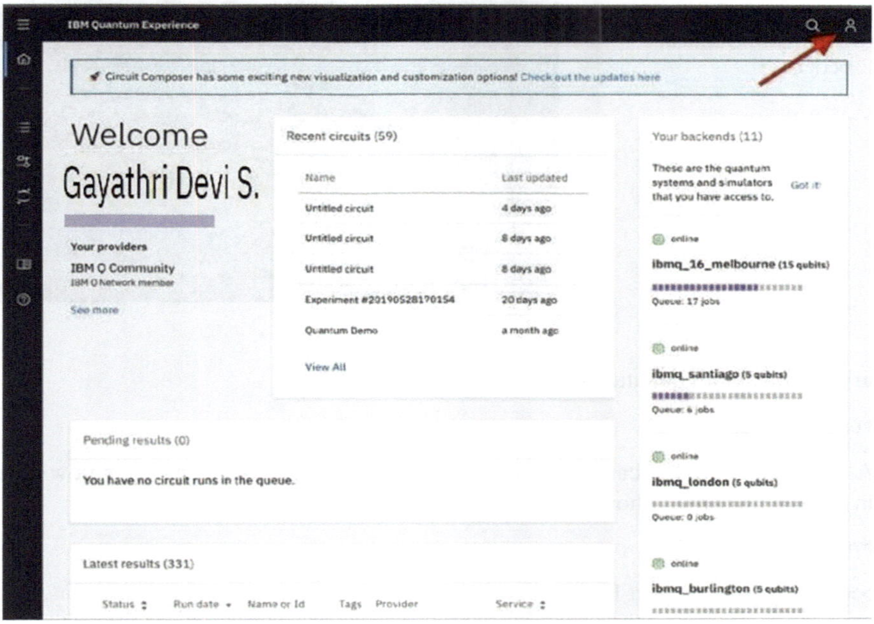

Fig. 2 Home page of IBM quantum experience

You'll be sent to the Home page once you've logged in. When you first log in to the IBM Quantum Experience site, this is the first page you'll encounter as shown in Fig. 2.

2.1 *Qiskit in Local Environment*

Step 1: Create an account by going to https://quantum-computing.ibm.com/ and signing up.

Step 2: Select "My Account" from the user icon in the top right corner.

Step 3: To copy your API token to the clipboard, click the "Copy token" (blue button) as shown in Fig. 3.

Step 4: Create a new file in the name of qiskitrc inside '~/.qiskit/qiskitrc'

$ vim ~/.qiskit/qiskitrc Copy & paste the below lines and replace with the one obtained from Step 3.

[ibmq]

token = <YOUR-IBM-API-TOKEN>

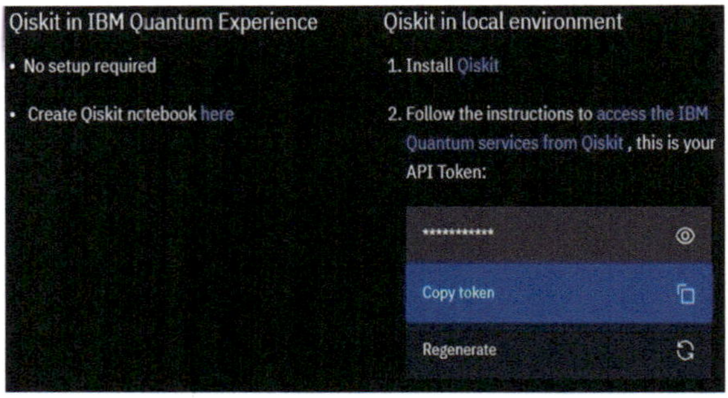

Fig. 3 Home screen of Qiskit in IBM Quantum Experience

url = https://auth.quantum-computing.ibm.com/api

verify = True

Alternatively, the file can be automatically generated by executing the below lines in Python prompt as shown below.

#python

>>> from qiskit import IBMQ

>>> IBMQ.save_account ('<YOUR-IBM-API-TOKEN>')

After executing the above lines confirm if the file 'qiskitrc' is created inside ~/.qiskit/ folder. Now all are set to simulate our first Quantum program. In addition to pre-built programs, upload and manage our own Qiskit Runtime programs. We can also run this tutorial in IBM Quantum Lab which is shown in Fig. 4.

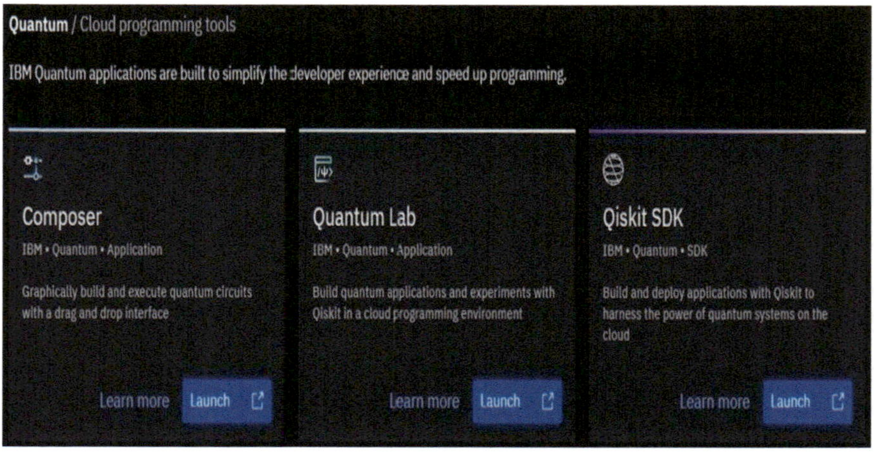

Fig.4 Quantum/cloud programming tools

3 IBM Quantum Composer

IBM Quantum Composer is a graphical quantum programming tool that allows us to construct quantum circuits and operate them on real quantum hardware or simulators by dragging and dropping operations. We can design a circuit by dragging and dropping operations onto the qubit wires, just like in the IBM Quantum Composer. It's a strong, interactive viewer when used to display an existing circuit [11]. It supports infinite side-scrolling of lengthier circuits, as well as clicking and seeing the internals of composite gates [12]. The dashboard is shown in Fig. 5.

3.1 *Visualise the States of Qubits*

Visualise how modifications to our circuit affect the state of qubits with an interactive q-sphere, histograms depicting measurement probabilities, or statevector simulations. Run on quantum computing hardware. To further understand the impacts of device noise, run the circuits on genuine quantum hardware.

3.2 *Generate Code Automatically*

Instead of creating code by hand, use Quantum Composer to build OpenQASM or Python code that functions exactly like the circuit we constructed.

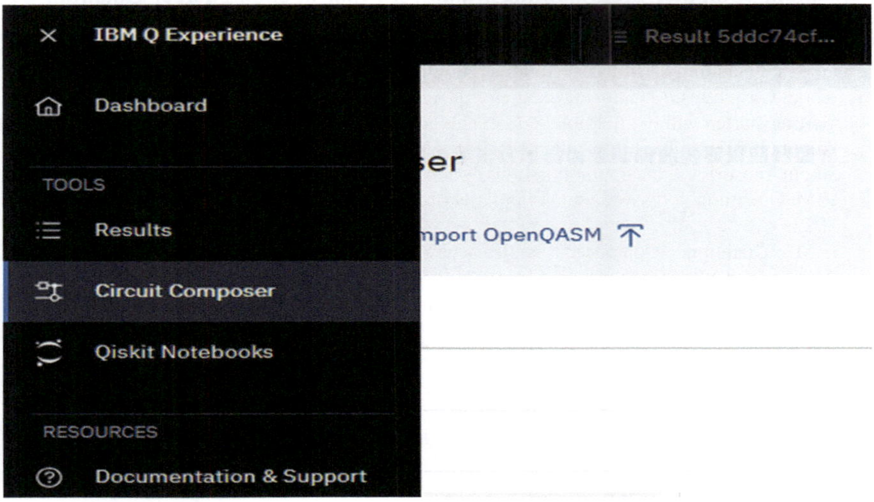

Fig. 5. Dashboard of IBM Q Experience

3.3 Make Our Very Own Quantum Circuit

Quantum Composer is a set of tools from IBM that let us construct, visualise and run quantum circuits on quantum systems or simulators. To access additional tools and tasks, go to the More options tab on each window.

3.4 Window for IBM Quantum Composer

View our files, jobs or documentation using the tool panels on the side. Click the icon for the open tab to close the side panel as shown in Fig. 6. We can open a new circuit, manage and inspect our saved circuits, personalise our workspace, seek help and more with the menu bar. Change the run settings before running our circuit on a quantum system or a simulator. The building blocks of quantum circuits are quantum gates and operations bars. Using the graphical circuit editor, drag and drop these gates and other actions. Colours are used to categories different types of gates together. Classical gates, for example, are dark blue, while phase gates are light blue and non-unitary operations are grey.

4 IBM Quantum LAB

In IBM Quantum Lab, a cloud-enabled Jupyter notebook environment, we may program and prototype in Python. There is no need to install anything. Qiskit is a quantum computer programming framework that is open source. Quantum Lab

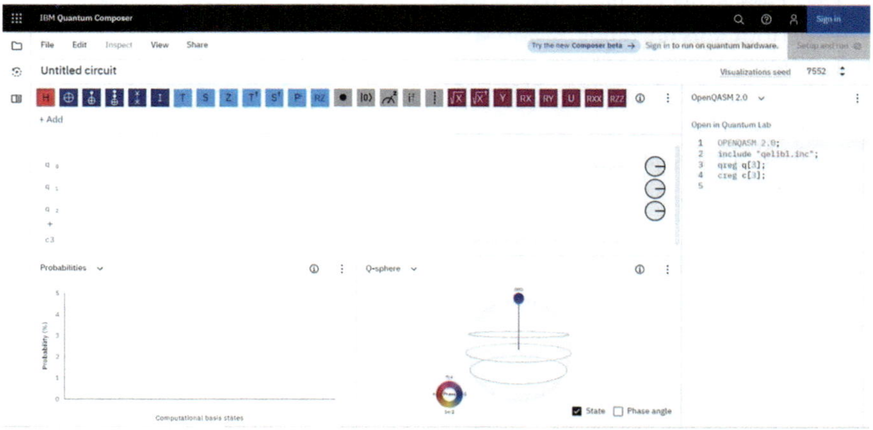

Fig. 6 IBM Quantum composer's window

allows us to create scripts that mix Qiskit code, equations, visuals and narrative text in a customised Jupyter Notebook environment, all without having to install anything. Use real quantum hardware or simulators to run our code. We can view, manage and store our files from anywhere. Use this step-by-step guide to code your own quantum circuit or go straight to Quantum Lab [5].

4.1 Quantum Circuits

The object at the heart of Qiskit is the quantum circuit. Here's how we create one, which we will call qc.

In [36]:

```
from qiskit import QuantumCircuit
qc = QuantumCircuit()
```

This circuit is currently completely empty, with no qubits and no outputs.

4.2 Quantum Registers

To make the circuit less trivial, we need to define a register of qubits. This is done using a QuantumRegister object. For example, let's define a register consisting of two qubits and call it qr.

In [37]:

```
from qiskit import QuantumRegister
qr1 = QuantumRegister(2,'a1')
```

Give it a label like 'a1' is optional. Now we can add it to the circuit using the add_register method, and see that it has been added by checking the qregs variable of the circuit object. This guide uses Jupyter Notebooks. In Jupyter Notebooks, the output of the last line of a cell is displayed below the cell:

In [38]:

```
qc.add_register(qr1)

qc.qregs
```

Out[38]: QuantumRegister(2, 'a1')]

Now our circuit has some qubits, we can use another attribute of the circuit to see what it looks like: draw() .

In [40]:

```
qc.draw('mpl')
```

Out[40]:

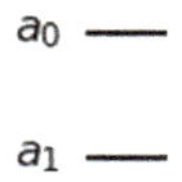

4.3 *Qiskit Backends*

The quantum backend systems have been provisioned for our purpose (as shown in Fig. 7). It not only lists the available backends, but also provides information about each one, such as its state. The state of the device indicates whether it is online or in maintenance mode, how many qubits (quantum bits) it has, and how many experiments are waiting to be done on it. It also has a colour bar graph to show queue wait periods, as shown in the screenshot below between ibmq 16 melbourne and ibmq rome. Please keep in mind that the quantum devices mentioned for you may not be the same as those listed here.

In [43]:
```
from qiskit import IBMQ
IBMQ.load_account()
provider = IBMQ.get_provider()
for backend in provider.backends():
    print(backend)
```

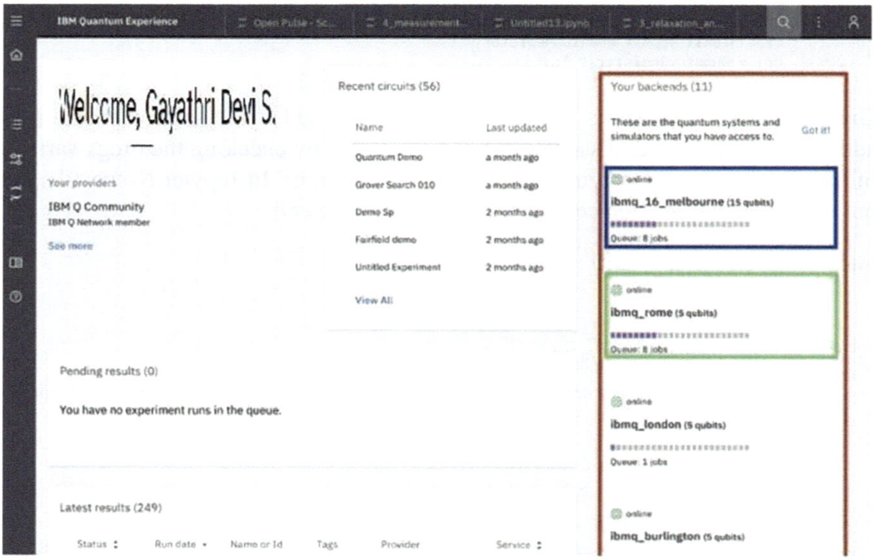

Fig. 7 Provisioned backend simulators and device

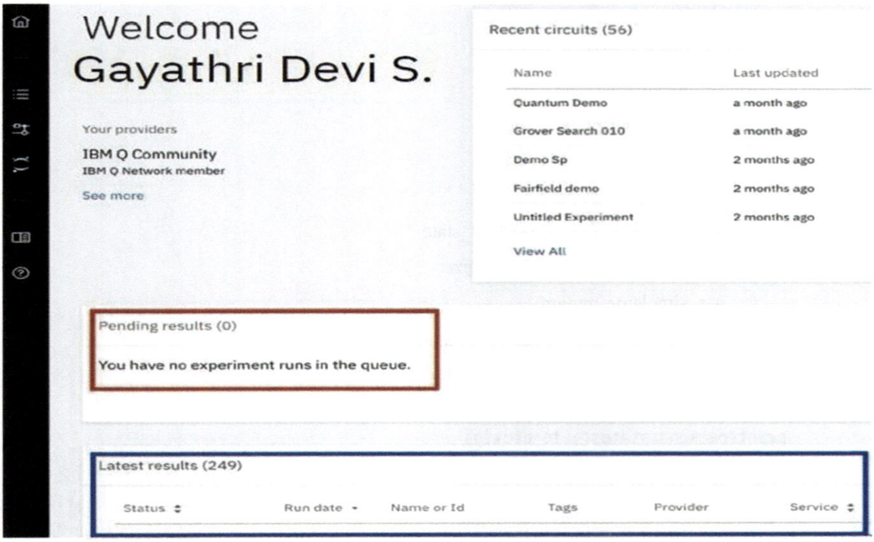

Fig. 8 Pending results and latest results

Out [43]:
```
ibmq_qasm_simulator
ibmq_armonk
ibmq_santiago
ibmq_bogota
ibmq_lima
ibmq_belem
ibmq_quito
simulator_statevector
simulator_mps
simulator_extended_stabilizer
simulator_stabilizer
ibmq_manila
```

We can see the device's basic status information. Figure 8 is what we see before we expand the device information. In the right-hand square, we go over the configuration, connectivity and error rates of the devices in greater depth.

In addition, as seen in the accompanying screenshot, the work ID is listed so that we can call back the details from that task at a later time.

4.4 Qiskit Simulators

In [45]:
```
from qiskit import Aer
sv_sim = Aer.get_backend('aer_simulator')
```

A list of all possible simulators in Aer can be found using

In [46]:
```
for backend in Aer.backends():
    print(backend)
```

Out [46]:

aer_simulator

aer_simulator_statevector

aer_simulator_density_matrix

aer_simulator_stabilizer

aer_simulator_matrix_product_state

aer_simulator_extended_stabilizer

aer_simulator_unitary

aer_simulator_superopqasm_simulator

statevector_simulator unitary_simulator

pulse_simulator

```
print(backend.status().to_dict())
```

```
{'backend_name': 'ibmq_qasm_simulator', 'backend_version': '0.1.547', 'operational':
{'backend_name': 'ibmq_armonk', 'backend_version': '2.4.34', 'operational': True, 'p
{'backend_name': 'ibmq_santiago', 'backend_version': '1.4.1', 'operational': True, '
{'backend_name': 'ibmq_bogota', 'backend_version': '1.6.37', 'operational': True, 'p
{'backend_name': 'ibmq_lima', 'backend_version': '1.0.35', 'operational': True, 'pen
{'backend_name': 'ibmq_belem', 'backend_version': '1.0.41', 'operational': True, 'pe
{'backend_name': 'ibmq_quito', 'backend_version': '1.1.28', 'operational': True, 'pe
{'backend_name': 'simulator_statevector', 'backend_version': '0.1.547', 'operational
{'backend_name': 'simulator_mps', 'backend_version': '0.1.547', 'operational': True,
{'backend_name': 'simulator_extended_stabilizer', 'backend_version': '0.1.547', 'ope
{'backend_name': 'simulator_stabilizer', 'backend_version': '0.1.547', 'operational'
{'backend_name': 'ibmq_manila', 'backend_version': '1.0.29', 'operational': True, 'p
```

4.5 GHZ Circuit

This circuit was designed by Greenberges, Micheal Home, Anton Zeilinger in 1980.

In [11]:
```
from qiskit import *
from qiskit import QuantumCircuit
def ghz_circuit(num_qubits):
    circ = QuantumCircuit(num_qubits)
    circ.h(0)
    for j in range(1,num_qubits):
        circ.cx(0,j)
    return circ
ghz_circuit(3).draw()
```

Out [11]:

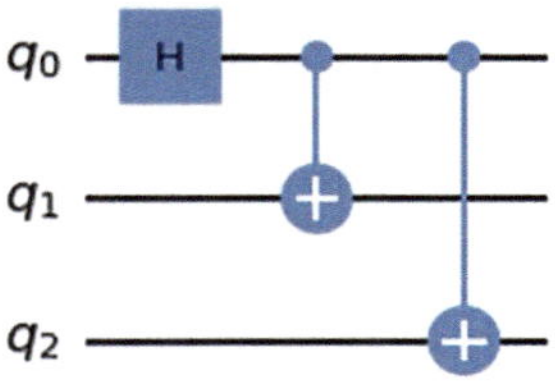

5 Running Jobs on IBM Real Quantum Machines

The Dashboard in Fig. 9 shows you a summary of the IBM Quantum Services you've scheduled. The most crucial aspect is that it displays 'Your Systems'. Select 'Your Systems' from the filter drop-down next to the search box after clicking 'View All'. You are given an overview of the systems that are available to you. These are a few systems that only have a few qubits [13]. However, these are sufficient to get you started.

We may get a more detailed look of a system's architecture and configuration by clicking on it. The figure in the bottom right, for example, depicts how the system's qubits are connected. When dealing with a real quantum computer, we must deal with the fact that we can only entangle connected qubits with each other.

Figure 10 shows the available real quantum systems in IBM. Depending on our requirements, we choose the system. The below code executes the job in least busy real quantum system.

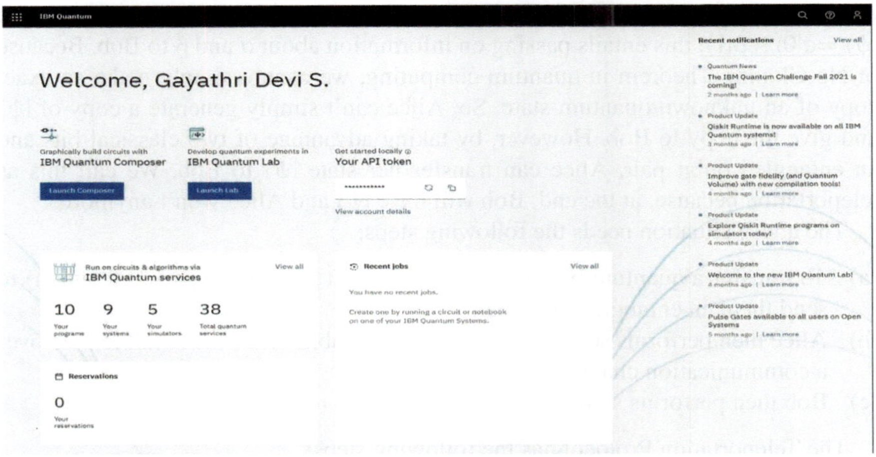

Fig. 9 Dashboard of IBM Quantum Services

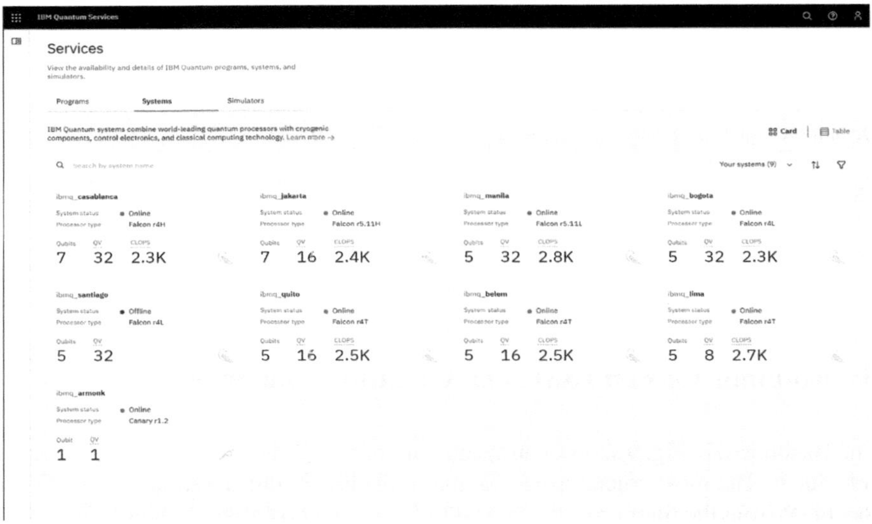

Fig. 10 IBM Quantum Services

```
from qiskit.providers.ibmq import least_busy
backend = least_busy(provicer.backends(filters=lambda x: x.configuration().n_qubits
                                               not x.configuration().simulator and x.status
job_exp=execute(qc,backend=backend,shots=8192)
```

6 The Quantum Teleportation Protocol

Alice wants to send quantum information to Bob. She wants to send the qubit state $|\psi\rangle = \alpha|0\rangle + \beta|1\rangle$, this entails passing on information about α and β to Bob. Because of No Cloning Theorem in quantum computing, we cannot simply make an exact copy of an unknown quantum state. So, Alice can't simply generate a copy of $|\psi\rangle$ and give the copy to Bob. However, by taking advantage of two classical bits and an entangled qubit pair, Alice can transfer her state $|\psi\rangle$ to Bob. We call this as teleportation because, at the end, Bob will have $|\psi\rangle$ and Alice won't anymore.

The above situation needs the following steps:

(a) To transfer a quantum bit, Alice and Bob must use a third party (Telamon) to send them an entangled qubit pair.
(b) Alice then performs some operations on her qubit, sends the result to Bob over a communication channel.
(c) Bob then performs some operations on his end to receive Alice's qubit.

The Teleportation Protocol has the following steps:

Step 1: Initialising the state to be teleported. We will do this on Alice's qubit q0.

Step 2: Creating entanglement between two qubits, we will use qubits q1 and q2 for this. Recall that Alice owns q1 and Bob owns q2.

Step 3: Applying a Bell measurement on Alice's qubits q0 and q1.

Step 4: Applying classically controlled operations on Bob's qubit q2, depending on the outcomes of the Bell measurement on Alice's qubits.

6.1 Simulating the Teleportation Protocol

Step 1:

A third-party Telamon creates an entangled pair or Bell Pair of qubits and gives one to Bob and one to Alice.

To create a Bell pair between two qubits is to transfer one of them to the X-basis {|+>,|->}, using a H gate, and apply CNOT gate onto the other qubit controlled by the one in the X-basis.

Let's say Alice owns q1 and Bob owns q2 after they part ways.

Step 2:

Alice applies a CNOT gate to q1, controlled by $|\psi\rangle$ (the qubit she is trying to send Bob), i.e. q0. Then Alice applies a Hadamard gate to $|\psi\rangle$.

Step 3:

Alice applies a measurement to both qubits that she owns q1 and $|\psi\rangle$ and stores this result in two classical bits. She then sends these two bits to Bob.

Step 4:

Bob, who already has the qubit q2, then applies the following gates depending on the state of the classical bits.

00➔ Do nothing

01➔ Apply X gate

10➔ Apply Z gate

11➔ Apply ZX gate

Note that this transformation of information is purely classical.

Alice's qubit has now teleported to Bob.

At the end of this protocol, Alice's qubit has now teleported to Bob.

6.2 Applications of Teleportation

Quantum teleportation is a technique that uses entangled states to transport quantum information from a source to a destination. The following are some of the applications of quantum teleportation [10]. Even if there are no quantum communication channels between two parties, quantum teleportation allows them to exchange unknown qubits. Teleportation is used in a variety of processing and communication tasks. Some of the real-world applications are listed below [9].

a. Network of quantum particles
b. Quantum computing system

In 2019, IBM unveiled the 'IBM Q', the first industrial-grade quantum computing system created with integrated commercial universal quantum systems for business and science applications.

7 Applications of Quantum Algorithms

A quantum algorithm is a quantum computation method that works on a realistic quantum computation model, the most common of which is the quantum circuit model of computation. A classical (or non-quantum) algorithm is a finite sequence of instructions or a step-by-step technique for solving a problem that can be performed on a traditional computer. A quantum algorithm, on the other hand, is a step-by-step method that may be carried out on a quantum computer. Although any classical algorithm can be run on a quantum computer, the phrase 'quantum algorithm' refers to algorithms that appear to be intrinsically quantum or employ a key property of quantum processing such as quantum superposition or entanglement. Few of the algorithms are mentioned here.

i. Deutsch–Jozsa algorithm

This algorithm was introduced by David Deutsch and Richard Jozsa in 1992. It is one of the most fundamental quantum algorithms. It demonstrated the superiority of quantum algorithms over classical algorithms. It is used to determine whether a function is constant or balanced when it has a single bit input and a single bit output [7].

ii. Grover's algorithm

Grover's method is a quantum algorithm for unstructured search proposed by Lov Grover in 1996. It quadratically accelerates unstructured searches and has applications beyond search. Grover's algorithm and amplitude amplification are powerful subroutines that can be employed in more complex quantum algorithms to provide quantum speedups for a variety of tasks. Only a few of these speedups are listed below:

- Finding the minimum of an unsorted list of N integers.
- Determining graph connectivity.
- Pattern matching, a fundamental problem in text processing and bioinformatics.
- Shor's factoring algorithm

Peter Shor devised Shor's algorithm in 1994. It's an integer factorization algorithm that runs in polynomial time on a quantum computer. Shor's algorithm discovers the prime factors of an integer N. On a sufficiently massive and stable quantum computer, Shor's method provides an efficient means of factoring arbitrarily large integers. The availability of Shor's algorithm is a major driver of post-quantum cryptography research towards new cryptographic methods that are resistant to quantum computers.

8 IBM Q Experience in Research Field

The IBM Q Experience has lately demonstrated its capacity to serve as an experimental platform open to anyone on the world. Some of the research works have been listed below which use the IBMQ Experience (see Table 1).

9 Conclusion

Business and technological advancements have always gone hand in hand, especially when it comes to information technology. Businesses have begun to look to quantum computers to solve challenges beyond the constraints of classical computers, which manipulate information using the same physical rules as atoms. More than a hundred businesses, universities and government agencies have already teamed up with IBM to investigate how the company's cloud-based quantum computers may assist tackle the industry's most pressing problems. We looked at the dashboard in this chapter, which has a lot of information to assist us figure out where we are. By examining the backend services, monitoring queue times we'll be able to keep track of our trials and understand where they're at. We also see about Circuit Composer or Qiskit notebooks to design an experiment. Finally, applications of teleportation and quantum algorithms are listed. This chapter is the start point for IBM Quantum Experience. Researchers in quantum computing field may take this to start their research work in IBMQ.

Table 1 List of quantum research fields which uses the IBMQ

S. no.	Title and author	Research area	Concept implemented
1	Fault-Tolerant Logical Gates in the IBM Quantum Experience by Harper and Flammia [6]	Quantum error detection	Error detecting code improves the fidelity of the fault-tolerant gates implemented in the code space as compared to the fidelity of physically equivalent gates implemented on physical qubits
2	Efficient quantum algorithm for solving travelling salesman problem: An IBM quantum experience by Srinivasan et al. [3]	Quantum optimization	A quantum algorithm to solve the travelling salesman problem using phase estimation technique
3	Quantum robots can fly; play games: an IBM quantum experience by Mahanti et al. [8]	Quantum robotics	A new quantum circuit to make the quantum robot fly. Demonstration of its applications in playing a game The quantum robot we present here shows the behaviour of 'fear', and its movement is deterministic in nature
4	An improved novel quantum image representation and its experimental test on IBM quantum experience by Su et al. [2]	Quantum image processing (QIP)	A quantum image control circuit was designed based on INCQI and quantum image preparation experiments were conducted on IBM Quantum Experience (IBMQ) to verify the feasibility and effectiveness of INCQI quantum image preparation
5	IBM Research team made IBM Quantum Experience available on the IBM Cloud [14]	Quantum computations	To accelerate innovation
6	Dr. James Wootton from the University of Basel, Switzerland [17]	IBM Quantum service	To develop games
7	Tara Tosic, a physics student at the École Polytechnique Fédérale de Lausanne (EPFL) [18]	IBM quantum service	Used for doing research in Arctic

(continued)

Table 1 (continued)

S. no.	Title and author	Research area	Concept implemented
8	Strilanc—an advisor of science [15]	Teleportation	Game changer in telecommunications
9	Colin Bell from HSBC and Dr. Darío Gil from IBM announced that they are excited to team with HSBC to explore applications of quantum technology to their business operations and help turn their aspirations into reality [19]	Quantum computations	HSBC uses IBM Quantum Experience to advance its net-zero goals, and to mitigate risks, including identifying and addressing fraudulent activity
10	Ryan Babbush, a scientist at Google [16]	Quantum simulation	Efficient quantum algorithms for emerging analog and digital quantum hardware
11	Solving Linear Systems of Equations by Using the Concept of Grover's Search Algorithm: an IBM Quantum Experience by Maji et al. [4]	Grover's search algorithm	Finding particular matrices that solve the set of equations and constructing corresponding quantum circuits using the basic quantum gates. We explicitly illustrate the whole process by taking 48 different sets of equations and solving them by using the concept of Grover's algorithm

References

1. A. Annunziata, IBM Quantum Summit 2020: Exploring the Promise of Quantum Computing for Industry (30 Sep 2020). https://www.ibm.com/blogs/research/2020/09/quantum-industry/
2. J. Su, X. Guo, C. Liu, S. Lu, L. Li, An improved novel quantum image representation and its experimental test on IBM quantum experience. Sci. Rep. **11** (2021). Article number: 13879
3. K. Srinivasan, S. Satyajit, B.K. Behera, P.K. Panigrahi, Efficient quantum algorithm for solving travelling salesman problem: an IBM quantum experience. https://doi.org/10.48550/arXiv.1805.10928
4. R. Maji, B.K. Behera, P.K. Panigrahi, Solving linear systems of equations by using the concept of Grover's search algorithm: an IBM Quantum Experience. Int. J. Theor. Phys. **60**, 1980–1988 (2021). https://doi.org/10.1007/s10773-021-04817-w
5. R. Loredo, *Learn Quantum Computing with Python and IBM Quantum Experience* (Packt Publisher, Sep 2020), ISBN. 9781838981006
6. R. Harper, S.T. Flammia, Fault-tolerant logical gates in the IBM Quantum Experience. Phys. Rev. Lett. **122**, 080504 (26 Feb 2019)
7. C. Shao, Y. Li, H. Li, Quantum algorithm design: techniques and applications. J Syst Sci Complex **32**, 375–452 (2019). https://doi.org/10.1007/s11424-019-9008-0
8. S. Mahanti, S. Das, B.K. Behera, P.K. Panigrahi, Quantum robots can fly; play games: an IBM quantum experience. Quantum Inf. Proc. **18** (2019). Article number: 219

9. T. Liu, The applications and challenges of quantum teleportation, in *Journal of Physics: Conference Series*, vol. 1634 (IOP Publishing, 2020), p. 012089. https://doi.org/10.1088/1742-6596/1634/1/012089
10. Y.B. Band, Y. Avishai, Quantum information. Quantum Mech. Appl. Nanotechnol. Inf. Sci. (2013)
11. https://quantum-computing.ibm.com/
12. https://quantumcomputingreport.com/ibm-quantum-experience-first-looks/
13. https://www.ibm.com/blogs/research/tag/quantum-experience/
14. https://en.wikipedia.org/wiki/IBM_Quantum_Experience
15. https://www.physicsforums.com/threads/what-are-some-potential-applications-of-quantum-teleportation.752454/ (applications of teleportation)
16. https://www.ibm.com/blogs/research/2017/12/approximate-quantum-computing-advantage-applications/
17. ScienceAtHome | Decodoku: Quantum Battleships
18. IBM Q in the Arctic: 76.4° North-IBM Research blog
19. HSBC Working with IBM to Accelerate Quantum Computing Readiness (29 Mar 2022)

Quantum Programming on Azure Quantum—An Open Source Tool for Quantum Developers

Kumar Prateek and Soumyadev Maity

Abstract Quantum computing has become a new buzzword in recent years. Although quantum computing techniques have been available in the literature for the past 40 or more years, the desire for real-time implementation of such quantum computing techniques has become possible due to the ongoing superspeed development of quantum computers by multinational corporations. Albeit, only 40 qubits quantum computer has been developed to date. Still, the pathway of big corporations reveals that by the end of this decade, a full-fledged quantum computer will be available in the market for everyone. Quantum computing uses quantum key distribution for quantum communication. It is expected that quantum computing and quantum communication will completely change the workflow of many industries. Studies are also predicting that the market demand for the quantum computing industry will be in multi-trillion dollars as early as 2030. Besides, the perspective of researchers has been drastically changing due to the plethora of opportunities brought forth by quantum computing for data processing and data encryption tasks. The quantum computer is deep-rooted in uncertainty principle, and probability theories thereby prohibit the copying and replicating of quantum information. Consequently, the guarantee of unconditional security for transmitted data is ensured, otherwise impossible. Generally, transmitted data are hacked due to attackers' generation of keys replica. We may note that despite quantum computing being in a nascent stage, it possesses the potential to change internet activities by speeding up many tasks. Many day-to-day activities of many industries like finance, healthcare, and security will unseal imperceptible abilities. Furthermore, many big corporations invest in developing quantum computers and open-source tools to enable the development of quantum programs running on quantum computers. Also, community-driven activities are accomplished to upgrade the skills of current software developers to make them ready with appropriate skills for the development of future quantum software, which will run on large bits quantum computers. In this direction, Microsoft Incorporation has not only developed a quantum development kit (QDK) but also provides cloud-based quantum comput-

K. Prateek (✉) · S. Maity
Department of Information Technology, Indian Institute of Information Technology Allahabad, Prayagraj, India
e-mail: pcl2017003@iiita.ac.in

R. Pandey et al. (eds.), *Quantum Computing: A Shift from Bits to Qubits*,
Studies in Computational Intelligence 1085,
https://doi.org/10.1007/978-981-19-9530-9_16

ing as a service, namely Azure Quantum, for developing and testing new quantum programs for the community. The newly designed quantum programs can now be simulated locally or run on the real quantum computer through Azure Quantum. Consequently, this chapter introduces the what, when, and why's of quantum computing. Also, this chapter presents all necessary tools (with detailed installation and execution steps) required by the quantum developer for the possible development of a quantum program.

1 Introduction

Quantum computing that uses quantum bits to perform computations, has recently become a buzzword in mainstream society [1]. It is estimated that quantum computer will become available to the mainstream public by 2030 [2]. This would enable the scientist to perform activities usually considered impossible these days. Quantum computing can revolutionize all existing technologies and systems. It may even enable the world to find discoveries. Quantum computers will transform artificial intelligence (AI) [3] because quantum computing can process an enormous amount of data at unparallel speeds and can simulate neural networks of enormous sizes [4]. Even quantum computers using quantum computing techniques can calculate certain optimization problems by simulating them simultaneously instead of calculating each solution sequentially as it is performed using classical computing techniques. Quantum computing can find the answer to any optimization problem significantly faster than a classical computer, thereby revolutionizing machine learning, which can solve complex problems considered impossible these days [5]. Quantum computing and deep learning reinforce each other and can solve data classification and data recognition at a much faster speed. Also, the fast data classification [6], and data recognition would enable the detection of emotions [7] from the vocal sample of any person, thereby can revolutionize the interview preparation strategy for the mass public that aspires to get a job in multinational companies. Not only that, the gaming industry can create new characters that interact with the player in a hyper-realistic manner [8]. Electric vehicles are very popular among the mainstream public since the government provides subsidies. These days cooperative societies are campaigning to encourage everyone to use electric vehicles to reduce their negative impact on the environment and reduce pollution. Another use case of quantum computing lies in the design of an energy-efficient battery used in electric vehicles [9]. Quantum computing can provide better simulation techniques to find the ageing of battery cells, thereby increasing the mileage capacity of vehicles between subsequent recharges [10]. Furthermore, traditional computing is inefficient for solving problems like traffic optimization involving travelling salespeople [11]. Specifically, determining the shortest possible routes between multiple places within any city and returning to the starting point using quantum computing techniques could be very simple. Not only that, but the emerging challenges of real-time routing and navigation can be solved using live feed from roads and railways combined with quantum computing techniques.

The development of a new drug requires the simulation of the molecule's structure to determine the behaviour of each molecule and the interaction of the individual molecule with neighbouring molecules [12]. The modelling of the molecule with high accuracy is extremely difficult to achieve these days by employing a classical computer. Also, it is impossible to simulate a molecule with only a few atoms. However, quantum computing can solve these problems as these molecules demonstrate quantum properties, which will transform the development of new drugs and cures for any disease. The quantum computing techniques can visualize better the interaction between a drug sample and a target patient and will reduce development costs significantly [13]. Henceforth, the development of better vaccines and medications will become much faster, enhancing the discovery process of innovative solutions for new and serious types of diseases like cancer, heart disease, and brain disorders [14]. Besides, the greatest impact of quantum computing could be seen in the finance industry [15]. Quantum computing and financial modelling can be used in tandem to model the entire financial market as a quantum process. Even quantum computing research has been in progress to improve the Monte Carlo Model [16], which predicts the probabilities of many outcomes with its own set of risks. Quantum computing can test more outcomes with high accuracy in minimum time, improving customer portfolio management by performing activities like credit scoring, prediction modelling, and fraud detection [17]. Also, quantum computing can solve the evolving weather forecasting problems by calculating different weather variables simultaneously instead of performing sequential calculations [18]. Quantum computing can even streamline the process used nowadays in the supply chain industry by resolving supply chain problems—the availability of manufacturing components and their pricing relationship with different vendors efficiently in many industries such as the food chain [19] and automotive. Even the components dataset enables the identification of manufacturing processes that have contributed to operational failures [20].

The main use case of quantum computing is in the field of cyber security, which is popularly known as quantum cryptography or quantum key distribution [20–25]. The potential of quantum cryptography can break any encryption standards that rely on hard mathematical problems. In fact, Shor's algorithm [26] has demonstrated that the computational time of hard mathematical problems can be reduced from millions of years to a few seconds. Quantum cryptography endangers the widely used RSA encryption techniques [27]. In fact, for any classical adversary, it would take 300 trillion years to break the 617 digits of the RSA-2048 encryption standard. However, with the availability of a fully working quantum computer, the RSA-2048 encryption technique could be broken in 10 s [28]; which can pose national security threats to many developing countries. Recently, in the literature, many works utilizing quantum key distribution have been proposed in evolving fields like the internet of things [29], vehicular-ad-hoc networks [30, 31], cloud computing [32–34] and many more [35–39]. The many use cases will demand many quantum developers who can write and execute different quantum programs. Therefore, in view of this, the specific contribution of this chapter is described subsequently.

This chapter explores quantum computing in detail, thereby providing quantum enthusiasts with an opportunity to learn what is quantum computing?, the concepts of quantum computing?, why do we need quantum computing?, the history and background of quantum computing?, basics of mathematics that are required for quantum computing like matrices, advance matrices concepts?, why will we use quantum computing?. Also, basics of quantum computing like qubits, why do we need qubits?, difference between bits and qubits, what is multiple qubits?, what is Dirac notation? why Pauli measurement is used?. Moreover, this chapter describes a detailed study on an open-source tool: Azure Quantum, which comprises a quantum development kit (QDK) provided by Microsoft Inc. under the aegis of the azure quantum service. Specifically, what is Azure Quantum?, What Azure Quantum offers for the quantum community, i.e., what are the solutions offered by Azure Quantum?, which model of cloud computing service is provided by Azure Quantum?, Who is Azure Quantum for? is inspected. Following this, the details of QDK are thoroughly presented, describing what is QDK?, what QDK offers to quantum developers?, the programming language supported by QDK?, and where to run QDK like whether we should run QDK locally or we should run QDK on the web-based computing platform. Also, steps to run QDK locally? and steps to run QDK on a web-based interactive computing platform are presented, along with a discussion on how to update the QDK? and how one can contribute to QDK. Furthermore, this chapter also investigates the AQ workspace, a prerequisite for developing quantum software. How to create and install AQ workspace is explored with a discussion on the required steps for creating AQ workspace within the local environment and creating AQ workspace with Jupyter notebook. The steps to develop quantum software are also presented. The core of any quantum software is quantum programs. Therefore, examples of different ways to write any quantum programs (supported in AQ) such as writing Q$^\sharp$ and standalone program, writing Q$^\sharp$ and python program, writing Q$^\sharp$ and .NET program along with individual program requirement is described. Generally, libraries are used to keep any quantum program high level. Consequently, details of important libraries supported and available with respect to AQ are presented. Besides, resource estimators for any quantum program are explored, detailing steps to estimate resources for any quantum program on a local simulator and steps to estimate resources for any quantum program on a real quantum computer. In addition, how to run the quantum programs on the local simulator and how to run the quantum programs on a real quantum computer provided through AQ (cloud-based quantum service) are also displayed. A discussion is presented on each component of the workflow required for running any quantum programs on cloud-based quantum services—AQ. At last, after properly summarizing implementation details on available open-source tools, Azure Quantum (AQ) enables quantum developers to design quantum applications.

The organization of the rest of the chapter is as follows: First, Sect. 2 describes Azure Quantum—a cloud service provided by Microsoft Inc. Afterwards, Sect. 3

describes the overall structure of the quantum program detailing the mathematics involved and how to write any quantum program. Section 4 details the available simulator provided by QDK. Subsequently, the Azure Quantum workspace and ways to develop quantum software are discussed in Sects. 5 and 6, respectively. Finally, Sect. 7 concludes the chapter.

2 Azure Quantum

Azure quantum is a cloud service provided by Microsoft Inc [40] and contains varying types of quantum technologies and quantum solutions. Specifically, it enables an open ecosystem for quantum developers to access different quantum solutions, hardware, and software. Besides, AQ lets the developer utilize the benefits of quantum computing through a full-stack open cloud computing facility through quantum hardware, quantum software, and pre-built quantum solutions.

2.1 Who is Azure Quantum For?

AQ is designed for quantum developers, enthusiasts, or individuals who require quantum computation during their product development. Particularly, AQ is for everyone who desires to learn, test, and design a new quantum application, including the following stakeholder:

2.1.1 Industry

Any industry in security, power and energy, information technology (IT), finance, and many more areas may adopt AQ for incorporating quantum solutions to innovate, design, and develop their new product and new protocols for meeting the goals like efficiency and unconditional security. AQ brings quantum resources to one place so the industry can design and develop any new product.

2.1.2 Developer

Developers can write the quantum program and create quantum applications using the programming language of one's choice within AQ. AQ enables quantum developers to learn quantum programming to develop new applications comprising quantum search and quantum machine learning to name a few.

2.1.3 Researchers

AQ allows budding and expert researchers to test new quantum schemes and protocols. Also, AQ motivates researchers of different domains such as cyber security, machine learning, and chemistry to incorporate quantum computing in one's work to harness the benefits of quantum computing. The researchers can test, build, and deploy any new quantum solution in their respective domains at scale.

2.1.4 Quantum Enthusiast

Quantum enthusiasts can now use a set of quantum technologies, test and learn quantum technologies covering the quantum computing basics, quantum programming language: $Q^{\sharp}$ and many more by performing real-time implementation through AQ.

2.2 *What AQ Offers?*

AQ offers a platform for everyone to design, develop, and compile the different quantum programs, quantum algorithms, quantum schemes, and quantum protocols. All offerings of AQ can be categorized as follows:

2.2.1 Quantum Computing Based Solutions

AQ enables quantum developers and researchers to learn quantum computing to experiment and design any new type of prototype on various quantum hardware provided through the platform. Currently, the availability of such development scenarios through cloud-based platforms allows the quantum developer to analyze their product at hand, thus making such quantum solutions ready for the future quantum machine much before the arrival of such quantum machines.

2.2.2 Optimization Solutions

AQ allows the development of new solutions tailored to reduce the cost of operations on several optimization problems in application areas ranging from fleet management, fleet scheduling, finance, and energy trading. AQ also allows faster implementation of optimization problems when compared to other classical optimization techniques on devices like CPU, GPU, FPGA, and custom silicon. In addition, AQ allows the implementation of quantum-inspired optimization (QIO) algorithm on devices like GPU, CPU, FPGA, and silicon custom, which guarantees speedup with respect to existing classical approaches. QIO uses the benefits of quantum computing and simulates the quantum effects on a classical computer. Besides, the partner and

Microsoft Inc. have designed state-of-art QIO algorithms, and AQ provides access to numerous QIO algorithms so that one can test, deploy, and use them instantly during product development.

2.3 What is QDK?

QDK stands for the quantum development kit. It provides a development environment to develop quantum programs. The Microsoft QDK is open-source in nature and allows development in AQ. It facilitates quantum developers to work online and offline and supports quantum programming language, namely Q$^{\sharp}$.

2.4 What is Offered by QDK?

QDK is well known for providing a quantum development environment for quantum programs. Specifically, QDK provides a tool that assists quantum developers in writing individual quantum programs, thereby leading to quantum software development. The Microsoft QDK provides the following:

2.4.1 Ready to Use Libraries

Libraries are generally used to make the code high level. Here, two types of libraries are provided by QDK, namely standard and domain-specific libraries. Standard libraries assist developers in implementing common patterns in many quantum algorithms. In contrast, domain-specific libraries cater to the specific needs of the quantum program to be used in specific fields like machine learning, chemistry, finance, and many more.

2.4.2 Quantum Computing Simulators

Simulators are the best way to emulate real-world environments. The quantum computing simulator allows instances of the quantum program to execute and produces results without the availability of real quantum hardware. The quantum computing simulator does not consider noise, whereas actual quantum hardware involves a certain amount of noise.

2.4.3 Noise Simulators

The noise simulator produces a result as soon as instances of the quantum program are executed under the influence of noise and considers stabilizer representation.

2.4.4 Resource Estimators

Resource Estimators estimate the real-world cost of any quantum solution or individual quantum program. It calculates the number of quantum bits required by a particular quantum program, the time taken by a particular quantum program, and many more.

Besides, QDK provides extension support for visual studio (VS), visual studio code (VS Code), and Jupyter notebook, allowing interoperability with .NET languages and python. Even python packages are available to design new AQ optimization solutions by the budding quantum developer. Not only this, but QDK can also be integrated with Qiskit and Cirq, thus enabling quantum solutions designed and developed in another programming language to run in AQ.

2.5 Programming Language Supported by QDK

A special type of programming language is designed, namely Q♯—a quantum programming language for writing quantum algorithms without involving the underlying circuit at the start of writing the quantum algorithm. Specifically, Q♯ enables the merging of classical algorithms with quantum counterparts and brace control flow of algorithm execution similar to classical computing algorithm. Consequently, algorithms are expressed understandably with respect to representation using a circuit model comprising quantum gates in a fixed sequence. Q♯ facilitates requirements of specifying quantum bit (whether logical or physical) during runtime when the quantum algorithm gets executed. Similarly, runtime determines the mapping, which enables execution of quantum algorithm involving quantum bit state transfer, remapping, i.e., Q♯ allows mapping to be carried over until the topology of the target device is known thus facilitating the quantum developer to write quantum programs easily.

QDK can be executed locally on the client's (quantum developer) local computer. Below are the steps that need to be followed to set up or install the required tools that enable the development of quantum computing and optimization solutions for AQ (to later run in AQ) in one's local environment. QDK can be used locally or on the cloud. The detailed description is as follows.

2.5.1 Use QDK Locally with Docker

Quantum developers can download the QDK docker image and run its local docker installation. Docker image can be downloaded at [41].

2.5.2 Use QDK in Cloud

Quantum developers can write a new quantum program using QDK that can be executed in the cloud. Quantum developers can use three procedures independently, as mentioned below, to write a quantum program using QDK and Q$^\sharp$ without the need to install anything locally [42].

- **Use QDK in Jupyter Notebook in Azure Quantum Portal:** The azure quantum portal allows quantum developers to upload, execute or store jupyter notebooks.
- **Use QDK with Binder:** Quantum developers can use a binder to run and execute a quantum program without installing QDK locally. The binder allows the execution and sharing of Q$^\sharp$ console application jupyter notebooks online for free.
- **Use QDK with Docker:** Quantum developers can use the QDK docker image in the cloud. Cloud service providers like Amazon web services or Google Cloud support uploading docker images.

3 Quantum Program

The quantum program consists of classical and quantum codes interchangeably. The quantum portion of the quantum program can be described using the advanced mathematics concept, which is discussed in detail as follows:

3.1 Mathematics Required for Quantum Computing

The details of all concepts required to understand quantum computing and write quantum programs are described herewith.

3.1.1 Linear Algebra for Quantum Computing

Popularly, linear algebra is known as the language of quantum computing. Specifically, linear algebra is quite helpful for describing quantum operations and quantum states and assists the quantum enthusiast in the visualization process of quantum computer workflow, i.e., how quantum computers react when a series of instructions are fed into them by the quantum developer through quantum code. Linear algebra uses vectors and matrices to describe quantum operations and quantum states.

3.1.2 Vector and Matrices in Quantum Computing

A quantum bit can be in a superposition state, zero state, or one state at any time. In linear algebra, the vector is used to describe quantum states and is generally represented by a single column matrix like matrix A. The element of matrix $A = \binom{a}{b}$, i.e., a and b represent the probability of collapsing the particular qubit to state 0 and state 1, respectively. All the single column matrices whose sum of the square of individual elements results in 1, i.e.,

$$|a|^2 + |b|^2 = 1. \tag{1}$$

are known to be valid quantum state vectors; otherwise, the quantum state vector is considered invalid.

The Pauli matrices, let us say $\mathrm{Y} = \begin{pmatrix} 0 & 1 \\ 1 & 0 \end{pmatrix}$, are used to represent quantum operations. The matrix Y must be a unitary matrix, i.e., the valid quantum operations are only represented through the unitary matrix. Whenever any quantum operation $\mathrm{Y} = \begin{pmatrix} 0 & 1 \\ 1 & 0 \end{pmatrix}$ is applied to the quantum bit in state $\mathrm{A} = \binom{1}{0}$, then it results in a new quantum state. The new quantum state is represented using matrix multiplication of two matrices that represent the state A of quantum bit and quantum operation Y, respectively, i.e., $\begin{pmatrix} 0 & 1 \\ 1 & 0 \end{pmatrix} \otimes \binom{1}{0}$.

3.1.3 Representing 2-Qubit States in Quantum Computing

many quantum programs use more than one qubit. Therefore, the need to represent 2-qubit cannot be overlooked. Generally, two qubits are represented using a tensor product of individual qubit states. We cannot simply multiply the state of two individual qubits because each qubit is represented using vector space. The tensor product of two-qubit states results in a four-dimensional matrix as follows: $\binom{a}{b} \otimes \binom{c}{d} = \begin{pmatrix} ac \\ ad \\ bc \\ bd \end{pmatrix}$.

Each element of the 4*1 matrix denotes the probability that the qubit is collapsed to that state. Here, the elements ac, ad, bc, and bd represent the probability that the 2-qubit is collapsed to 00, 01, 10, and 11 states, respectively. Similar to a single qubit, the 2-qubit must also satisfy the square sum of each element result equivalent to 1. Here,

$$|ac|^2 + |ad|^2 + |bc|^2 + |bd|^2 = 1. \tag{2}$$

3.1.4 Qubit

The quantum bit (qubit) acts as the fundamental information in quantum computing. In classical computing, any bit can possess the value of either 0 or 1 or a superposition of 0 & 1. The 2D column vector of the unit norm represents the state of a qubit where the magnitude sum of a square of elements corresponds to 1. Also, any 2D column vector with elements as a real number or with a norm 1 depicts a valid quantum state. Besides, the quantum state vectors $\binom{1}{0}$ & $\binom{0}{1}$ plays an important role in quantum computing as this vector forms a basis to describe the qubit's state. Henceforth, any quantum state $\binom{a}{b}$ can be represented using the sum of basis state vectors a$\binom{1}{0}$ + b $\binom{0}{1}$. Also, in quantum computing, the quantum developer follows the standard convention of $\binom{1}{0}$ & $\binom{0}{1}$ to represent 0 & 1 of classical computing. Although, contrary choices like $\binom{0}{1}$ & $\binom{1}{0}$ can also be chosen for representing 0 & 1. Therefore, only two out of many infinite states of quantum bits correspond to the classical bit of 0 & 1.

3.1.5 Measuring a Qubit

The act of measuring a qubit collapse that bit to either 0 or 1 state, i.e., the qubit is collapsed to two classical states either $\binom{1}{0}$ or $\binom{0}{1}$. Informally, looking at a qubit is known as measuring the qubit. If a qubit is represented using $\binom{a}{b}$ quantum state vector then the classical state $\binom{1}{0}$ is obtained with probability $|a|^2$ and quantum state $\binom{0}{1}$ is obtained with probability $|b|^2$. Also, these probabilities, i.e.,

$$|a|^2 \quad + |b|^2 \quad = 1. \tag{3}$$

would sum up to 1 if the quantum states were valid quantum states. Since probabilities are calculated by squaring the magnitude, the overall sign of the quantum state vector becomes irrelevant while measuring any qubit. Hence negation of quantum state vector a → –a & b → –b has no impact on the probabilities of obtaining the state $\binom{1}{0}$ or $\binom{0}{1}$. Also, these negative phases are called global phases and are represented using the form $e^{i\phi}$ instead of $+ - 1$.

The measurement of the quantum bit does not always change the quantum state. if one measures quantum state $\binom{1}{0}$ or $\binom{0}{1}$ representing 0 & 1 of classical computing the state vector $\binom{1}{0}$ or $\binom{0}{1}$ does not damage that state; instead, the quantum state remains unchanged, thereby allowing replication of classical data on a quantum computer similar to replication in a classical computer. However, storing and copying quantum states (superposition of 0 &1) are prohibited due to the no-cloning theorem.

3.1.6 Bloch Sphere

Bloch Sphere is a representation representing the qubits in a 3D vector. Specifically, the Bloch sphere describes a 2D qubit state in the form of a 3D real-valued vector, thereby helping quantum developers visualize a single qubit state to understand the reasoning of multiple qubit states. In Bloch Sphere, the blue-headed arrow displays the direction where the quantum state points, whereas the rotation of the quantum state with respect to one cardinal axes is represented using the transformation of the arrow.

3.2 Ways to Write Quantum Program

A Quantum program can be written using Q$^\sharp$ programming language in the quantum development environment using Microsoft QDK. While writing a quantum program, the Q$^\sharp$ programming language can be used in a variety of ways, as described below [43].

3.2.1 Q$^\sharp$ and Standalone Program

Q$^\sharp$ and standalone quantum program can be configured using the jupyter notebooks, VS code, visual studio and any other editor/IDE. Q$^\sharp$ can be executed directly or use another language like C$^\sharp$ or F$^\sharp$. The details of setting up the configuration of the Q$^\sharp$ standalone program in a variety of ways are as follows:

- **Q$^\sharp$ and Jupyter Notebook**

  Step 1: Download miniconda [44], anaconda [45] as per specification of operating system of developer computer. Ideally, 64-bit installations are required.
  Step 2: Type **Conda init** command to initialize conda in quantum developer host operating system (Linux/Windows).
  Step 3: Now, open a new terminal and use **qsharp-env** command to create and activate a new conda environment along with specific packages to develop new quantum programs. Specifically, the command to create and activate a new conda environment are as follows:

  conda create -n qsharp-env -c microsoft qsharp notebook
  conda activate qsharp-env

  Step 4: Finally, quantum developers has to run **python -c "import qsharp"** command to verify the installation of Q$^\sharp$ standalone program execution environment.

- **Set up Q♯ and python Environment**
  Here, we will demonstrate how to configure a python program integrated with the python development environment in the backend. This python program allows the calling of Q♯ operations to develop a quantum program. The detailed steps are described herewith.

  Step 1: Install the qsharp python package the installation of the qsharp python package brings forth all necessary functionalities required to simulate and compile operations using Q♯ in the program written in the python. Specifically, the installation of qsharp incorporates iQ♯ kernel.
  Step 2: Install miniconda or anaconda with the 64-bit installation procedure.
  Step 3: Afterwards, initialize conda with **Conda init** command to initialize conda in quantum developer host computer Powershell.
  Step 4: Now, create and activate a new environment with the name **qsharp-env** using the following command:

  conda create -n qsharp-env -c microsoft qsharp notebook
  conda activate qsharp-env

  The command execution, as mentioned above, will create the required packages to run the quantum program successfully. Finally, quantum developers have to run **python -c "import qsharp"** command to verify the installation of Q♯ standalone program execution environment.

3.2.2 Write a Q♯ Standalone Program and Run it on Local Quantum Simulator

Until now, we have seen the different ways to set up the configuration to run quantum programs. Now, we will see how to run the Q♯ program comprising Q♯ in our preferred development environment (jupyter notebooks/ VS code/ Visual studio/another editor with command prompt). The Q♯ program can be a standalone, python, or .NET program.

- **Run a Q♯ program in Jupyter Notebook locally**

  Step 1: In the installed environment (conda environment/python environment), start the jupyter notebook server by typing **Jupyter notebook**
  Step 2: As soon as jupyter notebook starts, choose **New** → Q♯. Subsequently, a new jupyter notebook encompassing Q♯ kernel will be created.
  Step 3: Write your quantum code in the created notebook. This chapter uses QuantumRandomNumberGenerator for demonstration purposes.
  Step 4: Run the created jupyter notebook after writing the code of QuantumRandomNumberGenerator. It will result in QuantumRandomNumberGenerator in the output cell.

Step 5: Finally, simulate the quantum program (here QuantumRandomNumberGenerator), which is just created. Type %Simulate QuantumRandomNumberGenerator. It results in either zero or one as the QuantumRandomNumberGenerator program produces a random result.

- **Run a Q♯ program in VS Code locally:** Any quantum developer can create and run a quantum program in visual studio (VS code) using the command line template provided by the VS code. First, the quantum developer must develop and create a new project in VS code. Afterwards, the quantum developer has to run the quantum application. The details for creating a new project and executing the application are mentioned herewith.

  Step 1: Click **(View → Command Palette →** Q♯**: Create New project)**
  Step 2: Click Standalone Console Application
  Step 3: Save the project at a preferred location with a suitable project name (let us say chapter).
  Step 4: Afterwards, enter the chapter and then **(Click create project)** which will result in successfully creating the project chapter. Now, the quantum developer can open, edit, and inspect the created project chapter. VS code provides a file, namely program.qs, to demonstrate the sample quantum code. This file is basically a Q♯ program which results in printing a message to the console.
  Step 5: Now, click **(Terminal → New Terminal)**
  Step 6: Type dotnet run in the terminal prompt to run the sample program. At the end of this step, the quantum developer obtains the output **(Hello quantum world!)** and has completed the execution of the sample program in VS code. Alternatively, the quantum developer can write the QuantumRandomNumberGenerator program within a new file and follow steps 4–6 to obtain the output as either 0 or 1 in VS code.

- **Run a Q♯ program in VS (windows only) locally:** Contrary to VS code, the quantum developer has to follow different steps if he wants to execute a Q♯ program on the visual studio (VS) (windows only). First, the quantum developer has to create a new Q♯ application. Afterwards, the quantum developer has to run the quantum application to produce the output. The details for creating a new Q♯ application and steps to perform the execution of the created Q♯ application are mentioned herewith.

  Step 1: Open visual studio **(File → New → Project)**
  Step 2: Select **(Q♯ application → Next)**
  Step 3: Click Create after entering the name and preferred location for the Q♯ application.
  Step 4: Since the project is created, the quantum developer can inspect the project by going through the file, namely program.qs. The file contains a Q♯ program which prints a message to the outer console. Alternatively, the quantum developer can write a new Q♯ program or can extend the already existing program.qs file by adding more print statements.

Step 5: Now, Select **(Debug → Start without debugging)**
The execution of the above steps results in printed **(Hello quantum world!)** in an outer console window if the quantum developer has not changed anything in the file program.qs. However, if the file has been incorporated with the QuantumRandomNumberGenerator program, one may obtain either 0 or 1 as QuantumRandomNumberGenerator randomly produces the output.

4 Quantum Simulators

Quantum simulators play an important role in providing an execution environment for successfully implementing any newly designed quantum program. Specifically, the successful implementation of the particular quantum programs involving Hadamard gate (H), controlled-NOT gate (CNOT), and measurement (Measure). Besides, the quantum simulator also handles quantum bits (qubits) tracking and qubit management. A quantum simulator may be formally defined as software programs running on a classical computer that acts as a target machine for quantum programs. It also enables the quantum developer to test and execute a newly designed quantum program. Additionally, it allows the quantum developer to understand the qubit's reaction when qubits are subjected to different quantum primitive operations within any quantum program. Each quantum program can be simulated using different classes of a quantum simulator. These classes of quantum simulators differ in their approach to providing the varying implementation of quantum primitives like H, CNOT, and Measure. All the classes of the quantum simulator are included in the quantum development kit (QDK). The details of each class of quantum simulator are described herewith.

4.1 Full State Simulator

The full state simulator is available in QDK and acts as a quantum machine to allow simulation of the quantum program at the local classical computer of the quantum developer. The full state simulator allows the debugging of execution of such quantum programs, which requires up to 30 qubits only. The steps to invoke and run a full state simulator with different host programs such as $Q^\sharp$ with $C^\sharp$, $Q^\sharp$ with python, $Q^\sharp$ in the command line, $Q^\sharp$ in Jupyter notebooks. The full state simulator uses the namespace, namely Microsoft.Quantum.Simulation.Simulators in any quantum program. Algorithm 1 presents an instance example of a sample full state simulator class.

Algorithm 1 Sample of full state simulator class instance

```
using (var sim = new QuantumSimulator())
    {
        var res = myOperation.Run(sim).Result;
    }
```

- **Invoking the full state simulator with $C^{\sharp}$ program:** Here, the quantum developer has to create a **QuantumSimulator** class instance and subsequently call the created instance in the **Run** method along with different quantum operations as detailed above. Besides, the full state simulator behaviour may be changed within any quantum program using commands like **throwOnReleasingQubitsNotInZeroState, randomNumberGeneratorSeed, disableBorrowing**.
- **Invoking the full state simulator with python program:** The quantum developer can use the method namely **simulate()** which exists in $Q^{\sharp}$ python library.
  result = myOperation.simulate()
- **Invoking the full state simulator from command line:** The quantum developer who wants to invoke a full state simulator to execute a $Q^{\sharp}$ code using the command line has to run the below-mentioned command. By default, the target machine is set to a full state simulator. So, any quantum developer can use any of the two commands described below.

  dotnet run
  dotnet run -s QuantumSimulator
- **Invoking the full state simulator from Jupyter Notebooks:** The quantum developer can type **%simulate myOperation** to run the full state simulator for used $Q^{\sharp}$ operation within any quantum program.

4.2 *Sparse Simulator*

The namespace used for the sparse simulator is Microsoft.Quantum.Simulator. Simulators. A sparse simulator is widely used to simulate quantum programs with a small number of states in superposition (sparse states). Unlike the full state simulator, it uses sparse representation for quantum state vectors. Due to sparse representation, the memory footprint used for representing quantum states gets reduced, allowing performing a simulation with an increased number of qubits. Ideally, sparse simulators represent such quantum states that feature most of their amplitude coefficient as zero in their computational basis, i.e., the representation of quantum states that feature a sparse computational basis should be performed using the sparse simulator for efficiency. Algorithm 2 presents an instance example of a sample sparse simulator class.

Algorithm 2 Sample of sparse simulator class instance

```
using (var sim = new SparseSimulator())
  {
     var res = RunMyOperation.Run(sim).Result;
  }
```

- **Invoking the sparse simulator with C♯ program:** Here, the quantum developer has to create a **SparseSimulator** class instance and subsequently call the created instance in the **Run** method along with different quantum operations. Like a full state simulator with C♯ program, here, the sparse simulation behaviour may be changed within any quantum program using different commands like **throwOnReleasingQubitsNotInZeroState, randomNumberGeneratorSeed, numQubits and disableBorrowing**.
- **Invoking the sparse simulator with python program:** The quantum developer can use the method namely **simulate_sparse()** which exists in Q♯ python library.
  result = RunMyOperation.simulate_sparse()
- **Invoking the sparse simulator from command line:** The quantum developer who wants to invoke a sparse simulator to execute a Q♯ code using the command line has to run the below-mentioned command.
  dotnet run -s SparseSimulator
- **Invoking the sparse simulator from Jupyter Notebooks:** The quantum developer can type **%simulate_sparse RunMyOperation** to run the sparse simulator for used Q♯ operation within any quantum program.

4.3 Toffoli Simulator

The Toffoli simulator is available in QDK, and it also acts as a quantum machine to allow the simulation of the quantum program at a local classical computer at the quantum developer end. Unlike the full state simulator, the Toffoli simulator allows the debugging and execution of quantum programs involving millions of qubits, thereby widely used for the evaluation purpose of boolean functions using oracles. However, being a special-purpose simulator, the Toffoli simulator only supports the gates, namely CNOT, X and multi-controlled X gate, and classical computations. A sample Toffoli simulator class instance is presented through Algorithm 3.

Algorithm 3 Sample of toffoli simulator class instance

```
var sim = new ToffoliSimulator();
var res = myOperation.Run(sim).Result;
```

- **Invoking the toffoli simulator with C$^\sharp$ program:** Here, the quantum developer has to create a **ToffoliSimulator** class instance and subsequently call the created instance in the **Run** method along with different quantum operations as the first parameter.
  var June = new ToffoliSimulator();
  var July = myOperation.Run(June).Result;

- **Invoking the sparse simulator with python program:** The quantum developer can use the method, namely **toffoli_simulate()** which exists in Q$^\sharp$ python library.
  result = myOperation.toffoli_simulate()()

- **Invoking the sparse simulator from command line:** The quantum developer who wants to invoke the toffoli simulator to execute a Q$^\sharp$ code using the command line has to run the below-mentioned command.
  dotnet run -s ToffoliSimulator

- **Invoking the sparse simulator from Jupyter Notebooks:** The quantum developer can type **%toffoli myOperation** to run the toffoli simulator for the used Q$^\sharp$ operation within any quantum program.

Besides, the Toffoli simulator is well known for supporting measurement operation only on a Pauli Z basis, resulting in either 0 or 1 probability. Also, the rotation operation represented by **R** command and exponentiated Pauli operation represented by **Exp** command have been supported by the Toffoli simulator. The functions such as **DumpMachine** & **DumpRegister** has also been supported by Toffoli simulator. Additionally, each instance of the Toffoli simulator uses 65,536 quantum bits. However, the quantum developer can use parameter **qubitcount** to set the number of quantum bits if their developed quantum program requires more than 65,536 qubits. Adding each qubit brings forth the need for one extra byte of memory in any quantum program.

4.4 Noise Simulator

The namespace used for the noise simulator is Microsoft.Quantum.Experimental. Currently, QDK provides a noise simulator as a the preview feature to conduct a simulation of open quantum systems. Open quantum systems are quantum systems that involve noise from the environment or in which qubits interact with the environment. As the noise simulator is still in the preview feature, it only allows the simulation of quantum programs with only CNOT, H and phase gates under the effect of the noisy environment. Additionally, the noise simulator can only be used for Q$^\sharp$ within c$^\sharp$ host program, Q$^\sharp$ within the python host program, and with Q$^\sharp$ standalone program. Algorithm 4 presents an instance example of a sample noise simulator class.

Algorithm 4 Sample of noise simulator class instance

```
import qutip as qt
import qsharp
import qsharp.experimental
import qsharp.experimental.enable_noisy_simulation()
namespace NoisySimulation
{
  open Microsoft.Quantum.Intrinsic;
  open Microsoft.Quantum.Measurement;
  open Microsoft.Quantum.Canon;
  open Microsoft.Quantum.Diagnostics;
    @EntryPoint()
        static void Main(string[] args)
          operation DumpPlus() : Unit
          {
          use q = Qubit();
          H(q);
          DumpMachine();
          X(q);
          Reset(q);
          }
      }
```

4.5 Resource Estimator

The resource estimator's responsibility includes estimating the resources required to successfully execute the quantum program comprising Q$^{\sharp}$ operations on the real quantum computer or local quantum simulator. The resource estimator can estimate the resources for any quantum program with Q$^{\sharp}$ operations requiring thousands of qubits, i.e., the resource estimator presents the top-level analysis of quantum programs. The resource estimator can be described/assumed as a kind of simulation target invoked in quantum programs. Algorithm 5 presents an instance example of a sample resource estimator class.

- **Invoking the resources estimator with C$^{\sharp}$ program:** Here, the quantum developer has to create a **ResourcesEstimator** class instance and subsequently call the created class instance in the **Run** method along with different quantum operations as its first parameter. Since ResourcesEstimator class does not implement **IDisposable** interface hence does not require to employ 'using' statement in C$^{\sharp}$ program. Also, as it is seen from the code that ResourcesEstimator incorporates **ToTSV()** method, quantum developers can save the produced output to a new file or analyze it from the console. The output of ToTSV() is depicted through Table 1. The ResourcesEstimator is developed on top of QuantumTraceSimulator; therefore, by default, the ResourcesEstimator counts **T** gates only. However, the configuration can be changed using QCTraceSimulator instance.

Algorithm 5 Sample of resource estimator class instance

```
using System;
using Microsoft.Quantum.Simulation.Simulators;
using Microsoft.Quantum.Simulation.Simulators.QCTraceSimulators;
namespace Quantum.MyProgram
{
  class Driver
    {
      static void Main(string[] args)
      {
        ResourcesEstimator estimator = new ResourcesEstimator();
        MyOperation.Run(estimator).Wait();
        Console.WriteLine(estimator.ToTSV());
      }
    }
}
```

Table 1 Resource estimator output

Metric	Sum	Max
CNOT	1000	1000
QubitClifford	1000	1000
R	0	0
Measure	4002	4002
T	0	0
Depth	0	0
Width	2	2
QubitCount	2	2
BorrowedWidth	0	0

- **Invoking the resources estimator with python program:** The quantum developer can use the method namely **estimate_resources()** which exists in Q$^\sharp$ python library.
  result = myOperation.estimate_resources()
- **Invoking the resources estimator from command line:** The quantum developer who wants to invoke the resources estimator to execute a Q$^\sharp$ code using the command line has to run the below-mentioned command.
  dotnet run -s ResourcesEstimator
- **Invoking the sparse simulator from Jupyter Notebooks:** The quantum developer can type **%estimate myOperation** to run the resources estimator for the used Q$^\sharp$ operation within any quantum program.

4.6 Quantum Trace Simulator

The QuantumTraceSimulator assists the quantum developer in debugging classical code used within the quantum program. Also, it estimates the resource required for running the instance of any quantum program comprised of $Q^\sharp$ operations. The difference between QuantumTraceSimulator and ResoucesEstimator is that ResoucesEstimator dispenses only a limited set of metrics, whereas QuantumTraceSimulator can estimate resources with more number of metrics. Algorithm 6 presents an instance example of a quantum trace simulator class.

- **Invoking the Quantum Trace Simulator with $C^\sharp$ program:**
  Here, the quantum developer has to create a **QCTraceSimulatorr** class instance and subsequently call the created class instance in the **Run** method along with different quantum operations as its first parameter. Since ResourcesEstimator class does not implement **IDisposable** interface hence does not require employing using statement in $C^\sharp$ program. Also, QDK supports configuration flag change through **QCTraceSimulatorConfiguration** subsequently passing to **QCTraceSimulator**. The list of supported configurations is presented through Table 2.

Algorithm 6 Sample of QC trace simulator class instance

```
using Microsoft.Quantum.Simulation.Core;
using Microsoft.Quantum.Simulation.Simulators;
using Microsoft.Quantum.Simulation.Simulators.QCTraceSimulators;
namespace Quantum.MyProgram
{
  class Driver
    {
        static void Main(string[] args)
        {
            QCTraceSimulator sim = new QCTraceSimulator();
            var res = MyQuantumProgram.Run(sim).Result;
            System.Console.WriteLine("Press any key to continue...");
            System.Console.ReadKey();
        }
    }
}
```

5 Azure Quantum Workspace

This section describes creating an Azure Quantum workspace and steps to get started with $Q^\sharp$ and an Azure Quantum notebook.

Table 2 QC trace simulator configuration command

Configuration	Description
Distinct inputs checker	Checks for potential conflicts with shared qubits
Invalidated qubits use checker	Checks if the program applies an operation to a qubit that is already released
Primitive operations counter	Count the number of primitives used by every operation invoked in a quantum program
Depth counter	Gather counts that represent the lower bound of the depth of every operation invoked in a quantum program
Width counter	Count the number of qubits allocated and borrowed by each operation in a quantum program

5.1 Create an Azure Quantum Workspace

Azure Quantum workspace is a collection of all assets (tools and libraries) needed to execute a quantum application or optimization application. If a quantum developer wants to execute a quantum application or any optimization problems on real quantum hardware, then S/he needs an Azure Quantum workspace [46]. The quantum developer has to use the Azure Quantum portal and needs Azure Quantum credits to run Azure Quantum workspace to use Azure Quantum workspace. For first-time users of the Azure Quantum workspace, the Azure Quantum portal automatically provides 500USD worth of free Azure Quantum credits so that any developer can run quantum applications on real quantum hardware. Besides, the Azure Quantum portal provides two options for creating a new workspace: quick create and advanced create. Both options create a new Azure Quantum workspace. However, they differ from each other in the facility of managing resource groups and storage accounts. Specifically, quick create automatically creates the required resource group & storage account and all quantum hardware providers, namely Quantinuum, IonQ, and Microsoft QIO providers. Also, it automatically adds Azure Quantum workspace credits for accessing the quantum hardware, whereas advance create provides the option to manually configure resource groups and storage accounts with the facility to set flags for categorized resources. Besides, the following steps need to be followed to create a new workspace:

- **Step 1:** With user account's credentials sign in to (**Azure Portal**) [47]
- **Step 2:** Now, the developer has to Select (**Create a resource → Azure Quantum →Azure quantum service**)
- **Step 3:**Afterwards, Select (**Azure Quantum → Create**) . It results in a form to create a workspace.
- **Step 4:** Now, the quantum developer has to associate a subscription Azure Quantum credits to the new workspace.
- **Step 5:** Finally, Select (**Quick Create/Advance Create**).

In the quick create, the quantum developer has to enter a name for the workspace and select the region for the workspace. Finally, the quantum developer has to click create, whereas in advance create, the quantum developer has to choose an existing resource group or create a new one. Afterwards, the quantum developer has to enter a name for the workspace and select the region for the workspace. Finally, it also provides an option to create a storage account for the workspace and allows to add available providers manually. Additionally, the quantum developer can exercise the modification by clicking the button modify.

5.2 *Get Started with Q♯ and an Azure Quantum Notebook*

The quantum developer can use a jupyter notebook in the Azure Quantum workspace. Below are the steps that any quantum developer needs to follow for writing any quantum application incorporating Q♯ language on Azure Quantum notebook.

- **Step 1:** With own account's credentials log in to **(Azure Portal)**. Henceforth, the developer has to select **Azure Quantum workspace**
- **Step 2:** Select **(Jupyter notebook)**
- **Step 3:** Search for **(Hello, World:Q♯ notebook tile)** in the sample gallery. Afterwards, choose the quantum hardware provider (either Quantinuum or IonQ).
- **Step 4:** Select **Copy to my notebooks**. It results in copying the sample notebook under **my notebook**.
- **Step 5:** Finally, execute the sample notebook under **(My notebook)**.

If the quantum developer has been using the IonQ platform, then he has to check **My notebook → Hello-world-q-sharp-ionq notebook**. Afterwards, he has to select **Run all**. Subsequently, the quantum developer can navigate to many cells and execute the shell individually. The sample notebook comprises of hello world program, which executes a quantum random number generator and presents the histogram of results. In the sample notebook, the python 3 kernel is used by default. However, while creating a notebook in the Azure Quantum workspace, the quantum developer can select one out of two kernels that are compatible with Q♯ namely Azure Quantum Q♯ kernel and python 3 (ipykernel). Also, it may be noted that the responsibility of the first cell of the sample notebook includes loading of subscription information of the quantum developer account to establish a connection with the Azure Quantum service. The responsibility of the second cell of the sample notebook includes fetching the target hardware within the Azure Quantum workspace. The target may be a quantum simulator or real quantum hardware provided by different providers. The third, fourth and subsequent cells define the quantum program, set the target quantum simulator or real hardware, and submit the program, plots, and present the result. The steps are the same for the quantum target machine, similar to the IonQ target machine for any quantum program.

5.3 Submit a Quantum Circuit with Qiskit Using an Azure Quantum Workspace

The Azure Quantum workspace allows submission of a qiskit circuit to available quantum targets such as Quantinuum & IonQ. The quantum developer can create a new notebook with the qiskit program in the Azure Quantum workspace. The quantum developer needs to have an active Azure Quantum workspace before creating a new notebook. First, the quantum developer has to log in to the Azure Quantum portal. Afterwards, S/he has to select the created Azure Quantum workspace. Subsequently, from the left pane of the Azure Quantum portal, one has to select **Notebooks** → **My Notebooks** → **Add New**. Finally, the developer has to assign a name to the created file and click **create file**. After successful file creation, the quantum developer has to load the required imports, connect to Azure Quantum services, define a circuit, list all targets, select a target to run the quantum program, and estimate the cost for that quantum program.

6 Ways to Develop a Quantum Software

The quantum developer has to follow the below-mentioned steps to complete the development of quantum software [48].

- **Step 1: Write the quantum code:**
  The quantum developer should write the quantum code using Q♯ programming language available through QDK.
- **Step 2: Use libraries to keep your code high level:**
  The QDK includes quantum libraries that help quantum developers keep the quantum code high level, facilitating the developer to focus on the logic of quantum code. The quantum developer has to not worry about quantum code's background and implementation details if s/he is using the libraries.
- **Step 3: Integrate with classical software:**
  The quantum code can be integrated with classical code encompassing classical libraries to fast-track the solutions of some particular real-world problems.
- **Step 4: Run quantum code on simulator:**
  Once the development of quantum code gets complete, the quantum developer can execute the quantum code on the available variety of simulators provided through QDK. Each simulator has limitations with the capability to support the number of qubits. However, these simulators are still very important to visualize the behaviour of the quantum program.
- **Step 5: Estimate the required resources:**
  Before running the quantum code on the real hardware, the quantum developer can use a resources estimator to find the number of qubits required for the developed program and calculates the execution time of the specific quantum program.

- **Step 6: Run Code on quantum hardware:**
  The quantum developer can run the developed quantum code on real quantum hardware through Azure Quantum. The Azure Quantum provides different quantum hardware platforms enabled by providers like IonQ and Quantinuum.

7 Conclusion

The demand for the quantum developer is expected to rise exponentially in coming years with the evergrowing impact of quantum computing techniques in the real world. Many industries relating to the internet of things, vehicular ad-hoc networks, autonomous vehicles, and many more have started looking for ways to incorporate quantum computing techniques in their existing infrastructure. Furthermore, the threat brought forth by quantum cryptography will require many quantum cryptographers who can design and analyze new encryption schemes on quantum computing platforms like Azure Quantum. Consequently, this chapter describes the Azure Quantum, Azure Quantum workspace, quantum programs, ways to develop quantum software, and many more detailing steps required to be followed by any new quantum developer. The presented discussion will help the software developer who is aspiring to transition into a quantum developer and any quantum enthusiast who wants to use Azure Quantum for the possible development of a quantum program that will solve real-world problems. This chapter motivates future readers and developers to quick start one's journey towards developing quantum applications so that they can upgrade themselves in a new role—quantum software engineer.

References

1. T.A. Shaikh, R. Ali, Quantum computation for big information processing, in *Blockchain, Big Data and Machine Learning* (CRC Press, 2020), pp. 315–338
2. S.S. Biswas, Quantum computers: a review work. Adv. Comput. Sci. Technol. **10**(5), 1471–1478 (2017)
3. B. Rawat, N. Mehra, A.S. Bist, M. Yusup, Y.P.A. Sanjaya, Quantum computing and AI: impacts and possibilities. ADI J. Recent Innov. **3**(2), 202–207 (2022)
4. M. Schuld, I. Sinayskiy, F. Petruccione, The quest for a quantum neural network. Quantum Inf. Process. **13**(11), 2567–2586 (2014)
5. Y. Li, M. Tian, G. Liu, C. Peng, L. Jiao, Quantum optimization and quantum learning: a survey. IEEE Access **8**, 23568–23593 (2020)
6. T. Hur, L. Kim, D.K. Park, Quantum convolutional neural network for classical data classification (2021), arXiv preprint arXiv:2108.00661
7. W. Deng, H. Liu, J. Xu, H. Zhao, Y. Song, An improved quantum-inspired differential evolution algorithm for deep belief network. IEEE Trans. Instrum. Meas. **69**(10), 7319–7327 (2020)
8. F.S. Khan, N. Solmeyer, R. Balu, T.S. Humble, Quantum games: a review of the history, current state, and interpretation. Quantum Inf. Process. **17**(11), 1–42 (2018)
9. M. Carrega, A. Crescente, D. Ferraro, M. Sassetti, Dissipative dynamics of an open quantum battery. New J. Phys. **22**(8), 083085 (2020)

10. S. Ghosh, T. Chanda, A. Sen, Enhancement in the performance of a quantum battery by ordered and disordered interactions. Phys. Rev. A **101**(3), 032115 (2020)
11. S. Yarkoni, F. Neukart, E.M.G. Tagle, N. Magiera, B. Mehta, K. Hire, M. Hofmann, Quantum shuttle: traffic navigation with quantum computing, in *Proceedings of the 1st ACM SIGSOFT International Workshop on Architectures and Paradigms for Engineering Quantum Software* (2020), pp. 22–30
12. M. Zinner, F. Dahlhausen, P. Boehme, J. Ehlers, L. Bieske, L. Fehring, Quantum computing's potential for drug discovery: early stage industry dynamics. Drug Discov. Today **26**(7), 1680–1688 (2021)
13. M. Ruzycka-Ayoush, P. Kowalik, A. Kowalczyk, P. Bujak, A.M. Nowicka, M. Wojewodzka, I.P. Grudzinski, Quantum dots as targeted doxorubicin drug delivery nanosystems in human lung cancer cells. Cancer Nanotechnol. **12**(1), 1–27 (2021)
14. Y. Cao, J. Romero, A. Aspuru-Guzik, Potential of quantum computing for drug discovery. IBM J. Res. Dev. **62**(6), 6–1 (2018)
15. A. Bouland, W. van Dam, H. Joorati, I. Kerenidis, A. Prakash, Prospects and challenges of quantum finance (2020), arXiv preprint arXiv:2011.06492
16. P. Rebentrost, B. Gupt, T.R. Bromley, Quantum computational finance: Monte Carlo pricing of financial derivatives. Phys. Rev. A **98**(2), 022321 (2018)
17. A. Ganapathy, Quantum computing in high frequency trading and fraud detection. Eng. Int. **9**(2), 61–72 (2021)
18. J. Zhang, X. Qiu, X. Li, Z. Huang, M. Wu, Y. Dong, Support vector machine weather prediction technology based on the improved quantum optimization algorithm. Comput. Intell. Neurosci. (2021)
19. Y. Li, Y. Zhao, J. Fu, L. Xu, Reducing food loss and waste in a two-echelon food supply chain: a quantum game approach. J. Clean. Prod. **285**, 125261 (2021)
20. S. Yarkoni, A. Alekseyenko, M. Streif, D. Von Dollen, F. Neukart, T. Bäck, Multi-car paint shop optimization with quantum annealing, in *2021 IEEE International Conference on Quantum Computing and Engineering (QCE)*. (IEEE, 2021), pp. 35–41
21. C.H. Bennett, G. Brassard, Quantum cryptography: public key distribution and coin tossing, in *Proceedings of the IEEE International Conference on Computers, Systems, and Signal Processing* (IEEE, New York, 1984), pp. 175–179
22. C.H. Bennett, Quantum cryptography using any two nonorthogonal states. Phys. Rev. Lett. **68**(21), 3121–3124 (1992)
23. D. Bruß, Optimal eavesdropping in quantum cryptography with six states. Phys. Rev. Lett. **81**(14), 3018–3021 (1998)
24. C.H. Bennett, G. Brassard, N.D. Mermin, Quantum cryptography without Bell's theorem. Phys. Rev. Lett. **68**(5), 557–559 (1992)
25. A.K. Ekert, Quantum cryptography based on Bell's theorem. Phys. Rev. Lett. **67**(6), 661–663 (1991)
26. R. LaPierre, Shor algorithm, in *Introduction to Quantum Computing* (Springer, Cham, 2021), pp.177–192
27. N. Kilber, D. Kaestle, S. Wagner, Cybersecurity for quantum computing (2021), arXiv preprint arXiv:2110.14701
28. L. Zhang, A. Miranskyy, W. Rjaibi, G. Stager, M. Gray, J. Peck, Making existing software quantum safe: lessons learned (2021), arXiv preprint arXiv:2110.08661
29. L. Ankur, Karambir, Using quantum key distribution and ECC for secure inter-device authentication and communication in IoT infrastructure, in *Proceedings of 3rd International Conference on Internet of Things and Connected Technologies (ICIoTCT)* (2018), http://dx.doi.org/10.2139/ssrn.3166511
30. K. Prateek, F. Altaf, R. Amin, S. Maity, A privacy preserving authentication protocol using quantum computing for V2I authentication in vehicular Ad Hoc networks. Secur. Commun. Netw. (2022)
31. Z. Chen, K. Zhou, Q. Liao, Quantum identity authentication scheme of vehicular Ad-Hoc network. Int. J. Theor. Phys. **58**, 40–57 (2019). https://doi.org/10.1007/s10773-018-3908-y

32. G. Sharma, S. Kalra, Identity based secure authentication scheme based on quantum key distribution for cloud computing, in *Peer-to-Peer Networking Application* (2018), pp. 220–234. https://doi.org/10.1007/s12083-016-0528-2
33. G. Murali, R.S. Prasad, Secured cloud authentication using quantum cryptography, in *2017 International Conference on Energy, Communication, Data Analytics and Soft Computing (ICECDS)* (Chennai, 2017), pp. 3753–3756. https://doi.org/10.1109/ICECDS.2017.8390166
34. Y. Dong, S. Xiao, H. Ma, L. Chen, Research on quantum authentication methods for the secure access control among three elements of cloud computing. Int. J. Theor. Phys. https://doi.org/10.1007/s10773-016-3132-6
35. K. Prateek, S. Maity, Post-quantum blockchain-enabled services in scalable smart cities, in *Quantum Blockchain, An Emerging Cryptographic Paradigm* (2022), p. 263
36. J. Ahn, J. Chung, T. Kim, B. Ahn, J. Choi, An overview of quantum security for distributed energy resources, in *2021 IEEE 12th International Symposium on Power Electronics for Distributed Generation Systems (PEDG)*. (IEEE, 2021), pp. 1–7
37. L.P. Raghav, R.S. Kumar, D.K. Raju, A.R. Singh, Optimal energy management of microgrids using quantum teaching learning based algorithm. IEEE Trans. Smart Grid **12**(6), 4834–4842 (2021)
38. Z. Wang, Z. Guo, G. Mogos, Z. Gao, Quantum key distribution by drone. J. Phys.: Conf. Ser. **2095**(1), 012080 (IOP Publishing, 2021)
39. M.H. Adnan, Z. Ahmad Zukarnain, N.Z. Harun, Quantum key distribution for 5g networks: a review. State Art Future Direct. Future Internet **14**(3), 73 (2022)
40. J. Hooyberghs, Azure quantum, in *Introducing Microsoft Quantum Computing for Developers* (Apress, Berkeley, 2022), pp. 307–339
41. Download Docker image, https://github.com/microsoft/Quantum/tree/master/.devcontainer
42. Q♯ user guide, https://docs.microsoft.com/en-us/azure/quantum/install-overview-qdk
43. Ways to run a Q♯ program, https://docs.microsoft.com/en-us/azure/quantum/user-guide/host-programs?tabs=tabid-python
44. Download Miniconda, https://docs.conda.io/en/latest/miniconda.html
45. Download Anaconda, https://docs.anaconda.com/anaconda/install/windows/
46. Azure Quantum Workspace, https://azure.microsoft.com/en-in/services/quantum/
47. Azure Portal, https://azure.microsoft.com/en-in/features/azure-portal/
48. M. Mariia, Quantum software development using the QDK (2021), https://devblogs.microsoft.com/qsharp/quantum-software-development-using-the-qdk/

Survey of Open-Source Tools/Industry Tools to Develop Quantum Software

Dhaval Mehta, Amol Ranadive, Jigna B. Prajapati, and Rajiv Pandey

Abstract Quantum computing is an advanced-level computing paradigm based on the fundamental principles on which nature operates, i.e. quantum mechanics and its ability to perform complex computations and high potential to set trends in the new era in computing technology. As quantum computing is slightly different from traditional computing, special types of software are needed for implementing Quantum Computing. There is a list of quantum software projects which are exclusively available on GitHub and/or maintained by the Quantum Open Source Foundation and there are proprietary tools offered by leading companies. With this in mind, we have compiled and curated detailed information & insights on some of the best quantum computing software tools which are widely used now-a-days. This chapter focuses on Microsoft Quantum Development Kit, IBM Quantum Tools, Amazon Braket, Google Quantum tools, and other popular open-source tools.

1 Introduction

As Quantum Computing is slightly different than traditional computing, hence special types of software is needed for implementing the concept of Quantum Computing. There is a list of quantum software projects available exclusively on

D. Mehta
School of Engineering and Technology, Navrachana University, Vadodara, Gujarat, India

A. Ranadive
School of Business and Law, Navrachana University, Vadodara, Gujarat, India

J. B. Prajapati (✉)
Acharya Motibhai Patel Institute of Computer Studies, Ganpat University, Mahesana, Gujarat 384012, India
e-mail: jignap15@gmail.com

R. Pandey
Amity Institute of Information Technology (AIIT), Amity University, Lucknow Campus, Uttar Pradesh, India

R. Pandey et al. (eds.), *Quantum Computing: A Shift from Bits to Qubits*,
Studies in Computational Intelligence 1085,
https://doi.org/10.1007/978-981-19-9530-9_17

GitHub maintained by the Quantum Open-Source Foundation and the proprietary tools provided by leading companies [1].

We can divide the quantum software broadly into two categories [2–6]:

- **The software which runs quantum algorithms**: This type of Quantum software development kit enables users to develop and test quantum algorithms for a problem.
- **The software which enables quantum computers to perform**: Quantum computers might face performance problems because of random errors and these types of software have capabilities to correct such errors.

This chapter focuses on the basic functionality of the tools for developing applications/solving real-life problems using Quantum Computing.

2 Quantum Computer Vendors' Toolkits

This type of toolkits/software is exclusively provided by the manufacturer of quantum computers. At present, four major companies are dealing in it and their popular tools are depicted in Fig. 1 [1].

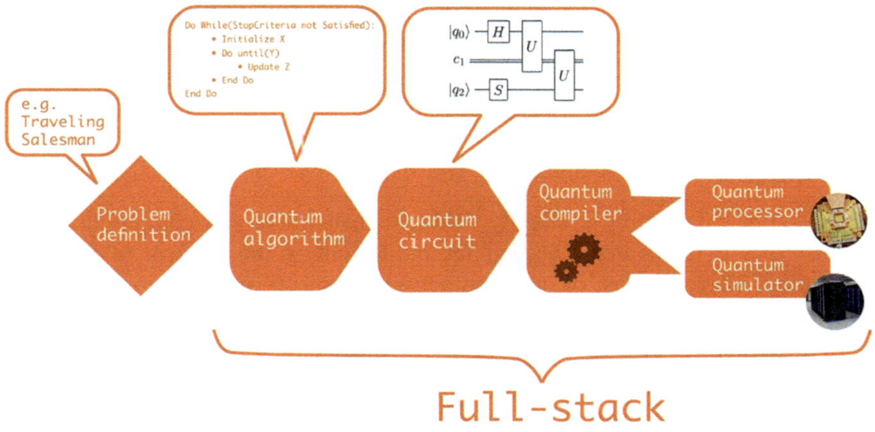

Fig. 1 Visualization of a typical quantum algorithm workflow on a gate-model quantum computer [1]. "Visualization of a typical quantum algorithm workflow on a gate-model quantum computer" by Fingerhuth et al. (2018), retrieved from https://doi.org/10.1371/journal.pone.0208561, used under Creative Commons Attribution-License

2.1 *Microsoft Quantum Development Kit*

The Microsoft Quantum Development Kit is provided by Microsoft to carry out work in the field of quantum computing. This can be integrated with the Visual Studio development environment. Even Azure Quantum, a cloud-based service, can also be used. It uses Q# as a programming language for designing quantum circuits and optimizing or solving complex problems. It's an open ecosystem that enables users to access various quantum software, hardware, and solutions from Microsoft and its partners [7].

2.1.1 Azure Quantum Solutions

Azure Quantum suggests two main alternatives of quantum solutions as follows:

(i) *Quantum Computing*:

This solution facilitates learners to study, experiment, and prototype with various quantum hardware providers and make them ready for the future of scaled quantum machines. In comparison with other solutions, the user is not separated to single hardware technology and helps the user from a full-stack point of view for protecting the long-term investments [8].

(ii) *Optimization*:

This alternative is more economical as it builds solutions that help users to decrease the cost of operation in many fields like finance, energy cost, fleet management, scheduling, and much more [8].

Using Azure Quantum and the Quantum Development Kit toolset, the user can devise quantum algorithms and optimization solutions, and then apply these quantum solutions on the existing Azure platform to get real-world outputs even before the development of a general-purpose quantum computer [8].

(iii) *Unique features*:

- Its' an open ecosystem that offers solutions to access different quantum software, hardware, and solutions from Microsoft and associates.
- It's a reliable, extensible, and secure platform which will continue to accommodate rapidly evolving quantum in future.
- It allows users to run on readymade solutions for traditional and accelerated computing resources for quantum impact.
- It provides a unified development interface by which users can easily interact.

2.1.2 Q# and the Quantum Development Kit

Q# is a quantum programming language that is a part of The Microsoft Quantum Development Kit (QDK) which is an open-source development kit for Azure

Quantum that provides the facility to the user to work both in online mode with the service and offline mode. Q# is a high-level programming language that offers the user to concentrate on the problem at the algorithm level and application level to build quantum programs [8].

The Quantum Development Kit (QDK)

The QDK provides the following tools which help users during the quantum software development process:

- Ready-to-use Libraries: QDK provides two libraries, i.e., standard libraries which implement examples that are common for many quantum algorithms and domain-specific libraries like chemistry or machine learning. These two keep the user's code high level [8].
- Quantum computing simulators enable the user to run an instance of the program and see the output without the actual use of hardware access [8].
- Noise simulators enable the user for simulating the behavior of the program under the effect of noise and the stabilizer representation [8].
- A resource estimator that provides approximate cost to run the solutions in real world, e.g., how many qubits are required, what types of other resources are required, and how long time a program will take [8].
- The Quantum Development Kit consists of extensions for IDEs like Visual Studio and Visual Studio Code and can be easily integrated with Jupyter Notebooks. The Quantum Development Kit supports compatibility with Python and other NET languages like C#. A user can also devise optimization solutions with the Azure Quantum optimization Python package. As quantum systems advance, code can also be endured [8].
- In addition to that, QDK provides a facility to integrate with Qiskit and Cirq, and the users who have already worked on other languages can also run their programs on Azure Quantum without editing [8].

2.1.3 Q# Language

In quantum computing, a use majorly deals with code rather than with circuits and hence robust programming language is required. The Q# quantum programming language is part of Microsoft's Quantum Development Kit and facilitates use with rich IDE and provides tools for program visualization and analysis. The Q# language also permits a user to write code for classical as well as for quantum computing. In addition to that, it helps general classical control flow during the execution of a program which allows a clean sign of adaptive algorithms that are difficult to communicate directly in the circuit model of a static sequence of quantum gates. Q# can be used as standalone, in notebooks, and also at the command line or can be used with a host language such as Python or C# [9].

The Q# language has many features like, it doesn't state whether qubits are logical or physical rather. It can be decided at the runtime when the code is executed. Same way, the mapping of a qubit variable in a program to an actual logical or physical qubit is determined at the runtime, and that mapping may be postponed until the topology and other details of the target device are known. The runtime is accountable for determining a mapping that permits the algorithm to execute, with any qubit state transfer and remapping essential during execution. Following is sample code, written using Q# language [9]:

```
namespace HelloQuantum {

    open Microsoft.Quantum.Canon;
    open Microsoft.Quantum.Intrinsic;


    @EntryPoint()
    operation HelloQ() : Unit {
        Message("Hello quantum world!");
    }

}
```

The output of the above code will be

```
Hello, quantum world!
```

2.1.4 Workflow of the Quantum Software Development on Azure Quantum

Following is the diagram which shows the phases from which a quantum program goes through from the problem identification to the solution on Azure Quantum and the tools offered by the Quantum Development Kit for each phase. Figure 2 shows the workflow of the quantum software development stepwise [8].

In the first phase, the user can write a program in Q# programming language using various development tools like Visual Studio, Visual Studio Code, or Jupyter Notebooks. In the second phase, the user can focus on the main logic of the algorithm while the quantum libraries will help for high-level implementation work. In the third stage, the user can also integrate the Q# program code with Python and .NET framework for getting the benefits of advancement that occurred in the classical computing field in the last many years. Once the code is written, the user can select the quantum simulator to run the code and can see how the code works without the use of actual hardware. Once the code is run on the simulator then in the next phase the usercan get the estimation of hardware and other resources to run the code. In the last phase, a user can select appropriate quantum hardware to run the code. Here users can use the code for all steps of the workflow. At some point, a user may have to twist some portions of the code because of the current hardware limitations. But in the long run, a user can switch between various simulators and hardware providers without any code alterations.

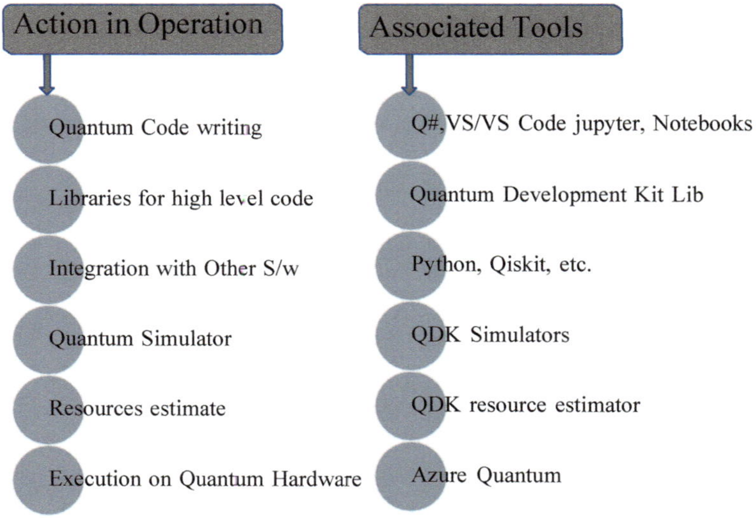

Fig. 2 Workflow of the quantum software development on Azure Quantum [8]. Adapted from "What is Azure Quantum?" Accessed from https://docs.microsoft.com/en-us/azure/quantum/overview-azure-quantum

2.1.5 Quantum Cloud Solutions

Once the program is error-free and the validation process is over and fit to run on the hardware then the program can be submitted to Azure Quantum. The following diagram illustrates the standard workflow after the program is submitted. Figure 3 shows quantum cloud solutions [8].

Azure Quantum offers the most overwhelming and different quantum resources available today from industry leaders. Azure Quantum currently affiliates with the following providers which offer users to run your Q# quantum code on real hardware, and the option to test code on simulated quantum computers.

2.1.6 Quantum Computing Providers

Following are some well-known vendors who provide quantum computing facilities. Users can select the provider according to the requirements of the problem [8].

- The world's largest integrated quantum computing company Quantinuum had combined Cambridge Quantum's advanced software development with Honeywell Quantum Solutions' high-fidelity hardware to speed up quantum computing. It has a trapped-ion system with high reliability, fully linked qubits, and the capacity to perform mid-circuit measurements.

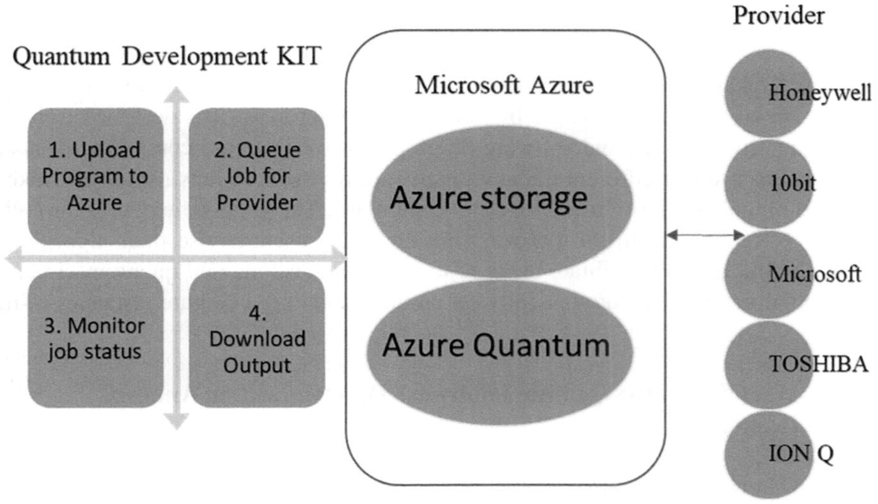

Fig. 3 Quantum cloud solutions available on Azure Quantum [8]. Adapted from "What is Azure Quantum?" Accessed from https://docs.microsoft.com/en-us/azure/quantum/overview-azure-quantum

- IONQ provides the world's most powerful quantum computer which is a dynamically reconfigurable trapped-ion quantum computer for up to 11 fully connected qubits, which allows users to execute a two-qubit gate between any pair, features minimal gate errors, and with the best performance.
- Quantum Circuits, Inc. provides a fast and high-reliable system with powerful real-time response to enable error correction.

2.1.7 Optimization Providers

For optimization solutions, the following are the list of providers [8]:

- 1QBit offers iterative heuristic algorithms which look for techniques to solve QUBO problems.
- Microsoft QIO is a set of multiple targets of hardware that express the optimization problem differently which is inspired by the quantum research of the last many years.
- Toshiba SBM is also known as Toshiba Simulated Bifurcation Machine which is a GPU-powered ISING machine that optimizes large-scale combinatorial optimization problems at high speed.

2.2 *Amazon Braket*

Amazon Braket (introduced in August 2020) is a fully managed quantum computing service based on cloud provided by Amazon Web Services. It is designed to help rapid scientific research and software development for quantum computing. It also helps the user to test different quantum hardware for the solutions. Amazon Braket has a set of built-in algorithms, tools, and documents. These can be accessed through Jupyter Notebooks from the Bracket console. Once users devise their algorithm, Amazon Bracket will facilitate users to test their algorithms and quantum circuits via simulators that will automatically set up the mandatory compute instances [10].

2.2.1 Workflow of the Quantum Software Development on Amazon Braket

Figure 4 shows the diagram of the workflow on Amazon Braket.

In the first phase, the user can write a program using fully manager Jupyter Notebooks or in the local environment. Then the program is validated and tested on local simulators or fully managed high-performance simulators. In the next phase, the program is run on quantum computers of the user's choice. Here users can get the flexibility to combine the classical and quantum computing resources for hybrid algorithms. Then these results are analyzed for measuring the performance.

Amazon also provides the option of Amazon Quantum Solutions Lab which facilitates users to work in a collaborative manner where users can work with industry experts in the domain of quantum computing, machine learning, and high-performance computing. The programs assist the user in research and identifying the applications of quantum computing in various fields. Amazon Quantum Solution Lab experts also help organizations to be ready for the future of quantum computing.

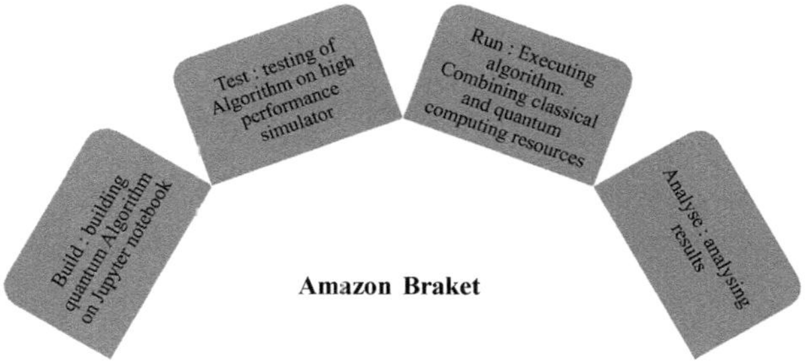

Fig. 4 Workflow of the quantum software development on Amazon Braket [10] Adapted from "About Amazon Braket." Accessed from https://aws.amazon.com/braket/

2.2.2 Amazon Braket Features

Following are the unique features of Amazon Braket [11]:

1. **Developer Tools**

- *Hardware-agnostic developer framework*

To shorten and streamline the process of designing and executing quantum algorithms, Amazon Braket SDK can be used. It has been intended to be technology agnostic, eliminating the need to code against different quantum programming tools for each type of quantum hardware. The SDK provisions a unified development framework that empowers the user or developer to build quantum algorithms and run them on any compatible Quantum Hardware provided as a part of the Amazon Braket service. As new quantum technologies appear and are supplemented to the Amazon Braket service, the development experience remains consistent and it also ensures that the existing designs and quantum algorithms can be tested on these novel systems [11].

- *Fully managed executions of quantum–classical algorithms with Hybrid Jobs*

Amazon Braket Hybrid Jobs allows users to initiate, monitor, and run hybrid quantum–classical algorithms. Once the user provides code and selects the Quantum Processing Unit (QPU) or simulator to run, Amazon Braket accelerates the classical computing, executes the code, and releases the resources once the job is finished. It also allows users to define custom metrics for the algorithms, which are automatically documented by Amazon CloudWatch and exhibited in real time in the Amazon Braket console as the code executes. This gives users live reports of how the algorithm is advancing, so users can alter the algorithm if needed. In addition to that, Hybrid Jobs provides prioritized access to your chosen QPU to help your algorithm execute quickly and predictably, enabling you to improve the quality and reproducibility of results [11].

- *Develop variational quantum algorithms with PennyLane*

Amazon Braket supports PennyLane which is an open-source software framework. This is developed on the concept of quantum differentiable programming. This framework helps the user to build and to run hybrid (quantum classical), or variational, algorithms. This approach facilitates users to train quantum circuits in the same way, the user would train a machine learning model to find solutions to computational problems in the field of quantum chemistry, quantum machine learning, and optimization. PennyLane is performance-optimized for Amazon Braket and gives interfaces to familiar machine learning tools, like PyTorch and TensorFlow, to train quantum circuits smoothly and speedily [11].

- *Fully managed Jupyter Notebooks*

Fully managed Jupyter Notebooks are another great feature where the user has a choice of using their development environment or use of fully managed Jupyter

Notebooks to build quantum algorithms and manage experiments. Jupyter notebook creation is easy where pre-configured with a suite of quantum computing developer tools, including the Amazon Braket SDK, PennyLane, and Ocean can be generated. Users can select the notebook instance type to match the performance requirements and configure security settings such as encryption for stored data [11].

- *Pre-built algorithms and tutorials*

In Amazon Braket notebooks, the Amazon Braket SDK is pre-installed along with tutorials. Users can easily start by selecting a pre-built algorithm in a single place [11].

2. **Simulators**

- *Choice of simulation tools*

With Amazon, Braket offers options of four circuit simulators to run and test quantum algorithms. These include the local simulator which is included in the Amazon Braket SDK along with three fully managed simulators. The local simulator can run on a client machine or within an Amazon Braket managed notebook and facilitates users to simulate the quantum circuits with and without noise. The fully managed simulators are SV1 (a general-purpose state vector simulator) DM1 (a density matrix simulator) for supporting noise modeling and TN1 (a tensor network simulator) which concentrates on certain larger scale structured quantum circuits [11].

- *Consistent experience*

Amazon Braket facilitates users to run a circuit on simulators with the help of a single API call. A request to run the code on a simulator works similarly as a request to run on quantum hardware, by changing a single line of code, a user can switch from running on a simulator to an actual quantum computer [11].

- *Choice of result types*

Users can select various result types for simulation jobs, like individual samples, custom observables, separate amplitudes, or the full state vector. Amazon Braket simulators can compute exact results or return measurement samples that match the behavior of quantum computers [11].

3. **Quantum Computers**

- *Simplified access to quantum computers*

Amazon Braket offers safe access to different types of quantum computing technologies. The user has not any upfront commitment or needs to sign any contract as the user will pay the bill only for what is used [11].

- *Choice of quantum processing units (QPUs)*

Amazon Braket allows using both annealing and gate-based quantum computers. By following the gate-based quantum computing paradigm, users can approach trapped-ion technology from an organization like IonQ and superconducting quantum processors from Oxford Quantum Circuits and Rigetti. On the other hand, users can solve quantum annealing problems using the latest QPUs from D-Wave. This helps the user to examine different technologies, evaluate the compute performance of different machines for the problem, and select the hardware system that is best suited to the application [11].

- *Amazon Quantum Solutions Lab*

As of date, Quantum Computing lies in its infancy stage, and still many uncertainties prevail resulting in challenges that are yet to be solved. The Amazon Quantum Solutions Lab is a collaborative research and professional services program that works with quantum computing experts that assist in more effectively exploring quantum computing and evaluating the current performance of this budding technology. Additionally, it also offers the users or developers to work with some of its qualified technology and consulting partners in the AWS Partner Network (APN) that specifically focus on applications for quantum computing and can help address those niche requirements [11].

4. **Management and Security**

- *Management Console*

As a part of Amazon Web Service, Amazon Braket can also be accessed through the AWS Management Console, a centralized and user-friendly web interface for Amazon Web Services, which facilitates users to make secure login using your AWS account or AWS Identity and Access Management (IAM) credentials. Users can do many things like to accomplish and monitor Amazon Braket resources, such as notebooks and tasks, and access detailed information about quantum circuit simulators and QPUs from the console [11].

- *User access management, security, and monitoring*

Amazon Braket provides smooth interoperability with other AWS services such as Amazon CloudWatch, AWS CloudTrail, Amazon EventBridge, and AWS IAM to enable users to monitor the loads, generate the notifications when the tasks are finished, and manage access controls and permissions from the security point of view. The simulation and quantum task results are delivered to the preferred Amazon Simple Storage Service (S3) bucket for storage and evaluation and giving full control to the user over the data [11].

2.3 *Google Quantum AI*

Google Quantum AI advancing is excellent in quantum computing and developing the tools for researchers to operate beyond classical capabilities. It provides Cirq, an open-source framework for programming quantum computers. Open Fermion, a library for compiling and analyzing quantum algorithms to simulate fermionic systems, including quantum chemistry and TensorFlow Quantum, a library for hybrid quantum–classical machine learning [12].

An open-source framework for programming quantum computers by Google Quantum AI.

2.3.1 Open-Source Frameworks

- **Cirq**

Cirq programming language is a Python software-based library that enables the user to write, manipulate, and optimize quantum circuits, and then allows it to run on quantum computers and quantum simulators. As in today's scenario, the details of the hardware are dynamic for getting excellent results and Cirq provides valuable abstractions for working with noisy intermediate-scale quantum computers. It also facilitates users to visualize data in the graphical format [13].

Following is the sample code which is written in Cirq [14]:

```
import cirq

# Pick a qubit.
qubit = cirq.GridQubit(0, 0)

# Create a circuit
circuit = cirq.Circuit(
    cirq.X(qubit)**0.5,  # Square root of NOT.
    cirq.measure(qubit, key='m')  # Measurement.
)
print("Circuit:")
print(circuit)

# Simulate the circuit several times.
simulator = cirq.Simulator()
result = simulator.run(circuit, repetitions=20)
print("Results:")
print(result)
```

Output

```
Circuit:
0, 0): ——X^0.5——M('m')——
Results:
m=11000111111011001000
```

- **OpenFermion**

OpenFermion is a Python software-based library that is designed for enabling the simulation of fermionic models and quantum-based chemistry problems on quantum hardware. It has an interface to common electronic structure packages and streamlines the translation between a molecular specification and a quantum circuit for resolving and reviewing the electronic structure problem on a quantum computer. The package is developed in such a way that is robust and can be easily extended and maintains high software standards in the terms of documentation and testing. Plugins of OpenFermion provide users with an effective and low overhead for translating electronic structure calculations into quantum circuit calculations [15].

- **TensorFlow Quantum (TFQ)**:

TFQ is a quantum machine learning library for rapid prototyping of hybrid quantum–classical machine learning models. Research in quantum algorithms and applications can be useful to Google's quantum computing frameworks within TensorFlow. TensorFlow Quantum concentrates on quantum data and building hybrid quantum–classical models. It facilitates users to utilize quantum computing algorithms and logic designed in Cirq and provides quantum computing basics corresponding with existing TensorFlow APIs, along with high-performance quantum circuit simulators [16].

2.3.2 IBM Quantum

IBM Quantum is an online service that allows users to access cloud-based quantum computing services. IBM Quantum provides various tools like IBM's prototype quantum processors, tutorials for quantum computation, and interactive textbooks. This is the first initiative of the industry to develop universal quantum computers for science, engineering, and business purposes. The main objective of this attempt includes the expansion of the whole quantum computing technology stack and studying the applications to make quantum mostly usable and accessible. IBM Quantum includes Quantum Composer and Quantum Lab [17].

IBM Quantum Composer

IBM Quantum Composer is a graphical user interface-based quantum programming tool that facilitates users to drag and drop tasks to build quantum circuits in interactive mode. IBM Quantum Composer has a variety of tools that can be customized and permits the user to build, visualize, and run code on simulators or real quantum hardware.

It also facilitates users the following:

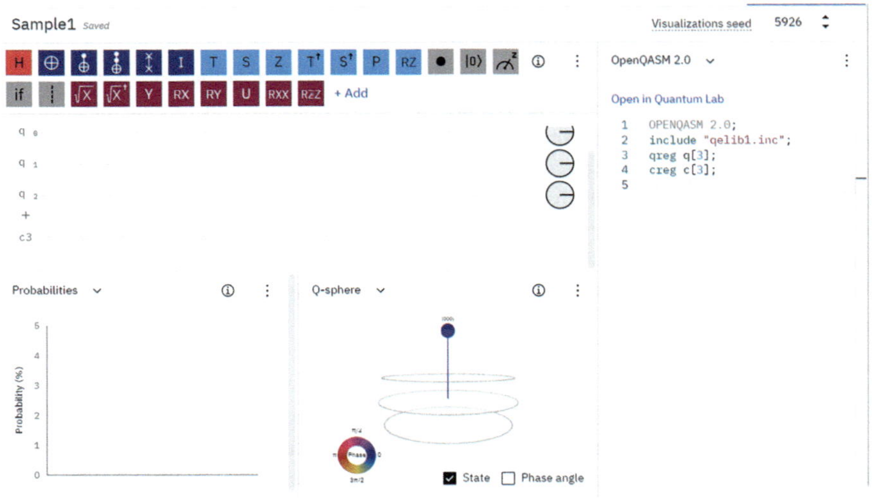

Fig. 5 About IBM quantum composer [18]

- **Visualize qubit states**: It allows the user to check how changes to the circuit affect the position of qubits, in the form of an interactive q-sphere, or histograms displaying measurement possibilities or state vector simulations [18].
- **Run-on quantum hardware**: The user can run the code on actual quantum hardware to know the outcomes of device noise [18].
- **Automatically generate code**: It is also capable of generating the code automatically and hence in place of writing code by hand, the user can get the advantage of this feature. The automatically generated code (in OpenQASM or Python) runs the same way as the circuit which is created with Quantum Composer [18].
- **Quantum system details**: The user can check the status, topology, measurement data, and access details of the IBM quantum systems [18].

Figure 5 shows the snapshot of the Dashboard of IBM Quantum Composer.

IBM Quantum Lab

IBM Quantum Lab facilitates users to write code using Qiskit language (which is an open-source framework to write codes for quantum computers) and text description in a customized Jupyter Notebook environment. For that, no installation is required to access Qiskit from any typical web browser using the Internet. Users can run code on simulators or using real quantum hardware. Users can also manage the files (up to 1 GB) from anywhere. Figure 6 is the snapshot of IBM Quantum Lab [19].

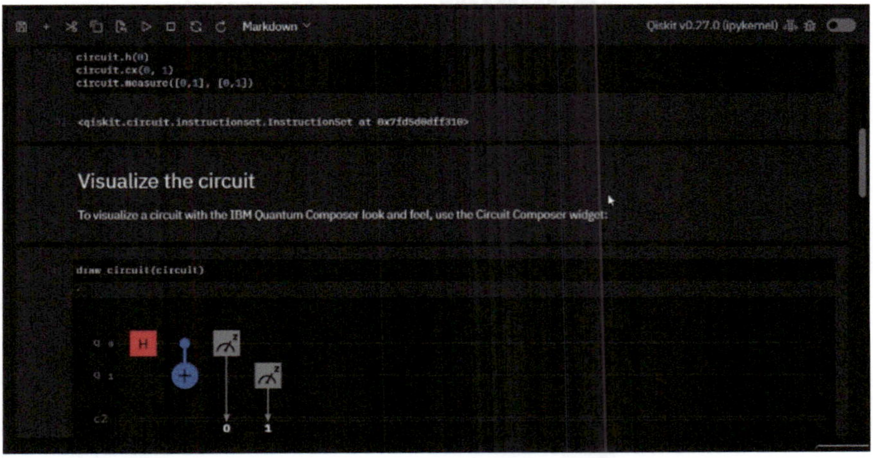

Fig. 6 About IBM quantum lab [19]

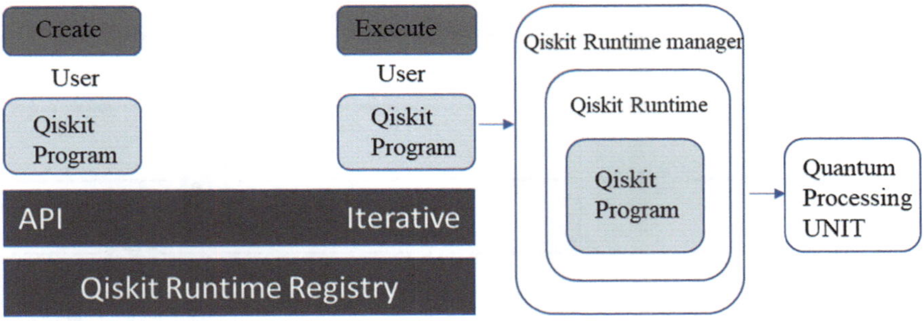

Fig. 7 Runtime architecture of Qiskit [22]

About Qiskit Language

Qiskit [quiss-kit] is an open-source SDK for writing code for quantum computers. It is available for all three major operating systems like Windows, Linux, and MacOS and capability to work at pulses, circuits, and application module levels. It provides a circuit library that consists of a complete set of quantum gates and different pre-built circuits [20]. The transpiler is a unique feature that converts Qiskit code into an improved circuit using a backend's native gate set, which facilitates users to write down code for any quantum processor or processor architecture with minimal effort [21]. Figure 7 represents the Qiskit runtime architecture.

Following is the sample code which is written in Qiskit:

```
from qiskit import QuantumCircuit

qc = QuantumCircuit(5,5)

qc.x(0)

for q in range(4):

    qc.cx(0,q+1)

qc.measure_all()

qc.draw()
```

Output

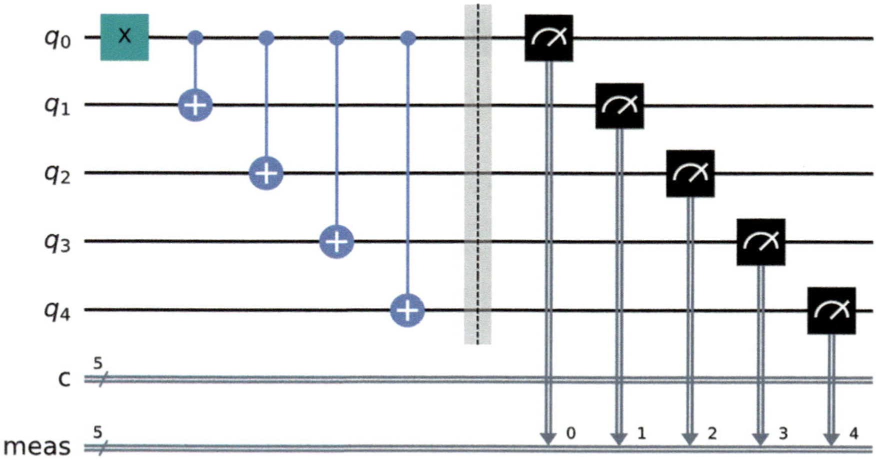

3 Survey of Open-Source Software

Open-source software is always playing a crucial role in the development and testing of quantum algorithms. Some of the tools are supported by leading vendors and those are discussed in previous sections. This section focuses on the evaluation of each project by looking at the various features like documentation, licenses, programming language, accordance with norms of software engineering, and the type of the project [1].

Table 1 shows the name of open-source software, small description, programming language, license of usages, and supported OS for developing the algorithm.

Table 1 Overview of Open-Source Software

Name	Description	Programming language	License	Supported OS	References
Cliffords.jl	Efficient calculation of Clifford circuits in Julia	Julia	MIT	Windows, Mac, Linux	[23]
Dimod	Shared API for Ising/quadratic unconstrained binary optimization samplers	Python	Apache-2.0	Windows, Mac, Linux	[24]
dwave-system	Basic API for easily incorporating the D-Wave system as a sampler in the D-Wave Ocean software stack	Python	Apache-2.0	Mac, Linux	[21]
FermiLib	Open-source software for analyzing fermionic quantum simulation algorithms	Python	Apache-2.0	Windows, Mac, Linux	[25]
Forest (pyQuil and Grove)	Simple yet powerful toolkit for writing hybrid quantum–classical programs	Python	Apache-2.0	Windows, Mac, Linux	[26, 27]
ProjectQ	An open-source software framework for quantum computing	Python, C++	Apache-2.0	Windows, Mac, Linux	[28, 29]
PyZX	Python library for quantum circuit rewriting and optimization using the ZXcalculus	Python	GPL-3.0	Windows, Mac, Linux	[30]
QGL.jl	A performance-orientated QGL compiler	Julia	Apache-2.0	Windows, Mac, Linux	[31]
Qbsolv	Decomposing solver that finds a minimum value of a large quadratic unconstrained binary optimization problem by splitting it into pieces	C	Apache-2.0	Windows, Mac, Linux	[32]
Qiskit Terra and Aqua	Quantum Information Science Kit for writing experiments, programs, and applications	Python, C++	Apache-2.0	Windows, Mac, Linux	[33, 34]

(continued)

Table 1 (continued)

Name	Description	Programming language	License	Supported OS	References
Qiskit.js	Quantum Information Science Kit for JavaScript	JavaScript	Apache-2.0	Windows, Mac, Linux	[35]
Qrack	Comprehensive, GPU-accelerated framework for developing universal virtual quantum processors	C++	GPL-3.0	Mac, Linux	[36]
Quantum Fog	Python tools for analyzing both classical and quantum Bayesian networks	Python	BSD3-Clause	Windows, Mac, Linux	[37]
Quantum++	A modern C++ 11 quantum computing library	Python, C++	MIT	Windows, Mac, Linux	[38, 39]
Qubiter	Python tools for reading, writing, compiling, simulating quantum computer circuits	Python, C++	BSD3-Clause	Windows, Mac, Linux	[40]
Quirk	Drag-and-drop quantum circuit simulator for your browser to explore and understand small quantum circuits	JavaScript	Apache-2.0	Windows, Mac, Linux	[41]
reference-qvm	A reference implementation for a Quantum Virtual Machine in Python	Python	Apache-2.0	Windows, Mac, Linux	[42]
ScaffCC	Compilation, analysis, and optimization framework for the Scaffold quantum programming language	C++, Objective C, LLVM	BSD2-Clause	Mac, Linux	[43, 44]
Strawberry Fields	Full-stack library for designing, simulating, and optimizing continuous variable quantum optical circuits	Python	Apache-2.0	Windows, Mac, Linux	[45, 46]
XACC	eXtreme-scale accelerator programming framework	C++	Eclipse PL1.0	Windows, Mac, Linux	[47]

(continued)

Table 1 (continued)

Name	Description	Programming language	License	Supported OS	References
XACC VQE	Variational quantum eigen solver built on XACC for distributed and shared memory systems	C++	BSD3-Clause	Windows, Mac, Linux	[48]

The following table shows the name of open-source software and its features for supporting like to develop a quantum algorithm, quantum circuits, compilers, simulator, QPU and backend and full-stack support (see Table 2).

Table 2 Overview of open-source software in align with quantum software development work

Name	Quantum computing paradigm	Quantum algorithms	Quantum circuits	Quantum compiler	Quantum computer simulator	QPU backend	References
Cliffords.jl	Discrete gate model	No	Yes	No	Yes	No	[23]
FermiLib	Discrete gate model	Yes	No	No	No	No	[25]
Forest (pyQuil and Grove)	Discrete gate model	Yes	Yes	Yes	Yes	Yes	[26, 27]
ProjectQ	Discrete gate model	Yes	Yes	Yes	Yes	Yes	[28, 29]
PyZX	Discrete gate model	No	No	Yes	No	No	[30]
QGL.jl	Discrete gate model	No	No	Yes	No	No	[31]
Qiskit Terra and Aqua	Discrete gate model	Yes	Yes	Yes	Yes	Yes	[32, 33]
Qiskit.js	Discrete gate model	Yes	Yes	Yes	Yes	Yes	[34]
Qrack	Discrete gate model	No	Yes	Yes	Yes	No	[35]
Quantum Fog	Discrete gate model	Yes	Yes	No	No	No	[36]
Quantum++	Discrete gate model	No	Yes	No	Yes	No	[37, 38]
Qubiter	Discrete gate model	Yes	Yes	Yes	Yes	Yes	[36]
Quirk	Discrete gate model	Yes	Yes	No	Yes	No	[36]

(continued)

Table 2 (continued)

Name	Quantum computing paradigm	Quantum algorithms	Quantum circuits	Quantum compiler	Quantum computer simulator	QPU backend	References
reference-qvm	Discrete gate model	No	Yes	No	Yes	No	[37]
ScaffCC	Discrete gate model	No	No	Yes	No	No	[38, 39]
Strawberry fields	Continuous gate model	Yes	Yes	Yes	Yes	No	[40, 41]
XACC	Discrete gate model	Yes	Yes	Yes	Yes	Yes	[42]
XACC VQE	Discrete gate model	Yes	No	No	No	No	[43]

References

1. M. Fingerhuth, T. Babej, P. Wittek, Open source software in quantum computing. PLoS One **13**(12), e0208561 (2018). https://doi.org/10.1371/journal.pone.0208561
2. N. Otanasap, Quantum Computer and Applications for the Society 5.0. Sau J. Sci. Technol. **7**(1), 22–32 (2021)
3. F. Da Silva, G. Núñez Reyes, The era of platforms and the development of data marketplaces in a free competition environment (2022)
4. C. Dilmegani, Cloud Quantum Computing & Top cloud QC vendorsin 2021 (22 Feb 2022). AIMultiple. https://research.aimultiple.com/quantum-computing-cloud/
5. Tools of Quantum Computing. Quantum Computing Report. https://quantumcomputingreport.com/tools/. Accessed 28 Mar 2022
6. Open-Source Quantum Software Projects. https://github.com/qosf/awesome-quantum-software. Accessed 28 Mar 2022
7. Azure Quantum Documentation. https://docs.microsoft.com/en-us/azure/quantum/. Accessed 28 Mar 2022
8. What is Azure Quantum? https://docs.microsoft.com/en-us/azure/quantum/overview-azure-quantum. Accessed 28 Mar 2022
9. The Q# programming language user guide. https://docs.microsoft.com/en-us/azure/quantum/user-guide/. Accessed 28 Mar 2022
10. About Amazon Braket. https://aws.amazon.com/braket/. Accessed 28 Mar 2022
11. Amazon Braket Features. https://aws.amazon.com/braket/features/. Accessed 28 Mar 2022
12. Google Quantum AI. https://quantumai.google/. Accessed 28 Mar 2022
13. About Cirq. https://quantumai.google/cirq. Accessed 28 Mar 2022
14. Getting started with Cirq. https://quantumai.google/cirq/start. Accessed 28 Mar 2022
15. About OpenFermion. https://quantumai.google/openfermion. Accessed 28 Mar 2022
16. About TensorFlow Quantum. https://www.tensorflow.org/quantum. Accessed 28 Mar 2022
17. IBM Quantum. https://quantumcomputing.ibm.com/composer/docs/iqx/. Accessed 28 Mar 2022
18. IBM Quantum Composer. https://www.ibm.com/quantum-computing/tools/composer/. Accessed 28 Mar 2022
19. IBM Quantum Lab. https://quantum-computing.ibm.com/lab. Accessed 28 Mar 2022

20. About Qiskit Langauge. https://qiskit.org/textbook/ch-appendix/qiskit.html. Accessed 28 Mar 2022
21. A. Condello, contributors, dwave-system (2018). https://github.com/dwavesystems/dwave-system. Accessed 28 Mar 2022
22. Quantum Lab. https://quantumcomputing.ibm.com/lab/docs/iql/runtime/. Accessed 28 Mar 2022
23. B. Johnson, M. da Silva, contributors, Cliffords.jl (2018). https://github.com/BBN-Q/Cliffords.jl. Accessed 28 Mar 2022
24. A. Condello, contributors, Dimod (2018). https://github.com/dwavesystems/dimod. Accessed 28 Mar 2022
25. I. Kivlichan, C. Gidney, R. Babbush, D. Steiger, contributors, FermiLib (2018). https://github.com/ProjectQ-Framework/FermiLib. Accessed 28 Mar 2022
26. W. Zeng, S. Heidel, M. Harrigan, N. Tezak, P. Karalekas, contributors, PyQuil (2018). https://github.com/rigetticomputing/pyquil. Accessed 28 Mar 2022
27. A. Polloreno, J. Otterbach, K. McKiernan, J. Lin, N. Rubin, N. Tezak, W. Zeng, contributors, Grove (2018). https://github.com/rigetticomputing/grove.
28. D. Steiger, T. Haener, contributors, ProjectQ (2018). https://github.com/ProjectQFramework/ProjectQ
29. D.S. Steiger, T. Häner, M. Troyer, ProjectQ: an open source software framework for quantum computing. Quantum **2**, 49 (2018). https://doi.org/10.22331/q-2018-01-31-49
30. A. Kissinger, J. van de Wetering, contributors, PyZX (2018). https://github.com/Quantomatic/pyzx
31. C. Ryan, D. Riste, contributors, QGl.jl (2018). https://github.com/BBN-Q/QGL.jl. Accessed 28 Mar 2022
32. A. Douglass, A. Condello, B. Ellert, M.W. Booth, contributors, Qbsolv (2018). https://github.com/dwavesystems/qbsolv. Accessed 28 Mar 2022
33. D. Rodriguez, J. Gambetta, contributors, Qiskit Terra (2018). https://github.com/Qiskit/qiskit-terra. Accessed 28 Mar 2022
34. R. Chen, M. Pistoia, S. Hu, M. Marques, S. Wood, contributors, Qiskit Aqua (2018). https://github.com/Qiskit/aqua. Accessed 28 Mar 2022
35. J. Perez, contributors, Qiskit.js (2018). https://github.com/Qiskit/qiskit-js. Accessed 28 Mar 2022
36. D. Strano, contributors, Qrack (2018). https://github.com/vm6502q/qrack. Accessed 28 Mar 2022
37. R.R. Tucci, Quantum Bayesian nets. Int. J. Mod. Phys. B **9**(03), 295–337 (1995). https://doi.org/10.1142/S0217979295000148
38. V. Gheorghiu, Quantum++: a modern C++ quantum computing library. PLoS One **13**(12), e0208073 (2018). https://doi.org/10.1371/journal.pone.0208073
39. V. Gheorghiu, contributors, Quantum++ (2018). https://github.com/vsoftco/qpp. Accessed 28 Mar 2022
40. R. Tucci, contributors, Qubiter (2018). https://github.com/artiste-qb-net/qubiter. Accessed 28 Mar 2022
41. C. Gidney, contributors, Quirk (2018). https://github.com/Strilanc/Quirk. Accessed 28 Mar 2022
42. N. Rubin, contributors, reference-qvm (2018). https://github.com/rigetticomputing/reference-qvm. Accessed 28 Mar 2022
43. A. JavadiAbhari, S. Patil, D. Kudrow, J. Heckey, A. Lvov, F.T. Chong, et al., ScaffCC, in *Proceedings of CF14, 11th, ACM Conference on Computing Frontiers* (2014), p. 1. https://doi.org/10.1145/2597917.2597939
44. A. Javadi-Abhari, contributors, ScaffCC (2018). https://github.com/epiqc/ScaffCC. Accessed 28 Mar 2022

45. N. Killoran, J. Izaac, N. Quesada, V. Bergholm, M. Amy, C. Weedbrook, Strawberry fields: A software platform for photonic quantum computing. Quantum **11**(3), 129 (2019)
46. J. Izaac, contributors, Strawberry Fields (2018). https://github.com/XanaduAI/strawberryfields. Accessed 28 Mar 2022
47. A. McCaskey, contributors, XACC (2018). https://github.com/eclipse/xacc. Accessed 28 Mar 2022
48. A. McCaskey, contributors, XACC-VQE (2018). https://github.com/ORNL-QCI/xaccvqe. Accessed 28 Mar 2022

Simulating Quantum Principles: Qiskit Versus Cirq

Rajiv Pandey, Pratibha Maurya, Guru Dev Singh, and Mohd. Sarfaraz Faiyaz

Abstract Quantum computing is a computing paradigm for addressing computer issues of classical systems by harnessing all the possibilities given by quantum physics concepts. In these computers, information is represented by quantum states and utilizes different quantum phenomena such as quantum superposition, entanglement, and interference provided by quantum physics. Quantum computations are based on the fundamental notion of reversible computing. Quantum algorithms are built based on quantum computational complexity. Various quantum algorithms have been developed so far, with the general conclusion that exploiting quantum physics effects results in a significant speedup over conventional algorithms. In this chapter, we have implemented some important quantum principles using different open-source software development kit for dealing with quantum computers with an aim to provide an understanding on how we can create and execute quantum programs on prototype quantum devices or simulate them on a local machine. The chapter deals in the implementation details of quantum principles along with providing other prime representations like Visual Circuit, State vector, Q-sphere, and Visual Probabilistic result.

Keywords Quantum computing · Quantum principles · Qiskit · Cirq

Note: It is suggested that this chapter may be read in relation to Chap. 7 titled "Evolutionary Analysis: Classical Bits to Quantum Qubits" of this volume, to get a fair understanding of the simulations.

R. Pandey · P. Maurya (✉) · G. D. Singh
Amity Institute of Information Technology, Amity University Uttar Pradesh, Lucknow Campus, Lucknow, India
e-mail: pmaurya@lko.amity.edu

R. Pandey
e-mail: rpandey@lko.amity.edu

Mohd. S. Faiyaz
Amity School of Engineering and Technology, Amity University Uttar Pradesh, Lucknow Campus, Lucknow, India

R. Pandey et al. (eds.), *Quantum Computing: A Shift from Bits to Qubits*,
Studies in Computational Intelligence 1085,
https://doi.org/10.1007/978-981-19-9530-9_18

1 Introduction

Quantum computing is the technique of solving computer problems by implementing all the possibilities provided by quantum physics' principles [1–3]. Traditional or "Classical" computers only utilize a tiny portion of these capabilities. They compute in the same way as humans do. There are more results about the amazing things we could achieve if we just had a powerful enough quantum computer [4, 5]. The most important is that we would be able to model phenomena of quantum mechanics in physics, chemistry, and biology that are impossible to replicate with ordinary computers [6]. Information is stored in bits in traditional computing, which have discrete outcome of 0 and 1. The underlying principle of quantum information underpins quantum computers. Information is represented in these computers via quantum states and the use of quantum phenomena given by quantum mechanics (such as quantum superposition, entanglement, interference, the no-cloning theorem, decoherence, and so on) [7–14]. Quantum systems can appear in a variety of ways at the physical layer (atom energy levels, spin, etc.). A generic quantum system is a m-dimensional quantum system ($m = 2$ for a qubit system), and so a quantum register (a set of n quantum states) in superposition allows us to represent m^n alternative classical values at the same time. The quantum states are naturally characterized as entangled systems in quantum circuit calculations; consequently, the state of one quantum system relies on the other.

The fundamental principle of reversible computing [15] underpins quantum calculations. Theoretically, the original state may be retrieved from the result state of a reversible computation. Reversible circuits may also be developed for classical systems in which the number of inputs and outputs of a reversible gate must be equal and the mapping of a specific input to a specific output must be one to one. These criteria must also be followed by a quantum computer system; as a result, the input quantum states of a quantum circuit emerge reversibly through unitary operations.

Quantum algorithms make use of quantum computational complexity foundations. So far, various quantum algorithms have been created with the common aim to harness quantum mechanics effects to achieve a large speedup above classical algorithm. Aside from that (as proved in the prime factoring problem), it is inferred that quantum algorithms can tackle a variety of issues that are now intractable using classical techniques [16–19].

In this chapter, we have implemented some important quantum principles using two different open-source platforms Qiskit and Cirq and analyzed these principles from different perspectives. Qiskit is an open-source (Apache 2.0 licenced) project. It is a software development kit (SDK) that allows you to access for both quantum computing simulators and the actual quantum hardware in the IBM Quantum Experience for free. All you must do is register for an API key.

Cirq is a Python framework that allows to "create, update, and invoke Noisy Intermediate-Scale Quantum (NISQ) circuits." The initial version of the program was made available in July 2018. Cirq includes a quantum circuit library as well as a high-performance local quantum circuit simulator.

The chapter has been organized as follows. Section 2 deals with the first quantum principle of superposition and explores the principle on both quantum development platforms. Sections 3 and 4 describe the other two quantum principles like entanglement and interference. The chapter is concluded in Sect. 5.

2 Superposition

One of the characteristics that distinguishes a qubit from a conventional bit is its ability to be in superposition, it is one of the basic concepts of quantum physics. In classical mechanics, a wave expressing a musical note can be regarded as a superposition of many waves with various frequencies. A quantum state in superposition, similarly, may be seen as a linear transformation of other independent quantum states [20, 21]. This quantum state in superposition results in the formation of a new genuine quantum state.

Qubits can exist in a combination of both the $|0\rangle$ and $|1\rangle$ base states as shown in Fig. 6. When qubit is measured (only observables may be measured), it collapses to one of its eigenstates, and the value obtained reflects that state. A measurement, for example, will cause a qubit in a superposition state of equal weights to collapse to one of its two base states, $|0\rangle$ and $|1\rangle$, with an equal chance of 50% (as shown in Figs. 6 and 8). $|0\rangle$ is the state that, when measured and so collapsed, always yields the value 0. Similarly, $|1\rangle$ will always equal 1.

Figure 1 shows the visual programming of superposition on IBM's composer using Quantum Assembly Language (QASM). It is machine-independent language used to describe quantum computing in a circuit architecture.

Figure 2 shows the visual circuit of superposition on IBM's composer using QASM. Two quantum registers (qreg/q) and two classical registers (c) (for taking measurements from quantum registers), one Hadamard gate (H) and measurement gates (Z) are used to make this circuit.

Figure 3 represents state vector and Q-sphere of superposition while Fig. 4 represents the visual probabilistic result showing the probability of 00 and 01 states with almost of 50% of 1024 shots for superposition.

```
OPENQASM 2.0;
include "qelib1.inc";

qreg q[2];
creg c[2];
h q[0];
measure q[0] -> c[0];
```

Fig. 1 Visual programming of superposition on IBM's composer

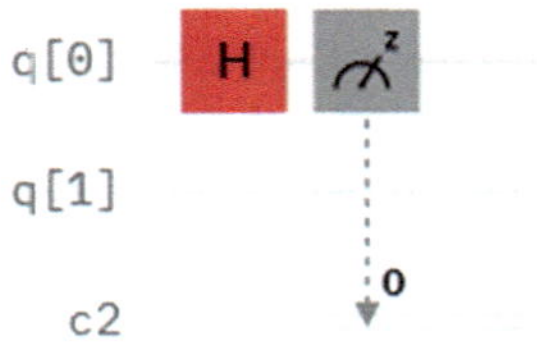

Fig. 2 Visual circuit of superposition on IBM's composer

Fig. 3 State vector and Q-sphere of superposition on IBM's composer, representing the position and velocity of Hadamard gate

Fig. 4 Visual probabilistic result of superposition on IBM's composer

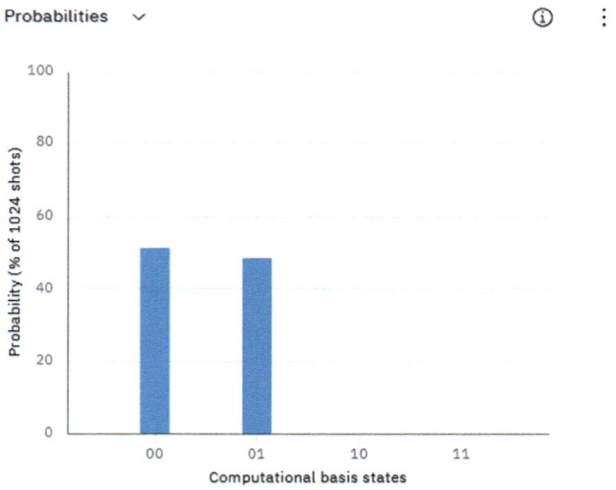

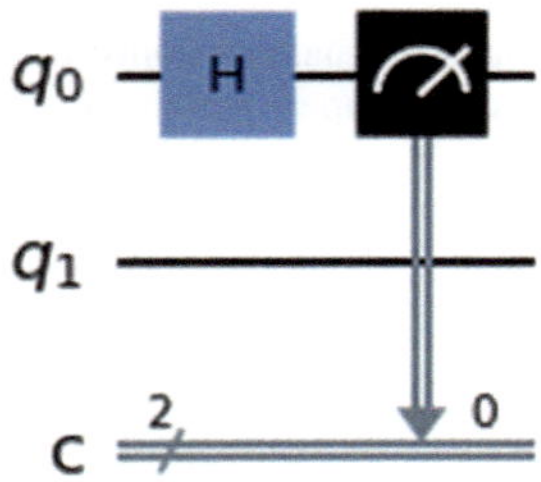

Fig. 5 Visual circuit of superposition on IBM's Qiskit

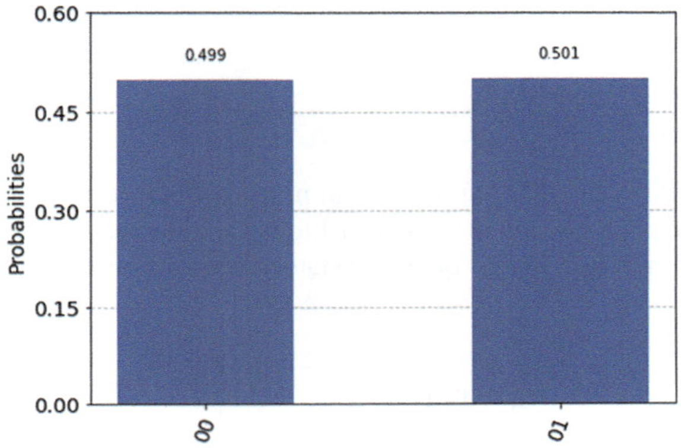

Fig. 6 Visual probabilistic result of superposition on IBM's Qiskit

Figure 5 illustrates the visual circuit of superposition on IBM Qiskit which is obtained by applying one Hadamard gate and measurement on $\mathbf{q}_0$ register and Fig. 6 gives Visual Probabilistic Result on same development platform with the probability of almost 50% of both 00 and 01.

Figures 7 and 8 represent same phenomena on Google's Cirq.

The circuit in Fig. 7 is made by applying one Hadamard gate and measurement on $\mathbf{q}_0$ register.

Quantum superposition differs substantially from classical wave superposition. A quantum computer with n qubits can be in a superposition of two states: $|000\ldots0\rangle$ to

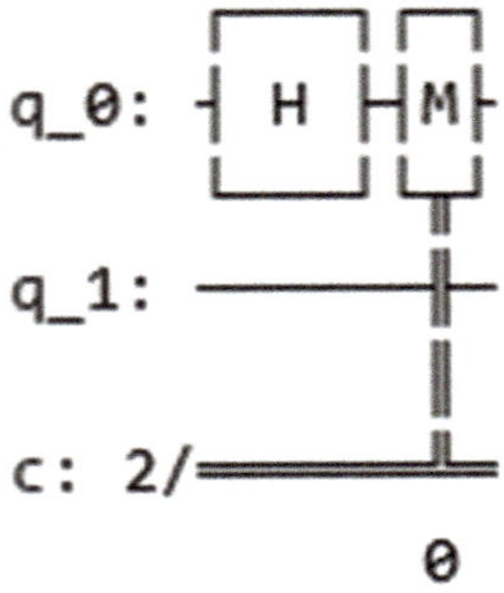

Fig. 7 Visual circuit of superposition on Google's Cirq

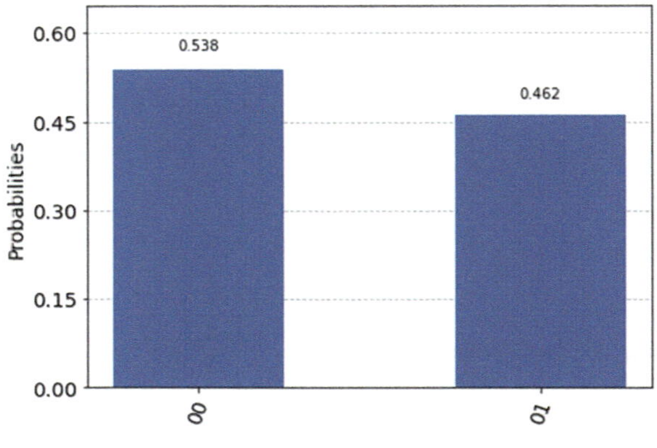

Fig. 8 Visual probabilistic result of superposition on Google's Cirq, showing the probability of 00 and 01 states with almost of 50%

$|111\ldots1\rangle$. Playing n musical notes with all various frequencies, on the other hand, can only result in a superposition of n frequencies. Combining classical waves scales linearly, whereas quantum state superposition scales exponentially.

3 Entanglement

Entanglement is another counter-intuitive phenomenon in quantum physics. When the quantum state of one particle cannot be characterized separately of the quantum state of all the other particles, they are said to be entangled. Although the pieces of the system are not in a distinct state, the quantum state of a system as a whole may be defined.

When two qubits become entangled, they form a unique relationship. The outcomes of measurement will reveal the entanglement. The measurements on the various qubits might result in a 0 or a 1 [22]. However, the result of one qubit measurement will always be connected with the outcome of the other qubit measurement (as shown in Fig. 12). Even if the particles are separated by a great distance, this is always the case. The Bell states are examples of such states.

In Fig. 9, three quantum and three classical registers, applying Hadamard gate in q_0 register followed by two controls—NOT or C-NOT (cx) gates (in the first cx gate, q0 register acts as "Control" and q1 register acts as "Target"; in the second CX gate, q1 register acts as "Control" and q2 register acts as "Target"; and three measurements in the q0 to q2 register) have been used.

Figure 10 describes the visual circuit for entanglement and consists of three quantum registers and three classical registers. On the first register, the Hadamard gate is used. Two cx gates are used (in the first cx gate, q0 register acts as "Control" and q1 register acts as "Target"; in the second cx gate, q1 register acts as "Control" and q2 register acts as "Target"; and three measurements are used in the q0 to q2 register).

```
OPENQASM 2.0;
include "qelib1.inc";

qreg q[3];
creg c[3];
h q[0];
cx q[0], q[1];
cx q[1], q[2];
```

Fig. 9 Visual programming of entanglement on IBM's composer

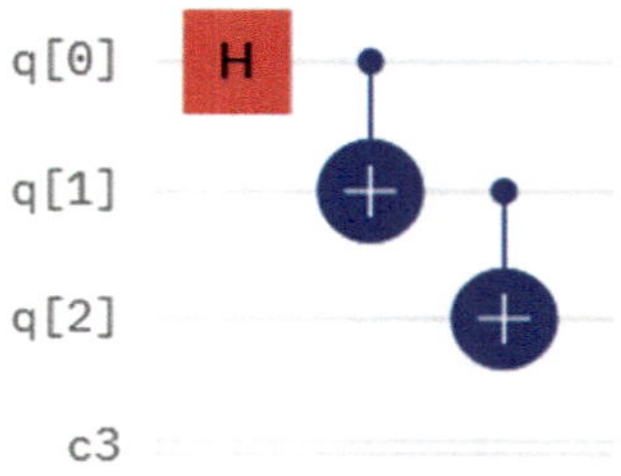

Fig. 10 Visual circuit of entanglement on IBM's composer

Figure 11 represents the position and velocity of quantum gates applied in the system for entanglement. Figure 12 shows the entangled probability of 000 and 111 states with almost of 50% of 1024 shots on IBM's composer.

In Fig. 13, the circuit is made by using three quantum registers and three classical registers. The Hadamard gate is applied on first register: two cx gates are used (in first cx gate $\mathbf{q}_0$ register is acting as "Control" and $\mathbf{q}_1$ is acting as "Target" and in second cx gate $\mathbf{q}_1$ register is acting as "Control" and $\mathbf{q}_2$ is acting as "Target") then three measurements are used in $\mathbf{q}_0$ to $\mathbf{q}_2$ register.

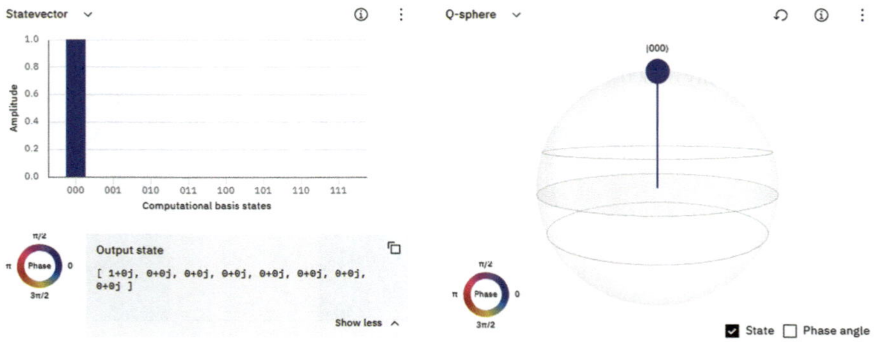

Fig. 11 State vector and Q-sphere demonstration of entanglement on IBM's composer

Fig. 12 Visual probabilistic result of entanglement on IBM's composer

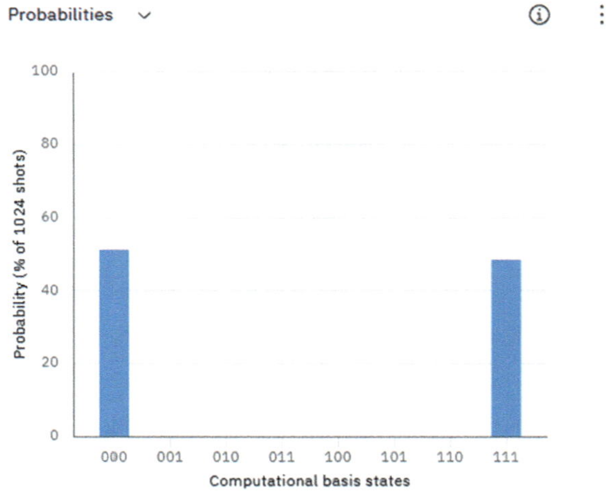

Fig. 13 Visual circuit of entanglement on IBM's Qiskit

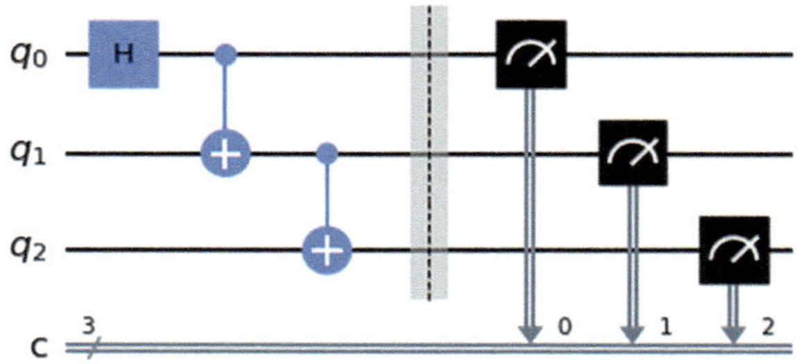

Figures 14, 15 and 16 represent entanglement on Google Cirq.

The circuit is implemented by using three quantum registers and three classical registers. The Hadamard gate is used on the first register. Two cx gates are used (in

Fig. 14 Visual probabilistic result of entanglement on IBM's Qiskit, represents the entangled probability of 000 and 111 states

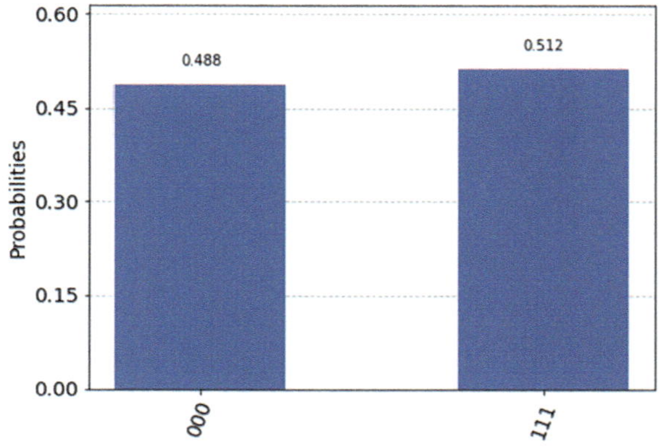

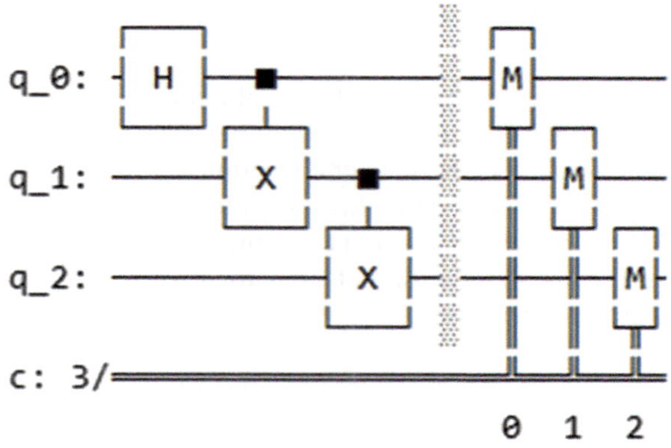

Fig. 15 Visual circuit of entanglement on Google's Cirq

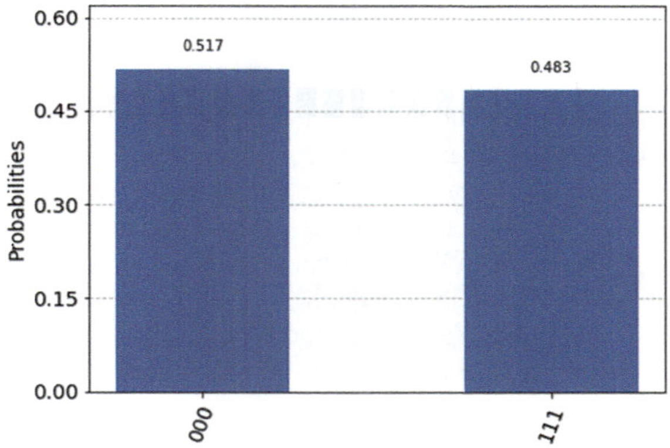

Fig. 16 Visual probabilistic result of entanglement on Google's Cirq, represents the entangled probability of 000 and 111 states

the first cx gate, q_0 register acts as "Control" and q_1 register acts as "Target"; in the second cx gate, q_1 register acts as "Control" and q_2 register acts as "Target"; and three measurements are used in q_0 to q_2 register).

Two particles are generated in a way that the system's total spin is zero. If the spinning of one of the particles is measured along a specific axis and observed to be counter-clockwise, it is guaranteed that the spin of other particle (across the same axis) will be clockwise. This appears unusual since one of the entangled particles appears to "feel" that a measurement is being done on the other entangled particle and "knows" what the result should be, but this is not the case. This occurs in the absence of any exchange of information between entangled particles. They may be billions of kilometers apart and the connection would still exist.

4 Interference

Controlling the chance that a set of qubits falls into certain measurement states is a basic principle in quantum computing. The ability to influence the measurement of a qubit into a specific state or collection of states is enabled by quantum interference, a result of superposition (as shown in Figs. 17, 18 and 19) [23].

It is the case in which interference from environmental noise harms the quantum particle as well as the potential for particle wave functions to either enhance or degrade one another.

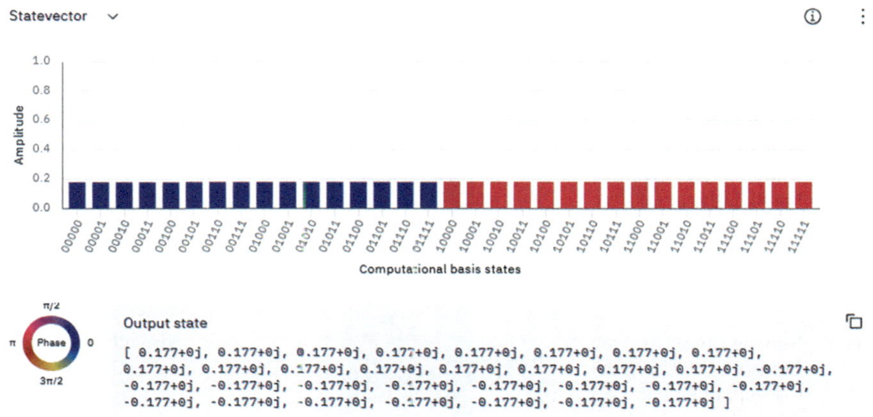

Fig. 17 Demonstrating state vector representation

Fig. 18 Demonstrating Q-sphere representation

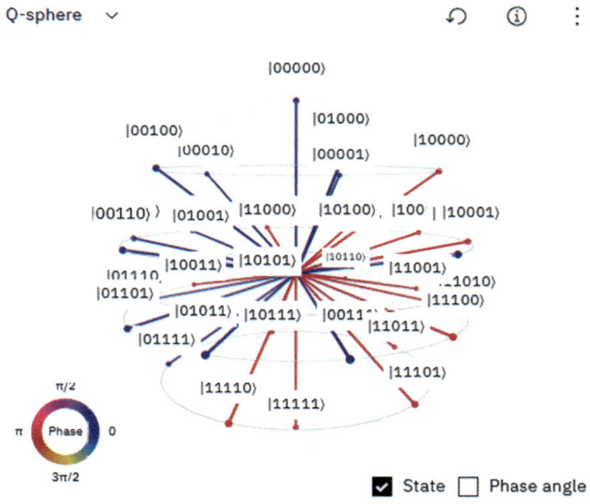

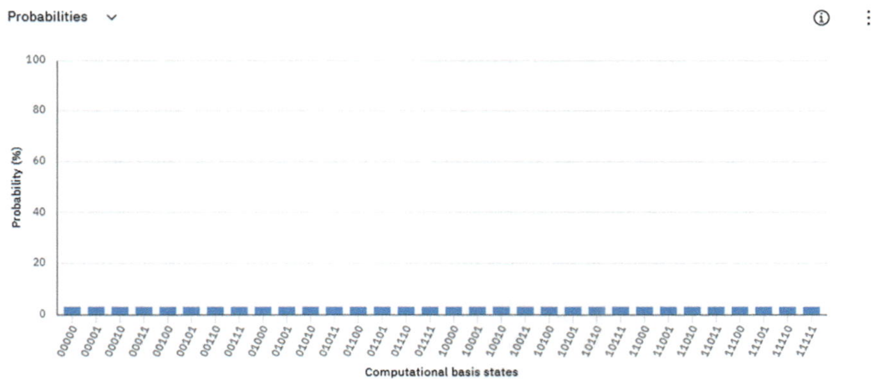

Fig. 19 Probability representation of quantum interference, result of superposition

It is the case in which interference from environmental noise harms the quantum particle as well as the potential for particle wave functions to either enhance or degrade one another.

Consider the qubit in superposition $|\psi\rangle = \frac{(1,-1)}{\sqrt{2}}$ and observe that $H|\psi\rangle = |1\rangle$ to fix concepts. Specifically, if the Hadamard gate is applied to $|\psi\rangle$, it will be measured in the pure state $|1\rangle$ with theoretical certainty. This is the purest kind of quantum interference.

In Fig. 20, the visual programming of interference has been shown by five quantum registers and five classical registers and applying different quantum gates in the registers.

Fig. 20 Visual programming of interference on IBM's composer

```
OPENQASM 2.0;
include "qelib1.inc";

qreg q[5];
creg c[5];
h q[0];
h q[1];
h q[2];
h q[3];
x q[4];
h q[4];
barrier q[0], q[1], q[2], q[3], q[4];
cx q[1], q[4];
cx q[2], q[4];
barrier q[0], q[1], q[2], q[3], q[4];
h q[0];
h q[1];
h q[2];
h q[3];
measure q[0] -> c[0];
measure q[1] -> c[1];
measure q[2] -> c[2];
measure q[3] -> c[3];
```

Figure 21 shows the visual circuit for interference, which is made by using five quantum registers and five classical registers. The Hadamard gate is applied on first four registers and in fifth register x gate (NOT gate) is used with Hadamard gate. Two cx gates are used (in first cx gate $\mathbf{q}_1$ register acting as "Control" and $\mathbf{q}_4$ acting as "Target"; in second cx gate $\mathbf{q}_2$ register is acting as "Control" and $\mathbf{q}_4$ is acting as "Target") then four Hadamard gates and four measurements are used in $\mathbf{q}_0$ to $\mathbf{q}_4$ registers (Figs. 21 and 24).

It consists of five quantum registers and five classical registers as shown in Fig. 24. The Hadamard gate is applied on first four registers and in fifth register x gate (NOT Gate) is used with Hadamard gate. Two cx gates are used (in first cx gate $\mathbf{q}_1$ register is acting as "Control" and $\mathbf{q}_4$ is acting as "Target" and in second cx gate $\mathbf{q}_2$ register is acting as "Control" and $\mathbf{q}_4$ is acting as "Target") then four Hadamard gates and four measurements are used in $\mathbf{q}_0$ to $\mathbf{q}_4$ registers.

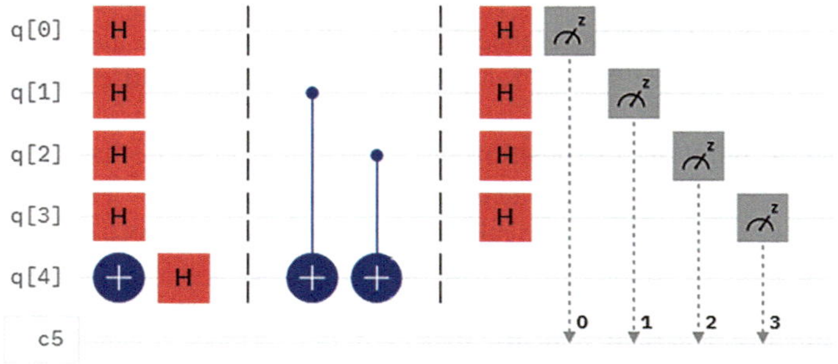

Fig. 21 Visual circuit of interference on IBM's composer

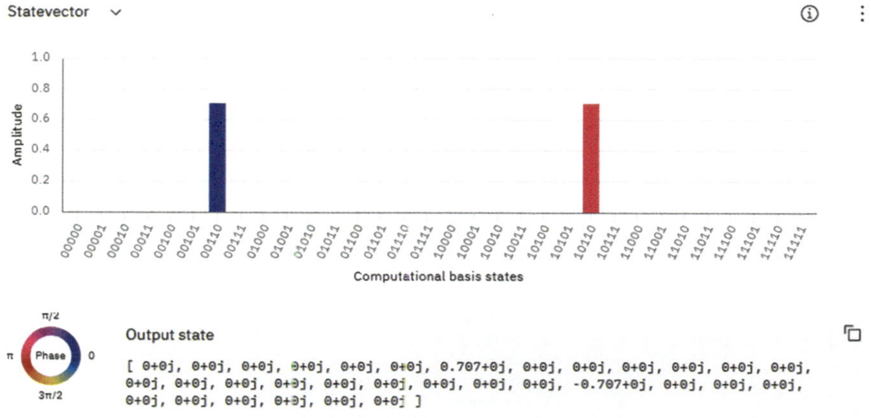

Fig. 22 State vector of interference on IBM's composer

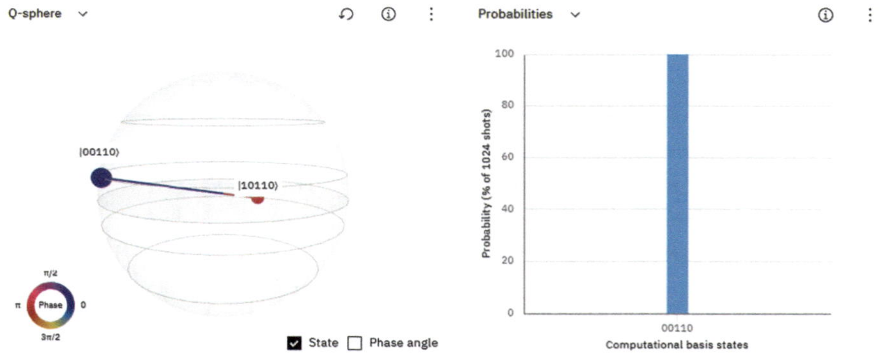

Fig. 23 Q-sphere and visual probabilistic result of interference on IBM's composer

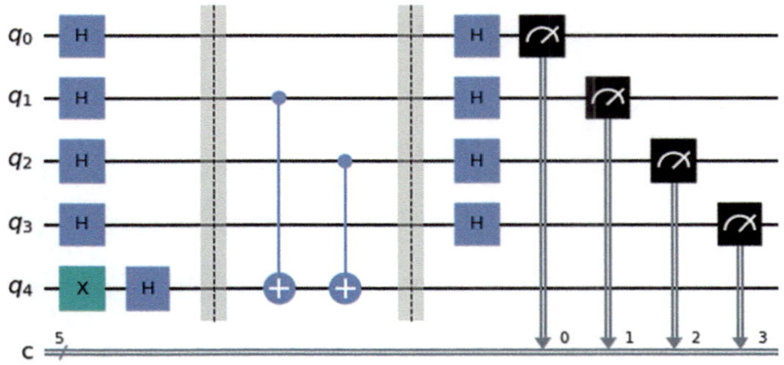

Fig. 24 Visual circuit of interference on IBM's Qiskit

Figures 25, 26 and 27 represent the different results for Interference on IBM's Qiskit.

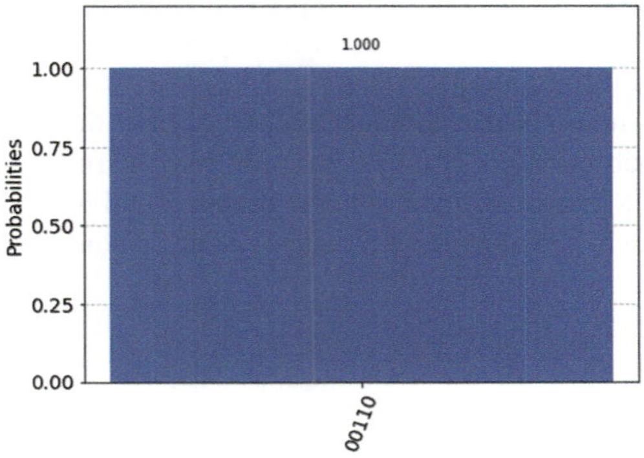

Fig. 25 Visual probabilistic result of interference on IBM's Qiskit

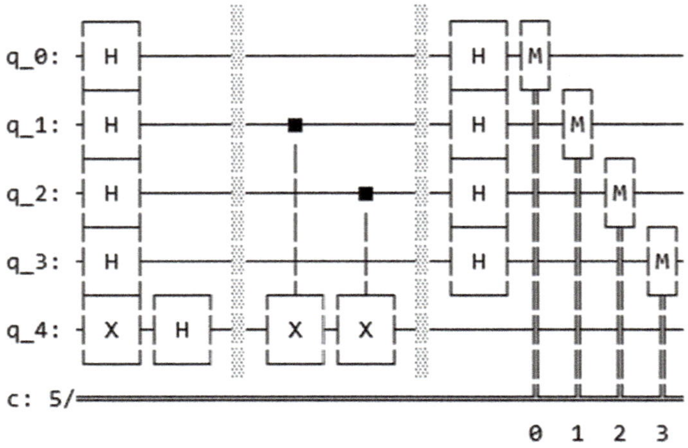

Fig. 26 Visual circuit of interference on Google's Cirq

Fig. 27 Visual probabilistic result of interference on Google's Cirq

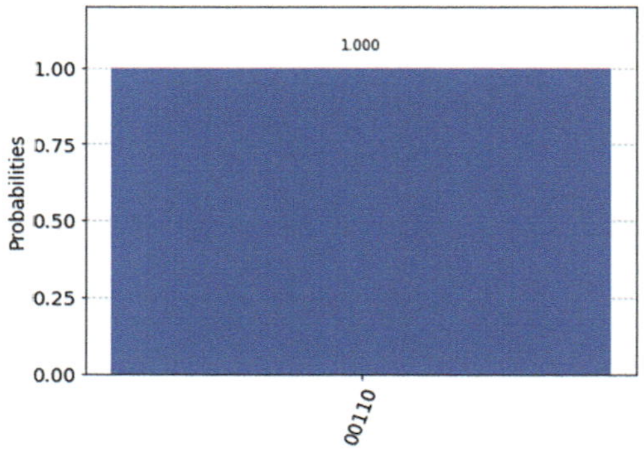

However, even though $|\psi\rangle$ can be measured with equal probability in states $|0\rangle$ and $|1\rangle$, this does not imply that $H|\psi\rangle = H|0\rangle$ with probability 1/2 and $H|\psi\rangle = H|1\rangle$ with probability 1/2. In reality, neither $H|0\rangle$ nor $H|1\rangle$ corresponds to a pure state. Furthermore, for $H|\psi\rangle = H|0\rangle$ and $H|\psi\rangle = H|1\rangle$ with equivalent probability, $|\psi\rangle$ should be measured incidentally before the use of the Hadamard gate. An inadvertent observation of a system qubit may disrupt quantum interference in this way. This is known as quantum decoherence, and it may be a significant source of inaccuracy when dealing with physical quantum computers.

5 Conclusion

Quantum computing provides several benefits over traditional computing. The ability of quantum computing to solve problems significantly quicker than traditional computers is one of its key advantages. This is due to a phenomenon known as "quantum parallelism," which enables a quantum computer to simultaneously investigate several potential solutions by combining the characteristics of a particle and a wave. Quantum computers are also immune to noise and capable of operating in hostile environments, making them ideal for use in machine learning and artificial intelligence applications. The ability of quantum computers to outperform flaws in conventional computing systems is another benefit. Finally, the ability of quantum computers to store and analyze huge volumes of data is crucial for the development of machine learning and artificial intelligence models. This chapter emphasizes on the significance of quantum compute and discusses in detail the important quantum principles.

Open Source Platforms for Quantum Computing

IBM: https://qiskit.org/.

Google: https://quantumai.google/cirq.

References

1. R.P. Feynman, Simulating physics with computers. Int. J. Theor. Phys. **21**, 467–488 (1982). https://doi.org/10.1007/BF02650179
2. T. Hey, Quantum computing: an introduction. Comput. Control Eng. J. **10**(3), 105–112 (1999)
3. R. Feynman, Quantum mechanical computers. Found. Phys. **16**, 507–531 (1986)
4. Y. Kanamori, S.-M. Yoo, F.T. Sheldo, A short survey on quantum computers. Int. J. Comput. Appl. (2006)
5. M.A. Nielson, I.L. Chuang, *Quantum Computation and Quantum Information* (Cambridge University Press, Cambridge, 2000)
6. D. Wecker, et al., Gate-count estimates for performing quantum chemistry on small quantum computers. Phys. Rev. A **90**(2), 022305–, 08 (2014)
7. X.M. Hu et al., Experimental creation of superposition of unknown photonic quantum states. Phys. Rev. A **94**, 033844 (2016)
8. S. Sami, I. Chakrabarty, A note on superposition of two unknown states using Deutsch CTC model. Modern Phys. Lett. A **31**, 1650170 (2016)
9. R. Jozsa, N. Linden, On the role of entanglement in quantum-computational speed-up. Proc. R. Soc. Lond. Ser. A Math. Phys. Eng. Sci. **459**(2036), 2011–2032 (2003)
10. D. Hucul, I.V. Inlek, G. Vittorini, C. Crocker, S. Debnath, S.M. Clark, C. Monroe, Modular entanglement of atomic qubits using photons and phonons. Nat. Phys. **11**, 37–42 (2014)
11. Y.S. Kim et al., Protecting entanglement from decoherence using weak measurement and quantum measurement reversal. Nat. Phys. **8**, 117 (2012)
12. L.M. Duan, M.J. Madsen, D.L. Moehring, P. Maunz, R.N. Kohn Jr., C. Monroe, Probabilistic quantum gates between remote atoms through interference of optical frequency qubits. Phys. Rev. A. **73**, 062324 (2006)
13. S. Mavadia, V. Frey, J. Sastrawan, S. Dona, M.J. Biercuk, Prediction and real-time compensation of qubit decoherence via machine learning. Nat. Commun. **8**, 14106 (2017)

14. P.W. Shor, Scheme for reducing decoherence in quantum computer memory. Phys. Rev. A **52**, 2493 (1995)
15. C.H. Bennett, Logical reversibility of computation. IBM J. Res. Dev. **17**, 525–532 (1973)
16. P.W. Shor, Polynomial-time algorithms for prime factorization and discrete logarithms on a quantum computer. SIAM J. Comput. **26**(5), 1484–1509 (1997)
17. T. Monz et al., Realization of a scalable Shor algorithm. Science **351**, 1068–1070 (2016)
18. R. Van Meter, Architecture of a quantum multicomputer optimized for Shor's factoring algorithm. Ph.D. Thesis, Keio University, 2006, arXiv: quant-ph/0607065
19. L.M.K. Vandersypen et al., Experimental realization of Shor's quantum factoring algorithm using nuclear magnetic resonance. Nature **414**, 883–887 (2001)
20. R.P. Poplavskii, Quantum computers, in *Thermo Dynamical Models of Information Processing* (1975), pp. 465–501
21. P. Benioff, The computer as a physical system: a microscopic quantum mechanical Hamiltonian model of computers as represented by turning machines. J. Stast. Phys. (1980)
22. D. Bacon, Too entangled to quantum compute one-way. Physics **2**, 38 (2009)
23. D.R. Simon, On the power of quantum computation, in *Proceedings of 35th Annual Symposium on Foundations of Computer Science*, Sante Fe, NM (1994), pp. 116–123

Future Direction and Applications

Quantum Machine Learning in Prediction of Breast Cancer

Jigna B. Prajapati, Himanshu Paliwal, Bhupendra G. Prajapati, Surovi Saikia, and Rajiv Pandey

Abstract Machine learning (ML) is the most promising subset of artificial intelligence. Quantum computing is prevalent for fast problem-solving approaches. The complex problems are classified and solved using huge multi-dimensional space. The various algorithms can interfere in multi-dimensional space and resolve the problems. Quantum Machine Learning provides the platform for various mining processes with to the point developments in quantum computing. Quantum computing & Machine learning both are very complex. Quantum Machine learning focuses on quick problem-solving synthesis with a quantum framework using different algorithms. Machine Learning functions by supervised, unsupervised, and semi-supervised learning mechanisms. ML uses label and unlabeled data to implement different classification, clustering, and decision trees for complex problems. Quantum computing comprises quantum counterparts for various computational complexity. Quantum Machine Learning provides a profound sympathetic approach for various subjects to derive new dimensioned results. There are several serious life-threatening diseases such as cancer, hepatotoxicity, cardiotoxicity, nephrotoxicity, etc. require prompt and precise detection at the early stages of progression. The need of the hour is to develop rapid, accurate, and more efficient strategies for various disease predictions which are also cost-effective and non-invasive in nature. Breast cancer is also

---

J. B. Prajapati (✉)
Acharya Motibhai Patel Institute of Computer Studies, Ganpat University, Mahesana, Gujarat 384012, India
e-mail: jignap15@gmail.com; jigna.prajapati@guni.ac.in

H. Paliwal · B. G. Prajapati
Shree S K Patel College of Pharmaceutical Education and Research, Ganpat University, Mahesana, Gujarat 384012, India
e-mail: bhupendra.prajapati@guni.ac.in; bhupen27@gmail.com

S. Saikia
Translation Research Laboratory, Department of Biotechnology, Bharathiar University, Coimbatore, Tamil Nadu 641046, India

R. Pandey
Amity University, Noida, Delhi NCR, India

R. Pandey et al. (eds.), *Quantum Computing: A Shift from Bits to Qubits*,
Studies in Computational Intelligence 1085,
https://doi.org/10.1007/978-981-19-9530-9_19

such a disease that early screening is challenging owning to hereditary predisposition. Quantum computation techniques emerged with Machine learning as the promising approach in the past decade concerning the prediction of breast cancer. The quantum computes can be utilized for assisting cancer detection by employing quantum neural networks, quantum simulators, Super Vector Machine (SVM); Artificial Neural Networks (ANN), Dimensionality Reduction Algorithms etc. are used on the pre-processed dataset for the derived prediction of breast cancer. This book chapter will focus on current trends of Quantum Machine leaning for the prediction of breast cancers by solving complex computational problems using above stated algorithms. This chapter discusses the Molecular Classification of Breast Cancer as Luminal-A, Luminal-B, Normal-like, HER2 enriched, and Basal-like with Breast Cancer Diagnostic Techniques. It covers the study of Brest cancer prediction using Quantum Neural Network, Dimensionality Reduction Algorithms, and Support vector machines (SVM). It includes comparative discussions about different algorithms for breast cancer prediction.

1 Quantum Machine Learning

1.1 Introduction

Quantum Machine Learning (QML) is a consolidative method that combines Quantum Physics (QP) and Machine Learning (ML) [1]. Quantum Machine Learning provides the platform for applied researchs on various mining processes with to-the-point developments in quantum computing [2]. Quantum computing & Machine learning both are very complex in their nature. Quantum Machine learning focuses on quick problem-solving synthesis with a quantum framework using different algorithms [3]. Machine Learning functions by supervised, unsupervised, and semi-supervised learning mechanisms. Machine learning uses labeled and unlabeled data to implement different classifications, clustering, and decision trees for complex problem-solving [4]. Quantum computing comprises quantum counterparts for various computational complexity. It is very hard to manage theoretical advances in quantum computing for large and heterogeneous domains. The gap in quantum counterparts in reference to a machine learning approach emergent interdisciplinary problem-solving. Quantum Machine Learning provides a profound sympathetic approach for various subjects to derive new dimensioned results. Quantum Machine Learning is in focus for "risk investment with balancing the past & future risk aspect" as per management perspective. The neural networks are the extended part of QML, which includes a comparison of various physical and learning systems as well as methodological and structural systems as shown in Fig. 1. The chapter will be focusing on the primary features of quantum machine learning and types of machine learning. It will be comprised of an overview of the Quantum machine learning approaches which are useful in predicting breast cancer. The quantum computing

Fig. 1 Basic structure of QML

techniques, such as Quantum Neural networks, Dimensionality Reduction Algorithms, Support Vector Machines, etc. have been showing great potential in predicting breast cancer types.

1.2 Machine Learning

Over the past few years, machine learning has become a fast-growing segment as ML applies to a various broad range of domains, such as computational biology, computer vision, and computer security, to name a few. In today's industry, data analysis (DA) is very crucial. The ML and DA practice statistical methods to inspect collected or available data and provide learning based on the computer in the possible shortest time even though handling large data. ML algorithms (MAA) are divided into three classes based on their learning styles: supervised learning (SL), unsupervised learning (UL), and semi-supervised learning (SSL) [5].

1.2.1 Supervised Learning

Each data point is labeled in supervised learning (Fig. 2). The label is also known as an outcome, a result, or a predictor variable in the conventional approach [6]. A regression problem can arise if labels have a continuous numerical range. Labels are part of a fixed, finite system of numerical values or qualitative forms in classification [7].

1.2.2 Semi-supervised Learning

The other approach to learning is semi-supervised learning (Fig. 2) which uses both labeled and unlabeled instances. Labels are often difficult to come by, although data instances are numerous. The semi-supervised strategy uses labeled instances to learn the pattern, then uses unlabeled examples to refine the decision limit among the classes [8].

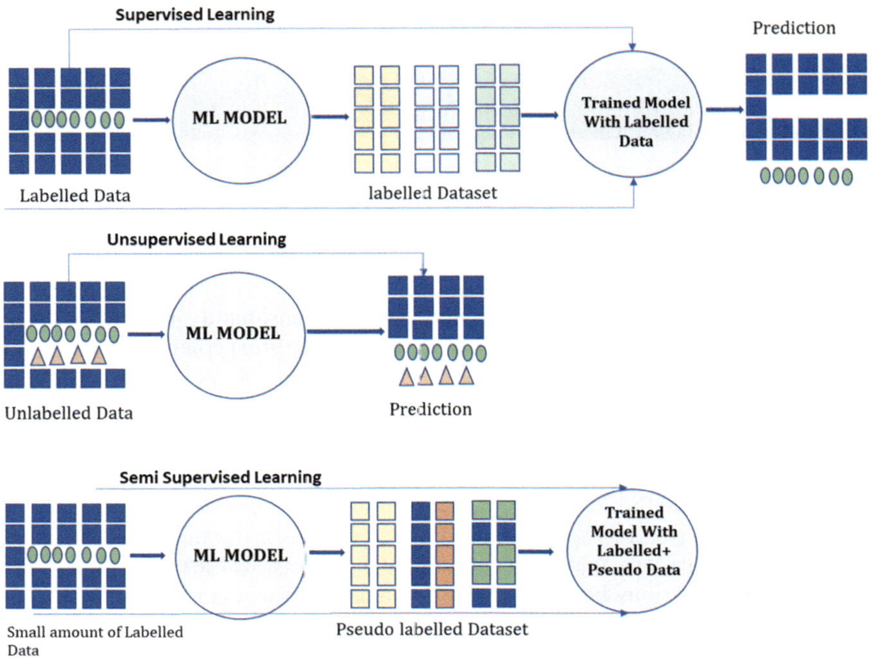

Fig. 2 ML Types: supervised learning (SL), unsupervised learning (UL), and semi-supervised learning (SSL)

1.2.3 Unsupervised Learning

Another approach to learning is Unsupervised learning (Fig. 2). There is no label on training instances. The learning process automatically recognizes the classes, which frequently results in a decision boundary [9].

The major limitation of applying ML algorithms includes the time taken for computations and storage when handling a large quantum of data. Thus, the need for time is to work on the ability of quantum computers that are sufficiently clever to decrease the overall storage and computation time required [10].

1.3 Deep Quantum Learning

DL is quite a competent and operative tool for ML. With its performance across a wide range of applications, deep learning (DL) transformed the ecosphere of ML modeling. Long training times have been a serious barrier to the development of neural networks with multiple layers of hidden nodes. Deep learning solves this challenge by employing computational techniques such as layer learning and immensely parallel systems. New possibilities opened up for creating very sophisticated ML models utilizing DL, which was previously impossible due to a lack of high-performance computing resources. Usually, a huge volume of training data is required for deep learning. Processing such a large volume of data is a computationally challenging operation, which prompted researchers to look into quantum benefits for DL systems.

Current advances in the development of programmable photonic circuits and quantum annealers [11, 12] have paved the way for the creation of DL algorithms that can now be implemented in a quantum framework. The most recent research on Boltzmann machine formulation in a quantum paradigm [13] showed that quantum deep learning is on its way. Quantum DL algorithms can be implemented on quantum annealers, which are special-purpose quantum information processors that are much easier to build than larger quantum computers.

The representation of information in DL is distributed: the layers in the network correspond to various degrees of abstraction. Different levels of abstraction are achieved by increasing the number of layers of QML and the number of nodes in the layers. DL architectures can generalize more effectively with less human intervention, building the layers of abstraction required for difficult learning tasks automatically [14]. Different concepts are learned from other concepts on a lower level of abstraction, forming a hierarchy. As a result, high-level notions develop at the conclusion of the learning pipeline. The loss function is frequently nonconvex [14]. Deep learning often combines unsupervised and supervised learning.

Deep neural networks (DNN) are powerful ML tools. It is well suited to spurring the development of deep quantum learning (DQL) methods. DQL networks can be built using special-purpose quantum information processors like quantum annealers and programmable photonic circuits [15, 16].

The Boltzmann machine (BM) is the simplest DNN to quantize. The typical BM is made up of bits with configurable interactions. It is skilled by altering those interactions until the thermal statistics of the bits, as defined by a Boltzmann Gibbs distribution (see Fig. 1b), match the data statistics [17]. The BM is quantized by expressing the neural network as a set of interacting quantum spins, which corresponds to an Ising tunable model. The user can then read out the output qubits to get an answer by setting the input neurons in the BM to a fixed state and allowing the system to thermalize. DQL is distinguished by the fact that it does not necessitate the use of a big, general-purpose quantum computer. Second, quantum information processing offers novel deep learning models that are inherently quantum. The quantum BM may be transformed into a variety of quantum systems by adding more quantum connections [18, 19].

Totaling a tunable transverse interaction to a tunable Ising model is recognized worldwide for full quantum computing [18]: this model can run any algorithm that a general-purpose quantum computer can run with the right weight assignments. Such universal deep quantum learners are capable of recognizing and classifying patterns that a traditional computer is incapable of. Quantum BM, unlike conventional BM, produces a quantum state. As a result, deep quantum networks can be trained to generate quantum states that represent a wide range of systems. This characteristic, which is not present in classical machine learning, enables it to function as a quantum associative memory [20]. Quantum Boltzmann training can be used for more than just identifying quantum states and developing better models for classical data.

2 Quantum Machine Learning Algorithms

A lot of machine learning tasks require linear algebra to figure out matrix operations by describing data in matrices, Quantum computing (QC) can speed up various linear algebra operations, which vastly improves traditional. Numerical approaches for optimization are a popular subject of study that aims to progress the computations of optimization procedures. Quantum optimization, a branch of QC, tries to optimize the aforementioned techniques even further, similar to classical optimization. Quantum Gradient Descent (QGD) and Quantum Approximate Optimization Algorithm are two well-known approaches of this type (QAOA). Quantum neural networks (QNN) such as Quantum Boltzman Machines use these technologies effectively [21]. The goal of quantum machine learning (QML) is to implement machine learning algorithms (MLLs) in a quantum environment. By using quantum properties like superposition and quantum entanglement to solve a variety of issues with great efficiency.

2.1 Quantum Machine Learning Approaches

For a varied range of machine learning models, such as neural networks (NN), graphical models, support vector machines (SVM), and so on, several quantum methods have been offered. Some implicit methodologies are very popular in QML with classical algorithms like Quantum Support Vector Machines (QSVM) & Quantum Principle Component Analysis (QPCA). QSVM is used for linear classification and QPCA is a popular approach for estimating clustering and density [22]. Deep Learning (DL) is an emerging sub-discipline of ML [23]. The quantum computers are working with DL applications with substantial storage and time. Quantum Boltzmann Machines, Quantum Generative Adversarial Networks [24], Quantum Convolutional Neural Networks and Quantum Variational Autoencoders are connected with the same [25, 26]. Another rational subset of ML is Reinforcement Learning (RL) [27]. RL can be described as learning as time continues by exploring the environment. RL is defined as continuous learning through exploration of the environment as time passes [28]. Table 1 shows various quantum machine learning approaches.

Table 1 Quantum machine learning approaches

Algorithm	Grover	Speedup	Quantum data	References
K-medians	Yes	Quadratic	No	[29]
Hierarchical clustering	Yes	Quadratic	No	[29]
Regression	No	–	Yes	[30]
K-means	Optional	Exponential	Yes	[22]
Principal components	No	Exponential	Yes	[31]
Associative memory	Yes	–	No	[32]
Nearest neighbors	Yes	Quadratic	No	[33]
Neural networks	Yes	–	No	[34]
Support vector machines	Yes	Quadratic	No	[35]

2.2 Grover Search Algorithm

Grover's Search Algorithm (GAA) is one of the most widely used quantum search algorithms [36, 37]. This algorithm identifies a set of elements that meet a set of criteria. For this task, an oracle is capable of recognizing the elements to meet the requirements. Assuming the group has N items, & oracle is O(M) [38] times for traditional calculations to obtain all elements that satisfy the abovementioned requirements.

This algorithm can achieve the same result by using quantum mechanics ($\sqrt{}$M). By utilizing a specific characteristic of quantum computing known as parallel processing, the method can make multiple calls to the oracle at the same time.

The search elements for N number, GAA uses dimension N to represent, with n = *log*M (base 2) qubits. Each $el \in N$ with index y is denoted by an orthonormal vector, as $|y\rangle$ in the qubits's space. The objective is to identify the particular element's index z with sustaining search criteria.

$$\text{If } y = z, \ \text{Uz}|y\rangle = -|y\rangle \tag{1}$$

$$\text{If } y \neq z, \ \text{Uz}|y\rangle = |y\rangle \tag{2}$$

The Eqs. (1) and (2) can as

$$\text{U}_z = \text{I} - 2|z\rangle\langle z| \tag{3}$$

Here, I for identity operator. The GAA concurrently uses the last oracle operator and the Grover diffusion operators, as

$$U_p = 2|\mathrm{p}\rangle\langle \mathrm{p}| - \mathrm{I} \quad (4)$$

here $|\mathrm{p}\rangle = \frac{1}{\sqrt{\mathrm{n}}} \sum_{\mathrm{y}=1}^{\mathrm{n}} \mathrm{y}$.

The (3) and (4) utilized as with the qubits' state are being initialized to the $|\mathrm{p}\rangle$ state. Thus, the Uz and Up are consecutively applied iteratively for r(N) number of times. Further assessed the eigenvalue (λz)f, which may conclude the said index (z).

2.3 *Quantum Reinforcement Learning*

Reinforcement learning (RL) is a common category of learning approach in addition to supervised and unsupervised learning [27]. In contrast to SL and UL, RL assesses input–output pairings using a scalar value termed reward and uses a trial-and-error policy to interact with the environment in order to learn a mapping from states to actions [39, 40]. It has been widely used in artificial intelligence (AI), notably robotics, thus far [41]. It is because it performs well in online adaptation and has a strong learning ability for any complex nonlinear system. When dealing with practical applications, such as the exploration method and poor learning speed; in particular, when dealing with complex difficulties, there may be some complicated problems. It can be seen, particularly when the state-action space expands in size and the number of parameters to learn grows exponentially with the number of dimensions.

For RL optimization, different learning paradigms are combined. Smith suggested a new model based on self-organizing maps (SOM) and benchmark Q-learning to describe and generalize in model-less RL [42]. Furthermore, to adapt fuzzy inference techniques to issues with large/continuous state-action spaces, Watkins' Q-learning is introduced. In practice, various RL approaches have been improved [43–45]. Rigatos and Tzafestas used quantum computing theory to gain the benefit of parallelization in the fuzzy-Logic control algorithm (FCA), which has the goal of speeding up fuzzy inference.

To improve the performance of conventional evolutionary algorithms (EA), quantum inspired evolutionary algorithms (QIEA) have been created, Hogg and Portnoy [46] proposed a quantum technique for solving over-constrained satisfiability and asymmetric traveling salesman problems in combinatorial optimization. Dynamic programming has been studied using the quantum search approach. Dong et al. [47] proposed the Quantum Reinforcement Learning (QRL) concept, which is based on the state superposition principle and parallelism, which are fundamental concepts in quantum computing.

2.4 Quantum Annealing

Quantum annealing (QA), which is used in the study of quantum–mechanical fluctuations in statistical mechanics [48], is the quantum counterpart of simulated annealing (SA) [49, 50]. The same is popular as the Quantum Stochastic Optimization algorithm. Quantum annealing is applied for hard optimization problems using the principle of quantum adiabatic evolution. QA is applied in different fields such as graph theory [51], communications [52–54], computer science [55], machine learning [56, 57], finance (Nicholas %J arXiv preprint arXiv:0.06800 [52]), aeronautics and any others [58]. Simulated annealing random walk of a statistical-mechanical system is active as the cost function for a given problem [59]. This can control the potential energy profile of space and some noise avoidance which may get stuck [60]. Decreased temperature is sufficiently slow to lead to the zero-temperature equilibrium state, the lowest-energy state [61]. SA is used effectively in many real-world applications due to its broad applicability, decent performance, and very simple implementation in most circumstances. SA is used to find an approximate answer within a measured computation time when a problem demands an endlessly long period to derive the exact solution Finding the ground state of a classical Ising Hamiltonian H0 can be thought of as finding the minimization of a cost function in an optimization problem. Cost functions having a significant number of local minima can be found in a variety of practical issues [61].

2.5 Quantum Neural Networks

Quantum neural networks (QNNs) are neural network models that incorporate quantum mechanics principles [62, 63]. From a computational standpoint, one type of quantum neural network combines artificial neural network models with quantum computing properties to generate more durable and efficient models [64, 62]. Often these QNN models attempt to replace conventional binary or McCulloch-Pitts neurons with quantum mechanically manifested qubits also known as "qurons". Subhash and Ron [47, 65] have established the resemblance of the neural activation function with the quantum mechanical Eigenvalue equation to propose a quantum neural model. Narayanan and Menneer presented a photonic implementation of a quantum neural network that uses many theories to collapse into the desired states when quantum measurements are applied [66]. Great efforts have been devoted to finding the quantum version of the perceptron. However, the advances in this direction were impeded because the characteristic neural nonlinear activation functions seldom follow the mathematical structure of quantum theory due to the inherent linear operations in a quantum system. After much effort, Schuld, Sinayskiy, and Petruccione used the quantum phase estimation algorithm to implement the activation function. Apart from this, lots of quantum-inspired models have come up to implement a fuzzy logic-based neural network. Elizabeth Behrman and Jim Steck suggested a new quantum computing configuration with configurable mutual interactions that consist of a number of qubits. The interaction strengths are based on the training set and the classical backpropagation technique. Dan Ventura and Tony Martinez proposed quantum associative memory in 1999 [32]. For a circuit-based quantum computer, the authors suggested an approach to imitate an associative memory. The memory states are envisioned as a superposition of quantum states and the closest to a particular input is then retrieved using a quantum search using an exponential memory storage capacity.

2.6 Support Vector Machine

SVM is a prominent SL technique that can be applied to solve classification as well as regression issues [67]. It is widely used in ML for classification difficulties. SVM Algorithm is aimed at the optimal line or decision boundary for the categorization of n-dimensional space into groups so that additional data sets can be conveniently inserted into the relevant category in the future to have the best decision boundary as a hyperplane [68, 69]. It selects the extreme points and vectors that aid in the formation of the hyperplane. Such cases which are extreme in nature are known as support vectors. Thus, the associated algorithm is referred to as a Support Vector Machine [20]. This is helpful in both text and hypertext categorization. In both inductive and transudative environments, their use can greatly reduce the labeled training data. Support vector machines are used in several shallow semantic parsing approaches. The quantum support vector machine with least-square linear regression [70] is an important amplitude encoding-based quantum machine learning technique [4]. There are two types of support vector machines, Linear support vector machine and a Nonlinear support vector machine.

Linear support vector machine—It is used for linearly separable data, which means that if a dataset can be classified into two groups using only a single straight line, it is considered linearly separable data, and the classifier employed is the Linear support vector machine classifier.

Non-linear support vector machine—It's for non-linearly separated data, which implies that if a dataset can't be classed using a straight line, it's nonlinear data, and the classifier employed is called Nonlinear support vector machine classifiers.

Figure 3 illustrates this kind of classification problem. In mathematics, this classification problem is formulated as finding the maximum margin between the hyperplane and the closest data points to it, so that the classifier hyperplane makes these two classified classes have the maximum margin. Thus, SVM is also called maximum margin classifiers. Imagine that this is a sectional view and one is looking at it from the side, then this data set is linearly separable in some dimensions. The development of support vector machines started with optimal margin hyperplanes that separate two classes with the highest expected generalization performance. Soft margins allow noise training instances in cases where the two classes are not separable. The data

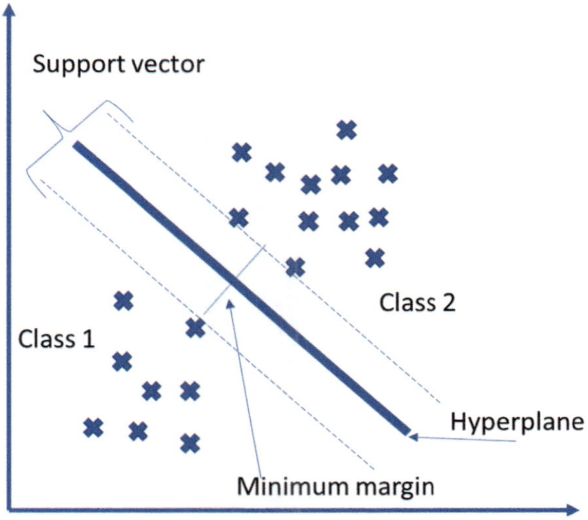

Fig. 3 SVM classification. Adapted and redrawn from Zhang and Ni [71]

points in the parallel dashed lines are called support vectors. In addition, the line in the middle of these two classes is actually a hyperplane, which is corresponding to the equation: $\vec{w} \cdot \vec{x} + b = 0$ where b is constance and w is hyperplane. The positive class: $\vec{x}$ in $\vec{w} \cdot \vec{x} + b \geq 1$, The negative class: $\vec{x}$ in $\vec{w} \cdot \vec{x} + b \leq 1$, The problem is transformed into an optimization problem, to estimate the parameters $\vec{w}$ and b such that: $\max\limits_{\vec{w},b} \frac{2}{|\vec{w}|} st.yi(\vec{w} \cdot \vec{x} + b \geq 1)$ where 2/w is the margin between two classes. After estimating $\vec{w}$ and b, the new vector can be determined by $\vec{x}_0$.

3 Breast Cancer

3.1 Introduction

Cancer has been one of the most unpredictable and life-threatening diseases which has adversely impacted the lives of humans for a long back [72]. The previous century has witnessed a large number of cancer patients which may lead to frightening scenarios in the future [73]. The condition has gotten worse in recent times due to the emergence of a pandemic due to coronavirus in 2019 because it has negatively affected the diagnosis and treatment of cancer [74, 75]. According to a report by National Cancer Registry Programme (India) in 2020, cancer is one of the leading causes of death with about 9% of total deaths caused by any other non-communicable diseases [76]. Figure 4 illustrates the different types of cancers classified according to different organs in humans.

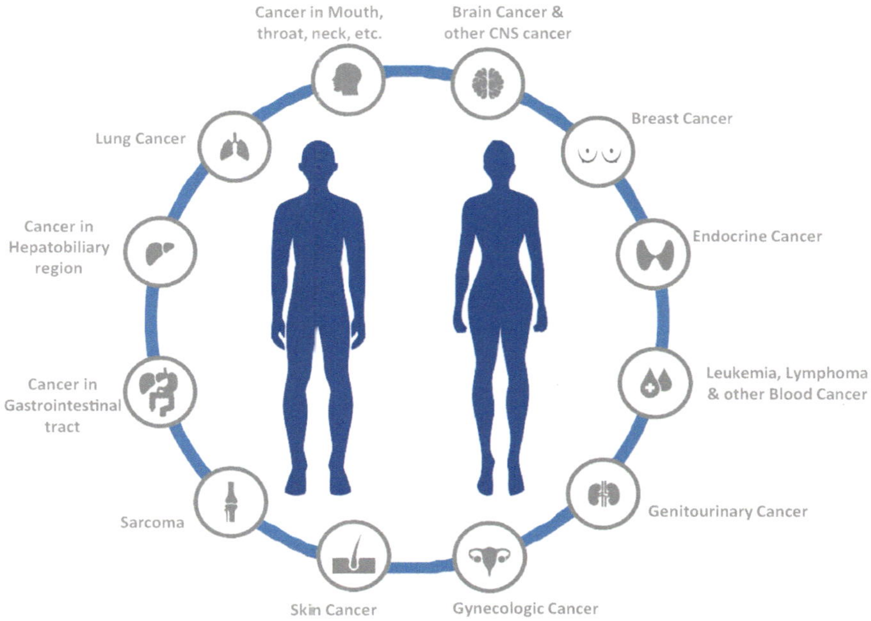

Fig. 4 Organ-based classification of various types of cancers affecting humans

The different types of cancers also add complexity to the specificity and accuracy of the method of diagnosis and effectivity of the treatment [77, 78]. Breast, lung, colon, and rectal cancers are a few of the most common types of cancer, especially in women [79]. The cancer is generated due to several sequences of mutations in the genes which results in alterations in the function of the cells [80]. The gene mutations may be due to some chemical compounds and cigarette smokers may inhale carcinogenic chemicals which may cause lung cancer. In addition to this, there is a number of environmental chemicals reported to show carcinogenicity by inducing gene mutations or genetic defects in the cytoplasm or nucleus of the cell [81–83]. Infectious agents (bacteria, viruses, etc.) and radiations are other leading carcinogenic factors [84].

Breast cancer is one of the types of cancer which takes place in the breast and is known to be the major cause of cancer-related deaths in young women (<45 years). It is mostly seen in women but rarely men are also reported to have breast cancer [85]. However, the incidences of breast cancer in younger females are very low in comparison with overall breast cancer cases but breast cancer in such a young age group is prognostically more critical type. The invasive breast cancer in younger individuals shows significantly higher mitotic activity and the worst types of breast cancer like basal and triple-negative are observed more frequently [86]. The genetic predisposition of breast cancer leading to the expression of biomarkers, such as human epidermal growth factor receptor 2 (HER2), proliferation markers, etc. are considered to be an imperative feature, especially in young females. There is a need for exhaustive research targeted at understanding the biology of breast cancer for proactive diagnosis and customized therapeutic alternatives for the specific age group, cancer progression, severity of cancer, etc. [87].

3.2 Molecular Classification of Breast Cancer

Breast cancer is classified on a molecular basis in order to develop an understanding of the nature of the tumour which facilitates the precise diagnosis and choice of treatment [88, 89]. The conventional ways of classification based upon pathologic characteristics or estimation of biomarkers do not furnish the complete clinical profile of the individual types of breast cancers [90, 91]. There has been uncertainty with regard to carcinogenesis and progression of breast cancer but some of the high-throughput methodologies have generated pivotal information about genetic mutations and the biological process leading to breast cancer. Furthermore, the information allows discernment of novel treatment approaches and types of patients which might influence the management of breast cancer [92–95].

A large number of cases of breast cancer patients are positive for estrogen receptor (ER) and progesterone receptor (PR) [96]. The luminal group tumours are produced due to positive expression of ER responsive genes which codes for characteristic luminal epithelial cell proteins. The luminal tumours are classified into the luminal subtypes, i.e., luminal-A and luminal-B [97, 98]. The molecular classification of breast cancer with luminal and other subtypes is included below:

3.2.1 Luminal-A

Carey et al., have reported that about 50–60% of total breast cancer cases are of luminal-A subtype [99]. Luminal-A is identified with a high quantity of ER and a lesser quantity of genes corresponding to proliferation. It is known to show low nuclear pleomorphism, low proliferation, and low histological grade with considerably good prognostic features. It is indicated by the expression of luminal epithelium cytokeratins, such as CK-8 and CK-18. Additionally, the other luminal-linked markers, like ER1, ZIP6 (zinc transporter), etc. are also expressed in luminal-A. Basically, it is ER- and PR-positive, but negative HER2 and low proliferating cell nuclear antigen (Ki67) [99, 100]. Individuals with luminal-A type breast cancer display a decent prognosis and the rate of relapse is considerably lesser than others. The primary treatment option is hormonal therapy which shows relapses commonly in bones, liver, lungs, and central nervous system [101, 102].

3.2.2 Luminal-B

This subtype of tumour is found in about 15–20% of total cases of breast cancers. Luminal- B displays aggressive features, rapid proliferation, and a comparatively worse prognosis [103]. The information derived from the immunohistochemistry indicates that the luminal-B subtype comprises positive ER, negative HER2, and high Ki67, however not all luminal-B tumours abide by this explanation as few of them are negative for both ER and HER2, while about 30% of HER2-positive tumours are also categorized as luminal-B type tumour. These tumours show expression of growth receptor signaling genes, viral oncogene homolog (v-MYB), lysosome-associated transmembrane protein 4-beta (LAPTMB4) gamma-glutamyl hydrolase (GGH), etc. [104, 105].

The rate of relapse is high with a low survival rate in comparison with luminal-A [106]. The survival rate displayed in this tumour subtype is comparable to risk-bearing tumours (basal-like and HER2-positive subtypes). Furthermore, the studies have shown that the luminal-B tumour does not show any therapeutic benefits from the endocrine therapy and treatment with paclitaxel and doxorubicin. Nevertheless, the luminal-B retaliated considerably better with the neoadjuvant chemotherapy with regards to increased overall pathological response rates [107–109].

3.2.3 Normal-Like

The normal-like tumours are found in about 5–10% of all cases of breast cancers. These tumours are negative for ER, PR, HER2, CK5, and EGFR epidermal growth factor receptor (EGFR) therefore they are also categorized as triple-negative tumours. It is believed to demonstrate the gene expression of adipose tissue with in-between prognosis between luminal- types and basal-like tumours. There is not much clarity about this subtype concerning its clinical potential and few researchers declared that this normal-like breast cancer subtype is merely an output of intensive contamination of normal tissues [110, 111].

3.2.4 HER2 Enriched

This subtype is found in about 15–20% of the total cases of breast cancer and is accompanied by highly aggressive clinical outcomes. HER2 enriched breast cancer is indicated by overexpression of HER2 gene and other HER2-related genes. Almost half of the HER2 enriched tumours also comprise ER, usually with reduced ER expression [112]. These tumours are extremely proliferative with 75% having high histological grade and more than 40% with p53 mutations. HER2 enriched tumours are likely to show poor prognosis, metastatic behaviour in brain and visceral organs, and resistance to hormonotherapy. On the other hand, they are highly sensitive to a chemotherapeutic agent, such as doxorubicin [113–115].

3.2.5 Basal-Like

The basal-like subtype accounts for about 8–37% of overall cases of breast cancers. They are similar to tripe-negative tumours due to the absence of ER, PR, and HER2 expression in their immunohistochemical classification. Interestingly, the basal-like classification is suitable only for research purposes, and most characteristics of triple-negative are primary biomarkers in clinical studies. They are considered to be high histological grade cancer with reduced tubule formation and necrotic and fibrotic zones. These are known for very poor prognostic behaviour, solid tumour growth patterns, aggressive clinical responses, and high metastasis in the brain and lungs [116, 117]. Figure 5 represents the molecular classification of breast cancer, divided based on ER, PR, cancer grade, and prognosis.

3.3 Diagnosis of Breast Cancer

The diagnosis of breast cancer is vital for finding out the individuals who may be facilitated with additional surveillance approaches for early treatment interventions. But there is still ambiguity around the use of different strategies for the screening of breast cancer. This may give rise to the situation where the chance of estimating a high-risk breast cancer in women may be lost [118, 119]. Apart from identifying

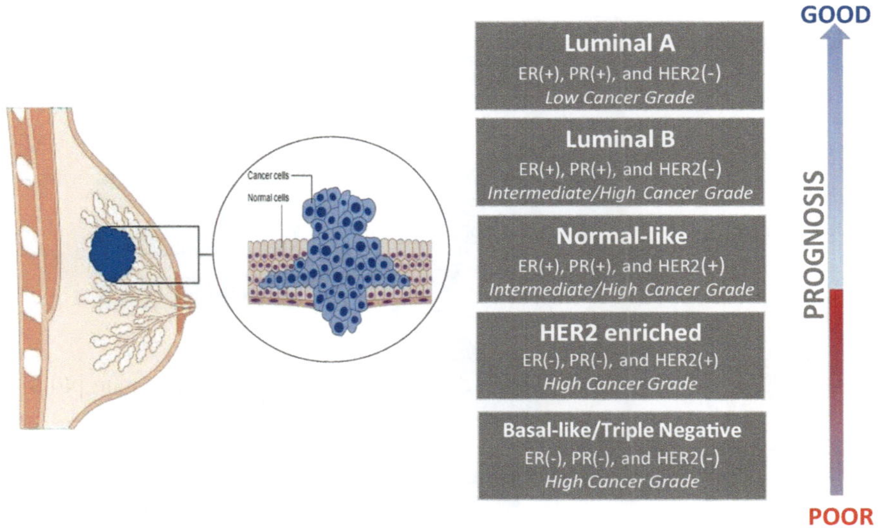

Fig. 5 The molecular classification of Breast Cancer is divided based on ER, PR, Cancer grade, and Prognosis. ER and PR in the figure represent the estrogen receptor and progesterone receptor, respectively

high-risk groups, screening modalities, and screening guidelines, several risk factors can be considered before proceeding with a screening strategy. The assessment of risk factors is useful in providing recommendations about diagnosis techniques and understanding an individual's preference [120]. Some of the important breast cancer risk factors are provided below:

(a) **Hormonal factors**: Estrogen exposure, breast cancer cell differentiation, high estrogen level
(b) **Reproductive factors**: Early menarche, late first birth, late menopause, and prolonged use of the postmenopausal combination
(c) **Factors related to the bone**: High bone density, estrogen receptors in bone, and high hip bone mineral density
(d) **Physical and Lifestyle factors**: Age, obesity, family history, lack of physical activity, alcoholism, and smoking habits.
(e) **Other factors**: High breast density, chest radiation therapy, and the presence of atypical ductal hyperplasia or lobular hyperplasia [120–122].

3.4 *Breast Cancer Diagnostic Techniques*

Breast cancer detection is critical for planning efficient treatment strategies and there is a number of advanced techniques discovered for early and rapid detection. Figure 6 displays the classification of breast cancer diagnostic techniques. The generally used techniques for breast cancer detection are included below:

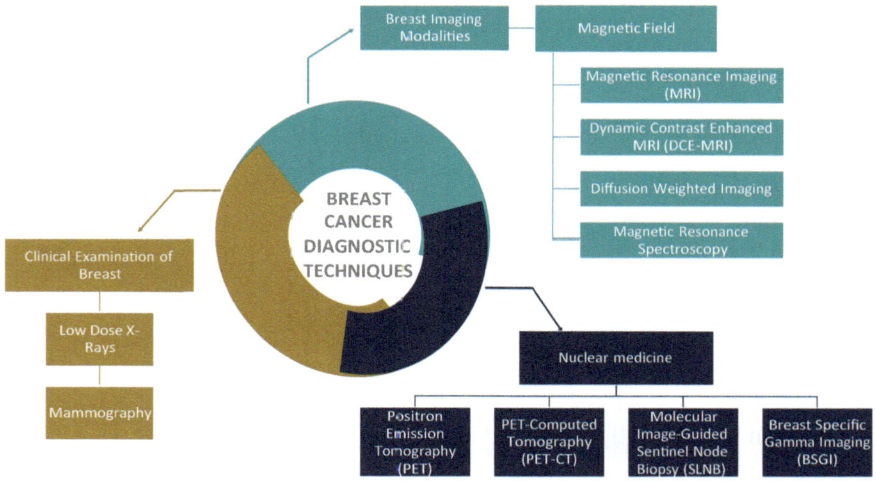

Fig. 6 Illustration of various types of breast cancer diagnostic techniques

(A) **Clinical Breast Examination**

Mammography

The use of mammography for screening of breast cancer in women above 50 years o age is recommended but there is still a disparity in using it for younger females [123]. In recent times, digital mammography has substituted film mammography in older women (>50 years) as well as in women below 50 years, especially those with dense breasts. The scientific reports recommend that mammography can also decrease mortality due to breast cancer, although the associated benefits and risks need to be weighed carefully [124, 125].

The introduction of mammography for the screening of breast cancer has potentiated early detection, and further allowed a reduction in the number of cases of tumours with worse prognostic behaviour [126]. The frequency of mammography is adjusted from recommendations derived from information obtained by decision modeling. According to this, biennial screening was more effective for women within the 50–74 age group with avoidance of about 7 breast cancer deaths per 1000 women screened. While, annual screening is recommended for the age group of 40–74 and avoids 10 deaths per 1000 women screened [127].

(B) **Imaging Techniques**

Magnetic Resonance Imaging (MRI)

It is mainly employed for the detection of breast cancer in high-risk individuals owing to their high screening sensitivity. It is not recommended for the average-risk individuals because of its lack of specificity and is comparatively more expensive than mammography [128]. MRI can be used for the diagnosis of breast cancer by estimating the size of the cancer and its loci. In some high-risk cases, the combination of MRI screening with the annual mammogram is suggested. MRI is associated with limitations due to its tendency to furnish false positive outcomes, especially in average-risk women, and the inability to detect calcium deposits [129].

Dynamic contrast-enhanced magnetic resonance imaging (DCE-MRI)

DCE-MRI is primarily used in patients with newly screened breast cancer for estimating the intensity of disease in a specific region and serves as the base for the planning of therapy. It works by probing the vasculature at different time points after intravenous administration of contrast agents and signals can be delineated through quantitative enhancement kinetic features. The technique offers a non-invasive and 3-D approach to visualizing the disease severity, angiogenic features, the appearance of the lesion, estimating alterations in angiogenic features, and overall response prediction. Despite its superior screening potential, there are several restrictions with DCE-MRI, such as intertwined features of malignant and benign tumours, the inability to manage the neoadjuvant type microscopic disease, and the unpredictable tumour development in the clinical studies [130, 131].

Diffusion-Weighted Imaging (DWI)

Diffusion-weighted MRI is one of the quick screening techniques which can be used for the measurement of water mobility in the tissue indicating the intercellular microenvironment. This technique imparts decreased diffusivity and tumour cells are visualized as contrastingly intense in comparison with the surrounding tissue. DWI is majorly an unenhanced method for the screening of breast cancer due to which the strategy is more economic and quite safe when compared to previous imaging modalities [132]. It enables the generation of exclusive tissue microstructure details which are useful in the description of breast lesions. The unique features result in greatly enhancing the accuracy of diagnosis and assist in planning the treatment regimen. Further, the technique is also useful in enhancing the positive predictive value for imaging breast cancer and characterizing the response to neoadjuvant chemotherapy. However, there is still a lack of a systematic standardized approach for using DWI for breast cancer detection [133, 134].

Magnetic Resonance Spectroscopy (MRS)

It is comparatively an accurate technique for the diagnosis of breast cancer which has an extremely enhanced capability to screen for breast cancer in individuals. It can be employed as a detection tool for at-risk individuals and routine oncological assessments for an increased rate of detection and aggressive primary treatment. Proton MRT (1H-MRS) is a non-invasive technique for the *in vivo* estimation of tissue metabolism and displaying the application in enhancing the specificity of MR breast cancer diagnosis and examining the responsiveness of tumours to neoadjuvant chemotherapies [135, 136].

(C) **Nuclear Medicine**

There has been a lot of interest in the use of nuclear imaging approaches for breast cancer detection and specially designed devices enable the utilization of low doses of radiotracers with adequate sensitivity to even tiny lesions. The patients with newly detected cancer are subjected to primary chemotherapeutic agents and there is a need for enhanced treatment techniques. There is a wide variation in the breast density among breast cancer patients, therefore more anatomic screening techniques are needed. General nuclear medicine is the primary contributor to the complete extent of clinical investigations like early screening of lesions and heterogeneity, contemplating the responses to the therapy, and monitoring the progression of the disease. The advancements in sophisticated instruments like high-resolution dedicated breast devices coupled with the diagnostic versatility of conventional cameras have augmented the application of nuclear medicine in clinical settings. Nuclear medicine for breast cancer screening comprises techniques (Fig. 6) such as positron emission tomography (PET), PET-computed tomography (PET-CT), molecular image-guided sentinel node biopsy (SLNB), and breast specific gamma imaging (BSGI) [137, 138].

3.5 QML in Breast Cancer

3.5.1 Quantum Neural Network in Breast Cancer

Daskin has presented a simple neural network, a smaller number of qubits, and quantum gates. They have defined a quantum circuit for a relevant classical neural net that involves fewer nodes in the hidden layer. A periodic neural network using backpropagation is described through the gradient descent for the breast cancer dataset. The implementation using ML has sped up rather than the classical approach Quantum Machine learning is promising and important to derive various results in the process of fast & speedy computation algorithms [139]. Mishra et al. have implemented a quantum framework using deep learning & supervised learning. They have carried out the implementation with a quantum neural network for the application of cancer detection. They have focused on ten qubits to learn label datasets and optimize circuits for fewer errors within defined quantum neural network functioning [140]. Azevedo et al. define that potential quantum computing can be achieved using machine learning (ML). They explore the quantum approach using ML with to have benefited from speed, complexity & storage. They introduce hybrid classical-quantum neural networks using transfer learning (TL). They focused on heatmaps for mammograms within networks during evaluating different performance metrics [141]. Kim et al. have presented a novel approach of TL (transfer learning) for medical image classification. They have mitigated lacking labeled data issues in the medical domain. They used the modality of bridge transfer Instead of the direct transfer of labeled data. They have experimented with using CNN to achieve a high classification performance. The small number of labeled target medical images has been compared to various transfer learning approaches [142].

Shie et al. have gone through several studies for TL using CNN for implementation and analysis of medical-related images. They have analyzed the brain, eye & breast images as clinical objects during the phase of image acquisition using various methods like X-rays, magnetic resonance, or any other. They have focused on the Sample size and depth to derive adequate result. Feature Extracting in TL is generally very popular for small datasets while it needs a very sub sequence approach with larger datasets. The image preprocessing as translation needs artificially generated data for the CNN model. This is the approach that exchanges computational stress for the TL approach [143]. Although there is no validated proof of which method works best for a given clinical problem [144], suggest that AlexNET is the most commonly CNN model used for brain magnetic resonance images [145, 146] and breast X-rays [147, 148], while DenseNet for lung X-rays [149] and shallow CNN models for skin

Table 2 Various types of learning techniques and their results in prediction

Type of learning	Results	References
CNN	62% accuracy, 75% training accuracy	[151]
KNN, LibSVM, decision trees, random forest, and naive Bayes	89.3–64.7% for classifying each class benign/malignant; 75.8–78.3% for classifying dense/fatty tissue and 71.0–83.1% for identification of a finding	[152]
CNN, RNN	76.9% accuracy, AUC of 74.9%, the sensitivity of 84.88%, and specificity of 64.91%	[153]
ANN, dimensionality reduction	Accuracies 99%	[154]
Principal component analysis) and backpropagation neural network	Accuracy of 96.58%	[155]
SVM, ANN and Naïve Bayes	Accuracy of 98.82%, the sensitivity of 98.41%, specificity of 99.07%	[156]
SVM	Accuracy 99.51%	[157]
LSVM	Accuracy 95.61%	[158]
St-SVM	Accuracy 97.71%	[158]

and dental X-rays [150]. Table 2 summarizes various types of learning techniques and their key results in the prediction of breast cancer.

3.5.2 Dimensionality Reduction Algorithms in Breast Cancer

Ahuja et al. [154] have discussed the biopsied cells' digitized images for breast cancer diagnosis using ML. They worked to increase classification accuracies by noise deduction. They have applied dimensionality reduction (DRT) and outlier removal using Decision Trees, DRT- Linear & Quadratic Discriminant Analysis, Logistic Regression, Gaussian Naive, Kernel Naïve & Support Vector Machines. The derived maximum accuracy when noise removal by dimensionality reduction. They presented an effective approach to improving discrimination accuracies using various ML algorithms [154]. Hasan et al. have discussed large volumed dataset biomedicine science. They focused on PCA (Principal Component Analysis) for dimensionality Reduction for enhancing interpretability [159]. Ritchie et al. have highlighted the challenges for the identification & characterization of genes for human diseases. They introduce MDR (multifactor-dimensionality reduction) using dimensionality reduction to identify associated disease risk. The MDR method is nonparametric & model-free directly applicable to studies. MDR is found enough correct to identify interaction among relatively small samples as breast cancer case–control data. MDR identified four polymorphisms from three different estrogen-metabolism genes for four-locus interaction associated with multifactorial disease [160]. Gupta et al. have discussed about the accurate diagnosis challenges for medical science. They stated that the numbers of the medial test are not deriving accurate results sometimes. They highlighted computational diagnostic techniques using artificial intelligence and machine learning. They worked on Breast cancer diagnosis with a higher degree of accuracy by dimensionality reduction approach. They focused on neural networks for the diagnostic evaluation of breast cancer, namely, benign and malignant with optimal performance to a greater extent. They derived a maximum accuracy of 96.58% with PCB (principal component analysis) and Backpropagation neural network [155].

Liu et al. have discussed about the dimensionality reduction for breast cancer with ultrasound image-based computer-aided diagnosis (CAD). They show better performance of the correntropy algorithm for robust feature selection (CRFS) in noise corrupted data. During their study, they found it tough to have labeled data which is very expensive & time-consuming while it was easy to have unlabeled data. They used semi-supervised learning with iterated Laplacian regularization (Iter-LR) and augment the classification accuracy of the breast ultrasound CAD. It was associated with texture feature modeled by Iter-LR-based semi- supervised CRFS (Iter-LR-CRFS) with dimensions reduction approach of breast cancer data. They worked with Iter-LR-CRFS, CRFS, and PCB (principal component analysis) and derived that Iter-LR-CRFS significantly perform well among all [161]. Daisy et al. have discussed the improvement of accuracy prediction & computational overhead in classification. They compared FCBF (correlation-based feature selection), MFBF (Multi thread-based feature selection) and DDC-DIC (decision dependent -decision independent correlation). It was used for redundancy checking to improve prediction accuracy and minimize computation time. They have worked on Insurance company data and breast cancer data using the Weka tool [162].

Omondiagbe has discussed Breast cancer, its phases, and its implications. They highlighted the accurate diagnosis of breast cancer with CAD (computer-aided detection) systems using machine learning. They have investigated SVM (Support Vector Machine) ANN (Artificial Neural Networks) and Naïve Bayes using the Wisconsin Diagnostic Breast Cancer (WDBC) Dataset. They followed the study by feature selection, extraction, and dimensionality reduction. They have shown the combination of high dimensionality reduction & LDA (linear discriminant analysis) for breast cancer diagnosis. They derived the accuracy of 98.82%, a sensitivity of 98.41%, specificity of 99.07%, and area under the receiver operating characteristic curve of 0.9994 [156]. Jamal et al. have discussed about the prediction of Breast cancer using machine learning algorithm as SVM (Support Vector Machine) and the Extreme Gradient Boosting technique. Before the implementation of these algorithms, they have extracted features using PCB (Principal Component Analysis) & K-Means is used for dimensionality reduction. They have shown the comparison among four models based on two-dimensionality reduction on Wisconsin Breast Cancer Dataset. They have derived the results as k-means perform well compare to PCB by measuring accuracy, sensitivity, and specificity [163].

3.5.3 Support Vector Machines (SVM) in Breast Cancer

Noble WS has discussed the support vector machine for common biomedical engineering [164], and Suthaharan has discussed Machine learning models and algorithms for big data classification using SVM [165]. Widodo & Yang has discussed Support vector machines in machine condition monitoring and fault diagnosis [166], Pisner & Schnye has introduced the SVM for neuroimaging analysis [167], and Joachims has discussed SVM light as an implementation of Support Vector Machines (SVMs) in C [168]. Liu et al. have discussed the Support Vector Machine (SVM) classification algorithm which was used to diagnose breast cancer. They have also compared SVM to other ML techniques in the prediction of breast cancer using the UCI ML repository [169]. Akay has discussed about Breast cancer & curable cancer types. They have focused on breast cancer diagnosis by feature selection & abstraction using SVM. Wisconsin breast cancer datasets (WBCD) have been used to make experiments for breast cancer diagnosis. The parameters such as classification accuracy, sensitivity, specificity, predictive values, and ROC curves are analyzed for better performance. They derived the classification accuracy (99.51%) for the SVM model which is very high compared to others [157]. Azar et al. have discussed about analysis of six types of SVMs for the diagnosis of classical Wisconsin breast cancer issues. They have worked on various classifications such as proximal support vector machine, Lagrangian support vector machines, finite Newton method for Lagrangian support vector machine, Linear programming support vector machines, and smooth support vector machine. They have stated that the derived results of SVM classifiers are fast, simple, and efficient for breast cancer diagnosis. They found the lowest accuracy of LSVM at 95.61%, and the highest accuracy of St-SVM at 97.71%) in the diagnosis of breast cancer [158]. Ozer et al. have highlighted AI and ML in diverse fields of clinical medicine and bioengineering. They have studied that breast cancer is one of the malignancies worldwide with multiple underlying molecular etiologies. They introduce the SVM for robust classification. This data is high-dimensional big data in nature. They have shown the prediction of breast cancer subtypes using SVM models employment they also compare the performance of SVM and other therapeutic prediction models and conclude that SVMis a new solution for mentioned problems in biomedical, bioengineering, and clinical applications [170]. SVM & MLP BPN classifier was explored by Ghosh et al. in 2014 and compared them for the detection of breast cancer [171].

4 Future Scope of Quantum Machine Learning for Breast Cancer Prediction

Many industries especially healthcare is dynamic in nature where every day new challenges need to focus on. The health care industry is enabled by rapid technological support. One of the vibrant sectors of Quantum Machine Learning is working on fast and accurate results with fundaments of quantum computing and Machine Learning. Quantum computing is faster than classic computers in great capacity. It has a large storage volume rather than conventional data storage. It is popular for extremely complex calculations. One of the reasons for the increasing women's death is Breast cancer. The study of Breast cancer diagnosis, treatment, and prediction problems covers many aspects such as classification of breast cancer, multiple cancer cells, machine learning mammography images, breast cancer microarray gene expression data, breast cancer ultrasound images, machine learning quantum realm, and quantum federated learning. Quantum machine learning can play a vital role in the upcoming era for more mature technological support. It can provide a more dynamic solution for complex problems in associated Breast Cancer phases. Quantum machine learning is more scalable & can enhance the accurate prediction of breast cancer by various QML approached such as Quantum classification, by solving very complex problems, optimization of resource supports, solving nonlinear problems using sequential operations, using Quantum Cloud with emulators, simulators, or cloud processors, Quantum Cognition with working as the human brain for decision making, memory like human, Quantum Cryptography using decoding, Quantum Neural Networks (QNN) for automated control systems, Quantum Optics for the interaction of photons with particles, Optimization using quantum annealing, Machine Learning / Big data and using Simulation approaches.

5 Conclusion

This Book chapter first introduces the basics of quantum machine learning with Machine learning types & Deep Learning. There is a bird's eye view of Quantum Machine Learning Algorithms as Quantum Machine Learning Approaches, Grover Search Algorithm, Quantum Reinforcement Learning, Quantum Annealing, Quantum Neural Networks, and Support vector machine. It elaborates on Breast Cancer with Molecular Classification of Breast Cancer as *Luminal-A,* Luminal-*B,* Normal-like, HER2 enriched, and Basal-like with Breast Cancer Diagnostic Techniques. Several approaches to quantum computation for machine learning are performing well for the prediction of breast cancer with QNN (Quantum Neural Network), DRA (Dimensionality Reduction Algorithms) & SVM (Support vector machines). Breast cancer analysis, diagnosis & prediction using various QML algorithms generate very fast results with utilization optimized storage.

References

1. D. Quiroga, P. Date, R. Pooser, Discriminating quantum states with quantummachine learning, in *2021 IEEE International Conference on Quantum Computing and Engineering (QCE)*, 17–22 October 2021, pp. 481–482
2. M. Nivelkar, S.G. Bhirud, Optimized machine learning: training and classification performance using quantum computing, in *2021 IEEE 6th International Conference on Computing, Communication and Automation (ICCCA)*, 17–19 December 2021, pp. 8–13
3. P. Sangeetha, P. Kumari, Quantumalgorithms for machine learning and optimization, in *2020 2nd PhD Colloquium on Ethically Driven Innovation and Technology for Society (PhD EDITS)*, 8 November 2020, pp. 1–2
4. P. Rebentrost, M. Mohseni, S. Lloyd, Quantum support vector machine for big data classification. Phys. Rev. Lett. **113**(13), 130503 (2014). https://doi.org/10.1103/PhysRevLett.113.130503
5. G. Bonaccorso, *Machine Learning Algorithms* (Packt Publishing Ltd., 2017)
6. F. Schwenker, E. Trentin, Pattern classification and clustering: a review of partially supervised learning approaches. Pattern Recogn. Lett. **37**, 4–14 (2014). https://doi.org/10.1016/j.patrec.2013.10.017
7. M. Riedmiller, Advanced supervised learning in multi-layer perceptrons—from backpropagation to adaptive learning algorithms. Comput. Stand. Interf. **16**(3), 265–278 (1994)
8. S.S. Sawant, M. Prabukumar, A review on graph-based semi-supervised learning methods for hyperspectral image classification. Egypt. J. Remote Sens. Space Sci. **23**(2), 243–248 (2020). https://doi.org/10.1016/j.ejrs.2018.11.001
9. N. Li, M. Shepperd, Y. Guo, A systematic review of unsupervised learning techniques for software defect prediction. Inf. Softw. Technol. **122**, 106287 (2020). https://doi.org/10.1016/j.infsof.2020.106287
10. T.O. Ayodele, Types of machine learning algorithms. New Adv. Mach. Learn. **3**, 19–48 (2010)
11. R. Chatterjee, T. Yu, Generalized coherent states, reproducing kernels, and quantum support vector machines (2016). arXiv preprint arXiv:1612.03713
12. Z. Zhao, J.K. Fitzsimons, J.F. Fitzsimons, Quantum-assisted Gaussian process regression. Phys. Rev. A **99**(5), 052331 (2019)
13. J.D. Whitfield, M. Faccin, J.D. Biamonte, Ground-state spin logic. EPL (Europhys. Lett.) **99**(5), 57004 (2012)
14. Y. Bengio, Y. LeCun, Scaling learning algorithms towards AI. Large-Scale Kernel Mach. **34**(5), 1–41 (2007)
15. M. Denil,N. De Freitas, Toward the implementation of a quantum RBM (2011)
16. V. Dumoulin, I. Goodfellow, A. Courville, Y. Bengio, On the challenges of physical implementations of RBMs' (2014)
17. J. Biamonte, P. Wittek, N. Pancotti, P. Rebentrost, N. Wiebe, S. Lloyd, Quantum machine learning. Nature **549**(7671), 195–202 (2017). https://doi.org/10.1038/nature23474
18. J.D. Biamonte, P.J. Love, Realizable Hamiltonians for universal adiabatic quantum computers. Phys. Rev. A **78**(1), 012352 (2008)
19. S. Lloyd, B.M. Terhal, Adiabatic and Hamiltonian computing on a 2D lattice with simple two-qubit interactions. New J. Phys. **18**(2), 023042 (2016)
20. S.-Y. Xia, Z.-Y. Xiong, Y.-G. Luo, L.-M. Dong, A method to improve support vector machine based on distance to hyperplane, **126**(20), 2405–2410 (2015)
21. E. Farhi, J. Goldstone, S. Gutmann, A quantum approximate optimization algorithm (2014). arXiv preprint arXiv:1411.4028
22. S. Lloyd, M. Mohseni, P. Rebentrost, Quantum algorithms for supervised andunsupervised machine learning (2013). arXiv preprint arXiv:1307.0411
23. J. Schmidhuber, Deep learning in neural networks: an overview. Neural Netw. **61**, 85–117 (2015). https://doi.org/10.1016/j.neunet.2014.09.003
24. S. Lloyd, C. Weedbrook, Quantum generative adversarial learning. Phys. Rev. Lett. **121**(4), 040502 (2018). https://doi.org/10.1103/PhysRevLett.121.040502

25. I. Cong, S. Choi, M.D. Lukin, Quantum convolutional neural networks. Nat. Phys. **15**(12), 1273–1278 (2019). https://doi.org/10.1038/s41567-019-0648-8
26. C. Lavor, L.R.U. Manssur, R. Portugal, Grover's algorithm: Quantum database search (2003). arXiv preprint quant-ph/0301079
27. M. Kaloev, G. Krastev, Experiments focused on exploration in deep reinforcement learning, in *2021 5th International Symposium on Multidisciplinary Studies and Innovative Technologies (ISMSIT),* 21–23 October 2021, pp. 351–355
28. W.D. Smart, L.P. Kaelbling, Effective reinforcement learning for mobile robots, in *Proceedings 2002 IEEE International Conference on Robotics and Automation (Cat. No. 02CH37292)* (IEEE, 2002), pp. 3404–3410
29. E. Aïmeur, G. Brassard, S. Gambs, Quantum speed-up for unsupervised learning. Mach. Learn. **90**(2), 261–287 (2013). https://doi.org/10.1007/s10994-012-5316-5
30. A. Bisio, G. Chiribella, G.M. d Ariano, S. Facchini, P. Perinotti, Optimal quantum learning of a unitary transformation, **81**, 032324 (2010)
31. S. Lloyd, M. Mohseni, P. Rebentrost, Quantum principal componentanalysis, **10**, 631–633 (2014)
32. D. Ventura, T. Martinez, Quantum associative memory. Inf. Sci. **124**(1), 273–296 (2000). https://doi.org/10.1016/S0020-0255(99)00101-2
33. N. Wiebe, A. Kapoor, K. Svore, Quantum algorithms for nearest-neighbor methods for supervised and unsupervised learning (2014). arXiv preprint arXiv:1401.2142
34. A. Narayanan, T. Menneer, Quantum artificial neural network architectures and components. Inform. Sci. **128**, 231–255 (2000a). https://doi.org/10.1016/S0020-0255(00)00055-4
35. D. Anguita, S. Ridella, F. Rivieccio, R. Zunino, Quantum optimization for training support vector machines. Neural Netw. **16**(5–6), 763–770 (2003). https://doi.org/10.1016/s0893-6080(03)00087-x
36. K.-A. Brickman, P. Haljan, P. Lee, M. Acton, L. Deslauriers, C. Monroe, Implementation of Grover's quantum search algorithm in a scalable system, **72**(5), 050306 (2005)
37. P. Kwiat, J. Mitchell, P. Schwindt, A.G. White, Grover's search algorithm: an optical approach, **47**(2–3), 257–266 (2000)
38. C. Allauzen, M. Crochemore, M. Raffinot, Factor oracle: a new structure for pattern matching, in *International Conference on Current Trends in Theory and Practice of Computer Science* (Springer, 1999), pp. 295–310
39. D. Dong, C. Chen, H. Li, T. Tarn, Quantum reinforcement learning. IEEE Trans. Syst. Man Cybern. Part B (Cybern.) **38**(5), 1207–1220 (2008). https://doi.org/10.1109/TSMCB.2008.925743
40. K. Kashyap, S.D. Lalit, L. Gautam, From classical to quantum: a review ofrecent progress in reinforcement learning, in *2021 2nd International Conference for Emerging Technology (INCET).* 21–23 May 2021, pp. 1–5
41. C. Chen, D. Dong, Complexity analysis of Quantum reinforcement learning, in *Proceedings of the 29th Chinese Control Conference*, 29–31 July 2010, pp. 5897–5901
42. A.J. Smith, Applications of the self-organising map to reinforcement learning. Neural Netw. **15**(8–9), 1107–1124 (2002)
43. C. Chen, H.-X. Li, D. Dong, Hybrid control for robotnavigation-a hierarchical q-learning algorithm, **15**(2), 37–47 (2008)
44. M. Kaya, R. Alhajj, A novel approach to multiagent reinforcement learning: utilizing OLAP mining in the learning process. IEEE Trans. Syst. Man Cybern. Part C (Appl. Rev.) **35**(4), 582–590. https://doi.org/10.1109/TSMCC.2004.843188
45. S. Whiteson, Evolutionary function approximation for reinforcement learning, **7** (2006)
46. T. Hogg, D. Portnov, Quantum optimization. Inf. Sci. **128**(3–4), 181–197 (2000)
47. A.A. Ezhov, D. Ventura, Quantum neural networks, in *Future Directions for Intelligent Systems and Information Sciences* (Springer, 2000), pp. 213–235
48. C.M. Catherine, *Adiabatic Quantum Computation and Quantum Annealing: Theory and Practice* (Morgan & Claypool, 2014)

49. Y. Matsuda, H. Nishimori, H.G. Katzgraber, Ground-state statistics from annealing algorithms: quantum versus classical approaches. New J. Phys. **11**(7), 073021 (2009)
50. G.E. Santoro, R. Martoňák, E. Tosatti, R. Car, Theory of quantum annealing of an Ising spin glass. Science **295**(5564), 2427–2430 (2002)
51. W. Vinci,K. Markström, S. Boixo, A. Roy, F.M. Spedalieri, P.A. Warburton, et al., Hearing the shape of the Ising model with a programmable superconducting-flux annealer, **4**(1), 1–7 (2014)
52. N. Chancellor, S. Szoke, W. Vinci, G. Aeppli, P.A. Warburton, Maximum- entropy inference with a programmable annealer, **6**(1), 1–14 (2016)
53. Y. Otsubo, J.-I. Inoue, K. Nagata, M. Okada, Effect of quantumfluctuation in error-correcting codes, **86**(5), 051138 (2012)
54. Y. Otsubo, J.-I. Inoue, K. Nagata, M. Okada, Code-division multiple-access multiuser demodulator by using quantum fluctuations, **90**(1), 012126 (2014)
55. V. Choi, Adiabatic quantum algorithms for the NP-complete maximum—weightindependent set, exact cover and 3SAT problems (2010)
56. S.H. Adachi, M.P. Henderson, Application of quantum annealing to the training of deep neural networks (2015). arXiv preprint arXiv:1510.06356
57. M.H. Amin, E. Andriyash, J. Rolfe, B. Kulchytskyy, R. Melko, Quantum Boltzmann machine. Phys. Rev. X **8**(2), 021050 (2018)
58. K. Hukushima, K. Nemoto, Exchange Monte Carlo method and application to spin glass simulations, **65**(6), 1604–1608 (1996)
59. S. Kirkpatrick, C.D. Gelatt Jr, M.P. Vecchi, Optimization by simulatedannealing, **220**(4598), 671–680 (1983)
60. D. de Falco, D. Tamascelli, An introduction to quantum annealing, **45**(1), 99–116 (2011)
61. Morita, S., & Nishimori, H (2008). Mathematical foundation of quantum annealing. 49(12), 125210
62. M. Panella, G. Martinelli, Neural networks with quantum architecture and quantum learning. Int. J. Circ. Theory Appl. **39**(1), 61–77 (2011)
172 R. Zhou, H. Zheng, N. Jiang, Q. Ding, Self-organizing Quantum Neural Network (IEEE, 2006), pp. 1067–1072
64. A.J. da Silva, T.B. Ludermir, W.R. de Oliveira, Quantum perceptron over a field and neural network architecture selection in a quantum computer. Neural Netw. **76**, 55–64 (2016)
65. S.C. Kak, Quantum neural computing. Adv. Imag. Electron. Phys. **94**, 259–313 (1995)
66. A. Narayanan,T. Menneer, Quantum artificial neural network architectures and components, **128**(3–4), 231–255 (2000b)
67. S. Ding, Z. Zhu, X. Zhang, An overview on semi-supervised support vector machine. Neural Comput. Appl. **28**(5), 969–978 (2017)
68. S. Ding, X. Hua, J. Yu, An overview on nonparallel hyperplane support vector machine algorithms. Neural Comput. Appl. **25**(5), 975–982 (2014). https://doi.org/10.1007/s00521-013-1524-6
69. Y.-H. Shao,W.-J. Chen, N.-Y. Deng, Nonparallel hyperplane support vector machine for binary classification problems. Inform. Sci. **263**, 22–35 (2014). https://doi.org/10.1016/j.ins.2013.11.003
70. H. Xu, C. Guedes Soares, Manoeuvring modelling of a containership in shallowwater based on optimal truncated nonlinear kernel-based least square support vector machine and quantum-inspired evolutionary algorithm. Ocean Eng. **195**, 106676 (2020). https://doi.org/10.1016/j.oceaneng.2019.106676
71. Y. Zhang, Q. Ni, Recent advances in quantum machine learning, **2**(1), e34 (2020). https://doi.org/10.1002/que2.34
72. D.M. Hausman, What is cancer? Perspect Biol Med **62**(4), 778–784 (2019). https://doi.org/10.1353/pbm.2019.0046
73. P.S. Roy, B.J. Saikia, Cancer and cure: a critical analysis. Indian J. Cancer **53**(3), 441–442 (2016). https://doi.org/10.4103/0019-509x.200658

74. R.L. Siegel, K.D. Miller, H.E. Fuchs, A. Jemal, Cancer statistics, **72**(1), 7– 33 (2022). https://doi.org/10.3322/caac.21708
75. M.A. Zaimy, N. Saffarzadeh, A. Mohammadi, H. Pourghadamyari, P. Izadi, A. Sarli et al., New methods in the diagnosis of cancer and gene therapy of cancer based on nanoparticles. Cancer Gene. Ther. **24**(6), 233–243 (2017). https://doi.org/10.1038/cgt.2017.16
76. P. Mathur, K. Sathishkumar, M. Chaturvedi, P. Das, K.L. Sudarshan, S. Santhappan, et al., Cancer Statistics, 2020: Report From National Cancer Registry Programme, India, (6), 1063-1075 (2020). https://doi.org/10.1200/go.20.00122
77. R. Fisher, L. Pusztai, C. Swanton, Cancer heterogeneity: implications for targeted therapeutics. Br. J. Cancer **108**(3), 479–485 (2013). https://doi.org/10.1038/bjc.2012.581
78. C.E. Meacham, S.J. Morrison, Tumour heterogeneity and cancer cell plasticity. Nature **501**(7467), 328–337 (2013). https://doi.org/10.1038/nature12624
79. R.L. Siegel, K.D. Miller, A. Jemal, Cancer statistics, **66**(1), 7–30. (2016). https://doi.org/10.3322/caac.21332
80. K. Aizawa, C. Liu, S. Tang, S. Veeramachaneni, K.-Q. Hu, D.E. Smith, et al., Tobacco carcinogen induces both lung cancer and non-alcoholic steatohepatitis and hepatocellular carcinomas in ferrets which can be attenuated by lycopene supplementation, **139**(5), 1171–1181 (2016). https://doi.org/10.1002/ijc.30161
81. S.O. Antwi, E.C. Eckert, C.V. Sabaque, E.R. Leof, K.M. Hawthorne, W.R. Bamlet et al., Exposure to environmental chemicals and heavy metals, and risk of pancreatic cancer. Cancer Causes & Control **26**(11), 1583–1591 (2015). https://doi.org/10.1007/s10552-015-0652-y
82. M.G.K. Cumberbatch, A. Cox, D. Teare, J.W.F. Catto, Contemporary occupational carcinogen exposure and bladder cancer: a systematic review and meta-analysis. JAMA Oncol. **1**(9), 1282–1290 (2015). https://doi.org/10.1001/jamaoncol.2015.3209%JJAMAOncology
83. S.L. Poon, J.R. McPherson, P. Tan, B.T. Teh, S.G. Rozen, Mutation signatures of carcinogen exposure: genome-wide detection and new opportunities for cancer prevention. Gen. Med. **6**(3), 24 (2014). https://doi.org/10.1186/gm541
84. D.M. Parkin, The global health burden of infection-associated cancers in the year, **118**(12), 3030–3044 (2006). https://doi.org/10.1002/ijc.21731
85. Z. Anastasiadi, G.D. Lianos, E. Ignatiadou, H.V. Harissis, M. Mitsis, Breast cancer in young women: an overview. Updat. Surg. **69**(3), 313–317 (2017). https://doi.org/10.1007/s13304-017-0424-1
86. A. Metelková, A. Skálová, J. Fínek, Breast cancer in young women-correlation of clinical histomorphological, and molecular-genetic features of breast carcinoma in women younger than 35 years of age. Klin Onkol **30**(3), 202–209 (2017). https://doi.org/10.14735/amko2017202
87. H.A. Azim, A.H. Partridge, Biology of breast cancer in young women. Breast Cancer Res. **16**(4), 427 (2014). https://doi.org/10.1186/s13058-014-0427-5
88. F.K. Al-Thoubaity, Molecular classification of breast cancer: A retrospective cohort study. Ann. Med. Surg. (Lond.) **49**, 44–48. https://doi.org/10.1016/j.amsu.2019.11.021
89. N. Eliyatkın, E. Yalçın, B. Zengel, S. Aktaş, E. Vardar, Molecular classification of breast carcinoma: from traditional, old-fashioned way to a new age, and a new way. J. Breast Health **11**(2), 59–66 (2015). https://doi.org/10.5152/tjbh.2015.1669
90. E.A. Rakha, I.O. Ellis, Modern classification of breast cancer: should we stick with morphology or convert to molecular profile characteristics. Adv. Anat. Pathol. **18**(4) 255–267 (2011). https://doi.org/10.1097/PAP.0b013e318220f5d1
91. G. Viale, The current state of breast cancer classification. Ann. Oncol. **23**, x207–x210 (2012). https://doi.org/10.1093/annonc/mds326
92. A. Blagodatski, V. Cherepanov, A. Koval, V.I. Kharlamenko, Y.S. Khotimchenko, V.L. Katanaev, High-throughput targeted screening in triple-negative breast cancer cells identifies Wnt-inhibiting activities in Pacific brittle stars. Sci. Rep. **7**(1), 11964 (2017). https://doi.org/10.1038/s41598-017-12232-7
93. P.A. Fitzpatrick, N. Akrap, E.M.V. Söderberg, H. Harrison, G.J. Thomson, G. Landberg, Robotic mammosphere assay for high-throughput screening in triple-negative breast cancer. SLAS Discov. **22**(7), 827–836 (2017). https://doi.org/10.1177/2472555217692321

94. L. Huang, X. Yi, X. Yu, Y. Wang, C. Zhang, L. Qin, et al., High-throughput strategies for the discovery of anticancer drugs by targeting transcriptional reprogramming (review), **11** (2021). https://doi.org/10.3389/fonc.2021.762023
95. Z. Sun, B. Zhang, High-throughput screening (HTS) of natural products with triple- negative breast cancer (TNBC) organoids, **37**(15_suppl), e12558 (2019). https://doi.org/10.1200/JCO.2019.37.15_suppl.e12558
96. F.K. Al-thoubaity, Molecular classification of breast cancer: a retrospective cohort study. Ann. Med. Surg. **49**, 44–48 (2020a). https://doi.org/10.1016/j.amsu.2019.11.021
97. H.O. Habashy, D.G. Powe, T.M. Abdel-Fatah, J.M. Gee, R.I. Nicholson, A.R. Green et al., A review of the biological and clinical characteristics of luminal-like oestrogen receptor-positive breast cancer. Histopathology **60**(6), 854–863 (2012). https://doi.org/10.1111/j.1365-2559.2011.03912.x
98. C. Sotiriou, S.-Y. Neo, L.M. McShane, E.L. Korn, P.M. Long, A. Jazaeri et al., Breast cancer classification and prognosis based on gene expression profiles from a population-based study. Proc. Natl. Acad. Sci. U.S.A. **100**(18), 10393–10398 (2003). https://doi.org/10.1073/pnas.1732912100
99. L.A. Carey, Through a glass darkly: advances in understanding breast cancer biology, 2000–2010. Clin Breast Cancer **10**(3), 188–195 (2010). https://doi.org/10.3816/CBC.2010.n.026
100. J.J. Gao, S.M. Swain, Luminal a breast cancer and molecular assays: a review. Oncologist **23**(5), 556–565 (2018). https://doi.org/10.1634/theoncologist.2017-0535
101. V. Guarneri, P. Conte, Metastatic breast cancer: therapeutic options according to molecular subtypes and prior adjuvant therapy. Oncologist **14**(7), 645–656 (2009). https://doi.org/10.1634/theoncologist.2009-0078
102. H. Kennecke, R. Yerushalmi, R. Woods, M.C. Cheang, D. Voduc, C.H. Speers et al., Metastatic behavior of breast cancer subtypes. J Clin Oncol **28**(20), 3271–3277 (2010). https://doi.org/10.1200/jco.2009.25.9820
103. S. Dwivedi, P. Purohit, R. Misra, M. Lingeswaran, J.R. Vishnoi, P. Pareek, et al.,Chapter 5—Application of single-cell omics in breast cancer, in *Single-Cell Omics*, ed. by D. Barh, V. Azevedo (Academic Press, 2019), pp. 69–103. https://doi.org/10.1016/B978-0-12-817532-3.00005-0
104. Z.-H. Li, P.-H. Hu, J.-H. Tu, N.-S. Yu, Luminal B breast cancer: patterns of recurrence and clinical outcome. Oncotarget **7**(40), 65024–65033 (2016). https://doi.org/10.18632/oncotarget.11344
105. B. Tran, P.L. Bedard, Luminal-B breast cancer and novel therapeutic targets. Breast Cancer Res. **13**(6), 221 (2011). https://doi.org/10.1186/bcr2904
106. M.J. Ellis, Y. Tao, J. Luo, R. A'Hern, D.B. Evans, A.S. Bhatnagar et al., Outcome prediction for estrogen receptor-positive breast cancer based on postneoadjuvant endocrine therapy tumor characteristics. J. Natl. Cancer Inst. **100**(19), 1380–1388 (2008). https://doi.org/10.1093/jnci/djn309
107. R. Bhargava, S. Beriwal, D.J. Dabbs, U. Ozbek, A. Soran, R.R. Johnson et al., Immunohistochemical surrogate markers of breast cancer molecular classes predicts response to neoadjuvant chemotherapy: a single institutional experience with 359 cases. Cancer **116**(6), 1431–1439 (2010). https://doi.org/10.1002/cncr.24876
108. S. Paik, S. Shak, G. Tang, C. Kim, J. Baker, M. Cronin et al., A multigene assay to predict recurrence of tamoxifen-treated, node-negative breast cancer. N. Engl. J. Med. **351**(27), 2817–2826 (2004). https://doi.org/10.1056/NEJMoa041588
109. J.S. Parker, A. Prat, M. Cheang, M. Lenburg, S. Paik, C.J.C.R. Perou, Breast cancer molecular subtypes predict response to anthracycline/taxane-based chemotherapy, **69**(24 Suppl 3) (2009)
110. B. Weigelt, A. Mackay, R. A'Hern, R. Natrajan, D.S. Tan, M. Dowsett et al., Breast cancer molecular profiling with single sample predictors: a retrospective analysis. Lancet Oncol. **11**(4), 339–349 (2010). https://doi.org/10.1016/s1470-2045(10)70008-5
111. L. Yin, J.-J. Duan, X.-W. Bian, S.-C. Yu, Triple-negative breast cancer molecular subtyping and treatment progress. Breast Cancer Res. **22**(1), 61 (2020). https://doi.org/10.1186/s13058-020-01296-5

112. J.J. de Ronde, J. Hannemann, H. Halfwerk, L. Mulder, M.E. Straver, M.J. Vrancken Peeters, et al., Concordance of clinical and molecular breast cancer subtyping in the context of preoperative chemotherapy response. Breast Cancer Res. Treat **119**(1), 119–126 (2010). https://doi.org/10.1007/s10549-009-0499-6
113. U. Krishnamurti, J.F. Silverman, HER2 in breast cancer: a review and update. Adv. Anat. Pathol. **21**(2) (2014). https://journals.lww.com/anatomicpathology/Fulltext/2014/03000/HER2_in_Breast_Cancer__A_Review_and_Update.4.aspx
114. J.S. Ross, J.A. Fletcher, G.P. Linette, J. Stec, E. Clark, M. Ayers et al., The Her-2/neu gene and protein in breast cancer 2003: biomarker and target of therapy. Oncologist **8**(4), 307–325 (2003). https://doi.org/10.1634/theoncologist.8-4-307
115. J. Wang, B. Xu, Targeted therapeutic options and future perspectives for HER2-positive breast cancer. Signal Transduct. Target. Ther. **4**(1), 34 (2019). https://doi.org/10.1038/s41392-019-0069-2
116. B. Kreike, M. van Kouwenhove, H. Horlings, B. Weigelt, H. Peterse, H. Bartelink et al., Gene expression profiling and histopathological characterization of triple-negative/basal-like breast carcinomas. Breast Cancer Res **9**(5), R65 (2007). https://doi.org/10.1186/bcr1771
117. E.A. Rakha, S.E. Elsheikh, M.A. Aleskandarany, H.O. Habashi, A.R. Green, D.G. Powe et al., Triple-negative breast cancer: distinguishing between basal and nonbasal subtypes. Clin. Cancer Res. **15**(7), 2302–2310 (2009). https://doi.org/10.1158/1078-0432.Ccr-08-2132
118. E.R. Myers, P. Moorman, J.M. Gierisch, L.J. Havrilesky, L.J. Grimm, S. Ghate et al., Benefits and harms of breast cancer screening: a systematic review. JAMA **314**(15), 1615–1634 (2015). https://doi.org/10.1001/jama.2015.13183
119. M. Román, M. Sala, L. Domingo, M. Posso, J. Louro, X. Castells, Personalized breast cancer screening strategies: a systematic review and quality assessment. PLoS One **14**(12), e0226352 (2019). https://doi.org/10.1371/journal.pone.0226352
120. M. Escala-Garcia, A. Morra, S. Canisius, J. Chang-Claude, S. Kar, W. Zheng, et al., Breast cancer risk factors and their effects on survival: a Mendelian randomisation study. BMC Med. **18**(1), 327 (2020). https://doi.org/10.1186/s12916-020-01797-2
121. M.R. Ataollahi, J. Sharifi, M.R. Paknahad, A. Paknahad, Breast cancer and associated factors: a review. J. Med. Life **8**(Special Issue 4), 6–11 (2015)
122. Y.-S. Sun, Z. Zhao, Z.-N. Yang, F. Xu, H.-J. Lu, Z.-Y. Zhu et al., Risk factors and preventions of breast cancer. Int. J. Biol. Sci. **13**(11), 1387–1397 (2017). https://doi.org/10.7150/ijbs.21635
123. S. Raichand, A.G. Dunn, M.-S. Ong, F.T. Bourgeois, E. Coiera, K.D. Mandl, Conclusions in systematic reviews of mammography for breast cancer screening and associations with review design and author characteristics. Syst. Rev. **6**(1), 105 (2017). https://doi.org/10.1186/s13643-017-0495-6
124. M. Broeders, S. Moss, L. Nyström, S. Njor, H. Jonsson, E. Paap, et al., The impactof mammographic screening on breast cancer mortality in Europe: a review of observational studies. J. Med. Screen **19**(Suppl 1), 14–25 (2012). https://doi.org/10.1258/jms.2012.012078
125. C. van den Ende, A.M. Oordt-Speets, H. Vroling, H.M.E. van Agt, Benefits and harms of breast cancer screening with mammography in women aged 40–49 years: a systematic review. Int. J. Cancer **141**(7), 1295–1306 (2017). https://doi.org/10.1002/ijc.30794
126. H.G. Welch, P.C. Prorok, A.J. O'Malley, B.S. Kramer, Breast-cancer tumor size. Overdiag. Mammogr. Screen. Effect. **375**(15), 1438–1447 (2016). https://doi.org/10.1056/NEJMoa1600249
127. J.S. Mandelblatt, N.K. Stout, C.B. Schechter, J.J. van den Broek, D.L. Miglioretti, M. Krapcho, et al., Collaborative modeling of the benefits and harms associated with different U.S. breast cancer screening strategies. Ann. Intern. Med. **164**(4), 215–225 (2016). https://doi.org/10.7326/m15-1536
128. E. Warner, H. Messersmith, P. Causer, A. Eisen, R. Shumak, D. Plewes, Systematic review: using magnetic resonance imaging to screen women at high risk for breast cancer. Ann. Intern. Med. **148**(9), 671–679 (2008). https://doi.org/10.7326/0003-4819-148-9-200805060-00007
129. S. Radhakrishna, S. Agarwal, P.M. Parikh, K. Kaur, S. Panwar, S. Sharma et al., Role of magnetic resonance imaging in breast cancer management. South Asian J Cancer **7**(2), 69–71 (2018). https://doi.org/10.4103/sajc.sajc_104_18

130. L.W. Turnbull, Dynamic contrast-enhanced MRI in the diagnosis and management of breast cancer. NMR Biomed. **22**(1), 28–39 (2009). https://doi.org/10.1002/nbm.1273
131. J. Xiao, H. Rahbar, D.S. Hippe, M.H. Rendi, E.U. Parker, N. Shekar, et al., Dynamic contrast-enhanced breast MRI features correlate with invasive breast cancer angiogenesis. npj Breast Cancer **7**(1), 42 (2021). https://doi.org/10.1038/s41523-021-00247-3
132. N. Amornsiripanitch, S. Bickelhaupt, H.J. Shin, M. Dang, H. Rahbar, K. Pinker, et al., Diffusion-weighted MRI for unenhanced breast cancer screening, **293**(3), 504–520 (2019). https://doi.org/10.1148/radiol.2019182789
133. S.C. Partridge, N. Amornsiripanitch, DWI in the assessment of breast lesions. Top. Magn. Reson. Imaging: TMRI **26**(5), 201–209 (2017). https://doi.org/10.1097/RMR.0000000000000137
134. S.C. Partridge, E.S. McDonald, Diffusion weighted magnetic resonance imaging of the breast: protocol optimization, interpretation, and clinical applications. Magn. Reson. Imaging Clin. N. Am. **21**(3), 601–624 (2013). https://doi.org/10.1016/j.mric.2013.04.007
135. J.K.P. Begley, T.W. Redpath, P.J. Bolan, F.J. Gilbert, In vivo proton magnetic resonance spectroscopy of breast cancer: a review of the literature. Breast Cancer Res. **14**(2), 207 (2012). https://doi.org/10.1186/bcr3132
136. M.A. Bilal Ahmadani, S. Bhatty, Z.U. Abideen, M.S. Yaseen, T. Laique, J. Malik, Imaging in breast cancer: use of magnetic resonance spectroscopy.Cureus (2020). https://doi.org/10.7759/cureus.9734
137. W.A. Berg, Nuclear breast imaging: clinical results and future directions. J. Nucl. Med. **57**(Supplement 1), 46S (2016). https://doi.org/10.2967/jnumed.115.157891
138. L.R. Greene, D. Wilkinson, The role of general nuclear medicine in breast cancer. J. Med. Radiat. Sci. **62**(1), 54–65 (2015). https://doi.org/10.1002/jmrs.97
139. A. Daskin, A simple quantum neural net with a periodic activation function, in *2018 IEEE International Conference on Systems, Man, and Cybernetics (SMC)*, 7–10 October 2018, pp. 2887–2891
140. N. Mishra, A. Bisarya, S. Kumar, B.K. Behera, S. Mukhopadhyay, P.K. Panigrahi, Cancer detection using quantum neural networks: a demonstration on a quantum computer (2019)
141. V. Azevedo, C. Silva, I. Dutra, Quantum transfer learning for breast cancerdetection. Quant. Mach. Intell. **4**(1), 1–14 (2022b)
142. H.G. Kim, Y. Choi, Y.M. Ro, *Modality-Bridge Transfer Learning for Medical Image Classification* (IEEE, 2017), pp. 1–5
143. C.-K. Shie, C.-H. Chuang, C.-N. Chou, M.-H. Wu, E.Y. Chang, Transfer Representation-Learning for Medical Image Analysis (IEEE, 2015), pp. 711–714
144. M.A. Morid, A. Borjali, G. Del Fiol, A scoping review of transfer learning research on medical image analysis using ImageNet. Comput. Biol. Med. **128**, 104115 (2021)
145. S. Lu, Z. Lu, Y.-D. Zhang, Pathological brain detection based on AlexNet and transfer learning. J. Comput. Sci. **30**, 41–47 (2019). https://doi.org/10.1016/j.jocs.2018.11.008
146. S.-H. Wang, S. Xie, X. Chen, D.S. Guttery, C. Tang, J. Sun, et al., Alcoholism identification based on an AlexNet transfer learning model (original research), **10** (2019). https://doi.org/10.3389/fpsyt.2019.00205
147. P. Dauphin-Ducharme, K. Yang, N. Arroyo-Currás, K.L. Ploense, Y. Zhang, J. Gerson, et al., Electrochemical aptamer-based sensors for improved therapeutic drug monitoring and high-precision, feedback-controlled drug delivery. ACS Sens. **4**(10), 2832–2837 (2019). https://doi.org/10.1021/acssensors.9b01616
148. E.L. Omonigho, M. David, A. Adejo, S. Aliyu, Breast cancer: tumor detectionin mammogram images using modified AlexNet deep convolution neural network, in *2020 International Conference in Mathematics, Computer Engineering and Computer Science (ICMCECS)*, 18–21 March 2020, pp. 1–6
149. F. Yan, X. Huang, Y. Yao, M. Lu, M. Li, Combining LSTM and DenseNet for automatic annotation and classification of chest x-ray images. IEEE Access **7**, 74181–74189 (2019). https://doi.org/10.1109/ACCESS.2019.2920397

150. F. Nunnari, C. Bhuvaneshwara, A.O. Ezema, D. Sonntag, A study on the fusion of pixels and patient metadata in CNN-based classification of skin lesion images, in *Machine Learning and Knowledge Extraction*, ed. by A. Holzinger, P. Kieseberg, A.M. Tjoa, E. Weippl (Springer International Publishing, Cham, 2020), pp. 191–208
151. P.U. Hepsağ, S.A. Özel, A. Yazıcı, Using deep learning for mammography classification, 418–423 (2017)
152. J. Diz, G. Marreiros, A. Freitas, Applying data mining techniques to improve breast cancer diagnosis. J. Med. Syst. **40**(9), 203 (2016). https://doi.org/10.1007/s10916-016-0561-y
153. V. Azevedo, C. Silva, I. Dutra, Quantum transfer learning for breast cancer detection. Quant. Mach. Intell. **4**(1), 5 (2022a). https://doi.org/10.1007/s42484-022-00062-4
154. A. Ahuja, L. Al-Zogbi, A. Krieger, Application of noise-reduction techniques tomachine learning algorithms for breast cancer tumor identification. Comput. Biol. Med. **135**, 104576 (2021). https://doi.org/10.1016/j.compbiomed.2021.104576
155. K. Gupta, R.R. Janghel, Dimensionality reduction-based breast cancer classification using machine learning, in *Computational Intelligence: Theories, Applications and Future Directions—Volume I,* ed. by N.K. Verma, & A.K. Ghosh (Springer, Singapore, 2019), pp. 133–146
156. D.A. Omondiagbe, S. Veeramani, A.S. Sidhu, Machine learning classification techniques for breast cancer diagnosis. IOP Conference Series: Materials Science and Engineering **495**, 012033 (2019). https://doi.org/10.1088/1757-899x/495/1/012033
157. M.F. Akay, Support vector machines combined with feature selection for breast cancer diagnosis. Exp. Syst. Appl. **36**(2, Part 2), 3240–3247 (2009). https://doi.org/10.1016/j.eswa.2008.01.009
158. A.T. Azar, S.A. El-Said, Performance analysis of support vector machines classifiers in breast cancer mammography recognition. Neural Comput. Appl. **24**(5), 1163–1177 (2014). https://doi.org/10.1007/s00521-012-1324-4
159. B.M. Salih Hasan, A.M. Abdulazeez, A review of principal component analysis algorithm for dimensionality reduction. J. Soft Comput. Data Min. **2**(1), 20–30. https://publisher.uthm.edu.my/ojs/index.php/jscdm/article/view/8032
160. M.D. Ritchie, L.W. Hahn, N. Roodi, L.R. Bailey, W.D. Dupont, F.F. Parl et al., Multifactor-dimensionality reduction reveals high-order interactions among estrogen-metabolism genes in sporadic breast cancer. Am. J. Hum. Genet. **69**(1), 138–147 (2001). https://doi.org/10.1086/321276
161. X. Liu, J. Shi, S. Zhou, M. Lu, An iterated Laplacian based semi-superviseddimensionality reduction for classification of breast cancer on ultrasound images, in *2014 36th Annual International Conference of the IEEE Engineering in Medicine and Biology Society,* 26–30 August 2014, pp. 4679–4682
162. C. Deisy, B. Subbulakshmi, S. Baskar, N Ramaraj, Efficient dimensionalityreduction approaches for feature selection, in *International Conference on Computational Intelligence and Multimedia Applications (ICCIMA 2007)*, 13–15 December 2007, pp. 121–127
163. A. Jamal, A. Handayani, A.A. Septiandri, E. Ripmiatin, Y. Effendi, Dimensionality reduction using PCA and k-means clustering for breast cancer prediction. Lontar Komput. J. Ilm. Teknol. Inf **9**(3), 192–201 (2018)
164. W.S. Noble, What is a support vector machine? Nat. Biotechnol. **24**(12), 1565–1567 (2006)
165. S. Suthaharan, Support vector machine, in *Machine Learning Models and Algorithms for Big Data Classification* (Springer, 2016), pp. 207–235
166. A. Widodo, B.-S. Yang, Support vector machine in machine condition monitoring and fault diagnosis, **21**(6), 2560–2574 (2007)
167. D.A. Pisner, D.M. Schnyer, *Support Vector Machine. Machine Learning.* (Elsevier, 2020), pp. 101–121
168. T. Joachims, SVM light is an implementation of support vector machines (SVMs), in C. University of Dortmund, Collaborative Research Center on Complexity Reduction in Multivariate Data (SFB475); gmd.de/~thorsten/svm_light (2000)
169. H.X. Liu, R.S. Zhang, F. Luan, X.J. Yao, M.C. Liu, Z.D. Hu et al., Diagnosing breast cancer based on support vector machines. J. Chem. Inf. Comput. Sci. **43**(3), 900–907 (2003). https://doi.org/10.1021/ci0256438

170. M.E. Ozer, P.O. Sarica, K.Y. Arga, New machine learning applications to accelerate personalized medicine in breast cancer: rise of the support vector machines. Omics: J. Integrative Biol. **24**(5), 241–246 (2020)
171. S. Ghosh,S. Mondal, B. Ghosh, *A Comparative Study of Breast Cancer Detection Based on SVM and MLP BPN Classifier* (IEEE, 2014), pp. 1–4

Understanding of Argon Fluid Sensor Using Single Quantum Well Through K-P Model: A Bio-medical Application Using Semiconductor Based Quantum Structure

Gopinath Palai, Nitin Tripathy, Biswaranjan Panda, and Chandra Sekhar Mishra

Abstract Investigation of argon concentration in their fluid is made in this chapter. This assessment of argon fluid is realized through three layers of silicon-based quantum well structure. The principle of measurement of argon relies on the variation of electronic band-gap pertaining to the density of argon which ranges from 0 to 1.5 g/cm^3. The physics of the work depends on both nature of the structure including material properties, whereas mathematics of the research relies on the configuration of the proposed structure which is derived through Kronig–Penny model. As far as numerical results are concerned, the electronic band-gaps are 0, 0.12, 0.29, 0.48, and 0.65 eV pertaining to the density (concentration) of argon, 0, 0.5, 1.0, 1.3, 1.5 g/cm^3 respectively, that indicates electronic band-gap increases with the increasing of concentration. Further, the transmitted energy corresponding to the same concentration are 0, 0.12, 0.29, 0.48 and 0.65 eV respectively, which show that the transmitted energy and subsequently potential decreases linearly concerned to the concentration. Finally, the linear regression is used to estimate the relation between concentration and output potential, which is measured through the photo diode. More over the formula for this relation is concentration = $-2.254 \times$ potential + 3.366 with regression factor of 94.35. The outcomes of the chapter infer that one can determine the concentration of argon fluid by knowing the output potential.

Keywords Quantum well (QW) · K-P model · Energy band · Transmitted signal · Argon fluid sensor

G. Palai (✉) · C. Sekhar Mishra
Gandhi Institute for Technological Advancement (GITA), Bhubaneswar, Odisha, India
e-mail: gpalai_ece@gita.edu.in

N. Tripathy · B. Panda
IISER Berhampur, Berhampur, Odisha, India

R. Pandey et al. (eds.), *Quantum Computing: A Shift from Bits to Qubits*,
Studies in Computational Intelligence 1085,
https://doi.org/10.1007/978-981-19-9530-9_20

1 Introduction

A particular kind of potential well [1] with discrete energy values is known as a quantum well. Odoh and Njapba [2] made the first independent proposal for the abstraction of the quantum well in 1963 [3]. The quantum confinement phenomenon is what gives a quantum well its unique properties. A quantum well's primary objective is to compel a quantum particle, such as an electron, to move in a two-dimensional plane as opposed to a three-dimensional space. The quantum confinement effect happens if and only if the quantum well width is proportionate to the carriers, such as electrons and holes, possessing the de Broglie wavelength. Energy sub-bands are created through discrete energy levels of the carriers.

Numerous electrical devices are being built using the quantum well theory [4]. These are used in switches, modulators, photodetectors, and lasers. Electronics using quantum well (QW)s are much more efficient financially and operate more quickly than conventional devices, which is crucial in industries like telecommunications and technology. Many electronic gadgets now use quantum well devices (QWD)s instead of conventional electrical parts.

1.1 History

The study of classical electrical physics revealed a lot of things. It was ultimately revealed to be really outstanding and complex, nonetheless. In 1970, Fox and Ispasoiu [5] widened the semiconductor of the QW. The artificial super-latices were also produced by them. They had the completely original idea to alternate sheets of semiconductors with various band-gaps to create a heterostructure. Following this ground-breaking idea, interest in the physics of QW increased [6]. After then, a prodigious amount of diligent labour was finished.

Advancements in QWDs can be ascribed to technological improvements in crystal growth procedures since quantum devices have a unique structure. The designs of QWDs are highly pure and have few faults [5]. Thus, the suggested heterostructures of Tsu and Esaki yield semiconductor devices with extremely favorable properties. People saw that QWs and semiconductor technology had come to the peak of the list of hot issues. In 2000 [7], Zhores, Alferov and Herbert Kroemer shared the Nobel Prize in physics for their work in building semiconductor devices with a structure built of several semiconductors. Nowadays, people utilize digital devices like smartphones, PCs, smart watches, and other devices. These electrical devices are fully dependent on the effectiveness of parts like light-emitting diodes, transistors, etc. So QWD has made notable development in the fabrication and efficiency of those parts.

1.2 Fabrication

A QW structure is created via chemical vapour deposition or molecular beam epitaxy (MBE) with management over the layer width down to a single layer.

Thin films can sustain QW states as well, but they must be developed on semiconductor and metallic surfaces in particular [8]. By sandwiching a different thin layer between consecutive sheets of another sort of semiconductor with changing band-gaps, we may create a simple QW. Think about the following example: Two layers of AlGaAs around a thin sheet of GaAs with having large band-gap. We presume that only motion in the Z-axis results in material change. The potential consequently lies along the Z-axis (In XY-flat, there is no detention). A quantum potential well forms in GaAs regime through which restricted material's smaller band-gap than the AlGaAs nearby. The band energy varies during the building. Smaller energy carriers could become stuck below it since it can be thought of as a possible shift that a carrier may perceive [8]. The carriers might exist in several eigenstates within the QW. The potential well may contain less energy for particles in the conduction band than the AlGaAs area of that structure. Within the quantum potential well, low-energy particles in the conduction band can be captured. A similar strategy may be used to restrict the valence band holes on top of prospective valence band wells. The particle-in-a-box condition is comparable to it [8].

2 Quantum Well Structure

2.1 Modeling of Well (Infinite)

The modeling of a well is the fundamental prototype in a QW system. In the endless well prototype, the barriers or walls are presumed to be unlimited in length; however, in reality, no such infinite well exists [5]. The infinite well prototype is a practical paradigm which renders numerous perspectives on QWs. Taking the same QW that is oriented along the vertical (Z) axis, charge carriers are constrained along the z-axis but have unrestricted mobility in the xy-plane. Consider the QW to be from to d. The confined particle has 0 potential within the QW, whereas barrier of potential zone is limitless.

Schrodinger's equation for infinite well carriers prototype is:

$$-\frac{\hbar^2}{2m_w^*}\frac{\partial^2\psi(z)}{\partial z^2} = E\psi(z) \tag{1}$$

where, $^-h = (h/2\pi)$ and m_w^* = effective massof the particle within the well area of the carriers.

2.1.1 Levels of Energy and Solutions

Here, the solution cannot exist owing to the well's incredibly high barrier area. Using the best boundary conditions we can find,

$$\psi(0) = \psi(d) = 0 \tag{2}$$

It can be written as,

$$\psi_n(z) = \sqrt{\frac{2}{d}} \sin(k_n z) \text{ where, } k_n = 2\pi/d \tag{3}$$

Here, n represents as quantum number and k_n represents the wave vector connected for each state. The discrete energies are

$$E_n = \frac{\hbar^2 k_n^2}{2m_w^*} = \frac{\hbar^2}{2m_w^*}\left(\frac{n\pi}{d}\right)^2 \tag{4}$$

A crucial part of examining the physics underlying the QW structure, which is the result of quantum detention, is the basic infinite well prototype. The prototype precisely deduces the relationship that the well's energy varies inversely with its square. It denotes the well width and enables accuracy. In order to precisely create the semiconductor layer, a crucial component of band-gap engineering, in order to manage the well width.

Prototype well has the drawback of anticipating many energy levels than are actually possible since the walls of wells are constrained. The prototype also fails to account for the fact that, in actuality, wave functions do not decay exponentially to zero at the well's edge but rather go to zero when a carrier passes through a wall owing to quantum tunneling. This property, which is best exemplified by the limited well model, enables the design and production of superlattices as well as other novel QW devices.

2.2 Finite Potential Well

The finite potential well qualifies as a real model since it is realizable. Vo, the finite potential of the well's walls in a heterostructure. Here, with the walls' limited potential, the electron can subway through the blockade zone through which permission takes place.

Take a look at a QW with a Z-direction orientation, where the carriers can travel freely in the XY plane but are constrained in the Z-direction. Just the QW [6] from 0 to d is being taken. The carriers cannot obtain any potential within the well since V_O potential is in the blockade areas. The Schrodinger equation therefore holds true for carriers inside a well, such as an infinite potential well. But a boundary condition is present here. As a result, at the boundaries, the wave function and its slopes are continuous. For carriers inside the barrier zone, the Schrodinger's equation is

$$-\frac{\hbar^2}{2m_b^*}\frac{\partial^2\psi(z)}{\partial z^2} + V_0\psi(z) = E\psi(z) \tag{5}$$

Here m_b^*: Effective mass of a particle in carrier region with potential V_0 [5].

2.2.1 Solutions

The boundary conditions may be used to locate the solutions, and we are aware that the wave function is continuous at the wall's edge. Therefore, the following transcendental equations' solutions for the wave vector:

For odd

$$tan\left(\frac{k_n d}{2}\right) = -\frac{m_b^* k_n}{m_w^* k} \tag{6}$$

For even

$$tan\left(\frac{k_n d}{2}\right) = -\frac{m_w^* k}{m_b^* k_n} \tag{7}$$

Now see the exponential decay constant k is:

$$k = \frac{\sqrt{2m_b^*(V_0 - E_n)}}{\hbar} \tag{8}$$

The constant relies on E_n, also well's depth V_0 and the carrier's effective mass (m_b^*).

The aforementioned transcendental equation may be solved extremely quickly and easily using different approaches. The well's wave function resembles an infinite potential well in a sinusoidal fashion. However, the wall's barrier causes them to degrade dramatically.

3 Concept of Superlattices

Superlattices are constructed by the periodic arrangement of two types of materials with various band-gaps. Since alternating materials' various band-gaps, manifold QW structures (MQW), with well widths in the order of a few nanoparticles, finally develop. The barrier's thickness in the MQW structure is sufficiently narrow for charge carriers to surpass through it [6] (through quantum tunneling). Additionally, the barrier width between wells is sufficiently reduced in the superlattice structure enable to be contiguous. Multiple QW are periodic structures comprised of repeating QWs with barriers that prevent nearby wave functions from coupling [5].

4 Application

4.1 *Saturable Absorber*

Saturable absorption allows a QW to be deceived into not becoming an absorber. Saturable absorbs offer broad applications. A CO_2 that can produce pulses of up to 500 ps was mode-locked using p-type Germanium (Ge) [8]. Laser mode-locking at the time was accomplished using semiconductor absorbers (SESANs). The distributed Bragg semiconductors are used to grow the contemporary SESANs (DBRs). These SESANs are either MQW or SQW. The SESAMs were developed [9] through inherent simplicity of this form. In this case, the average and repetition rates were raised to 60 W and 160 GHz, respectively, for precise data [10]. The main benefit over all absorbers is that we may alter the absorber parameters across a wide range of values when employing SWSAMs. As an illustration, the saturation may be adjusted by adjusting the top reflector and altering the low-temperature growth.

4.2 *Thermoelectrics*

For energy harvesting, QWs exhibit thermoelectric behaviour. These are incredibly easy to make and have the capability of functioning at 300 K. Here wells are used to connect two reservoirs to the central hollow. The barriers can allow certain electron energy when employed as a filter. The temperature difference between the reservoirs and the cavities determines the output power and electron flow. Experimentally verified output power for a 1 K temperature differential is 0.18 W/cm^2. It roughly has twice the power of a quantum dot energy collector. But compared to quantum dots, the efficiency is somewhat lower [10]. The increased freedom enables the laser current. The distinction is that although QWs can transport up to a given level, quantum dots only allow electrons with a specified energy [11]. Utilizing the leftover heat from electrical circuits is the major objective.

4.3 Solar Cell Devices

The efficiency of solar cell is greatly improved by QWs. Trivial single-junction cells are predicted to have a theoretical maximum efficiency of roughly 34%. But productivity is not up to the mark due to its complex structure [12]. In p-i-n junction cells, it is a combination of different QWs, the photo-current can be enhanced above the dark current, which results in an improvement in net efficiency as compared to normal cells and relegate them to specialized uses. Further, it is fabricated with BEM (Beam epitaxial method) [13]. Additionally, adding nanoparticles to the lattice has been shown to significantly increase absorbance of signal onto lateral propagation channels contained inside MQW layer [14]. Apart from this, single and multi-solar cells relating to the QWD can be found from the literature [8, 14–18].

5 Mathematical Treatment for QW Structure

5.1 Bandgap Energy

Many scientists are presently engaged in the creation of QW. For example, to manufacture high-quality materials with few crystal dislocations, to develop better effectiveness QW based solar cells, and to boost the efficiency of light absorption. Scientists are using band-gap tunability to construct solar cells. The band-gap energy of QW determines the effective band-gap. Therefore, it aids in our estimation of the effective band-value gap's [8].

$$E_{G,\mathrm{Eff}} = {}^{E}{}_{\mathrm{G}}\mathrm{well,\ relaxed} + \Delta^{E}{}_{G}\mathrm{strain} + \Delta^{E}{}_{G}\mathrm{QSE} + \Delta^{E}{}_{G}\mathrm{QCSE} \quad (9)$$

Normally there are two effects on the band-gap by the cause of the strain of the material. The first one is the varying relative energy in the valence band and conduction band. Strain causes the energy change [8, 18].

$$\Delta E = -2a\left(\frac{C_{1!} - C_{!2}}{C_{11}}\right) \in \quad (10)$$

Here, C_{11} and C_{12}: elastic stiffness cofficients; $\in$: strain

In this case, heavy (hh) and light-hole (lh) degeneracy is the second effect of strain. When a material is crushed tightly, the heavy-holes transition to higher energy states [8, 19]. The difference in energy can be computed using the shear deformation potential.

$$\Delta_{hh} = b\left(\frac{C_{11} + 2C_{12}}{C_{11}}\right) \in \quad (11)$$

$$\Delta_{\dot{a}h} = 2b\left(\frac{C_{11} + 2C_{12}}{C_{11}}\right) \in \tag{12}$$

a: shear deformation potential $\in$
b: strain
C_{11} and C_{12}: coefficients

Here band-gap is persuaded by Stark effect. So the energy shift is [8]:

$$\Delta E_G^{\text{QCSE}} = \left(\frac{15 - \pi^2}{24\pi^2}\right)\left(\frac{m_e^* L_{\text{eff},CB}^4 + m_h^* L_{\text{eff},VB}^4}{\hbar^2}\right) \tag{13}$$

Here, q: elemental charge

$L_{\text{eff},\ CB}$ and $L_{\text{eff},\ VB}$: effective width of QWS in the conduction and valence band respectively
F: induced electric field due to piezoelectric and spontaneous polarization $\bar{h}$: Planck's constant

The QW thickness and QSE are inversely related. As a result, QSE drops as QW thickness rises [20]. When the QSEs decline, the $n = 1$ state descends and reduces the band-gap. The Kronig Penney model, together with the conduction band and valence band offsets in, may be used to compute quantum state.

5.2 Carrier Capture and Lifetime

By efficiently using the carriers, one can improve the performance of QW cells (QWSCs). The carrier combination is described as an electron and hole combination that cancels the charge of the electron and hole. In the case of quantum well, the carriers are optically generated [21]. The carrier lifetime is determined by tunneling and emission durations. These two lifetimes for thermionic emission and tunneling are dependent on the presence of barrier height. All of the physical means mentioned in the subsequent equation are expressed [22]:

$$\frac{1}{\tau_{tun.}} = \frac{n\pi\hbar}{2t_w^2 m_w^*} e^{-\frac{2}{\hbar}\int_0^{t_b}\left(\sqrt{2m_b^*(E_b - E(z)z)}\right) d_z} \tag{14}$$

$$\frac{1}{\tau_{therm.}} = \frac{1}{t_w}\sqrt{\frac{kT}{2\pi m_w^*}} e^{-\frac{E_b}{kT}} \tag{15}$$

$E(z)$: electric field

The lifetime can be calculated as

$$\frac{1}{\tau_{esc.}} = \frac{1}{\tau_{tun.}} + \frac{1}{\tau_{therm.}} \tag{16}$$

And probability can be written as

$$Ptot = PiN \tag{17}$$

Here,

$$P_i = \frac{\tau_{esc.}^{-1}}{\tau_{esc.}^{-1} + \tau_{rec.}^{-1}} \tag{18}$$

where,

$\tau_{rec.}$: recombination lifetime
N: the total number of QWs in the intrinsic region

For $\tau_{esc.} << \tau_{rec.}$, the probability is high for carrier recollection [23]. Solution of Energy is found using Kronig Penny Expression (Mathematics).

According to the theory, the electron's potential energy is zero close to the nucleus and reaches its maximum when it is located between two adjacent nuclei that are separated by the inter-atomic space 'a'. The electron's potential energy also varies periodically with the periodicity of the ion core (nucleus).

Schrodinger's time independent equation for one dimension is,

$$\frac{d^2\psi}{dx^2} + \frac{2m}{\hbar^2}(E - V)\psi = 0 \tag{19}$$

Applying Schrodinger's time independent equation (STIE) for the electron confined in the well gives,

As for the region (I), $V = 0$, the STIE becomes,

$$\frac{d^2\psi}{dx^2} = \frac{2m}{\hbar^2}(E)\psi = 0 \tag{20}$$

Also for the region (II), $V = V_0$, the STIE becomes,

$$\frac{d^2\psi}{dx^2} + \frac{2m}{\hbar^2}(E - V_0)\psi = 0 \tag{21}$$

For simplification purpose consider,

$$(i) \quad \frac{2m}{\hbar^2}(E) = \alpha^2$$

$$(ii) \quad -\frac{2m}{\hbar^2}(E - V_0) = \beta^2 \tag{22}$$

By putting the value of Eq. (22) in Eqs. (20) and (21) we got,

$$\frac{d^2\psi}{dx^2} + \alpha^2\psi = 0 \tag{23}$$

$$\frac{d^2\psi}{dx^2} - \beta^2\psi = 0 \tag{24}$$

Solution of the above mentioned Eqs. (23) and (24) is,

$$\psi = eikxUk(x) \tag{25}$$

where the $U_k(x)$ represents periodicity of the wave function ψ. Differentiating Eq. (25) twice with respect to x we got,

$$\begin{aligned}\frac{d^2\psi}{dx^2} &= e^{ikx}\frac{d^2U_{k(x)}}{dx^2} + ike^{ikx}\frac{dU_k(x)}{dx} - k^2U_ke^{ikx} + ike^{ikx}\frac{dU_{K(x)}}{dx}\\ &= e^{ikx}\frac{d^2U_{k(x)}}{dx^2} + 2ike^{ikx}\frac{dU_k(x)}{dx} - k^2U_ke^{ikx}\end{aligned} \tag{26}$$

After putting the value of Eq. (26) in Eqs. (23) and (24) and also by dividing thorough-out by e^{ikx} we get,

$$\frac{d^2U_{k(x)}}{dx^2} + 2ik\frac{dU_{k(x)}}{dx} + \left(\alpha^2 - k^2\right)U_{k(x)} \tag{27}$$

$$\frac{d^2U_{k(x)}}{dx^2} + 2ik\frac{dU_{k(x)}}{dx} - \left(\beta^2 + k^2\right)U_{k(x)} \tag{28}$$

The general solution of Eqs. (27) and (28) is,

$$U_1 = {}_{Ae}i(\alpha - k)x + {}_{Be} - i(\alpha + k)x \tag{29}$$

$$U_2 = {}_{Ce}(\beta - ik)x + {}_{De} - (\beta + ik)x \tag{30}$$

On solving Eqs. (29) and (30) we get,

$$\frac{P}{\alpha a}\sin\alpha a + \cos\alpha a = \cos ka \tag{31}$$

This Eq. (31) is Kronig Penny model. Where P is power of potential and a is the inter-atomic distance.

5.2.1 Case: 1

Where P tends to ∞ ($\lim_{p\to+\infty}$), Dividing 'P' throughout Eq. (31),

$$\frac{\sin\alpha a}{\alpha a}+\frac{\cos\alpha a}{P}=\frac{\cos ka}{P} \tag{32}$$

Applying P tends to ∞ ($\lim_{p\to+\infty}$): $\sin\alpha a = 0$;
The general solution is, $\alpha a = n\pi$,

$$\alpha=\frac{n\pi}{a} \tag{33}$$

Therefore, we can also write,

$$\alpha^2=\frac{n^2\pi^2}{a^2} \tag{34}$$

But from (22) we can write,

$$\alpha^2=\frac{2m}{\hbar}(E) \tag{35}$$

Equating the R.H.S. of Eqs. (34) and (35) we got,

$$\frac{n^2\pi^2}{a}=\frac{2mE}{\hbar^2} \tag{36}$$

By rearranging the Eq. (36) we got,

$$E=\frac{n^2\hbar^2}{8ma^2}\text{ where, }\left(\frac{\hbar}{2\pi}\right) \tag{37}$$

Here, n = order = 1, 2, 3, 4 ……… m = mass of electron a = inter-atomic distance.

In the above mentioned case 1 ($P \to \infty$) represent the nature of insulator. The Eq. (36) is the desired equation for insulator.

5.3 *Case: 2*

Where P is tends to 0 ($\lim_{p\to 0}$), in the Eq. (31),

$$\cos\alpha a=\cos ka$$
$$\alpha=k$$
$$\alpha 2=k2 \tag{38}$$

But, from the Eq. (22) we can write,

$$\alpha^2 = \frac{2m}{\hbar^2}(E) \tag{39}$$

From Eqs. (38) and (39) we got,

$$\frac{4\pi^2}{\lambda^2} = \frac{2mE}{\hbar^2} \tag{40}$$

After rearrangement of Eq. (40) we finally got that,

$$E = \frac{h^2}{2m\lambda^2} \text{ where, } \left(k = \frac{2\pi}{\lambda}\right) \tag{41}$$

After rearrangement of Eq. (41) we got,

$$\begin{aligned} E &= \frac{\frac{h^2}{\lambda^2}}{2m} \\ &= \frac{p^2}{2m} \end{aligned} \tag{42}$$

In the above mentioned case 2 ($P \rightarrow 0$) represent the nature of conductor. The Eq. (42) is the desired equation for conductor.

6 Semiconductor Based on Quantum Structure

One of the greatest inventions in the field of compound semiconductors we can consider as semiconductor quantum well structure. The ultra-thin semiconductor films are present alternatively in the quantum well structure where electrons and holes are cramped. As a result, new properties are shown by the materials due to the two-dimensional behaviour of the electron and holes. Esaki and Tsu had suggested this in 1969 [22]. In the initial level of development of QW technologies, the carrier transport performed in quasi-two-dimensional system. And many devices are discovered e.g. high electron mobility transistor (HEMT) by using modulation doping technology. By applying wave function engineering, a resonant tunneling transistor was made. There also many optical devices were invented e.g. quantum well laser, quantum dot laser, quantum cascade laser etc. [23].

The basics structure of quantum well is visualized in Fig. 1.

Figure 1 represents a heterostructure constructed from semiconductors AlGaAs (large band-gap) and GaAs (smaller bandgap) in a quantum well configuration. In the central GaAs region of length d, the conduction band energy is lower, and the valence band energy is higher. Therefore both electrons and holes can be confined in the GaAs

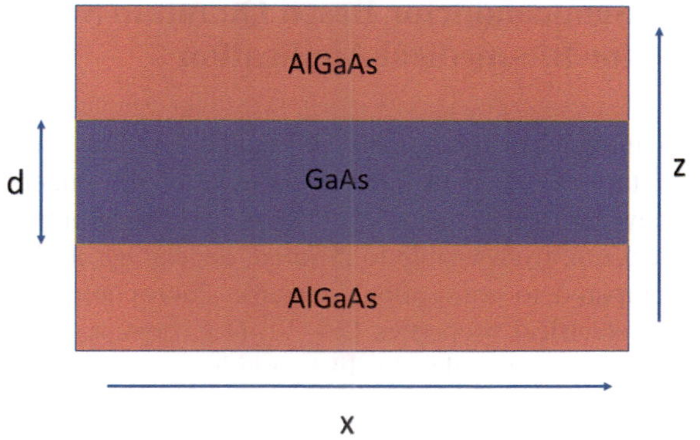

Fig. 1 Compound semiconductor based heterostructure

region. The basic concept of quantum well device is, a thin semiconductor layer having lower band-gap is designed between two thin semiconductor layer having large bandgap. The band-gap of Fig. 1 can be realized through Fig. 2.

Figure 2 showing the band-gap of AlGaAs-GaAs semiconductor based quantum well structure.

These quantum well devices are better option for many electronic devices [24]. As the width of the semiconductor material is harmonious with the de-Broglie wave length, as a result, the size quantization occurs. And also inside the small band-gap semiconductor, the electron and holes will be confined in which the potential energy is lowest. Since most often the thickness of the small band-gap [25] material is very less like 100°A, hence a smaller drive current will lead to population inversion and lasing a phenomenon observed in quantum lasers.

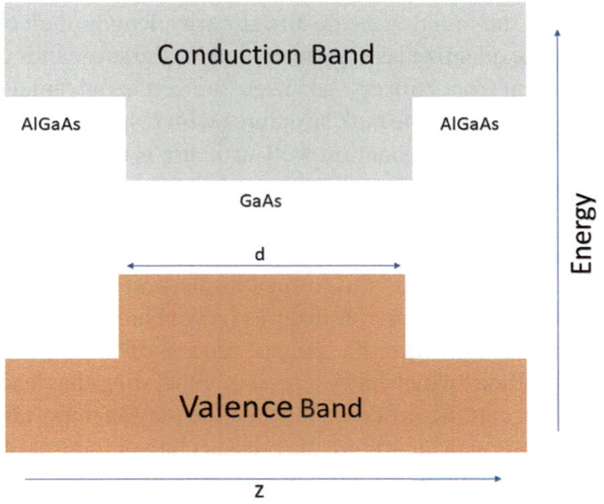

Fig. 2 The band structure diagram in a quantum well of GaAs in between AlGaAs. An electron in the conduction band or a hole in the valence band can be confined in the potential well created in the structure. The present states in the wells are sketched in the figure. These are "particle-in-a-box-like" states

7 Semiconductor Based Quantum Structure for Bio-medical Application

Quantum dots (QDs) colloidal nanocrystalline semiconductors exhibit distinctive photophysical characteristics because of the quantum confinement phenomenon. They emit different sorts of wave lengths throughout a broad range of the light spectrum, from visible to infrared, depending on their sizes and chemical makeup. Compared to conventional organic fluorophores, QDs exhibit distinctive optical and electrical properties [26–30] (e.g. organic dyes and fluorescent proteins). To biorecognition molecules like peptides, antibodies, nucleic acids, or small molecule legends, the enormous surface area of QDs is particularly advantageous. It is a revolution from electronics material science to biological applications as a result of these extraordinary properties [31–37]. The use of fluorescent labels for cellular labelling, intracellular sensors, deep-tissue and tumour targeting and imaging agents, sensitizers for photodynamic therapy (PDT), vectors for gene therapy, magnetic resonance imaging (MRI) contrast agents, and other uses are current and anticipated applications of QDs [37–41]. This study primarily describes the evolution of QD synthesis, surface modification, and toxicity, and briefly focuses on the advancements of QD applications in the bio-medical field [41–45].

8 Proposed Quantum Structure for Measurement of Argon Fluid

A quantum well is a kind of potential well which has discrete energy levels, where particles can move in hither and thither directions. The thickness of the quantum well is the order of the de-Broglie wavelength of electrons and holes. Moreover, the discrete quantized energy levels lead to the sub-bands of energy levels which are quite different from bulk crystal. There are certain advantages of the quantum well devices as compared to the bulk structures with respect to speed (faster), economical (cheap) and others. The quantum well structure is similar to the three layers structure of the photonic crystal. Besides this, many different types of quantum well structures have also been discussed in the literature. Again considering the sensor with respect to the medical, the concentrations of argon in their fluids have played a vital role. Basically, Argon based fluid is used for bio-medical purposes for destroying cancerous cells through cryosurgery procedures (cryoablation). Moreover, it is also used for plasma beam electrosurgery through the process of "argon-enhanced coagulation". However, the methods which carry a peril of generating that leads to clot the blood circulation. So concentrations of argon in the fluid play an important role. Keeping the importance of the argon in their fluid, the present chapter focuses on the realization of bio-medical sensor using single quantum well structure.

The proposed quantum well structure is a GaAs semiconductor based quantum structure, which is indicated in Fig. 3.

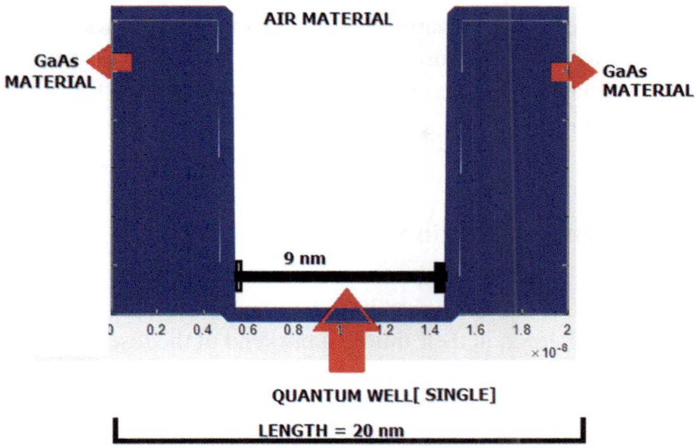

Fig. 3 GaAs based single quantum well structure

The single quantum well structure consists of three layers such that 1st and 3rd layer is made of GaAs material where middle layer is similar to the well which is called a quantum well. In this layer, different concentration of argon in their fluid is infiltrated. The thickness of well is about 9 nm where the width of the 1st and 3rd layers is taken at 5.5 nm each.

To realize the argon fluid sensor, Fig. 4 discusses the operational mechanism, which is indicated below.

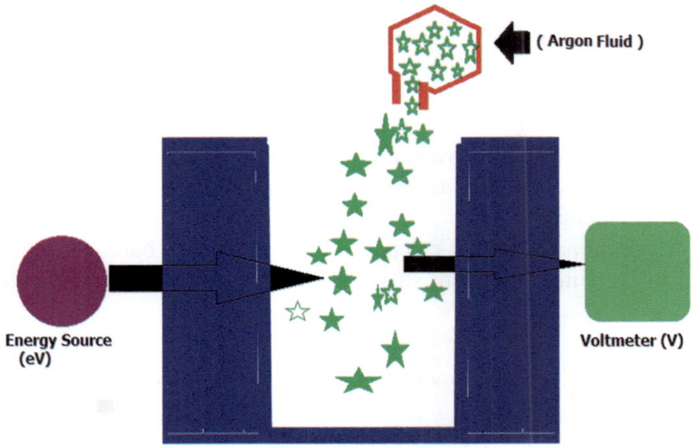

Fig. 4 Operational mechanism of argon fluid sensor

In Fig. 4; a signal of 1.42 eV energy is applied to the single well structure then said energy interacted with the structure of GaAs materials as well as different concentrations of argon. Finally, the energy transmitted from the structure and collected at the voltmeter in terms of 'Volt'.

9 Result and Interpretation

After analyzing Figs. 1 and 2, it is realized that the transmitted signal pertaining to each concentration of argon in their fluid is a backend of the research work. To carry out the same, the amount of signal lost in the middle layer has a vital role. Basically, the amount of loss relies on the band-gap of the entire three-layer structure with respect to the input source, which is taken at 1550 nm and such band-gap is a function of both nature of material and configuration of the structure. As far as the nature of the material is concerned, research deals with Si as first and third layers where the air is, in the second layer such that different concentrations of argon in their fluid have been infiltrated when input energy in terms of electron volt or eV is incident on the structure. The same amount of energy gets lost in form of reflection which is found in terms of the energy bandgap. Similarly focusing on the configuration of the structure, the thickness of first and third layers was taken at 5.5 nm whereas the middle layer took 9 nm. Again concentrating on the energy band of the structure, it is understood that the band-gap differs from different concentrations of argon in their fluid. For example; the band-gap for the concentration for 1.0 g/cm^3 is not the same as 1.3 g/cm^3 and is similar for others. As far as the mathematics of band-gap is concerned, it depends upon the kronig penny model which is derived in the preceding section. So, considering Eqs. (29)–(42), the bandgap is computed to each concentration i.e. for 0, 0.5, 1.0, 1.3, 1.5 g/cm^3 of argon in their argon fluid solution. The physics of the same depends upon the variation of the effective mass of e^- with their concentration, which is indicated in Table 1.

Table 1 represents the effective mass with respect to their different concentrations of argon. The outcome of the same result is indicated given in Fig. 5a–e for the above concentration respectively.

The aforementioned Fig. 5a–e represent the electronic band-gap energy of the three-layer of the single quantum structure. In this case, the vertical axis indicates

Table 1 Effective mass of the electrons in the quantum structure with concentration

Concentration (g/cm^3)	Effective mass (m_o^*/m)
0	1
0.5	0.7
1.0	0.48
1.3	0.45
1.5	0.42

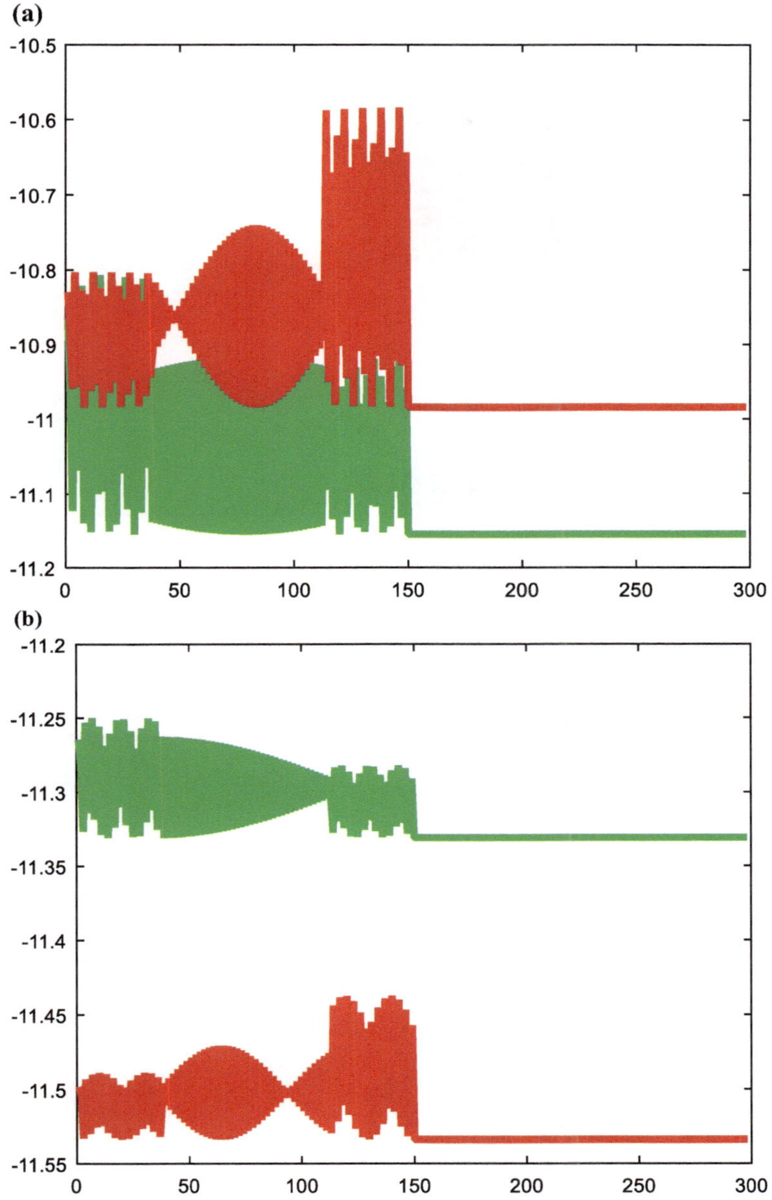

Fig. 5 **a** Electronic bandgap of quantum well for '0' (gm/cm^{-3}) concentration of argon. **b** Electronic bandgap of quantum well for '0.5' (gm/cm^{-3}) concentration of argon. **c** Electronic bandgap of quantum well for '1.0' (gcm^{-3}) concentration of argon. **d** Electronic bandgap of quantum well for '1.3' (gm/cm^{-3}) concentration of argon. **e** Electronic bandgap of quantum well for '1.5' (gm/cm^{-3}) concentration of argon

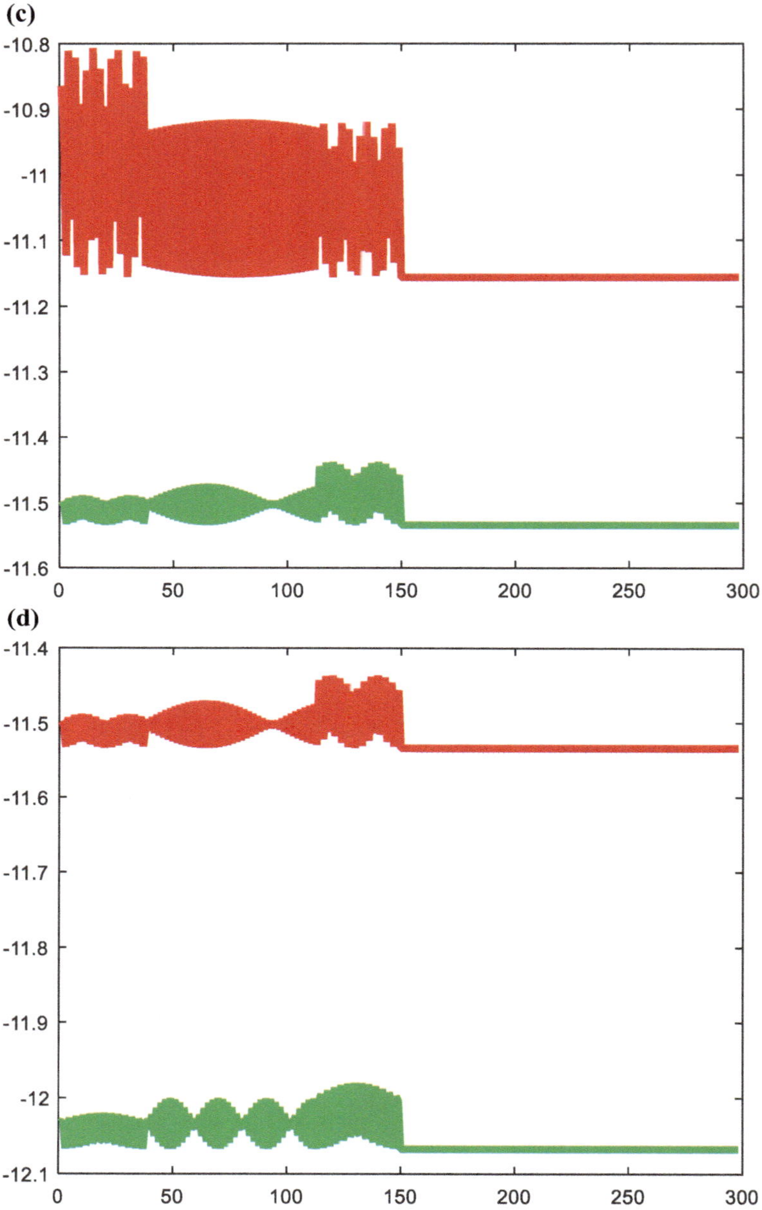

Fig. 5 (continued)

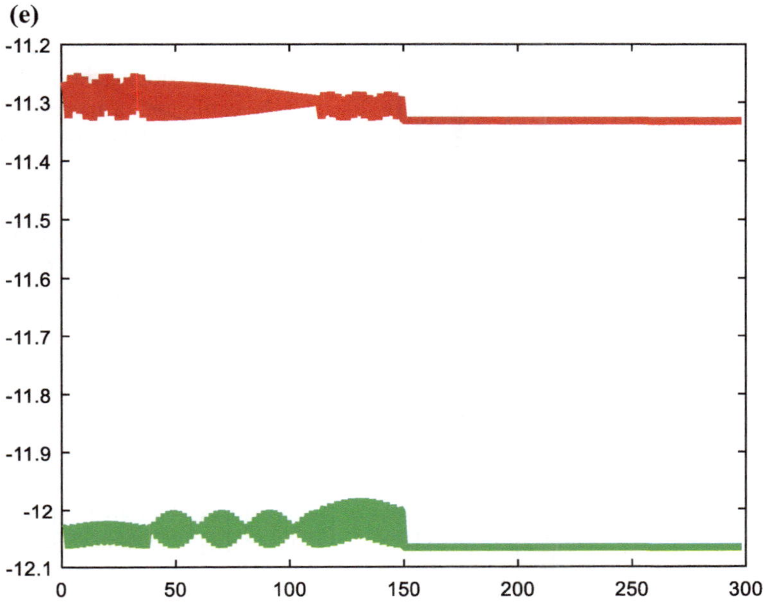

Fig. 5 (continued)

energy (in eV) and the horizontal axis represents the wave vector. Here the lower energy band (green color) represents the valence band whereas the red color represents the conduction band. Further, the bandgap is defined as the energy differences between the top of the valence band and the bottom of the conduction band which is found to differ from different figures. For example, a no-band gap is found in Fig. 5a because the valance band and conduction band overlap with each other. It indicates 0 bandgaps are found for the concentration 0 g/cm^3 argon of the argon fluid. Using the above definition of the electronic band-gap is computed for each concentration. The same energy is indicated in the Table 2.

In Table 2; the value of concentration and band-gap took along column 1 and column 2 respectively. Here it is found that the electronic bandgap increases with the increase in concentration. Then a graph is plotted between the electronic bandgap and concentration which is indicated in Fig. 6.

Table 2 Variation of electronic band-gap with different concentration

Concentration	Band-Gap (eV)
0	0
0.5	0.12
1.0	0.29
1.3	0.48
1.5	0.65

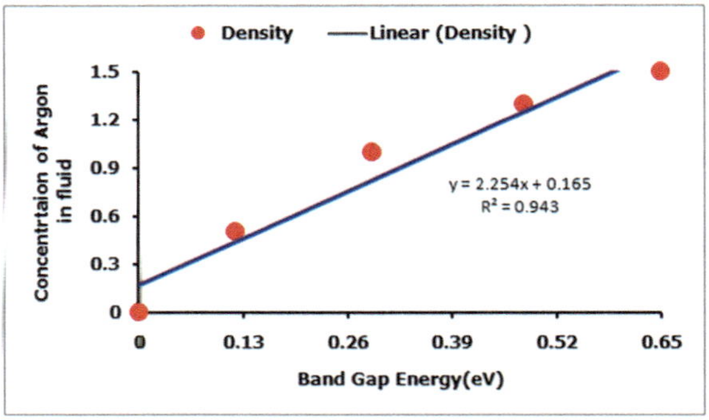

Fig. 6 Variation of bandgap energy with concentration

Table 3 Variation of output potential with concentration

Concentration	Output potential
0	1.42
0.5	1.3
0.7	1.13
1.3	0.94
1.5	0.77

In Fig. 6, bandgap energy is taken along the horizontal axis where the concentration is assumed along the vertical axis. From this figure, it is observed that band-gap energy varies linearly with the concentration of argon.

Then finally the transmitted energy (incident energy—electronic bandgap energy) can be obtained with respect to each concentration the same computed and then corresponding to the potential is detected at the photo detector. The entire result is shown indicated in Table 3.

Table 3 represents variation of concentration & corresponding output potential. The same is also shown in the graphical form in Fig. 7 with respect to the potential.

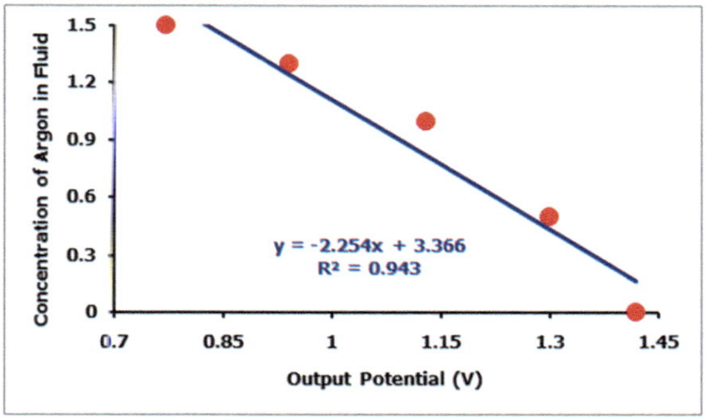

Fig. 7 Variation of output potential with respect to the concentration

In this curve, the transmitted potential is taken along the horizontal axis whereas the concentration is taken along the vertical axis. In this case, it is realized that output potential decreases linearly with respect to the argon concentration. So, potential decreases from 1.42 to 0.77 Vas the concentration increases from 0 to 1.5 g/cm^3. Apart from this, we have also fitted these two parameters through a linear trend line system and the equation (Mathematical Model) is inset in the figure. Besides this, it is understood that the linearity of the trend line is bound to 94.3%. Keeping the analysis it is understood that one can compute the argon concentration by knowing the output potential.

10 Conclusions

The assessment of the argon concentration in their fluid is made using silicon-based three layers of quantum structures through the principle of electronic bandgap variation with respect to the concentration of argon which ranges from 0 to 1.5 gm/cm^3. The KronigPenny mathematical model is a key factor to compute the electronic bandgaps [46]. The physics of the work relies on the variation of an effective mass whereas the mathematics of the research deals with the configuration of the structure. The numerical calculation shows that the electronic band-gap increases from 0 to 0.65 eV with respect to the concentration varies from 0 to 1.5 gm/cm^3, where transmitted potential decreases from 1.42 to 0.77 eV with the regression factor of 0.9430.

References

1. Quantum Well Infrared Photon Detectors|IRnova. www.ir-nova.se. Accessed 4 Sept 2018
2. E.O. Odoh, A.S. Njapba, A review of semiconductor quantum well devices. Adv. Phys. Theor. Appl **46**, 26–32 (2015)
3. H. Kroemer, A proposed class of hetero-junction injection lasers. Proceed. IEEE Inst. Electr. Electron. Eng. (IEEE) **51**(12), 1782–1783 (1963). https://doi.org/10.1109/proc.1963.2706, ISSN 0018-9219
4. Z.I. Alferov, R.F. Kazarinov, Authors Certificate 28448 (U.S.S.R) (1963)
5. M. Fox, R. Ispasoiu, *Quantum Wells, Superlattices, and Band-Gap Engineering*. Springer Handbook of Electronic and Photonic Materials (Springer, US, 2006), pp. 1021–1040. https://doi.org/10.1007/978-0-387-29185-7_42, ISBN 978-0-387-26059-4
6. B.R. Nag, *Physics of Quantum Well Devices* (Kluwer Academic Publishers, 2002). OCLC 754036669
7. S.H. Simon, *The Oxford Solid State Basics* (Oxford University Press, 2017). ISBN 978-0-19-968077-1, OCLC 1091723162
8. I. Sayed, S.M. Bedair, Quantum well solar cells: principles, recent progress, and potential. IEEE J. Photovoltaics **9**(2), 402–423 (2019). https://doi.org/10.1109/JPHOTOV.2019.2892079, ISSN 2156-3381. S2CID 67874610

9. D. Tang, H. Zhang, L. Zhao, X. Wu, Observation of high-order polarization-locked vector solitons in a fiber laser (PDF). Phys. Rev. Lett. **101**(15), 153904 (2008). arXiv:0903.2392, Bibcode:2008PhRvL.101o3904T, https://doi.org/10.1103/PhysRevLett.101.153904, PMID 18999601, S2CID 35230072. Archived from the original (PDF) on January 20, 2010
10. Scientists propose quantum wells as high-power, easy-to-make energy harvesters. Phys.org. Accessed 24 Oct 2013
11. B.R. Sothmann, R. Sánchez, A.N. Jordan, M. Büttiker, Powerful energy harvester based on resonant-tunneling quantum wells. New J. Phys. **15**(9), 095021 (2013). arXiv: 1309.7907, Bibcode:2013NJPh...15i5021S, https://doi.org/10.1088/1367-2630/15/9/095021, S2CID 119210320
12. K. Barnham, A. Zachariou, Quantum well solar cells. Appl. Surf. Sci. **113–114**, 722–733 (1997). Bibcode:1997ApSS..113..722B, https://doi.org/10.1016/S0169-4332(96)00876-8
13. S.M. Ramey, R. Khoie, Modeling of multiple-quantum-well solar cells including capture, escape, and recombination of photoexcited carriers in quantum wells. IEEE Trans. Electron. Dev. **50**(5), 1179–1188 (2003). Bibcode:2003ITED...50.1179R, https://doi.org/10.1109/TED.2003.813475
14. D. Derkacs, W.V. Chen, P.M. Matheu, S.H. Lim, P.K.L. Yu, E.T. Yu, Nanoparticle-induced light scattering for improved performance of quantum-well solar cells. Appl. Phys. Lett. **93**(9), 091107 (2008). Bibcode:2008ApPhL..93i1107D, https://doi.org/10.1063/1.2973988
15. M. Fox, R. Ispasoiu, S. Kasap, C. Safa, P. Capper (eds.), *Quantum Wells, Superlattices, and Band-Gap Engineering*. Springer Handbook of Electronic and Photonic Materials (Springer International Publishing, 2017), p. 1. https://doi.org/10.1007/978-3-319-48933-9_40, ISBN 978-3-319-48931-5
16. K. Barnham, I. Ballard, J. Barnes, J. Connolly, P. Griffin, B. Kluftinger, J. Nelson, E. Tsui, A. Zachariou, Quantum well solar cells. Appl. Surf. Sci., Proc. Eighth Int. Conf. Solid Films Surf. **113–114**, 722–733 (1997). Bibcode:1997ApSS..113..722B, https://doi.org/10.1016/S0169-4332(96)00876-8, ISSN 0169-4332
17. N.G. Anderson, Ideal theory of quantum well solar cells. J. Appl. Phys. **78**(3), 1850–1861 (1995). Bibcode: 1995JAP....78.1850A, https://doi.org/10.1063/1.360219, ISSN 0021-8979
18. H. Asai, K. Oe, Energy band-gap shift with elastic strain in GaxIn1−xP epitaxial layers on (001) GaAs substrates. J. Appl. Phys. **54**(4), 2052–2056 (1983). https://doi.org/10.1063/1.332252.ISSN0021-8979
19. S. Adachi, Material parameters of In1−xGaxAsyP1−yand related binaries. J. Appl. Phys. **53**(12), 8775–8792 (1982). https://doi.org/10.1063/1.330480.ISSN0021-8979
20. *Quantum Wells, Wires and Dots* (Wiley, 2006), pp. 243–270. https://doi.org/10.1002/0470010827.ch8, ISBN 978-0-470-01082-2
21. R.L. Anderson, Germanium-gallium arsenide heterojunctions [letter to the editor]. IBM J. Res. Dev. **4**(3), 283–287 (1960). https://doi.org/10.1147/rd.43.0283.ISSN0018-8646
22. J. Nelson, M. Paxman, K.W.J. Barnham, J.S. Roberts, C. Button, Steady-state carrier escape from single quantum wells. IEEE J. Quant. Electron. **29**(6), 1460–1468 (1993). Bibcode:1993IJQE29.1460N, https://doi.org/10.1109/3.234396, ISSN 0018-9197
23. R. King, D. Law, C. Fetzer, R. Sherif, K. Edmondson, S. Kurtz, ... & N.H. Karam, Pathways to 40%-efficient concentrator photovoltaics, in *Proceedings of 20th European Photovoltaic Solar Energy Conference* (2005), pp. 10–11
24. H. Fujii, K. Toprasertpong, Y. Wang, K. Watanabe, M. Sugiyama, Y. Nakano, 100-period, 1.23-eV bandgap InGaAs/GaAsP quantum wells for high-efficiency GaAs solar cells: toward current-matched Ge-based tandem cells. Prog. Photovoltaics Res. Appl. **22**(7), 784–795 (2014). https://doi.org/10.1002/pip.2454
25. S. Heckelmann, D. Lackner, C. Karcher, F. Dimroth, A.W. Bett, Investigations on AlxGa1-x as solar cells grown by MOVPE. IEEE J. Photovoltaics **5**(1), 446–453 (2015). https://doi.org/10.1109/jphotov.2014.2367869.S2CID41026351
26. N. Jain, J.F. Geisz, R.M. France, A.G. Norman, M.A. Steiner, Enhanced current collection in 1.7 eV GaInAsP solar cells grown on GaAs by metalorganic vapor phase epitaxy. IEEE J. Photovoltaics **7**(3), 927–933 (2017). https://doi.org/10.1109/jphotov.2017.2655035.OSTI1360894.S2CID20841656

27. I.E.H. Sayed, N. Jain, M.A. Steiner, J.F. Geisz, S.M. Bedair, 100-period InGaAsP/InGaP superlattice solar cell with sub-bandgap quantum efficiency approaching 80%. Appl. Phys. Lett. **111**(8), 082107 (2017). Bibcode:2017ApPhL.111h2107S, https://doi.org/10.1063/1.4993888, OSTI 1393377
28. O.O Emmanuel, S.N. Augustine, A review of semiconductor quantum well devices, **46** (2015). ISSN 2224-719X (Paper), ISSN 2225-0638 (Online)
29. T. Tsuyama, Development of various semiconductor quantum devices
30. L. Esaki, et al., Superlattice and negative differential conductivity insemiconductors. IBM J. Res. Dev. **14**(1), 61 (1970)
31. R. Dingle et al., Electron mobilities in modulation-doped semicon-ductor heterojunction superlattices. Appl. Phys. Lett. **33**, p665 (1987)
32. T. Mimura et al., A new field-effect transistor with selectively doped GaAs/n-Alx Ga1-x as heterojunctions. Jpn. J. Appl. Phys. **19**, L225 (1980)
33. L.L. Chang et al., Resonant tunneling in semiconductor double barriers. Appl. Phys. Lett. **24**, p593 (1974)
34. J.P. van der Ziel et al., Laser oscillation from quantum states in verythin GaAs–Al0.2Ga0.8As multilayer structures. Appl. Phys. Lett. **26**, 463 (1975)
35. Y. Arakawa et al., Multidimensional quantum well laser and temperature dependence of its threshold current. Appl. Phys. Lett. **40**(11), p939 (1982)
36. H.I. Alferov, The history and future of semiconductors heterostructure. Am. Inst. Phys., 1–13 (1998)
37. P. Bhattacharya, *Semiconductor Optoelectronic Devices*, 2nd edn. (Prentice Hall India, New Delhi, 2002), pp 79, 128
38. S. Das, R.K. Navak, T. Sahu, A.K. Panda, Electron mobility enhancement in barrier delta doped asymmetric double quantum well structures, in *Proceedings of the 17th International Workshop on the Physics of Semiconductor Devices*, ed. by V.K Jain, A. Verma (Spriger International Publishing, Switzerland, 2014), p. 255
39. R. Dingle, W. Wiegmann, C.H. Henry, Quantum states of confined carriers in very thin AlxGa1xAs-GaAs-AlxGa1-x as heterostructures. Phys. Rev. Lett. **33**, 827–830 (1974)
40. L. Esaki, R. Tsu, Superlattice and negative conductivity in semiconductors. IBM Research Report No. RC-2418 (1969)
41. L. Esaki, R. Tsu, Superlattice and negative differential conductivity in semiconductor lasers. IBM J. Res. Dev. **14**, 61–68 (1970)
42. T. Fuji, S. Yamakoshi, K. Nanbo, O. Wada, S. Hiyamizu, MBE growth of extremely high quality GaAs/AlGaAs, GRIN-SCH lasers with superlattice buffer layers. J. Vac. Sci. Technol. **2**, 259261 (1984)
43. P.A. Govind, K.N. Dutta, *Semiconductor Lasers*, 2nd edn. (Von Nostrand Reinhold, New York, 1993), pp. 163, 450
44. J.S. Harris Jr., From bloch functions to quantum wells. Int. J. Mod. Phys. **4**(6), 1149–1179 (1990)
45. V. Kirti, O.P. Vishwakarma, M.A. Ramrakhiani, Brilliant exposition of quantum well lasers. Int. J. Eng. Stud. **2**(1), 1–6 (2010)
46. B.D. Mallins, Ultrafast dynamics in InAs quantum DOT and GaInNAs quantum well semiconductor heterostrutures. Ph.D. Thesis. University of St. Andrews (2007), pp. 19–32

A Study on Quantum Cryptography and Its Need

Vandani Verma

Abstract With the increase of e-payments methods and the essential requirement to keep everything secret during online transactions, the demand of cryptography has increased a lot. In cryptography, factorization of huge integers or calculation of the exponents of a large prime number using hands or by normal computers is very hard but quantum computer can easily solve them within a lesser time. There is a need to move from classical cryptography to quantum cryptography to cope up with the advancement of technology and to maintain the demand of high level of information security. This chapter focuses on quantum cryptography, its need, and quantum key distribution that allows two parties to secretly share their information under the nose of an attacker with high computational power. This chapter also focusses on BB84, Ekert 91 protocol and the detection of eavesdropper in quantum key distribution based on photon polarization. Also, discuss the quantum masked authentication protocols in detail.

Keywords Quantum cryptography · QKD · BB84 · Ekert 91 · Masked authentication

1 Introduction

Cryptography uses codes to protect information and its transmission. It comprises of Alice and Bob, two cryptographic couples. An encryption algorithm is used to encrypt the original message using encryption key and through the conventional channel we deliver it to the recipient in the form of cipher text (an encrypted text, which is only readable to sender and recipient). Then the recipient decrypts the message by applying the encryption algorithm in the reverse manner. Its security implies

V. Verma (✉)
Amity Institute of Applied Sciences, Amity University-125, Uttar Pradesh, Noida 201313, India
e-mail: vandaniverma@yahoo.com

R. Pandey et al. (eds.), *Quantum Computing: A Shift from Bits to Qubits*,
Studies in Computational Intelligence 1085,
https://doi.org/10.1007/978-981-19-9530-9_21

(a) Confidentiality: It provides the assurance that message was not available to unauthorized parties.
(b) Integrity: It provide assurance that message has not been altered by third party and its source is genuine.
(c) Authentication: It provide verification of identity of the person.

It is based on mathematical computation, bit rate depends on computational power, and bit storage is 2^n, n bit string. Due to increase of computing power with time, it requires upgradation. Its communication range is millions of miles, with less expenditure cost for implementation but it is easily crackable using quantum computers. Suppose for example, for payments through internet, your credit card is protected with RSA encryption and the algorithm which comprises RSA, can easily be cracked using Shor's algorithm, which is a quantum algorithm. With quantum computers, mathematics-based cryptography can be cracked. Thus, giving rise to the need of Quantum Cryptography because of the following reasons: (a) In Cryptography coding depends on mathematics, though to factorize a large integer or to calculate exponents of a large prime is very tough using hands but using quantum computers we can do it easily. (b) The same key is used for both encryption and decryption in symmetric cryptography. So, if that key is leaked to any unauthorized party, then the whole communication can be decrypted and altered. (c) 128-bit keys are mostly used in asymmetric cryptography though they are very strong and use high number of calculations to break it but with the use of quantum computers that use quantum physics to operate it can be easily breakable. We need quantum cryptography as there is a demand of high level of security for information technology and quantum cryptography is secure even if the eavesdropper has high computational skills and uses quantum computers.

In Quantum Cryptography to perform encoding decoding tasks, we use quantum mechanics. It is 'provable secure' and resistant to cryptography. Average bit rate is 1 mbps and its bit storage is one n-bit string. 10 miles is the maximum transmission range. It has high expenditure cost but it is unbreakable. In 1935, an experiment was performed by Einstein et al. [1] through which the completeness of quantum mechanics is challenged. The basis for why quantum cryptography works is explained. This is known as EPR paradox. The following requirement of a complete theory is stated by EPR as "there should be an element in physical theory corresponding to every element in physical reality". There are two assumptions which are stated by EPR paper-(a) Locality: suppose there are two distant objects A and B then they don't have any direct effect on each other. The properties of B cannot be affected by the outcome of measurements of A. (b) Realism: a reality is existed that is independent of the observer and specific properties of the object are unchanged by different types of measurements. Though both are violated in quantum cryptography. Stephen Wiesner and Gilles Brassard started the work under quantum cryptography in 1970 using ideas introduced in EPR. He considered some fundamental rules of quantum physics: We cannot measure polarization of photons in incompatible base at the same time, we cannot describe the individual quantum processes distinctly, and security is preserved as we cannot measure the quantum state without disturbing the system and not possible to duplicate the unknown quantum state.

The idea of coding's of Quantum conjugates was introduced by Wiesner in New York, at Columbia University. After rejection from IEEE information theory society, in 1983 it was printed in the SIGACT news. It was shown that how we can transmit two messages simultaneously by encrypting them in conjugate noticeable (polarization of photons is linear and circular) so that only one of them is received or decoded at a time. Then Bennett and Brassard worked on the findings of Wiesner in 1984 and proposed a procedure of safe transmissions that is today known as the BB84 protocol. An important example of quantum cryptography is Quantum key distribution (QKD). It is used for industry, banks, Government, military.

Rest of the chapter is organized as follows: In Sect. 2, we discuss about the quantum key distribution in detail along with it no cloning theorem and security of QKD; Sect. 3, we discuss about the BB8 protocol and various security attacks; Sect. 4 discusses about the Ekert protocol. Now, we focus upon the mask authentication protocols in Sects. 5, 6 and 7. We also provide with the various attacks on quantum cryptographic algorithms in Sect. 8 and present our observations in Sect. 9 with concluding remarks.

2 Quantum Key Distribution (QKD)

An important example of quantum cryptography is Quantum key distribution (QKD). It is based on quantum mechanics theory through which eavesdropper's presence can be detected at the time, when two users are exchanging any one-time pad encryption key. Bennett and Brassard [2] invented QKD in 1984. QKD is unconditionally secure as it is bound by the laws of physics and the attacker cannot break the cipher text, no matter how much time or computer power is available, it cannot be cracked even with quantum computers. Since only one person cannot decide the secret key, agreement of both parties' matter therefore QKD is also called as Quantum key agreement and quantum key establishment. Photons are quantum carriers and are zero mass elementary particles of light which are unpolarized. A photon can spin in any of the directions—vertically, horizontally, diagonally. Unpredictable factor is the base of quantum physics and Heisenberg uncertainty principle defines unpredictable factor as it states that it is not possible to measure both position and momentum of a moving photon at the same time. Quantum cryptographers use a source of unpolarized light called light emitting diodes (LEDs) to create photon as they can generate a string of photons. Polarizing filters which are quantum measuring apparatus, are used to polarize the photons. They allow photons of light to pass through the filter even if the polarization of light is same as that of the filter. There are two polarizing filters—rectilinear (vertical/ horizontal) and diagonal (45° left/ 45° right).

Rectilinear polarizing filter is represented by $\oplus$. If photons are 45° left polarized and we measure it using a vertical polarizing filter, then only 50% of photons pass through the filter and polarization of these photon changes to vertical (Fig. 1).

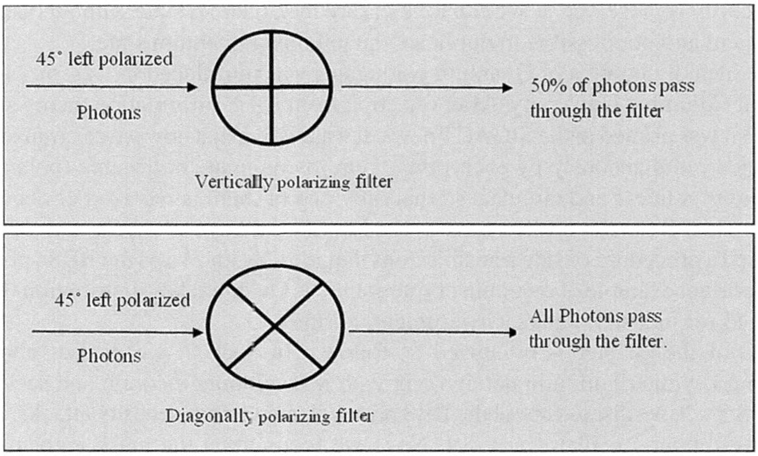

Fig. 1 Vertical and diagonal polarizing filter

Diagonally polarizing filter is represented by ⊗. When the photons are polarized at the same angle as the filter then all photons get through it.

If we estimate the angle of polarization of the particles using a filter, it changes the polarization of the photon and make it pass to the filter.

In this Alice and Bob require two channels to share their secret key—quantum channel (subjected to opponent) and another classical channel (which is authentic and subject to eavesdropper). The quantum channel used for transferring photons is optical fiber or open air e.g. satellite communication. The classical channel can be radio, phone, E-mail. **No cloning theorem**—A quantum state that is unrevealed, cannot be cloned (state whose parameters are not known) and the system will change if we take any measurement of an unknown system. Without disturbing the states, it is not possible to gain details about non-orthogonal quantum states. This is the main security of QKD, as the presence of an unauthorized party who is trying to gain any detail, can be detected by two communicating parties. Combination of QKD with one time pad encryption (crypto algorithm in which any random key is combined with original text. It is unbreakable), it provides long term security. Restriction of rate and distance in the links of QKD. For connecting multiple distant people together, quantum network is used which is composed of quantum links. It depends on QKD. For everyone who has entrance to the network, QKD provides them unconditional security. BB84, EPR, Ek91, B92, SSP, DPS, SARG04, COW, KMB04, S09 and S13 are some more protocols of quantum cryptography.

2.1 Security of Quantum Key Distribution

Though quantum cryptography is perfectly secure there are some loopholes in either the photon or in measurement of the polarizing filter. It is vulnerable to various side channel attacks and denial of service attack (the legal person is prevented by attackers from accessing the services). Without security proof, there will be an incomplete QKD system as there is no surety that the final key generated by us is secure. Attacks by Side channels: a natural example of this is supposed to be a group of soldiers that are using encrypted signals of the torch to communicate with the other group of soldiers. To know the position of the soldiers, attackers just need to physically identify the light of the torch, without decrypting any code in the torch signals. In this no encryption method gets failed but its physical implementation leads to side channel leaking information. They are attacks through which information is gained by computer systems without disturbing the cryptographic algorithms. Some main side channel attacks are

- Conventional side channel attack: security loopholes are on security threats which are linked to non-quantum aspects of QKD. They are leaking internal state's information of QKD hardware of Alice and Bob.
- Trojan horse attack: this eavesdropper monitors the reflected light and by this information the internal state is learned.
- Key elements used for the implementation of discrete variable QKD are Single particle detectors. Example avalanche photodiodes (APDs) which are operated on trigger modes. Many attacks occur on imperfection of APDs.
- As a single photon source is not commercially available, so instead of this, a weak laser diode is used. More than one photon is contained by this weak laser diode which constructs the condition for Eve to acquire knowledge of all the details about encrypted data.
- Laser source's intensity, quantum channel's length and time taken to detect a single photon. They are calibrated and take part in the safety of the system. So attack on the calibration procedure leads to security breach.
- Side channels on random number generators: in this Eve tries to gain knowledge about the randomness generated to run QKD protocols by Alice and Bob.

3 BB84 Protocol

Bennett and Brassard in [2] 1984 invented BB84. It is a cryptography which relies on laws of quantum mechanics. A quantum channel e.g. optical fiber and a classical channel e.g. phone, internet are used to join Alice and Bob. Pulses of polarized light are used and a single photon is contained by each pulse. Practically, we can also use the same link for both channels like we can use optical fiber which differs only in intensity of light pulse i.e. hundreds of photons per bit for the classical channel and one photon per bit for the quantum channel.

There are two types of bases

- Horizontal-vertical bases ⊖
  - Bit value 0: horizontal polarization ↔
  - Bit value 1: vertical polarization ↕
- Diagonal bases ⊗
  - Bit value 1: 45° right polarization ↙
  - Bit value 0: 45° left polarization ↘

Alice randomly chooses the basis as well as each photon's polarization and send this to Bob through a quantum channel. Here instead of transferring bits, the quantum channel transfers qubits and takes advantage of special properties of the qubit (we cannot copy a qubit and we cannot decide even if a qubit is created by Horizontal) vertical base or Diagonal base. For every photon, Bob chooses independently and randomly among one of the two basis. He either measures with the same bases as Alice to get a perfect correlated result or he measures with different bases and get uncorrelated result. Sometimes he is not able to measure the photons due to error in transmission or detection. Then a raw key is obtained by Bob which is a string of all the received bits. Then via the classical channel, Bob communicates with Alice about which bases he has used, and which photons are measured but he will not reveal anything about the result obtained. Alice and Bob then differentiate their chosen bases and keep the bits which correspond to similar bases. Due to randomness in the data, about 50% of the raw keys are discarded as they get correlated and uncorrelated results with same probability. This shorter key obtained is called sifted key. To check error rate which is due to technical imperfection in the system and eavesdropper, they randomly choose a few of persisting bits that are rejected afterwards. So remaining string of bits obtained after correcting error is called the secret key (Table 1).

Table 1 Process of BB84

Alice polarization state	↔	↕	↔	↘	↙	↕
Alice sequence	0	1	0	0	1	1
Randomly chosen encoding scheme by Alice or Alice's bases	(+)	(+)	(+)	(×)	(×)	(+)
Bob's bases	(+)	(×)	(+)	(×)	(+)	(+)
Measured state by Bob	↔	↘	↔	↘	↕	↕
Bit values by Bob	0	0	0	0	1	1
Correlated result	Yes	No	Yes	Yes	No	Yes
Remaining bits	1		0	0		1

3.1 Detecting: The Eavesdropper, Eve

Suppose an eavesdropper Eve tries to gain information and she intercepts the Quantum channel. She measures the photon with a polarizing filter and decodes the bit value for it and creates another photon for the same bit value and sends it to Bob. She must send a photon to Bob otherwise they will discard the measurement as he doesn't receive the qubit. So, Eve tries to transmit a qubit in its real state but due to the theorem of no cloning, copying the unknown quantum state is not possible. So, she will send it using the intercept resend strategy. As like Bob, she does not know which bases Alice used for measuring the states. So, in 50% of cases, she will guess the correct base and send the same base to Bob and her presence will not be noticed but in cases where she wrongly guesses the base, her presence gets noticed because of uncorrelated results between Alice and Bob. Rate of inaccuracy in the sifted key shows the existence of the eavesdropper. They can protect the transmission key using parity check (it is a process which ensures that during communication, only accurate data should be transmitted between nodes) in which they discuss the exact result and discard all the incorrect measurements.

3.2 Distilling a Secret Key

There are imperfections in the physical implementation and errors occur for many reasons like noise or losses in quantum channels, eavesdropping and wrong measurement using polarizing filter. Suppose Eve knows about 25% of the key element called as the estimation part of the protocol. Alice and Bob know that the key element contains some error rate e and a potential eavesdropper knows about 25% of details. So, they make use of the classical channel to take the bits that are secured and discard the rest and this is called as secret key distillation. It consists of reconciliation and privacy amplification. By reconciliation, they improve the transmission inaccuracy and wipe Eve's awareness of the key by reducing the size of the key by a procedure called privacy amplification. To maximize key length and to lower Eve awareness of the key, it is significant that only little information just to correct their errors is reveal during reconciliation. The final key is called the secret key which is used for cryptographic purposes by Alice and Bob.

4 Ekert 91 Protocol

A distinct perspective to quantum key distribution called quantum entanglement [it is unable to define the quantum state of one object without any mention of quantum state of another object, also we cannot make any conclusions on independent states of object] which is based on unique quantum correlation is developed by Ekert [3] in 1991. Ekert protocol is also called as the EPR protocol as there is a direct link to EPR paradox. In this Alice and Bob are sending particles using a central source that create entangled particle and send one to Bob and one to Alice. Due to the distance limitation factor in quantum cryptography, for its practical implementation, we need a central source like satellite that can send signals to multiple recipients at the same time. Quantum states called spin singlets are used. A protocol is set up by him, in which a pair of spin-1/2 particles (particles must fully rotate twice through 720° before it has same configuration) are emitted by the source in the singlet stage

$$\theta = \frac{1}{\sqrt{2}}(\downarrow\uparrow + \uparrow\downarrow)$$

It shows that the spin of particle A is pointing down and of particle B is pointing up, in state 1. Also the spin of particle A is pointing up and of particle B is pointing down, in state2. This can be called as superposition of states because we are unaware about the way in which either particle is spinning. Both Alice and Bob randomly choose between three bases which are represented by $p_a(a = 1, 2, 3)\, and\, q_b(b = 1, 2, 3)$. If in the z direction, the particles are travelling, then p_a and q_b are perpendicular to the trajectories of the particles and therefore located in the x–y plane; this is because light waves are recognized as transverse waves and electromagnetic oscillations are at 90° to the direction in which the light travels, the vectors p_a and q_b can be acquired by spinning vertical- horizontal bases about x-axis by angles

$\theta_1^p = 0, \theta_2^p = 45, \theta_3^p = 90°$ for Alice and $\theta_1^q = 45°, \theta_2^q = 90°, \theta_3^q = -45°$ for Bob.

Here orientation of Alice and Bob analyzers is given by p and q, respectively. The orientation of the analyzers or basis for every pair of particles is independently and randomly chosen by the user. After this, they publicly announce which basis they have chosen to measure every photon and then they separate these measurements into groups.

- Group A: In which, same bases are used by them to measure the particles
- Group B: In which, different bases are used by them to measure the particles.

If compatible bases are selected by Alice and Bob, then the measurement will be anti-correlated e.g., if Alice chose a spin down particle in state 1, then there is 100% probability that Bob will detect a spin up particle. It means that particles of Bob know how Alice has measured its particles and align themselves accordingly. Since particles of both Alice and Bob know one's presence, due to this rule of locality and realism in EPR, are disobeyed by quantum entanglement. Bob measurement outcome will be random, if non-compatible bases are chosen and Alice measures the particles in one of bases. Then they publicly announce about bases used by them to measure the particles and don't reveal any information about the outcome of the measurement and then discard all incompatible bases. This shorter key is called the sifted key. Within this, there are two results from every measurement.

- For spin up-bit value 1
- For spin down-bit value 0

It is almost impossible for the eavesdropper to get any information about the key, as entangled states are used. For example, measurement at 45° is decided by both Alice and Bob and suppose Alice discovered a spin down particle then there are chances that a spin up particle is detected by Bob but if the particle, in the process of detecting it, gets destroyed by the eavesdropper, Eve. She recreates a particle and sends it to Bob. But the orientation of that particle will be random. So, Bob will get error if the particle didn't correlate with Alice. Without leaking any information about the secret keys, the sender wants to detect the presence of the eavesdropper; therefore Ekert used 6 states.

$$E(p_a, q_b) = P_{++}(p_a, q_b) + P_{\pm}(p_a, q_b) + P_{-}(p_a, q_b) - P_{\mp}(p_a, q_b)$$

Above is the correlation coefficient of Alice and Bob's measurement where the probability of obtaining $\pm 1 along p_i$ and $\pm 1 along q_j$ is $P_{\pm}$. When analyzers are used with different orientations (incompatible basis)

$$S = E(p_1, q_1) + E(p_3, q_1) + E(p_3, q_3) - E(p_1, q_3)$$

The reason behind this choice are the combinations possible for p_a, q_b $p_1q_1, p_1q_2, p_1q_{3,} p_2q_1, p_2q_2, p_2q_{3,} p_3q_1, p_3q_2, p_3q_3$. If two pairs are orthogonal then correlation is equal to zero (50% identical and 50% opposite). Suppose p_1q_2 are orthogonal so we can't select both at the same time. So suppose if we select p_1. If two pairs of analyzers are of the same orientation p_2q_1 *and* p_3q_2, quantum mechanics predict total anti-correlation of the result between Alice and Bob $E\left(p_2, q_1\right) = E(p_3, q_2) = -1$. Therefore, the selected combinations are $p_1q_1, p_1q_{3,} p_3q_1, p_3q_3$. There is a negative sign with $E(p_1, q_3)$ this is due to CHSH inequality. Let a, rma' be the two possible orientations of one detector and b, rmb' are of the second detector.

$$E(a, b) - E\left(a, b'\right) = \int \left[A(a, \lambda).B(b, \lambda) - A(a, \lambda)B\left(b', \lambda\right)\right]p(\lambda)d\lambda$$

where λ is hidden the variable is assumed to be drawn from fixed distribution of possible states of source, $p(\lambda)$ is the probability of the source being in state λ, the integral over complete hidden space is 1 i.e $\int p(\lambda)d\lambda = 1$. Possible values of A and B are 1, 0, –1 therefore, |A| $\leq$ 1, |B| $\leq$ 1.| Possible values of A and B are 1, 0, –1 therefore, |A| $\leq$ 1, |B| $\leq$ 1.| Possible values of A and B are 1, 0, –1 therefore, |A| $\leq$ 1, |B| $\leq$ 1.| Possible values of A and B are 1, 0, –1 therefore, |A| $\leq$ 1, |B| $\leq$ 1.| Possible values of A and B are 1, 0, –1 therefore, |A| $\leq$ 1, |B| $\leq$ 1.

$$= \int [A(a, \lambda).B(b, \lambda) - A(a, \lambda)B\left(b', \lambda\right) + A\left(a', \lambda\right)B\left(b', \lambda\right)A(a, \lambda)B(b, \lambda)$$
$$- A\left(a', \lambda\right)B\left(b', \lambda\right)A(a, \lambda)B(b, \lambda)]\ p(\lambda)d\lambda$$

$$= \int [A(a,\lambda).B(b,\lambda)\left[1 \pm A(a',\lambda)B(b',\lambda)\right]p(\lambda)d\lambda$$
$$+ \int A(a,\lambda)B(b',\lambda)1 \pm A(a',\lambda)B(b,\lambda)]p(\lambda)d\lambda$$

By applying triangle inequality and using |A| $\leq$ 1, |B| $\leq$ 1 and $\left[1 \pm A(a',\lambda)B(b,\lambda)\right] and \left[1 \pm A(a',\lambda)B(b',\lambda)\right]$ are non-negative and since $\int p(\lambda)d\lambda = 1$

$$\Rightarrow \left|E(a,b) - E(a,b')\right| \leq \int\{\left[1 \pm A(a',\lambda)B(b',\lambda)\right] + \left[1 \pm A(a',\lambda)B(b,\lambda)\right]\}d\lambda$$

$$\Rightarrow \left|E(a,b) - E(a,b')\right| \leq 2 \pm \left[A(a',\lambda)B(b',\lambda) + A(a',\lambda)B(b,\lambda)\right]$$

$$\Rightarrow \left|E(a,b) - E(a,b')\right| \leq 2 \pm \left[E(a',b') + E(a',b)\right]$$

$$\Rightarrow \left|E(a,b) - E(a,b')\right| \pm \left|E(a',b') + E(a',b)\right| \leq 2$$

Quantum mechanics gave value of S as $2\sqrt{2}$. This is because according to quantum rules, expectation of entangled photons is E(a, b) = Cos [2($\varnothing_2 - \varnothing_1$)] but for spin ½ particles, argument of cosine is different in case of photons because half angle $\frac{\varnothing}{2}$ appears in the formula.

$$\begin{aligned} E(a,b) &= a.b = |a||.||b|| \text{ Cos } [\theta_2 - \theta_1] \\ &= \sqrt{a.a}\sqrt{b.b} \text{ Cos } [\theta_2 - \theta_1] \\ &= \text{Cos } [\theta_2 - \theta_1] \end{aligned}$$

$$E(p_1,q_1) = \text{Cos}\left[\theta_1^q - \theta_1^p\right] = Cos\left[45^\circ - 0^\circ\right] = \frac{1}{\sqrt{2}}$$

$$E(p_1,q_3) = \text{Cos}\left[\theta_3^q - \theta_1^p\right] = Cos\left[135^\circ - 0^\circ\right] = -\frac{1}{\sqrt{2}}$$

$$E(p_3,q_1) = \text{Cos}\left[\theta_1^q - \theta_3^p\right] = Cos\left[45^\circ - 90^\circ\right] = \frac{1}{\sqrt{2}}$$

$$E(p_3,q_3) = \text{Cos}\left[\theta_3^q - \theta_3^p\right] = Cos\left[135^\circ - 90^\circ\right] = \frac{1}{\sqrt{2}}$$

$$S = E(p_1,q_1) - E(p_1,q_3) + E(p_3,q_1) + E(p_3,q_3)$$

$$= \frac{1}{\sqrt{2}} + \frac{1}{\sqrt{2}} + \frac{1}{\sqrt{2}} + \frac{1}{\sqrt{2}} = 2\sqrt{2}$$

Table. 2 Obtaining the original message by applying XOR operations

Original message	1	1	0	0	1	0
Secret key	0	1	0	1	1	0
1st XOR operation	1	0	0	1	0	0
Bob's key	0	1	0	1	1	0
2nd XOR operation	1	1	0	0	1	0

We will get this S only if particles are not disturbed and their sifted key is secure. But there may be errors in the system due to imperfections, which are corrected using a classical channel. After obtaining the sifted key, they check how much of their data is anti-correlated. Anti-correlation means if the bit value obtained by Alice is 0 then Bob's bit value will be 1 and conversely. If the bit value obtained by both will be either 0 or 1, then there is some error in system. Error rate in the system is calculated by dividing the number of errors with size of the sample and multiplying it with 100%; through this we can confirm the presence of the eavesdropper and check the efficiency of the system. We can correct the error using XOR operations. Encrypted messages of Alice consist of original messages XORed with secret key. Then she sends the encoded text to Bob using internet and Bob's key is used to decrypt the message by applying XOR operation again (Table 2).

5 Zhang's Mask Authentication Protocol

It was proposed by Zhang [4] in 2009. In this we have three systems: Alice—as a user, Bob—as a service provider and a certifying authority that is semi trusted. It has three phases: set up phase, signature masking phase and verification phase. In the quantum signature masked authentication scheme, N EPR pairs are generated by Alice-$\{|\Phi_1>, |\Phi_2>, \ldots\ldots\ldots\ldots |\Phi_n>\}$, and each pair is in bell state. Then a particle sequence is given by Alice to Bob-$|\Phi_b>= (|\Phi_{b1}>, |\Phi_{b2}>, \ldots\ldots\ldots\ldots, |\Phi_{bn}>)$. For ensuring the security of the quantum channel, block transmission protocol is used for distribution of $|\Phi_b>$. The first sequence is transmitted from Alice to Bob then subsets of photons are measured by them. Then forth security of transmission for that first sequence is considered. If there is assurance that the channel is secure, then the second sequence from Alice is sent to Bob. For getting out the bell states, Bob carries out the bell basis measurement on the ordered N EPR pairs. Then the second eavesdropping test is carried out. By examining the error rate, whether the raw key is securely created or not is discovered. In this protocol, the photons are transmitted in batches of N EPR pairs. By measuring few photons in the first step where a particle sequence is held by Alice and Bob, the whole security of transmission is checked. This is the basic advantage of the block transmission protocol. Once we find that our quantum channel is secure it means the first particle sequence is not obtained by an eavesdropper Eve. Then no further information will be leaked to her, and she will not able to find our keys.

5.1 Setup Phase

Alice randomly selects their public key P_A–$P_A = (a_{1A}, b_{1A}, a_{2A}, b_{2A}, \ldots\ldots\ldots\ldots\ldots . a_{nA}, b_{nA})$

where $a_{iA}, b_{iA} \in \{0, 1\}, 0 < i \leq n$.

And send it to the certifying authority. After examining the qualification of Alice, the certifying authority generates the private key S_A for Alice

$$S_A = (c_{1A}, d_{1A}, c_{2A}, d_{2A}, \ldots\ldots\ldots\ldots\ldots . c_{nA}, d_{nA})$$

where $c_{iA}, d_{iA} \in \{0, 1\}, 0 < i \leq n$.

Then the secret key is calculated by the certifying authority namely K_A

$$K_A = P_A \oplus S_A = (e_{1A}, f_{1A}, e_{2A}, f_{2A}, \ldots\ldots\ldots\ldots e_{nA}, f_{nA})$$

where $e_{iA}, f_{iA} \in \{0, 1\}, 0 < i \leq n$ and $\oplus$ denotes XOR operation.

5.2 Signature-Masking Phase

Authentication message of Alice is represented by M

$$M = (m_1, m_2, \ldots\ldots\ldots\ldots, m_n)$$

where $m_i \in \{0, 1\}, 0 < i \leq n$.

According to S_A (i.e. secret key of Alice), an encrypted quantum state $|\Psi>$ is generated by Alice:

$$|\Psi >= \left|\Psi_{m_1 \oplus c_{1A}, d_{1A}} > \otimes \right|\Psi_{m_2 \oplus c_{2A}, d_{2A}} > \otimes \ldots\ldots\ldots\ldots \otimes \left|\Psi_{m_n \oplus c_{nA}, d_{nA}} >\right.$$

where for every $i = 1, 2, \ldots n$, a qubit, $|\Psi_{m_i \oplus c_{iA}, d_{iA}} >$ is in one of the following states

$$|\Psi_{0,0} >= |0>$$

$$|\Psi_{1,0} >= |1>$$

$$|\Psi_{0,1}>=\frac{|0>+|1>}{\sqrt{2}}$$

$$|\Psi_{1,1}>=\frac{|0>-|1>}{\sqrt{2}}$$

According to values of d_{iA}, quantum bases are selected. If $d_{iA}=0$, then $m_i \oplus c_{iA}$ is encrypted in Z-basis $\{|0>, |1>\}$ i.e. rectilinear basis. If $d_{iA}=1$, then $m_i \oplus c_{iA}$ is encrypted in X-basis $\{|+>, |->\}$ i.e. diagonal basis. Alice sends $|\Psi>$ to the certifying authority. Then the certifying authority applies $F^{[1]} to |\Psi>$ according to $K_A = P_A \oplus S_A$ and $|\Psi'>$ as a signature is obtained

$$F^{[1]} : |\Psi> \xrightarrow{K_A=P_A \oplus S_A} |\Psi'>$$

where $F^{[1]}$ is defined as

$$F^{[1]} = (X_1^{[1]}.X_2^{[1]} \ldots\ldots\ldots X_n^{[1]}) \otimes \left(Y_1^{[1]}.Y_2^{[1]} \ldots\ldots\ldots Y_n^{[1]}\right)$$
$$\Rightarrow F^{[1]} = X_1^{[1]}Y_1^{[1]} \otimes X_2^{[1]}Y_2^{[1]} \otimes \ldots\ldots\ldots\ldots\ldots \otimes X_n^{[1]}Y_n^{[1]}$$

where $\mathrm{X}_{\mathrm{i}}^{[1]}$ and $Y_i^{[1]}$ is defined as

$$\mathrm{X}_{\mathrm{i}}^{[1]} = X(e_{iA}), Y_i^{[1]} = Y(f_{iA})$$

where

$$X(1) = i\sigma_y = |0><1| - |1><0|$$

$$X(0) = I = |0><0| + |1><1|$$

$$Y(1) = H = \frac{1}{\sqrt{2}}(|0>+|1>) < 0| + \frac{1}{\sqrt{2}}(|0>-|1>) < 1|$$

$$Y(0) = \sigma_Z = |0><0| - |1><1|$$

At last, the certifying authority sends the signature $|\Psi'>$ to Bob along with all the decoy photon particles which are used for eavesdropping check.

5.3 Verification Phase

After receiving all the photons and states by Bob, states and position of the decoy particles are publicly announced by the certifying authority. Through the decoy particle eavesdropping method they check the security of the quantum channel. If the channel is secure, then Bob verifies Alice's originality by applying $F^{[2]} to|\Psi' >$ according to P_A and $|\Psi'' >$ is obtained

$$F^{[2]} : |\Psi' > \xrightarrow{P_A} |\Psi'' >$$

where $F^{[2]}$ is defined as

$$F^{[2]} = (X_1^{[2]}.X_2^{[2]} \ldots\ldots\ldots X_n^{[2]}) \otimes \left(Y_1^{[2]}.Y_2^{[2]} \ldots\ldots\ldots Y_n^{[2]}\right)$$
$$\Rightarrow F^{[2]} = X_1^{[2]}Y_1^{[2]} \otimes X_2^{[2]}Y_2^{[2]} \otimes \ldots\ldots\ldots\ldots\ldots \otimes X_n^{[2]}Y_n^{[2]}$$

where $\mathrm{X}_\mathrm{i}^{[2]}$ and $Y_i^{[2]}$ is defined as

$$\mathrm{X}_\mathrm{i}^{[2]} = X(a_{iA}), Y_i^{[2]} = Y(b_{iA})$$

where

$$X(1) = i\sigma_y = |0 >< 1| - |1 >< 0|$$

$$X(0) = I = |0 >< 0| + |1 >< 1|$$

$$Y(1) = H = \frac{1}{\sqrt{2}}(|0 > +|1 >) < 0| + \frac{1}{\sqrt{2}}(|0 > -|1 >) < 1|$$

$$Y(0) = \sigma_Z = |0 >< 0| - |1 >< 1|$$

On basis $(|0 >, |0 >, |0 > \ldots\ldots\ldots\ldots|0 >)$, Bob measures $|\Psi'' >$ and gets $(m'_1, m'_2, \ldots\ldots\ldots\ldots, m'_n)$ as the message. If $(m_1, m_2, \ldots\ldots\ldots\ldots, m_n) = (m'_1, m'_2, \ldots\ldots\ldots\ldots, m'_n)$ then the signature is valid otherwise invalid.

6 Quantum Signature Masked Authentication Protocol-Shi et al. [5]

It was proposed by Shi et al. [5] in 2015. They have taken the base as Zhang's scheme. But this scheme is quite complicated. In this we have three systems: Alice—as a user, Bob—as a service provider and a certifying authority that is semi trusted. It has three phases set up phase, signature masking phase and verification phase.

6.1 Setup Phase

In this Alice and Bob randomly select their public keys namely P_A *and* P_B respectively

$$P_A = (a_{1A}, b_{1A}, a_{2A}, b_{2A}, \ldots\ldots\ldots\ldots\ldots\ldots a_{nA}, b_{nA})$$

$$P_B = (a_{1B}, b_{1B}, a_{2B}, b_{2B}, \ldots\ldots\ldots\ldots\ldots\ldots a_{nB}, b_{nB})$$

where $a_{iA}, b_{iA}, a_{iB}, b_{iB} \in \{0, 1\}$,$0 < i \leq n$.

And sends it to the certifying authority. Then the certifying authority checks the qualification of Alice and Bob, it generates their private keys S_A *and* S_B and sends these keys through a secure quantum channel.

$$S_A = (c_{1A}, d_{1A}, c_{2A}, d_{2A}, \ldots\ldots\ldots\ldots\ldots\ldots c_{nA}, d_{nA})$$
$$S_B = (c_{1B}, d_{1B}, c_{2B}, d_{2B}, \ldots\ldots\ldots\ldots\ldots\ldots c_{nB}, d_{nB})$$

where $c_{iA}, d_{iA}, c_{iB}, d_{iB} \in \{0, 1\}$, $0 < i \leq n$.

Then secret keys are calculated by the certifying authority namely K_A, K_B *and* K_C

$$K_A = P_A \oplus S_A = (e_{1A}, f_{1A}, e_{2A}, f_{2A}, \ldots\ldots\ldots\ldots\ldots\ldots e_{nA}, f_{nA})$$

$$K_B = P_B \oplus S_B = (e_{1B}, f_{1B}, e_{2B}, f_{2B}, \ldots\ldots\ldots\ldots\ldots\ldots e_{nB}, f_{nB})$$

where $e_{iA}, f_{iA}, e_{iB}, f_{iB} \in \{0, 1\}$, $0 < i \leq n$ and $\oplus$ denotes XOR operation.

6.2 Signature-Masking Phase

Authentication message of Alice is represented by M

$$M = (m_1, m_2, \ldots\ldots\ldots\ldots, m_n)$$

where $m_i \in \{0, 1\}$, $0 < i \leq n$.

According to S_A (i.e. secret key of Alice), an encrypted quantum state $|\Psi>$ is generated by Alice:

$$|\Psi >= \left|\Psi_{m_1 \oplus c_{1A}, d_{1A}} > \otimes\right|\Psi_{m_2 \oplus c_{2A}, d_{2A}} > \otimes \ldots\ldots\ldots\ldots \otimes \left|\Psi_{m_n \oplus c_{nA}, d_{nA}} >\right.$$

where for every $i = 1, 2, \ldots n$, a qubit, $|\Psi_{m_i \oplus c_{iA}, d_{iA}} >$ is in one of the following states

$$|\Psi_{0,0}> = |0>$$

$$|\Psi_{1,0}> = |1>$$

$$|\Psi_{0,1}> = \frac{|0> + |1>}{\sqrt{2}}$$

$$|\Psi_{1,1}> = \frac{|0> - |1>}{\sqrt{2}}$$

Then according to one time pad encryption algorithm, Alice encode $s|\Psi>$ using $P_A \oplus S_A$, $E_{P_A \oplus S_A}\{|\Psi>\}$ and through the quantum channel, she send this to the certifying authority. The certifying authority obtains $|\Psi>$ by decoding $E_{P_A \oplus S_A}\{|\Psi>\}$ with $K_A = P_A \oplus S_A$.

Then the certifying authority applies $F^{[1]} to |\Psi>$ according to $K_A \oplus K_B$ and $|\mathrm{S}>$ as a signature is obtained

$$F^{[1]} : |\Psi> \xrightarrow{K_A \otimes K_B} |\mathrm{S}>$$

where $F^{[1]}$ is defined as

$$F^{[1]} = (X_1^{[1]}.X_2^{[1]} \ldots\ldots\ldots X_n^{[1]}) \otimes \left(Y_1^{[1]}.Y_2^{[1]} \ldots\ldots\ldots Y_n^{[1]}\right)$$
$$\Rightarrow F^{[1]} = X_1^{[1]}Y_1^{[1]} \otimes X_2^{[1]}Y_2^{[1]} \otimes \ldots\ldots\ldots\ldots\ldots \otimes X_n^{[1]}Y_n^{[1]}$$

where $\mathrm{X}_{\mathrm{i}}^{[1]}$ and $Y_i^{[1]}$ is defined as

$$\mathrm{X}_{\mathrm{i}}^{[1]} = X(e_{iA} \oplus e_{iB}),\ Y_i^{[1]} = Y(f_{iA} \oplus f_{iB})$$

where

$$X(1) = i\sigma_y = |0><1| - |1><0|$$

$$X(0) = I = |0><0| + |1><1|$$

$$Y(1) = H = \frac{1}{\sqrt{2}}(|0> + |1>) < 0| + \frac{1}{\sqrt{2}}(|0> - |1>) < 1|$$

$$Y(0) = \sigma_Z = 0 >< 0| - |1 >< 1|$$

Then according to one time pad encryption algorithm, the certifying authority encodes $|\mathrm{S}>$ using $P_A \oplus S_A$ $E_{P_A \oplus S_A}\{|\mathrm{S}>\}$ and through the quantum channel, he send this to the Alice. N EPR pairs are generated by Alice-$\{|\Phi_1>, |\Phi_2>$

, ... $|\Phi_n>\}$, and each pair is in a bell state i.e.

$$|\Phi_1>=\frac{1}{\sqrt{2}}\left(|0_{a_i}>|0_{b_i}>+|1_{a_i}>\left|1_{b_i}>\right)=\frac{1}{\sqrt{2}}\left(|+_{a_i}>|+_{b_i}>+|-_{a_i}>\left|-_{b_i}>\right)\right.\right.$$

where $i = 1, 2, 3 \ldots\ldots n$, ith two entangled states are denoted by a_i, b_i. Then a particle sequence is given by Alice to Bob using the block transmission protocol

$$|\Phi_b>=(|\Phi_{b1}>, |\Phi_{b2}>, \ldots\ldots\ldots\ldots, |\Phi_{bn}>)$$

And a particle sequence is kept by Alice

$$|\Phi_a>=(|\Phi_{a1}>, |\Phi_{a2}>, \ldots\ldots\ldots\ldots, |\Phi_{an}>)$$

To compute Alice's particle sequence,$|\Phi_a>$, randomly among one of two bases is chosen by her i.e. either Z-basis $\{|0>, |1>\}$ or X-basis $\{|+>, |->\}$ and the estimation result is recorded as$\{|p_1>, |p_2>, \ldots\ldots\ldots .|p_n>\}$ where $|p_i>\in\{|0>, |1>, |+>, |->\}$. These four states are encrypted by Alice into classical bits i.e.$|0>=00, |1>=01, |+>=10, |->=11$. Hence, $K_{AB}=\{k_1, l_1, k_2, l_2, \ldots\ldots\ldots .k_n, l_n\}$ is the final estimation result. Alice declares her measure basis to Bob, then with the same basis Bob measures the corresponding particles of$|\Phi_b>$. Same as Alice he then encrypts the four states into classical bits i.e.$|0>=00, |1>=01, |+>=10, |->=11$. At the end, Bob gets$K_{AB}$. A random sequence of length 2n is chosen by Alice

$$R_A=(g_1, h_1, g_2, h_2, \ldots\ldots\ldots\ldots\ldots .g_n, h_n)$$

where $g_i, h_i \in \{0, 1\}, 0 < i \leq n$. And calculates

$$R_A \oplus K_{AB}=(r_1, s_1, r_2, s_2, \ldots\ldots\ldots\ldots\ldots .r_n, s_n)$$

Alice decodes $E_{P_A \oplus S_A}\{|S>\}$ with K_A to get $|S>$ and applies $F^{[2]}\, to\, |S>$ according to $R_A \oplus P_B$ and $|S'>$ as the final signature is obtained

$$F^{[2]}: |S> \xrightarrow{R_A \oplus P_B} |S'>$$

where $F^{[2]}$ is defined as

$$F^{[2]}=(X_1^{[2]}.X_2^{[2]}\ldots\ldots\ldots X_n^{[2]}) \otimes \left(Y_1^{[2]}.Y_2^{[2]}\ldots\ldots\ldots Y_n^{[2]}\right)$$
$$\Rightarrow F^{[2]}=X_1^{[2]}Y_1^{[2]} \otimes X_2^{[2]}Y_2^{[2]} \otimes \ldots\ldots\ldots\ldots\ldots \otimes X_n^{[2]}Y_n^{[2]}$$

where $X_i^{[2]}$ and $Y_i^{[2]}$ is defined as

$$\mathrm{X}_{\mathrm{i}}^{[2]} = X(g_i \oplus a_{iB}), Y_i^{[2]} = Y(h_i \oplus b_{iB})$$

where

$$X(1) = i\sigma_y = |0 >< 1| - |1 >< 0|$$

$$X(0) = I = |0 >< 0| + |1 >< 1|$$

$$Y(1) = H = \frac{1}{\sqrt{2}}(|0 > +|1 >) < 0| + \frac{1}{\sqrt{2}}(|0 > -|1 >) < 1|$$

$$Y(0) = \sigma_Z = |0 >< 0| - |1 >< 1|$$

Credential certificate of Alice is $\{R_A \oplus K_{AB}, |S' >\}$ on M. For getting service from Bob, he will send $\{M, R_A \oplus K_{AB}, |S' >\}$ to Bob. Using signature $|S' >$ as final, Bob will get convinced that Alice is aware about the signature $|S >$.

6.3 *Verification Phase*

R_A is calculated by Alice by-$R_A \oplus K_{AB} \oplus K_{AB} = R_A$. Bob verifies Alice's originality by applying $F^{[3]}\, to\, |\mathrm{S}' >$ according to $P_A \oplus S_B \oplus R_A$ and $|\mathrm{S}'' >$ is obtained

$$F^{[3]} : \left|\mathrm{S}'\right\rangle \xrightarrow{P_A \oplus S_B \oplus R_A} \left|\mathrm{S}''\right\rangle$$

where $F^{[3]}$ is defined as

$$F^{[3]} = (X_1^{[3]}.X_2^{[3]} \ldots\ldots\ldots X_n^{[3]}) \otimes \left(Y_1^{[3]}.Y_2^{[3]} \ldots\ldots\ldots Y_n^{[3]}\right)$$
$$\Rightarrow F^{[3]} = X_1^{[3]} Y_1^{[3]} \otimes X_2^{[3]} Y_2^{[3]} \otimes \ldots\ldots\ldots\ldots\ldots \otimes X_n^{[3]} Y_n^{[3]}$$

where $\mathrm{X}_{\mathrm{i}}^{[3]}$ and $Y_i^{[3]}$ is defined as

$$\mathrm{X}_{\mathrm{i}}^{[3]} = X(a_{iA} \oplus c_{iB} \oplus g_i), Y_i^{[3]} = Y(b_{iA} \oplus d_{iB} \oplus h_i)$$

Where

$$X(1) = i\sigma_y = |0 >< 1| - |1 >< 0|$$

$$X(0) = I = |0 >< 0| + |1 >< 1|$$

$$Y(1) = H = \frac{1}{\sqrt{2}}(|0> + |1>) < 0| + \frac{1}{\sqrt{2}}(|0> - |1>) < 1|$$

$$Y(0) = \sigma_Z = |0>< 0| - |1>< 1|$$

On basis $(|0>, |0>, |0> \ldots\ldots\ldots\ldots |0>)$, Bob measures $|S''>$ and gets $(m'_1, m'_2, \ldots\ldots\ldots\ldots, m'_n)$ as message. If $(m_1, m_2, \ldots\ldots\ldots\ldots, m_n) = (m'_1, m'_2, \ldots\ldots\ldots\ldots, m'_n)$ then the signature is valid otherwise invalid.

6.4 Correctness

The initial quantum state $|\Psi>$ is generated by Alice using her secret key S_A. It passes through the following process during the signature and verification phase

$$Message\, M \xrightarrow{S_A} |\Psi> \xrightarrow{K_A \oplus K_B} |S> \xrightarrow{R_A \oplus P_B} |S'> \xrightarrow{P_A \oplus S_B \oplus R_A} |S''>$$

$$c_{iA} = e_{iA} \oplus e_{iB} \oplus g_i \oplus a_{iB} \oplus a_{iA} \oplus c_{iB} \oplus g_i$$

$$\Rightarrow c_{iA} = a_{iA} \oplus c_{iA} \oplus a_{iB} \oplus c_{iB} \oplus g_i \oplus a_{iB} \oplus a_{iA} \oplus c_{iB} \oplus g_i$$

$$d_{iA} = f_{iA} \oplus f_{iB} \oplus h_i \oplus b_{iB} \oplus b_{iA} \oplus d_{iB} \oplus h_i$$

$$\Rightarrow d_{iA} = b_{iA} \oplus d_{iA} \oplus b_{iB} \oplus d_{iB} \oplus h_i \oplus b_{iB} \oplus b_{iA} \oplus d_{iB} \oplus h_i$$

It can withstand forgery attack, impersonation attack.

7 Quantum Signature Masked Authentication Scheme-Fatahi and Afsheh [6]

It was proposed by Fatahi and Afsheh [6]. For avoiding complications he has only used private keys to calculate the secret key. In this we have three systems: Alice—as a user, Bob—as a service provider and a authorized center that is semi trusted. It has three phases set up phase, signature masking phase and verification phase.

7.1 Setup Phase

In this Alice and Bob randomly select their public keys namely $P_A\, and\, P_B$ respectively

$$P_A = (a_{1A}, b_{1A}, a_{2A}, b_{2A}, \ldots\ldots\ldots\ldots\ldots\ldots a_{nA}, b_{nA})$$

$$P_B = (a_{1B}, b_{1B}, a_{2B}, b_{2B}, \ldots\ldots\ldots\ldots\ldots\ldots a_{nB}, b_{nB})$$

where $a_{iA}, b_{iA}, a_{iB}, b_{iB} \in \{0, 1\}, 0 < i \leq n$.

And sends it to the authorized center, for eavesdropping check they will insert large number of decoy particles into the keys. The position and state of the decoy particles is publicly announced by Alice and Bob when the public keys are received by the authorized center. Then the authorized center uses the same basis as Alice and Bob to measure every decoy particle. Then he compares this result with Alice and Bob's measurement result. Through this the authorized center checks the qualification of Alice and Bob, generates their private keys $S_A and S_B$ and sends these keys through a secure quantum channel.

$$S_A = (c_{1A}, d_{1A}, c_{2A}, d_{2A}, \ldots\ldots\ldots\ldots\ldots\ldots c_{nA}, d_{nA})$$

$$S_B = (c_{1B}, d_{1B}, c_{2B}, d_{2B}, \ldots\ldots\ldots\ldots\ldots\ldots c_{nB}, d_{nB})$$

where $c_{iA}, d_{iA}, c_{iB}, d_{iB} \in \{0, 1\}, 0 < i \leq n$.

Then secret keys are calculated by the certifying authority namely $K_A, K_B\, and\, K_C$

$$K = S_A \oplus S_B = (e_1, f_1, e_2, f_2, \ldots\ldots\ldots\ldots\ldots\ldots e_n, f_n)$$

where $e_i, f_i \in \{0, 1\}$, $0 < i \leq n$ and $\oplus$ denotes XOR operation.

7.2 Signature-Masking Phase

Authentication message of Alice is represented by M

$$M = (m_1, m_2, \ldots\ldots\ldots\ldots, m_n)$$

where $m_i \in \{0, 1\}, 0 < i \leq n$.

According to S_A (i.e. secret key of Alice), an encrypted quantum state $|\Psi >$ is generated by Alice:

$$|\Psi >= \left|\Psi_{m_1 \oplus c_{1A}, d_{1A}} > \otimes \right|\Psi_{m_2 \oplus c_{2A}, d_{2A}} > \otimes \ldots\ldots\ldots\ldots \otimes \left|\Psi_{m_n \oplus c_{nA}, d_{nA}} >\right.$$

where for every $i = 1, 2, \ldots n$, a qubit, $|\Psi_{m_i \oplus c_{iA}, d_{iA}} >$ is in one of following states

$$|\Psi_{0,0} >= |0>$$

$$|\Psi_{1,0} >= |1>$$

$$|\Psi_{0,1} >= \frac{|0> + |1>}{\sqrt{2}}$$

$$|\Psi_{1,1} >= \frac{|0> - |1>}{\sqrt{2}}$$

A private key R_A is randomly selected by Alice

$$R_A = (g_1, h_1, g_2, h_2, \ldots\ldots\ldots\ldots\ldots g_n, h_n)$$

where $g_i, h_i \in \{0, 1\}, 0 \le i \le n$.

And send it to Bob in a secure quantum channel using decoy particle eavesdropping check method. Alice applies $F^{[1]} to |\Psi>$ according to R_A and $|\Psi'>$as the signature is obtained

$$F^{[1]} : |\Psi > \overset{R_A}{\rightarrow} |\Psi' >$$

where $F^{[1]}$ is defined as

$$F^{[1]} = (X_1^{[1]}.X_2^{[1]} \ldots\ldots\ldots X_n^{[1]}) \otimes \left(Y_1^{[1]}.Y_2^{[1]} \ldots\ldots\ldots Y_n^{[1]}\right)$$
$$\Rightarrow F^{[1]} = X_1^{[1]}Y_1^{[1]} \otimes X_2^{[1]}Y_2^{[1]} \otimes \ldots\ldots\ldots\ldots\ldots \otimes X_n^{[1]}Y_n^{[1]}$$

where $\mathrm{X}_{\mathrm{i}}^{[1]}$ and $Y_i^{[1]}$ is defined as

$$\mathrm{X}_{\mathrm{i}}^{[1]} = X(g_i), Y_i^{[1]} = Y(h_i)$$

where

$$X(1) = i\sigma_y = |0 >< 1| - |1 >< 0|$$

$$X(0) = I = |0 >< 0| + |1 >< 1|$$

$$Y(1) = H = \frac{1}{\sqrt{2}}(|0 > + |1 >) < 0| + \frac{1}{\sqrt{2}}(|0 > - |1 >) < 1|$$

$$Y(0) = \sigma_Z = |0 >< 0| - |1\rangle\langle 1|$$

Through a secure quantum channel, Alice sends $|\Psi' >$ to the authorized center. The authorized center applies $F^{[2]} to |\Psi' >$ according to $K = S_A \oplus S_B$ and $|\mathrm{S} >$ as a final signature is obtained

$$F^{[2]} : |\Psi' > \xrightarrow{K = S_A \oplus S_B} |\mathrm{S} >$$

where $F^{[2]}$ is defined as

$$F^{[2]} = (X_1^{[2]}.X_2^{[2]} \ldots\ldots\ldots X_n^{[2]}) \otimes \left(Y_1^{[2]}.Y_2^{[2]} \ldots\ldots\ldots Y_n^{[2]}\right)$$
$$\Rightarrow F^{[2]} = X_1^{[2]} Y_1^{[2]} \otimes X_2^{[2]} Y_2^{[2]} \otimes \ldots\ldots\ldots\ldots\ldots \otimes X_n^{[2]} Y_n^{[2]}$$

where $\mathrm{X}_\mathrm{i}^{[2]}$ and $Y_i^{[2]}$ is defined as

$$\mathrm{X}_\mathrm{i}^{[2]} = X(e_i), \; Y_i^{[2]} = Y(f_i)$$

where

$$X(1) = i\sigma_y = |0 >< 1| - |1 >< 0|$$

$$X(0) = I = |0 >< 0| + |1 >< 1|$$

$$Y(1) = H = \frac{1}{\sqrt{2}}(|0 > +|1 >) < 0| + \frac{1}{\sqrt{2}}(|0 > -|1 >) < 1|$$

$$Y(0) = \sigma_Z = |0 >< 0| - |1\rangle\langle 1|$$

At last, the authorized center sends the signature $|S >$ to Bob.

7.3 Verification Phase

Bob verifies Alice's originality by applying $F^{[3]}\, to\, |\mathrm{S} >$ according to $S_B \oplus R_A$ and $|\mathrm{S}' >$ is obtained

$$F^{[3]} : |\mathrm{S} > \xrightarrow{S_B \oplus R_A} |\mathrm{S}' >$$

where $F^{[3]}$ is defined as

$$F^{[3]} = (X_1^{[3]}.X_2^{[3]} \ldots\ldots\ldots X_n^{[3]}) \otimes \left(Y_1^{[3]}.Y_2^{[3]} \ldots\ldots\ldots Y_n^{[3]}\right)$$

$$\Rightarrow F^{[3]} = X_1^{[3]}Y_1^{[3]} \otimes X_2^{[3]}Y_2^{[3]} \otimes \ldots\ldots\ldots\ldots\ldots \otimes X_n^{[3]}Y_n^{[3]}$$

where $X_i^{[3]}$ and $Y_i^{[3]}$ is defined as

$$X_i^{[3]} = X(c_{iB} \oplus g_i),\ Y_i^{[3]} = Y(d_{iB} \oplus h_i)$$

where

$$X(1) = i\sigma_y = |0><1| - |1><0|$$

$$X(0) = I = |0><0| + |1><1|$$

$$Y(1) = H = \frac{1}{\sqrt{2}}(|0> + |1>) < 0| + \frac{1}{\sqrt{2}}(|0> - |1>) < 1|$$

$$Y(0) = \sigma_Z = |0><0| - |1><1|$$

On basis $(|0>, |0>, |0> \ldots\ldots\ldots\ldots |0>)$, Bob measures $|S'>$ and gets $(m'_1, m'_2, \ldots\ldots\ldots\ldots, m'_n)$ as message. If $(m_1, m_2, \ldots\ldots\ldots\ldots, m_n) = (m'_1, m'_2, \ldots\ldots\ldots\ldots, m'_n)$ then the signature is valid otherwise invalid.

7.4 Correctness

The initial quantum state $|\Psi>$ is generated by Alice using her secret key S_A. It passes through the following process during signature and verification phase

$$Message\ M \xrightarrow{S_A} |\Psi> \xrightarrow{R_A} |\Psi'> \xrightarrow{K=S_A \oplus S_B} |S> \xrightarrow{S_B \oplus R_A} |S'>$$

$$c_{iA} = g_i \oplus c_{iB} \oplus c_{iA} \oplus c_{iB} \oplus g_i$$

$$d_{iA} = h_i \oplus d_{iB} \oplus d_{iA} \oplus d_{iB} \oplus h_i$$

8 Attacks and Security

8.1 Intercept Resend Attack

In this, the quantum states i.e., photons which are transmitted by Alice are studied by Eve and then the replaced states of these are sent to Bob, because of no cloning theorem, copying of the unknown quantum state is not possible. The key that is shared by Alice to Bob causes errors in the BB84 protocol. As like Bob, Eve does not know which bases Alice used for measuring the states. So, Eve makes random guesses. As in fifty percent of cases, the corrected bases are guessed by Alice and send the same bases to Bob and her presence will not be noticed by Alice. But in cases when wrong guesses are made by her, her presence will be detected because of the uncorrelated result between Alice and Bob.

8.2 Man in the Middle Attack

In this attack, the communication between two parties is safely transferred and altered by the third party and the two parties are not aware about this. One example of this type is that in which an attacker named Eve intercepts the message and transfers this message to Bob, he will not figure out whether the message is from Alice or Eve. Therefore, Bob responds to the message with his encoded key. Eve replaces this key with her own key and transfers it to Alice, declaring it as Bob's key. Alice encodes the message using this key, trusting that its Bob's key. As it was Eve's key, therefore, Eve can easily decode the message, read it, change it and send the re-encoded message to Bob. While Bob or Alice is not aware about Eve but the full communication is controlled by the attacker, Eve.

8.3 Trojan Horses Attack

It is a type of malware that is impersonated as a legal software. Any hacker who is trying to get entrance to the system, can be enrolled by the Trojans and social engineering in charging and discharging Trojans into the system is used to trick the user. Once a Trojan is activated, it can authorize the attacker to observe you, and it makes the attacker to get backdoor entrance to your system and your sensitive data in the system is taken by the attacker. The actions that an attacker can perform after entering your system are: deleting, blocking, copying, and modifying data; also the performance of computer and its network is disturbed by the attacker.

8.4 Denial of Service Attack (DOS)

The two points i.e., from Alice to Bob in quantum key distribution are connected by optical fibers, by simply cutting and blocking the optical fiber, this attack can be produced. Due to this only quantum key distribution networks are established, in case of disturbance, it would passage communication through different links.

8.5 Security of Quantum Key Distribution [2]

Though quantum cryptography is perfectly secure there are some loopholes in either the photon or in measurement of the polarizing filter. It is vulnerable to various side channel attacks and denial of service attack (the legal person is prevented by attackers from accessing the services). Without security proof, there will be an incomplete QKD system as there is no surety that the final key generated by us is secure.

8.6 Attacks by Side Channels

A natural example of this is supposed to be a group of soldiers who are using encrypted signals of the torch to communicate with the other group of soldiers. To know the position of the soldiers, the attackers just need to physically identify the light of the torch, without decrypting any code in the torch signals. In this no encryption method gets failed but its physical implementation leads to side channel leaking information. They are attacks through which information is gained by computer systems without disturbing the cryptographic algorithms. Some main side channel attacks are.

- Conventional side channel attack: security loopholes are on security threats which are linked to non-quantum aspects of QKD. They are leaking internal state's information of QKD hardware of Alice and Bob.
- Trojan horse attack: in this the eavesdropper monitors the reflected light and by this information about the internal state is learned.
- Key elements used for the implementation of discrete variable QKD are Single particle detectors. Examples are avalanche photodiodes (APDs) which are operated on trigger modes. Many attacks occur on imperfection of APDs.
- As a single photon source is not commercially available, so instead of this, a weak laser diode is used. More than one photon is contained by this weak laser diode which constructs condition for Eve to acquire knowledge of all the details about encrypted data.
- Laser source's intensity, quantum channel's length and time taken to detect single photon. They are calibrated and take part in the safety of the system. So, attack on calibration procedure leads to security breach.

- Side channels on random number generators: in this Eve tries to gain knowledge about randomness generated to run QKD protocols by Alice and Bob.

9 Observation

In this section, we present year wise development in quantum cryptography and discuss the specialty of the protocol (some of them have been discussed in detail in Sects. 3–7).

Year	Authors	Protocol specialty
1984	Bennett and Brassard [2]	BB84: It was the first quantum key distribution mechanism to use four polarization states to transfer information
1991	Ekert [3]	EK91: It was the first Quantum Entanglement based protocol
1992	Bennett et al. [7]	BBM92: It was the first experimental QKD that use parity check and hashing algorithms to reconcile information
1992	Bennett [8]	BB92: The main distinction is that instead of four polarization states, just two low-intensity non-orthogonal coherent states are required in the BB92. i.e. (0^0, 45^0)
1996	Mu et al. [9]	Bit encoded in four non-orthogonal states characterized by quadrature phase amplitudes of a weak optical field with no units polarized photons
1998	Mayers and Yao[10]	Maiden device independent QKD
1999	Bechmann-Pasquinucci and Gisin [11]	The symmetry of this protocol makes security analysis much easier. It's the BB84 protocol plus an extra basis, meaning it has six states on the Poincare sphere: x, y, and z
2003	Inoue et al. [12]	It has a number of advantages, including easy configuration, efficient use of the time domain, and resistance to PNS attacks All photons are used to make the key, it's easy to set up, and it works well in the time domain
2003	Hwang [13]	To counteract the PNS attack in the context of substantial loss, presented a decoy-state technique
2004	Scarani et al. [14]	Use of non-orthogonal states offers several advantages, including easy configuration, efficient time domain use, and resistance to PNS attacks

(continued)

(continued)

Year	Authors	Protocol specialty
2004	Gisin et al. [15]	Time is used to encode the data. The presence of a spy can be monitored using an additional communication line. To work with high bit rates and weak coherent pulses. The system is simple in practice and resistant to PNS attack. As a result, no data will be lost
2009	Zhang [4]	Proposed One-way quantum identity authentication based on public key cryptography
2009	Khan et al. [16]	They used two mutually unbiased bases. The rate of index transmission error was introduced. Instead of employing two directions of one single base, the two parties employed two bases: one for encoding "0" and the other for encoding "1"
2012	Serna [17]	Uses QKD, a public cryptography. It is possible to use it for more than two parties. The one-photon protocol is still used. It permits vast key distribution amongst n–1 computers and a single key message distribution center. Because it does not use traditional channels, it is also immune to man-in-the-middle attacks. Because the qubits are exchanged many times, this protocol may be more difficult to implement
2012	Lo et al. [18]	Even if Alice and Bob's preparation routines aren't ideal, it works
2013	Serna [19]	It creates various secret keys for the transmitted qubits, meaning that no information is lost between the interlocutors. It varies from BB84 in that it uses asymmetric cryptography and a random seed. In this case, QKD becomes a zero-information-loss procedure. Only the traditional technique is different. This protocol can be used with any device that already exists
2015	Shi et al. [5]	Proposed new scheme that can solve Zhang's scheme security problem
2017	Abushgra and Elleithy [20]	Decoy states and parity check are included in this matrix. EPR authentication phase, in which the EPR string is utilized as the system's key
2018	Fatahi and Afsheh [6]	Introduced a novel quantum signature-masked authentication technique based on a private key that can withstand forgery, impersonation, and external attacks

(continued)

(continued)

Year	Authors	Protocol specialty
2021	Verma and Malhotra [21]	Proposed scheme for user authentication and is protected against forgery, impersonation and outside attack better than N. Fathahi, Zhang scheme

10 Conclusion

The declaration that quantum cryptography is an ultimate security is more solid than any other declaration on the security of earlier cryptography. In BB84, polarization states and bases are randomly chosen by user A and polarization states are then send to user B through a quantum channel. User B, randomly use his bases to measure the states and obtains the bit value from them. Then comparison of bases over the classical channel takes place through which they keep bits that correspond to the same bases and are free from eavesdropping. The left bits form a secret key. In Ek91, QKD is implemented using quantum entanglements. Here instead of Bob receiving particles from Alice, there is an in between source which creates entangled particles and sends one to Bob and other to Alice. Error is corrected using XOR operations. Ek91 use 6 bases while BB84 use 4 bases to measure the polarized particles. For everyone who have entrance to the network, QKD provides unconditional security to them but sometimes QKD links are vulnerable to quantum hacking and various types of attacks like denial-of-service attack and attacks by side channels like conventional side channel attacks, Trojan horses attack, attacks on single photon or multiple photon detectors, attack on random number generations or on the calibration procedures. Various protocols like B92, DPS, SARG04, COW, S09, are thus designed to provide safety against these attacks. Implementation of quantum cryptography will give good security in bank transactions, office purpose secret discussions, and for military purposes.

References

1. A. Einstein, B. Podolsky, N. Rosen, Can quantum-mechanical description of physical reality be considered complete? Phys. Rev. **47**(10), 77–80 (1935)
2. C.H. Bennett, G. Brassard, Quantum cryptography: public key distribution and coin tossing. Theoret. Comput. Sci. **560**(P1), 7–11 (2014). https://doi.org/10.1016/j.tcs.2014.05.025
3. A.K. Ekert, Quantum cryptography based on Bell's theorem. Phys. Rev. Lett. **67**(6), 661–663 (1991)
4. X. Zhang, One-way quantum identity authentication based on public key. Chin. Sci. Bull. **54**(12), 2018–2021 (2009). https://doi.org/10.1007/s11434-009-0350-9
5. W.M. Shi, Y.G. Yang, Y.H. Zhou, Quantum signature-masked authentication schemes. Optik (Stuttg) **126**(23), 3544–3548 (2015). https://doi.org/10.1016/j.ijleo.2015.08.277

6. N. Fatahi, H. Reza Afsheh, Secure electronic voting scheme by the new quantum signature-masked authentication (2018). https://doi.org/10.15406/paij.2018.02.00146
7. C. H. Bennett, F. Bessette, G. Brassard, L. Salvail, J. Smolin, Experimental quantum cryptography J. Cryptology **5**, 3–28 (1992)
8. C.H. Bennett, Quantum cryptography using any two nonorthogonal states. Phys. Rev. Lett. **68**(21), 3121–3124 (1992). https://doi.org/10.1103/PhysRevLett.68.3121
9. Y. Mu, J. Seberry, Y. Zheng, Shared cryptographic bits via quantized quadrature phase amplitudes of light. Opt. Commun. **123**(1–3), 344–352 (1996). https://doi.org/10.1016/0030-4018(95)00688-5
10. D. Mayers, A. Yao, Quantum cryptography with imperfect apparatus, in *Proceedings 39th Annual Symposium on Foundations of Computer Science (Cat. No.98CB36280)* (1998), pp. 503–509. https://doi.org/10.1109/SFCS.1998.743501
11. H. Bechmann-Pasquinucci, N. Gisin, Incoherent and coherent eavesdropping in the six-state protocol of quantum cryptography. Phys. Rev. A, **59**(6), 4238–4248 (1999). Accessed 14 May 2022. https://doi.org/10.1103/PhysRevA.59.4238
12. K. Inoue, E. Waks, Y. Yamamoto, Differential-phase-shift quantum key distribution using coherent light. Phys. Rev. A **68**(2), 022317 (2003). https://doi.org/10.1103/PhysRevA.68.022317
13. W.-Y. Hwang, Quantum key distribution with high loss: toward global secure communication. Phys. Rev. Lett **91**(5), 057901 (2003). https://doi.org/10.1103/PhysRevLett.91.057901
14. V. Scarani, A. Acin, G. Ribordy, N. Gisin, Quantum cryptography protocols robust against photon number splitting attacks for weak laser pulse implementations. Phys. Rev. Lett. **92**(5), 057901 (2004). https://doi.org/10.1103/PhysRevLett.92.057901
15. N. Gisin, G. Ribordy, H. Zbinden, D. Stucki, N. Brunner, V. Scarani, Towards practical and fast quantum cryptography (2004). Accessed 14 May 2022. https://arxiv.org/abs/quant-ph/0411022
16. M. M. Khan, M. Murphy, A. Beige, New Journal of Physics High error-rate quantum key distribution for long-distance communication. New J. Phys. **11**(16), 63043 (2009). https://doi.org/10.1088/1367-2630/11/6/063043
17. E. H. Serna, Quantum key distribution protocol with private-public key (Aug 2012). http://arxiv.org/abs/0908.2146
18. H.-K. Lo, M. Curty, B. Qi, Measurement-device-independent quantum key distribution (2012). Arxiv, 2012. Accessed 14 May 2022. https://doi.org/10.1103/PhysRevLett.108.130503
19. E.H. Serna, Quantum key distribution from a random seed (2013). ArXiv, 2013. https://doi.org/10.48550/arXiv.1311.1582
20. A.A. Abushgra, K.M. Elleithy, A shared secret key initiated by EPR authentication and qubit transmission channels. https://doi.org/10.1109/ACCESS.2017.2741899
21. V. Verma, K. Malhotra, A new secure quantum signature masked authentication scheme. Wirel. Pers. Commun. **120**(2), 1659–1674 (2021). https://doi.org/10.1007/s11277-021-08527-8

Evolution of Quantum Machine Learning and an Attempt of Its Application for SDN Intrusion Detection

Aakash R. Shinde and Shailesh P. Bendale

Abstract Even though Quantum Computers are still in their developmental phases, their technological implementation regarding their integration with classical computations is showing fascinating results around the world. Quantum Computers when compared with Classical Computational devices can be interpreted as a 'candle and light bulb', two different things accomplishing the same motive. Quantum Computers can provide us with tremendous processing power, which was unforeseen and has led us to question the ways in which they can transform existing technologies even raising concerns for current cryptic methods. Quantum systems have shown great potential in several fields one such is Quantum Machine Learning (QML) which even though a considerably novel field has benefitted from the integration of Quantum algorithms and Machine Learning algorithms, presenting exceptional results from various results. Scientists at Google were able to proclaim Quantum supremacy, this presented the Quantum computer's ability to perform extensively large computations in an extremely short time compared to a Super Computer, this could be beneficial for Machine learning algorithms to process huge amounts of data. Recent usage of a VPN for teleconferencing and a "work from home" scheme during the pandemic has caused a huge surge in network traffic forcing IT infrastructure providers to switch toward Software-Defined Network (SDN), Software Defined Network (SDN) is a prominent technology to provide betterment to the traditional network architecture. Machine Learning and Artificial Intelligence have been used extensively for various aspects of SDN, hence we attempted to explore QML and SDN interactions to assess their ability and the benefits we can achieve from this integration. This manuscript attempts to provide insights on the developments of QML over time and experimenting with SDN to provide a robust and efficient SDN system.

A. R. Shinde (✉)
MS in Quantum Information Sciences, University of Southern California, Los Angeles, CA, USA
e-mail: maxaakashshinde@gmail.com

S. P. Bendale
Head of Department (Computer Engineering), NBN Sinhgad School of Engineering, Maharashtra, Pune, India
e-mail: shailesh.bendale@sinhgad.edu

R. Pandey et al. (eds.), *Quantum Computing: A Shift from Bits to Qubits*,
Studies in Computational Intelligence 1085,
https://doi.org/10.1007/978-981-19-9530-9_22

Keywords Quantum machine learning · Software defined network · Quantum Computation · Artificial intelligence · Machine learning

1 Introduction

Quantum theory that once baffled several great physicists has made its way into the computational aspect by utilizing quantum mechanical phenomena such as quantum entanglement and quantum superposition to perform computations, and these systems that use quantum physics concepts for performing computations are termed Quantum Computers. Furthermore, quantum computers have theoretically been proven to be more efficient for solving computational problems with greater time complexity. Within a few years, quantum computers would be available commercially as per new promising research. Quantum Computers existing today are mostly in the research phase, in a laboratory environment that can be operated only under certain conditions among which a majority of them are accessible to users via systems connected to the cloud. Interesting results in several fields of quantum computers recently provide hope for staggering growth in the development of quantum computers, [1–9]. This has resulted in hefty investments in companies researching and developing quantum machines scalable for general users. Over $1.02 billion worth was invested into the quantum computing industry through the year 2021, which exceeds significantly considering the previous two years which was around $187.5 million invested. SDN (Software Defined Networks) has also seen similar growth, it has been estimated to increase by $25.97 billion from 2020 to 2025. SDN has proved its significance in network systems by creating customized network services and infrastructure. Considering the advantages provided by QC and SDN we tried to use technological prowess of QC improving the SDN systems providing a new prospect for QC application.

The dawn of quantum computing began in 1979 when a young physicist Paul Benioff submitted a paper titled "*The computer as a physical system: A microscopic quantum mechanical Hamiltonian model of computers as represented by Turing machines*", the paper submitted profound evidence of a probable coherent machine worth examining. This machine described a pure state changing over the action of a given Hamiltonian, which intern describes all the parts of a Turing machine as described by states defining the phase relation [10]. In 1980 Yuri Manin wrote a book in Russian named "*Computable and Uncomputable*" which was later translated a year after laying out the core idea for quantum computing [11]. Later in 1981, Richard Feynman conceptualized quantum computers in a lecture entitled "*Simulating Physics with Computers*" [12]. After the conceptualization of quantum computers by Benioff, Manin, and Feynman researchers started investigating algorithms that might work on quantum computers. In 1985 David Deutsch described and anticipated an algorithm for quantum computers that imposed the possibility of building a quantum computer in the later future [13]. Later along with Richard Jozsa, he generalized the algorithm that would run faster on a quantum computer

[14]. Building on the idea presented by Jozsa and Deutsch, Ethan Bernstein and Umesh Vazirani published a paper in 1993 proposing an algorithm that showed a clear quantum–classical separation allowing small errors [15]. Bernstein-Vazirani algorithm showcased the separation of small error allowances implying a non-deterministic quantum advantage. The problems proposed in Bernstein-Vazirani papers provide a solution for the problems with $O(n)$ time on a classical computer and that of $O(1)$ using a Bernstein-Vazirani circuit on a quantum computer, providing an efficient solution compared to a classical system. The 1993 paper of Bernstein-Vazirani even proposed a quantum version of Fourier transform, which in turn proved to be a building block for Peter Shor for developing his Shor's algorithm, one of the prominent factorization algorithms applicable to a QC. In 1994 Daniel Simon outlined a problem that could be solved much faster on a quantum computer rather than on a classical one, he proved so by showing that the upper bound of a quantum computer was much lesser than the lower bound of the classical computer [16]. During the same time spam Seth Lloyd who was working at Los Alamos suggested a method for building of a quantum computer which was published in the Science, this was addressed to be the first attempt for a practical approach towards a working quantum computer [17]. Studying the works of Deutsch, Bernstein-Vazirani and Simon, in 1994, Shor, a mathematical researcher at Bell Labs in New Jersey, came to realization for construction of a large number factorization algorithms into two prime factors. Only catch for the algorithm was that it needed a quantum computer to execute, these large number factorizations being exponentially hard for factorization into prime numbers on a classical system associate with public key cryptography implemented in RSA Algorithms which is the basis of all current internet communication [18]. In 2001, Shor's Algorithm was implemented by Issac Chaung on a nuclear magnetic resonance system to factor the number'15' as a demonstration [19]. After Shor's algorithm next came Lov Grover in May of 1996 contributing to the quantum computation by speeding up a search algorithm on a quantum computer. Grover's algorithm achieves quadratic speedup which means an algorithm that takes O (N) steps to execute on a classical computer will take O ($\sqrt{}$N) steps on a quantum computer to achieve the same goal [20]. Considering the work of Grover, Farhi and Gutmann created a framework for a continuous-time Hamiltonian version of Grover's algorithm [21]. Between 1999 and 2001, Yasunobu Nakamura built functioning and controllable superconducting qubits using Josephson junctions [22]. He created a two-level system manipulated by the user between its two states [23]. Manipulation and trapping of ions is another approach for implementation of a quantum computer, which was explored by Cirac and Zoller in 1995 where they proposed an ion trap as a physical system for quantum computational performance [24, 25].

Computational powers of a quantum computer rival that of a supercomputer as well, today its next to impossible for a complete representation of the molecular configuration of caffeine on a supercomputer but it can be represented using a 160-qubit quantum computer. For a classical computer, it would take around 10^{48} bits to represent the energy configuration of a single caffeine molecule at a single instance that roughly translates up to 10% of the total number of atoms on earth. This provides us with a glimpse of all that could be achieved with the computational power of a quantum computer. The major factor provided in this manuscript is a perspective on how quantum computers would affect artificial intelligence and machine learning, and how it would prove beneficial for other systems such as SDN. SDN is an enticing option as it contemplates a variety of problems and also it has been an epitome of networks with a programmable approach; it facilitates the network regarding the operations performed like the addition of rules, routing, or forwarding information from devices can be governed by one central controller. It interprets that the forwarding entities will have to perform those operations specified to them by this central controller only.

This manuscript reviews several research papers discussing the impact of Machine Learning, Artificial Intelligence, and Quantum Computation on one another and how these interdisciplinary fields have helped overcome several setbacks and aided in the development of these fields over several years. We also present our findings regarding the implementation of Quantum Machine Learning algorithms on the SDN dataset for anomaly detection. Novel contributions of this manuscript are as follows:

- Review major research in recent years regarding the integration of ML, AI, and quantum computers and address issues regarding the research.
- Summarize the most recent papers on the integration of ML, AI, and quantum computers, their development by interdisciplinary integration, and how they would affect those individual sectors in the future.
- Implementation of QML Algorithms on the SDN dataset for detecting Normal or an Attack on the SDN system.

This manuscript consists of three sections, Sect. 1, provides a general overview of major concepts regarding Quantum Computation and QML. This section summarizes a few major concepts of QC aiding in understanding their significance in the successive specified research. In Sect. 2, we discuss research in the fields of Artificial Intelligence, Machine Learning integrated with Quantum Computation. We explore the possibilities and how this research affects these sectors and helps us shape the future. This section helps in understanding the pros and cons of this integration, and also provides an insight into the current research and limitations in these specified integrated sectors, we also have proposed a few suggestions to overcome those limitations in the discussion based on research. Section 3, illustrates the information regarding the flow of experimentation, methods considered while experimenting, tools utilized, the process followed, and findings regarding the implementation of QML algorithms on the SDN dataset. Finally, we present our conclusions regarding the development and scope of improvement in the integrated fields of AI and ML, and QC, in Sect. 4. Also, in Sect. 5 we summarize the experimentation presenting our brief results and their implications on SDN.

2 Quantum Computing Terminologies

Quantum Information can be regarded for its structural importance in a quantum computer, and so for the considered as computations are to be performed using this information on a quantum computer. The information that is represented in these quantum computational devices works on the principals of quantum mechanics like quantum entanglement, quantum superposition, decoherence, the non-cloning theorem, quantum interference, etc. For example, a quantum circuit consists of states that are entangled with one another with a reliance on one another for performing computations [26–29]. Reversible computation is the basis for Quantum computation, as per the theory of reversible computation from the output state initial state can be recovered completely [30]. Utilizing the fundamentals of these quantum computational complexities, quantum algorithms are built. Multiple Quantum algorithms have been proposed using the principles of quantum mechanics resulting in remarkable growth in the speed of computation on quantum devices over their classical counterparts. Quantum algorithms have also been noteworthy in solving a few problems that seem to be farfetched over a classical computer or using a classical algorithm to solve them.

The following are major principles of quantum mechanics that are essential in understanding basic concepts used in quantum computers and quantum algorithms.

Quantum Superposition: Superposition is a fundamental principle in quantum mechanics that sets classical bits apart from 'qubits' or quantum bits [31]. Quantum Superposition is a quantum system feature allowing the particle to exist simultaneously in several different quantum states. Considering an example of an electron, electrons have a quantum feature called spin which is a kind of intrinsic angular momentum, where the electron can be in either spin up or spin down state. In a magnetic field the spin of an electron can be influenced and manipulated for computational purposes [32]. Until measured the state of the electron spin momentum cannot be determined, only after measuring the electron that might fall into either of the two states. This quantum phenomenon lets qubits be in the same state and in others at the same time to say 'up' and 'down' spin at the same time. Superposition not only puts up with the binary constraints of classical computers but also takes up the value for both simultaneously i.e., a qubit can be storing '0', '1' or resulting vector between '0' and '1' [33]. This property shown by a qubit being in either of the two states until measured is called quantum superposition.

Quantum Entanglement: Entanglement can be stated as a phenomenon of correlation among nonlocal measurements. Information is not sent from one to another using any device but allows one to obtain mutual information by exploiting the quantum physics phenomenon [34]. Any effect on one of the entangled qubits from its surrounding is automatically translated to another qubit in the entangled system. Entanglement establishes a strong correlation among two or more quantum particles, the particles are in such unison that the distance among them does not matter even theoretically placed at the ends of the universe. Einstein described this phenomenon of the interaction of particles as "spooky action at a distance" [33]. Due to Heisenberg's uncertainty principle, the amount of information extraction from a quantum state is reduced. Measuring the entangled state may lead to the termination of entanglement among particles. The preferred way of viewing an entanglement is through a physical resource, which is used up or "dissipated" as measurements are performed and exchanged for the type of communication we want to accomplish [34].

Quantum Decoherence: Loss of quantum coherence is specified as quantum decoherence. As long as there exists a definitive phase relation among different states, the system is said to be in coherence, and it can maintain coherence indefinitely if a system is perfectly isolated. Decoherence is the resultant vector of system interaction with the other environmental variables. Decoherence is a major setback for the quantum factorization algorithm as a coherence of a quantum wave function is essential for the algorithm. Decoherence causes environmentally induced super selection which destroys superposition between the states of a preferred pointer basis [35].

Qubits and Quantum Registers: Qubits, for a quantum computer are the basic units that can be physically interpreted by a two-level quantum mechanical system. Qubits are fundamental building blocks for a Quantum Computer and their ability to be in a superposition between '0' and '1' sets them apart from their classical counterpart. Mathematically a qubit can be represented using a unit vector in a two-dimensional complex Hilbert Space as:

$$|\psi\rangle = \alpha|0\rangle + \beta|1\rangle$$

where the two complex numbers represent the probability amplitude of the state and '0' and '1' is the basic states corresponding to the basic states of classical bits. As the absolute squares of the amplitudes equate to probabilities and must follow this $|\alpha|^2 + |\beta|^2 = 1$. The major difference between qubits and classical bits is the ability of qubits to be in a state of superposition of '0' and '1'. A quantum register consists of multiple qubits working together [36].

3 Past, Present, and Future of AI, ML, and QC

This section discusses various research done in the field of Quantum Computation, Artificial Intelligence, and Machine Learning integrated. It provides a historical finding of Artificial Intelligence, ML, and its evolution with Quantum Computation and new prominent research. Artificial Intelligence or AI was first termed at the Dartmouth College conference in 1956 [37], it subjugates numerous subfields that deal with understanding and abstraction of various human capacities noteworthy to be specified as intelligence and attempting to realize those capacities in machines for performing 'intelligent' tasks [38]. The resultant integration of AI and quantum computing can be classified into resultant into two perspectives first using ideas from quantum computing for solving certain problems in AI and conversely the second fortification of quantum application with help of AI [36]. Machine learning (ML) another aspect of AI stems from the pattern recognition perspective of determining results from the availed data which boldly emerges into systems like recognition of handwritten text and statistical learning. ML majorly deals with two prominent learning ways of supervised and unsupervised learning which are in close relation to data-mining and data analysis [39]. A broader view was erected for ML after the inclusion of the reinforcement learning field [40, 41].

ML is one of the prominent areas in which quantum computation and AI have shown promising results. ML faces several problems such as availability of data, extensive time for model training, and memory utilization, QC might provide solutions to such problems faced while working in the field of ML. Various research papers are devoted to the quantum generalization of computational learning theory [36]. The interest of this research specifies for provision of quantum algorithms better than their classical counterparts for certain usage like object detection and Boolean functions. Research in these related fields is specified in [15], and surveys are also provided in [42]. Decision trees formulate multiple decision problems, the Hamiltonian evolution-based quantum algorithm showed exponential speedup in comparison to classical counterparts in solving decision problems which were proposed by Farhi and Gutmann [43]. Grover's algorithm showcased the quadratic speed growth for search algorithms [44]. This led in 1999, Hoggs proposed how quantum search algorithms can be applied in AI but even after more than 10 years, any significant quantum searching algorithm in AI hasn't been reported [45]. Game theory is pretty prominently used in AI mostly for multi-agent systems and distributed AI. Benjamin and Hayden did introduce quantum games with multiple players [46]. Before this Lawrence introduced quantization on non-zero-sum, two-player games in [47]. Nelson, McEvoy, and Pointer were able to establish 'spooky action at a distance behavior' as quoted by Einstein, association in natural language [48], over which Bruza proposed a model describing 'spooky action at a distance behavior' similar to quantum entanglement by modeling word association in terms of tensor product [48]. In classical computation 'Bayesian' method is applied for statistical inference; statistical inference is at the heart of quantum theory because of its probabilistic nature of the quantum system [36]. Recently several quantum Bayes rules have made its way into physical literature for example [49]. Quantum generalization of Bayesian networks was introduced by Tucci for which instead of probabilities complex amplitudes are assigned to nodes that are used for the calculation of probabilities for some physical experiments [50]. Pearl introduced to casual Bayesian network augmenting it with local operations specifying probability distribution behavior in relation to external interventions [51]. Representation of local operations by superoperators which are the popular mathematical formalism of the dynamics of open quantum systems in a quantum casual system was proposed by Laskey [52]. Quantum state tomography [53] refers to a physical process that can produce a quantum state repeatedly; this resembles repetitive learning and would apply to quantum learning [36]. Similar to quantum state tomography, quantum process tomography theory was proposed by Chuang and Nielsen [54] and Poyatos et al. [55] for quantum operations. Extensive research can also be seen in the fields of Quantum neural networks [56] and Quantum genetic algorithms [57]. Advancements in Machine learning and Quantum Information Processing was tentatively established by the early theories of quantum signal processing [58], probabilistic aspects of quantum theory and quantum state estimation [59], and also by [60] which would lead to modern quantum metrology [61] including statistical analyses. Hamiltonian estimation identifies the optimal methods for estimation of the Hamiltonian parameters, where various estimation problems in Machine Learning can be given a perspective from Hamiltonian estimation [38]. This method was subsequently made an acquaintance with trusted quantum simulators resulting in an efficient and powerful estimation scheme [62], it also proves to be robust to moderate noises and imperfection of the simulators [62]. A restricted version working with simulators for this estimation method was released in [63]. Application of ML a technique like swarm optimization was first introduced and pioneered by the works of Hentschel and Sanders [64, 65] and later was built on by Sergeevich and Bartlett [66]. Later more well-known methods of differential evolution proved to be superior and efficient computationally [67]. Recently, combination of Hamiltonian estimation, Bayesian and sequential Monte Carlo-based estimation along with particle, and swarm optimization techniques were further proposed in [68]. Feedback-based learning and optimization were given attention and were explored by Chen where the application of a modified Q-learning algorithm for Reinforcement Learning on canonical control problems was discussed in [1]. The potential for Reinforcement Learning methods was showcased in context with optimal parameter estimation and also, typical optimal control scenarios, in [69]. The author Palittapongarnpim also provided a perspective on Machine Learning and Reinforcement Learning approaches for resolving quantum measurement and control problems. Reinforcement learning methods were used on projective simulation [70] for the task of protecting Quantum Computation from the local stray fields [71]. In a Measurement-Based Quantum Computation [72], by performing adaptive single-qubit projective measurements on a large entangled resource state the computation is driven [72, 73]. For a certain scenario if a resource state is exposed to a stray field, then qubits undergo a local rotation, for prevention of this scenario [71], introduced a learning agent playing the role of local probe qubit, learning ways for compensation of unknown fields. Work by Krenn [74] consisted of using a feedback-assisted search algorithm to identify previously unknown configurations which generate novel highly entangled states. This provided a perspective for designing novel quantum experiments that are able to automate. Combining the previous results with AI agents in quantum information protocols, these could act as 'lab robots' in future quantum laboratories [70]. An attempt to classify and characterize the way Machine Learning and give it an outlook for a quantum world was done in [75]. There, the basic object introduced is the database of labeled quantum or classical objects, which may come in copies. Such a database can, in general, then be processed to solve various types of tasks, using classical or quantum processing. Monràs addressed a broad perspective for quantum generalizations of supervised learning in its inductive form recently [76]. Key result provided by [77] is that quantum states are learnable with sample complexity scaling only linearly in the number of qubits, that is, logarithmically in the dimension of the density matrix. The comprehensive paper by Cheng [78] explores the problems of the learnability of quantum measurements. Quantum generalization of PAC learning was first introduced by Bshouty and Jackson [79] in which the quantum example oracle was defined to output coherent superposition that was defined. Cross considered the learning of linear Boolean functions under uniform distribution, ingenuity of his work is the assumption of noise allowing for evidence of a classical quantum learnability separation [80]. Recently Grilo and Kerenidis [81], showed that learning with errors which is an important topic in Computational Learning Theory is efficiently learnable for quantum examples. Gavinsky, hypothesized the Predictive Quantum (PQ) model where only one (or polynomially few) evaluation of the hypothesis is required considering a quantum Probability approximately correct learning setting [82]. In this setting author was able to identify a relational concept class which is not (polynomially) learnable in the classical case, but is PQ learnable under a standard quantum oracle under uniform distribution. Babbush [83] significantly generalized the scope of loss functions which led it to be embedded into quantum architectures on the observations that any polynomial unconstrained binary optimization can, with small overhead, be mapped onto a (slightly larger) Quadratic unconstrained binary optimization problem. Building on the work in [84], recent work by Sieberer and Lechner [85] showed ways to achieve programmable adiabatic architectures that allowed running of algorithms of which the weights were in superposition. In Wiebe [86] a quantum algorithm is devised that creates coherent encodings of the target distributions relying on quantum amplitude amplification, often attaining quadratic improvements in the number of training points and even exponential improvements in the number of neurons, in some regimes. In association rules mining [87], quadratic improvements have also been obtained in pure data mining contexts. Roughly speaking, it can identify correlations between objects in large databases. Quantum linear algebra is a collection of algorithms which solve certain linear algebra problems by directly encoding numerical vectors into state vectors. QLA includes algorithms for matrix inversion, principal component analysis [88, 89] and many others. These quantum sub-routines can result in speeding up numerous ML algorithms. Clader [90] brought to attention that standard matrix pre-conditioning techniques can also be applicable in the Quantum Linear Systems scheme. Zhao [91], demonstrated ways in which Quantum Linear Systems can be utilized for improving Gaussian process regression (GPR). Gaussian process regression is a process in which an initial distribution over possible latent functions is refined by taking into account the training set points, using Bayesian inference. The output of Gaussian process regression is, roughly speaking, a distribution over models which are consistent with the observed dataset [38]. Applications of QIP to reinforcement and other interactive learning problems has been comparatively less studied, when compared to quantum enhancements in supervised and unsupervised problems. One of the first proposals which provided a coherent view on learning agents from a physics perspective was that of Projective Simulation [70]. The Projective Simulation model has been combined with such additional learning machinery in an application to robotics and haptic skill learning [92–95]. As per the trends seen so far in the research discussed throughout this section, we can classify the research regarding AI, ML and QC into four criteria as follows:

i. Application of Machine Learning and Artificial Intelligence in Quantum Physics experimentation.
ii. Quantum Generalization of Machine Learning-based tasks.
iii. Quantum Enhancement application of Machine Learning and Artificial Intelligence implemented on Quantum Machines.
iv. Quantum Learning Agent and elements of Quantum Artificial Intelligence.

These areas classified proven to be helping interdisciplinary growth of fields over the imposed research. These research studies have proven to be of the betterment of both the fields in ways like designing quantum algorithms for problem-solving and increasing the efficiency of Artificial Intelligence algorithms and techniques, effective methods for formulating problems in Artificial Intelligence by using the concepts in the fields of quantum theory, development of Artificial Intelligence, and Machine Learning solutions for the present problems in the quantum world. Even so with the benefits and the long-run opportunities first class of the research in these fields is in the initial stage of development. Much more research in the field of efficiency of algorithms would be commendable as it is scarce. We can safely claim that Quantum computers would play an integral part in the field of Artificial Intelligence and Machine Learning systems once realized.

4 Experimentation

This section is primarily focused on the methods, tools, and processes to implement the QML algorithms for classifying an approach to an SDN as an attack or normal connection request based on the available parameters from the SDN dataset. Module created was tested and trained on a simulated environment on an Intel i7 processor with an 8 GB RAM and 4 GB Graphic Processor, device with a Windows 10 Operating System. With an ever-increasing use of Virtual Private Networks (VPN) by Tech Companies and users concerned with security, SDN has never been more important than now. Software-Defined Network (SDN) has been developed to reduce network complexity through control and managing the whole network from a centralized location. Today, SDN is widely implemented in many data centers' network environments. Nevertheless, emerging technology itself can lead to many vulnerabilities and threats that are still challenging for manufacturers to address [96, 97]. AI and ML have been extensively used when it concerns SDN for solving various of its problems, and various algorithms of ML have been used for creating Load balanced systems, AI Agents to control traffic, and Defense against DDoS attacks creating 'Intelligent Network Systems' [98]. We had previously attempted to utilize several AI algorithms on the dataset and realized that hardly any quantum algorithms or quantum techniques were proposed or used with the intention of intrusion detection in an SDN network. This is our attempt to access the possibility that these novel technologies implemented in this way of application would be able to create better and advanced systems in the future. This manuscript helps in expanding our research horizon, an exploration attempt on the collaboration of these two technologies. We also address the required developments necessary for a fully functional quantum system that would be necessary for such implantation on an industrial scale.

For this implementation, we utilized 'Qiskit' an Open-Source Quantum Development Kit [99], developed by IBM, a software to assist in the development of Quantum Circuits and Quantum Algorithms. Qiskit provides the complete set of tools required for creating essential quantum systems and utilizing quantum simulators, which in turn accelerate the development of quantum applications and provides an adaptable environment for the user, letting the user work with quantum computers at the level of circuits, pulses, and algorithms. Additionally, several domains specific application APIs exist on top of this core module [99]. This experiment was performed in two phases: first was Data Preparation and the second was Model Training, Fig. 1 is a flowchart attempting to explain the setup and the experimental process. To train a model for predicting we used a novel SDN intrusion dataset published by et. al. [96] which consisted of 84 columns and 68,424 rows of data for normal data and 84 columns and 136,743 rows of various types of intrusion attacks, which in total created a data size of 205,167 rows. During the data pre-processing phase, we decided to drop a few columns as they wouldn't provide any significant insight into the training model and also didn't have a major contribution to the intent of this experimentation, this, in turn, reduced the columns from 84 to 63. Later the data was subjected to several data size reduction techniques to hold the importance of columns and reduce them to effective qubit testable numbers, where column reduction method was used for this resulting in an efficient training of the quantum ML model. Reduction methods were used because the quantum simulator can simulate smoothly only for 4 qubits at max on the device used for experimentation and more qubits would have resulted in extensively heavy computational time, hence techniques like 'Truncated SVD along with t-SNE' and 'Principal Component Analysis (PCA)' were utilized for reducing 63 columns into variations of 2, 3 or 4. Employing truncated singular value decomposition (SVD), the Truncated SVD transformer can perform linear dimensionality reduction. The Truncated SVD estimator uses contradicting approach to PCA, before computing the singular value decomposition it does not center the data. This means it can work with sparse matrices efficiently [100]. t-distributed Stochastic Neighbour Embedding(t-SNE) is a tool utilized for visualizing high-dimensional data by converting similarities among data points to joint probabilities and oversees to minimize the Kullback–Leibler divergence between the high-dimensional data and the joint probabilities of the low-dimensional embedding [101, 102]. Using Singular Value Decomposition of the data to project data to a lower-dimensional space linear dimensionality is reduced by Principal Component Analysis (PCA). Before applying the SVD, the input data is centered but it is not scaled for each feature [103, 104]. Dependent on the shape of the input data and the number of components to extract, PCA uses the LAPACK implementation of the full SVD or a randomized truncated SVD by the method of et al. [105].

For training we used to function from Qiskit aqua algorithms like VQC [106] which were based on VQE, the Variational-Quantum-Eigensolver (VQE) [107–109] is a quantum/classical hybrid algorithm that can be used to find eigenvalues of

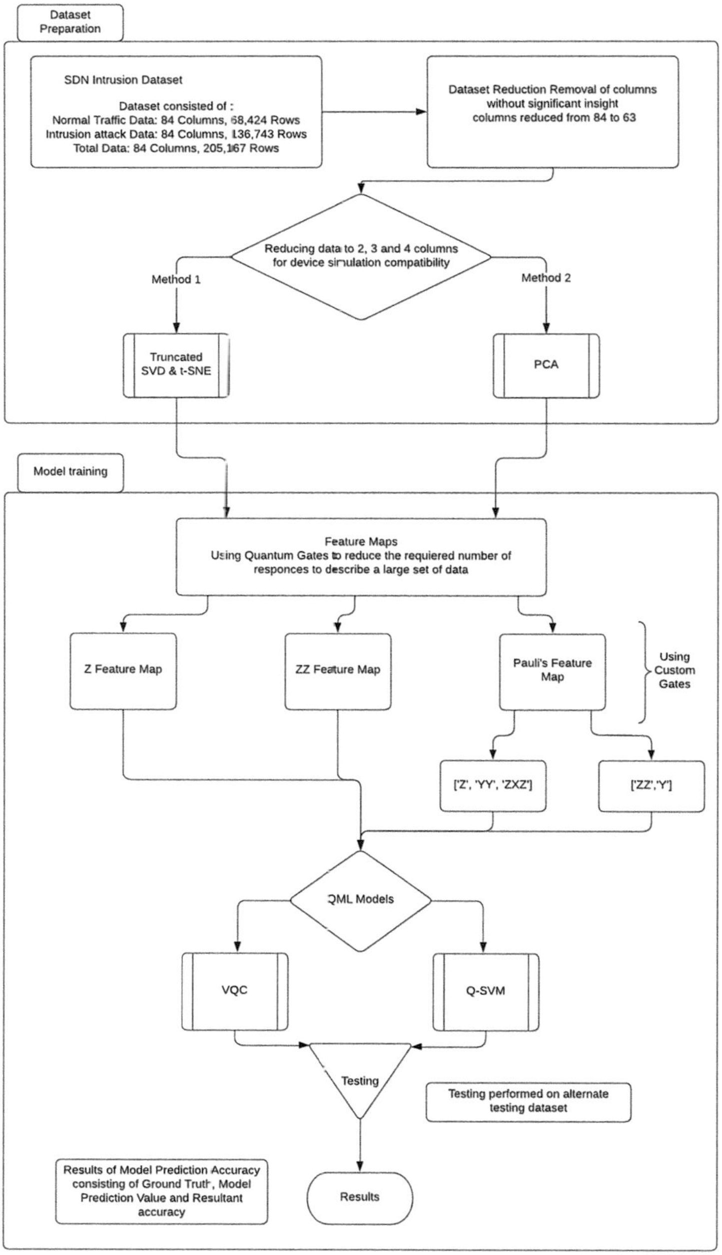

Fig. 1 Experimental process flowchart

an (often large) matrix 'H'. When this algorithm is used in quantum simulations, 'H' is typically the Hamiltonian of some system [110–112]. In this hybrid algorithm, a quantum subroutine is run inside of a classical optimization loop. Based on VQE, Variational Quantum Classifier (VQC) optimizes a parameterized quantum circuit using the variational method for solving problems in a quantum processor by providing a solution that cleanly separates the data [106]. Another algorithm used for classification was the Quantum-enhanced Support Vector Machine (Q-SVM), for solving classification problems where the QSVM algorithm needs to apply feature maps for which computing the kernel is essential. Classically computing kernels are inefficient which concludes that with the size of the problem the resources required for computation are expected to scale exponentially. By using a Quantum processor a direct estimation of the kernel in the feature space is provided by QSVM to solve the problem presented for classical computation [113]. Along with the Quantum Classification algorithms, Qiskit provides several important feature maps, A feature map reduces the required number of resources to describe a large set of data, supposing its relation with dimensionality reduction. With a greater number of variables involved a major problem stems while performing analysis of complex data, as a large amount of memory and computation power is required for analysis with a large number of variables, which may even lead to overfitting of the classification algorithm to training samples and with regards to new samples may generalize poorly [114, 115]. Z-Feature Map, ZZ-Feature Map and Pauli's Feature Map are the three feature maps utilized in this project for processing. After reducing the data size for computable time on a classical machine simulator Qiskit Aer [116] was used to simulate similar condition on the machine. Table 1, contains results generated over multiple simulations and tunings of feature maps, number of qubits, etc.

During the process, we were able to recognize major setbacks with the quantum systems especially the QML algorithms comparing them to the accuracies of their Classical counterparts [96]. One of the problems was the unavailability of Quantum Computers in hands of the general public freely, as using simulators is not only time-consuming for running these calculations for a long time but also not extensively useful for a higher number of qubits. As quantum computers are still developing the maximum number of qubits hasn't been able to reach the required number and a higher number of qubits quantum computers are still widely in the research phase. We were not able to train the data attributes in their original form but had to compress them using various methods as mentioned before this might have put a strain on the calculations and might have affected the accuracy of the overall model. Data points were humongous and hence were only able to run for a handle of sample from the original full dataset and iteration done to get average accuracy. Simulators on the local machine are handy to understand the concepts and working but the calculation time is steadily ranging for more than an hour to execute a model at a time, this could be said as a drawback of the classical machine but also prove a point of why actual devices are much better. Another big setback for the QML is the limited number of quantum algorithms that can be applicable for QML, hence a lot more effort is needed to not only create more quantum machine learning algorithms but also to encode multiple ones discussed before for easier use.

Table 1 Results of intrusion detection

Data reduction method	Train size	Test size	No. of qubits	Feature map	Optimizer	VQC accuracy %	QSVM accuracy %
Truncated SVD and TSNE	200	20	2	ZZ feature map	COBYLA	62.5	57.5
PCA	500	50	2	ZZ feature map	COBYLA	75	53.3
PCA	500	50	3	ZZ feature map	COBYLA	80	37.5
Truncated SVD and TSNE	200	20	4	ZZ feature map	COBYLA	54	31
PCA	500	50	4	ZZ feature map	COBYLA	72.6	35
Truncated SVD and TSNE	200	20	3	ZZ feature map	COBYLA	65	50
Truncated SVD and TSNE	500	50	2	ZZ feature map	COBYLA	61.9	52.3
PCA	500	50	2	Pauli's feature map = ['Z','YY','ZXZ']	COBYLA	62.3	51
PCA	500	50	3	Pauli's feature map = ['Z','YY','ZXZ']	COBYLA	69	32.4
Truncated SVD and TSNE	200	20	2	Pauli's feature map = ['ZZ','Y']	COBYLA	52	50
Truncated SVD and TSNE	500	50	3	Pauli's feature map = ['ZZ','Y']	COBYLA	63	55
PCA	500	50	2	Z feature map	COBYLA	60	0.45
PCA	500	50	3	Z feature map	COBYLA	85	55
PCA	500	50	4	Z feature map	COBYLA	82.5	67.5

5 Conclusion

This manuscript reviewed various aspects and research in the field of Quantum Computation and Artificial Intelligence, along with proof of concepts engaging with SDN and QML. It provides with latest trends and research regarding Quantum Computers and, a detailed discussion of their structure and functionality. Compared with the growth of Classical Computers in their early days we can safely say that Quantum Computers are in the transistor phase. Still being in such an early stage of its creation Quantum Computers have shown promising results as we discussed so far.

Various algorithms have been devised to be performed over Quantum Computer and have shown that they could outperform the traditional computers and their performing power. In the fields of Artificial Intelligence (AI) and Machine Learning (ML), the development is mutual as several ways are suggested so far for using the computational power of Quantum Computers for better performance and training of AI and ML models. Similarly, AI and ML were helpful in the development of models for the prediction of quantum states and the creation of AI Agents for improving the research in the field of Quantum Computation. Even though Quantum Computers seem to be lacking as compared to their Classical Counterpart on paper but it's not all true as the novel development in the early age of research is still able to produce such efficient results is considered outstanding. With better investment and more involvement of people in the Quantum Computation field better models and algorithms could be created with even more efficient results. Once we can overcome the current drawbacks posed for this system, we could not only be able to create exceptionally accurate models for computational systems like SDN which have to encounter ever-increasing traffic and require faster evaluation but also be able to predict much more complex systems like the climate, drugs simulation, newer material research and even might be able to find a solution for the climate crisis faced now.

References

1. R. Barends et al., Logic gates at the surface code threshold: Superconducting qubits poised for fault-tolerant quantum computing. Nature **508**(7497), 500–503 (2014). https://doi.org/10.1038/nature13171
2. J. Biamonte, P. Wittek, N. Pancotti, P. Rebentrost, N. Wiebe, S. Lloyd, Quantum machine learning, in *Nature*, vol. 549, no. 7671. (Nature Publishing Group, 2017), pp. 195–202. (13 Sept 2017). https://doi.org/10.1038/nature23474
3. S. Debnath, N.M. Linke, C. Figgatt, K.A. Landsman, K. Wright, C. Monroe, Demonstration of a small programmable quantum computer with atomic qubits. Nature **536**(7614), 63–66 (2016). https://doi.org/10.1038/nature18648
4. L. Dicarlo et al., Demonstration of two-qubit algorithms with a superconducting quantum processor. Nature **460**(7252), 240–244 (2009). https://doi.org/10.1038/nature08121
5. E. Farhi, J. Goldstone, S. Gutmann, H. Neven, Quantum algorithms for fixed qubit architectures (Mar 2017). Accessed 22 July 2020. http://arxiv.org/abs/1703.06199
6. B.L. Higgins, D.W. Berry, S.D. Bartlett, H.M. Wiseman, G.J. Pryde, Entanglement-free Heisenberg-limited phase estimation. Nature **450**(7168), 393–396 (2007). https://doi.org/10.1038/nature06257
7. T. Monz et al., Realization of a scalable Shor algorithm, in *Science (80-.)*, vol. 351, no. 6277 (Mar 2016), pp. 1068–1070. https://doi.org/10.1126/science.aad9480
8. N. Ofek et al., Extending the lifetime of a quantum bit with error correction in superconducting circuits. Nature **536**(7617), 441–445 (2016). https://doi.org/10.1038/nature18949
9. R. Van Meter, S.J. Devitt, Local and distributed Quantum Computation, in *Computer (Long. Beach. Calif.*, vol. 49, no. 9, (May 2016), pp. 31–42. https://doi.org/10.1109/MC.2016.291
10. P. Benioff, The computer as a physical system: a microscopic quantum mechanical Hamiltonian model of computers as represented by Turing machines. J. Stat. Phys. **22**(5), 563–591 (1980). https://doi.org/10.1007/BF01011339
11. Y. M.-S. Radio, undefined Moscow, and undefined 1980, Computable and uncomputable.

12. R.P. Feynman, Simulating physics with computers. Int. J. Theor. Phys. **21**(6–7), 467–488 (1982). https://doi.org/10.1007/BF02650179
13. D. Deutsch, Quantum theory, the Church-Turing principle and the universal quantum computer. Proc. R. Soc. London. A. Math. Phys. Sci. **400**(1818), 97–117 (1985). https://doi.org/10.1098/rspa.1985.0070
14. J. Ozsa2, Rapid solution of problems by Quantum Computation. Proc. R. Soc. London. Ser. A Math. Phys. Sci. **439**(1907), 553–558 (Dec. 1992). https://doi.org/10.1098/rspa.1992.0167
15. E. Bernstein, U. Vaziranit, Quantum complexity theory, in *Proceedings of the Annual ACM Symposium on Theory of Computing*, Jun. 1993, vol. Part F129585, pp. 11–20. https://doi.org/10.1145/167088.167097
16. D. Simon, On the power of Quantum Computation.
17. S. Lloyd, A potentially realizable quantum computer, in *Science (80-)*, vol. 261, no. 5128 (Sept. 1993), pp. 1569–1571. https://doi.org/10.1126/science.261.5128.1569
18. P.W. Shor, Algorithms for Quantum Computation: discrete log and factoring extended abstract.
19. L.M.K. Vandersypen, M. Steffen, G. Breyta, C.S. Yannoni, M.H. Sherwood, I.L. Chuang, Experimental realization of Shor's quantum factoring algorithm using nuclear magnetic resonance. Nature **414**(6866), 883–887 (2001). https://doi.org/10.1038/414883a
20. L.K. Grover, Quantum mechanics helps in searching for a needle in a haystack. Phys. Rev. Lett. **79**(2), 325–328 (1997). https://doi.org/10.1103/PhysRevLett.79.325
21. E. Farhi, S. Gutmann, An analog analogue of a digital Quantum Computation (Dec. 1996). Accessed 23 July 2020. http://arxiv.org/abs/quant-ph/9612026
22. Y. Nakamura, Y.A. Pashkin, J.S. Tsai, Coherent control of macroscopic quantum states in a single-Cooper-pair box. Nature **398**(6730), 786–788 (1999). https://doi.org/10.1038/19718
23. Y. Nakamura, Y.A. Pashkin, J.S. Tsai, Rabi oscillations in a Josephson-junction charge two-level system. Phys. Rev. Lett. **87**(24), 246601-1–246601-4 (Dec. 2001). https://doi.org/10.1103/PhysRevLett.87.246601
24. J.I. Cirac, P. Zoller, Quantum Computations with cold trapped ions. Phys. Rev. Lett. **74**(20), 4091–4094 (1995). https://doi.org/10.1103/PhysRevLett.74.4091
25. J.D. Hidary, Quantum Computing: an applied approach (2019)
26. L. Gyongyosi, S. Imre, A survey on quantum computing technology. Comput. Sci. Rev. **31**, 51–71 (2019). https://doi.org/10.1016/j.cosrev.2018.11.002
27. S. Imre, Quantum communications: Explained for communication engineers. IEEE Commun. Mag. **51**(8), 28–35 (2013). https://doi.org/10.1109/MCOM.2013.6576335
28. M. A. Nielsen and I. L. Chuang, *Quantum Computation and Quantum Information*. Cambridge University Press, 2010.
29. M. Vogel, Quantum Computation and quantum information, by M.A. Nielsen and I.L. Chuang. Contemp. Phys. **52**(6), 604–605 (2011). https://doi.org/10.1080/00107514.2011.587535
30. R.D. Van Meter, Architecture of a Quantum multicomputer optimized for Shor's factoring algorithm. (July 2006). Accessed 29 Jul 2020. http://arxiv.org/abs/quant-ph/0607065
31. Superposition and entanglement. https://www.quantum-inspire.com/kbase/superposition-and-entanglement/. Accessed 28 July 2020
32. Quantum superposition|Joint quantum institute. https://jqi.umd.edu/glossary/quantum-superposition. Accessed 28 July 2020
33. Quantum computing series, Part 4: superposition in quantum mechanics | by Scott Amyx|Medium. https://medium.com/@ScottAmyx/quantum-computing-series-part-4-superposition-in-quantum-mechanics-381b98180f62. Accessed 28 July 2020.
34. K. Shannon, E. Towe, O. Tonguz, C.R. Mar, On the use of quantum entanglement in secure communications: a survey, (pp. 1–16).
35. I.L. Chuang, R. Laflamme, P.W. Shor, W.H. Zurek, Quantum computers, factoring, and decoherence, in *Science (80-)*, vol. 270, no. 5242 (Dec. 1995), pp. 1633–1635. https://doi.org/10.1126/science.270.5242.1633
36. M. Ying, W. Scherer, Quantum Computation, quantum theory and AI. Artif. Intell. **174**(2), 162–176 (2019). https://doi.org/10.1007/978-3-030-12358-1
37. S. Russell, P. Norvig, *Artificial Intelligence A Modern Approach Third Edition*

38. V. Dunjko, H.J. Briegel, Machine learning & artificial intelligence in the quantum domain: a review of recent progress. Reports Prog. Phys. **81**(7) (2018). https://doi.org/10.1088/1361-6633/aab406
39. S. Shalev-Shwartz and S. Ben-David, *Understanding Machine Learning: From Theory to Algorithms*, vol. 9781107057135 (Cambridge University Press, 2013)
40. R.S. Sutton, A.G. Barto, *Reinforcement Learning: an introduction Second edition*, in progress.
41. E. Alpaydın, *Introduction to Machine Learning Second Edition Adaptive Computation and Machine Learning*
42. R. Bonner, *Quantum Computation and Learning*
43. E. Farhi, S. Gutmann, Quantum Computation and decision trees. Phys. Rev. A At. Mol. Opt. Phys. **58**(2), 915–928 (Aug 1998). https://doi.org/10.1103/PhysRevA.58.915
44. L.K. Grover, A fast quantum mechanical algorithm for database search, in *Proceedings Annual ACM Symposium Theory Computing*, vol. Part F129452 (May 1996), pp. 212–219. Accessed 30 July 2020. http://arxiv.org/abs/quant-ph/9605043
45. T. Hogg, Quantum search heuristics. Phys. Rev. A - At. Mol. Opt. Phys. **61**(5), 7 (2000). https://doi.org/10.1103/PhysRevA.61.052311
46. S.C. Benjamin, P.M. Hayden, Multi-player quantum games. Phys. Rev. A At. Mol. Opt. Phys. **64**(3), 4 (2000). https://doi.org/10.1103/PhysRevA.64.030301
47. J. Eisert, M. Wilkens, M. Lewenstein, Quantum games and quantum strategies. Phys. Rev. Lett. **83**(15), 3077–3080 (1999). https://doi.org/10.1103/PhysRevLett.83.3077
48. P.D. Bruza, K. Kitto, D. Nelson, C. McEvoy, Extracting spooky-activation-at-a-distance from considerations of entanglement. Lecture Notes Computer Science (including subseries Lecture Notes in Artificial Intelligence and Lecture Notes in Bioinformatics), vol. 5494, pp. 71–83 (Jan 2009). Accessed 30 July 2020. http://arxiv.org/abs/0901.4375
49. R. Schack, T.A. Brun, C.M. Caves, Quantum Bayes rule. Phys. Rev. A At. Mol. Opt. Phys. **64**(1), 143051–143054 (2000). https://doi.org/10.1103/PhysRevA.64.014305
50. R.R. Tucci, Quantum Bayesian nets. Int. J. Mod. Phys. B **09**(03), 295–337 (1995). https://doi.org/10.1142/s0217979295000148
51. J. Pearl et al., Causality: models, reasoning, and inference RE E EV V VI I IE E EW W WE (2000)
52. K.B. Laskey, *Quantum Causal Networks.*
53. K. Vogel, H. Risken, Determination of quasiprobability distributions in terms of probability distributions for the rotated quadrature phase. Phys. Rev. A **40**(5), 2847–2849 (1989). https://doi.org/10.1103/PhysRevA.40.2847
54. I.L. Chuang, M.A. Nielsen, Prescription for experimental determination of the dynamics of a quantum black box. J. Mod. Opt. **44–11**(12), 2455–2467 (1996). https://doi.org/10.1080/09500349708231894
55. J.F. Poyatos, J.I. Cirac, P. Zoller, Complete characterization of a quantum process: the two-bit quantum gate. Phys. Rev. Lett. **78**(2), 390–393 (1996). https://doi.org/10.1103/PhysRevLett.78.390
56. A. A. Ezhov, Pattern recognition with quantum neural networks, in Lecture Notes in Computer Science (including subseries Lecture Notes in Artificial Intelligence and Lecture Notes in Bioinformatics) vol. 2013 (2001), pp. 60–71. https://doi.org/10.1007/3-540-44732-6_7
57. A. Narayanan, M. Moore, Quantum-inspired genetic algorithms, in *Proceedings of the IEEE Conference on Evolutionary Computation* (1996), pp. 61–66. https://doi.org/10.1109/icec.1996.542334
58. C.W. Helstrom, Quantum detection and estimation theory. J. Stat. Phys. **1**(2) 231–252 (1969). (Kluwer Academic Publishers-Plenum Publishers). https://doi.org/10.1007/BF01007479
59. A. Holevo, *Probabilistic and Statistical Aspects of Quantum Theory.* Edizioni della Normale (2011)
60. S.L. Braunstein, C.M. Caves, Statistical distance and the geometry of quantum states. Phys. Rev. Lett. **72**(22), 3439–3443 (1994). https://doi.org/10.1103/PhysRevLett.72.3439
61. V. Giovannetti, S. Lloyd, L. MacCone, Advances in quantum metrology. Nat. Photonics **5**(4), 222–229 (2011). (Nature Publishing Group). https://doi.org/10.1038/nphoton.2011.35

62. N. Wiebe, C. Granade, C. Ferrie, D. Cory, Quantum Hamiltonian learning using imperfect quantum resources. Phys. Rev. A At. Mol. Opt. Phys. **89**(4), 042314 (2014). https://doi.org/10.1103/PhysRevA.89.042314
63. G. Wang, Quantum algorithm for linear regression. Phys. Rev. A **96**(1), 012335 (2017). https://doi.org/10.1103/PhysRevA.96.012335
64. A. Hentschel, B.C. Sanders, Efficient algorithm for optimizing adaptive quantum metrology processes. Phys. Rev. Lett. **107**(23), 233601 (2011). https://doi.org/10.1103/PhysRevLett.107.233601
65. A. Hentschel, B.C. Sanders, Machine learning for precise quantum measurement. Phys. Rev. Lett. **104**(6), 063603 (2010). https://doi.org/10.1103/PhysRevLett.104.063603
66. A. Sergeevich, S.D. Bartlett, Optimizing qubit Hamiltonian parameter estimation algorithm using PSO (2012). https://doi.org/10.1109/CEC.2012.6252948
67. N.B. Lovett, C. Crosnier, M. Perarnau-Llobet, B.C. Sanders, Differential evolution for many-particle adaptive quantum metrology. Phys. Rev. Lett. **110**(22), 220501 (2013). https://doi.org/10.1103/PhysRevLett.110.220501
68. M.P.V. Stenberg, O. Köhn, F.K. Wilhelm, Characterization of decohering quantum systems: machine learning approach. Phys. Rev. A **93**(1), 012122 (2016). https://doi.org/10.1103/PhysRevA.93.012122
69. The Deutsch-Jozsa Algorithm. The Math, the Code, and Why It Works|by Ryan Lam|Data Driven Investor|Medium. https://medium.com/datadriveninvestor/the-deutsch-jozsa-algorithm-f8611e2ff3ff. Accessed 26 July 2020
70. H.J. Briegel, G. De Las Cuevas, Projective simulation for artificial intelligence. Sci. Rep. **2**(1), 1–16 (2012). https://doi.org/10.1038/srep00400
71. M. Tiersch, E.J. Ganahl, H.J. Briegel, Adaptive quantum computation in changing environments using projective simulation. Sci. Rep. **5**(1), 12874 (2015). https://doi.org/10.1038/srep12874
72. R. Raussendorf, H.J. Briegel, A one-way quantum computer. Phys. Rev. Lett. **86**(22), 5188–5191 (2001). https://doi.org/10.1103/PhysRevLett.86.5188
73. H.J. Briegel, R. Raussendorf, Persistent entanglement in arrays of interacting particles. Phys. Rev. Lett. **86**(5), 910–913 (2001). https://doi.org/10.1103/PhysRevLett.86.910
74. M. Krenn, M. Malik, R. Fickler, R. Lapkiewicz, A. Zeilinger, Automated search for new quantum experiments. Phys. Rev. Lett. **116**(9), 090405 (2016). https://doi.org/10.1103/PhysRevLett.116.090405
75. E. Aïmeur, G. Brassard, S. Gambs, Machine learning in a quantum world, in Lecture Notes in Computer Science (including subseries Lecture Notes in Artificial Intelligence and Lecture Notes in Bioinformatics), vol. 4013 LNAI (2006), pp. 431–442. https://doi.org/10.1007/11766247_37
76. A. Monràs, G. Sentís, P. Wittek, Inductive supervised quantum learning. Phys. Rev. Lett. **118**(19), 190503 (2017). https://doi.org/10.1103/PhysRevLett.118.190503
77. S. Aaronson, *Quantum Machine Learning Algorithms: Read the Fine Print*
78. H.-C. Cheng, M.-H. Hsieh, P.-C. Yeh, The learnability of unknown quantum measurements. Quantum Inf. Comput. **16**(7–8), 615–656 (2015). Accessed 31 July 2020. http://arxiv.org/abs/1501.00559
79. N.H. Bshouty, J.C. Jackson, N.H. Bshouty, J.C. Jackson, N. Bshouty, Learning DNF over the uniform distribution using a quantum example oracle (1995)
80. A.W. Cross, G. Smith, J.A. Smolin, Quantum learning robust to noise. Phys. Rev. A - At. Mol. Opt. Phys. **92**(1) 2014. https://doi.org/10.1103/PhysRevA.92.012327
81. A.B. Grilo, I. Kerenidis, T. Zijlstra, Learning with errors is easy with quantum samples. Phys. Rev. A **99**(3) 2017. https://doi.org/10.1103/PhysRevA.99.032314
82. D. Gavinsky, Quantum predictive learning and communication complexity with single input, in *COLT 2010–23rd Conference on Learning Theory* (Dec 2008), pp. 207–217. Accessed 31 July 2020. http://arxiv.org/abs/0812.3429
83. R. Babbush, V. Denchev, N. Ding, S. Isakov, H. Neven, Construction of non-convex polynomial loss functions for training a binary classifier with quantum annealing (June 2014). Accessed 31 Jul 2020. http://arxiv.org/abs/1406.4203

84. W. Lechner, P. Hauke, P. Zoller, A quantum annealing architecture with all-to-all connectivity from local interactions. Sci. Adv. **1**(9), e1500838 (2015). https://doi.org/10.1126/sciadv.1500838
85. L.M. Sieberer, W. Lechner, Programmable superpositions of ising configurations. Phys. Rev. A **97**(5) 2017. https://doi.org/10.1103/PhysRevA.97.052329
86. N. Wiebe, A. Kapoor, K.M. Svore, Quantum deep learning. Quantum Inf. Comput. **16**(7–8), 541–587 (2014). Accessed: 31 Jul 2020. http://arxiv.org/abs/1412.3489
87. C.-H. Yu, F. Gao, Q.-L. Wang, Q.-Y. Wen, Quantum algorithm for association rules mining. Phys. Rev. A 94(4) (2015). https://doi.org/10.1103/PhysRevA.94.042311
88. A.W. Harrow, A. Hassidim, S. Lloyd, Quantum algorithm for linear systems of equations. Phys. Rev. Lett. **103**(15), 150502 (2009). https://doi.org/10.1103/PhysRevLett.103.150502
89. S. Lloyd, M. Mohseni, P. Rebentrost, Quantum principal component analysis. Nat. Phys. **10**(9), 631–633 (2014). https://doi.org/10.1038/NPHYS3029
90. B.D. Clader, B.C. Jacobs, C.R. Sprouse, Preconditioned quantum linear system algorithm. Phys. Rev. Lett. **110**(25), 250504 (2013). https://doi.org/10.1103/PhysRevLett.110.250504
91. Z. Zhao, J.K. Fitzsimons, J.F. Fitzsimons, Quantum assisted Gaussian process regression. Phys. Rev. A **99**(5) (2015). https://doi.org/10.1103/PhysRevA.99.052331
92. S. Hangl, V. Dunjko, H.J. Briegel, J. Piater, Skill learning by autonomous robotic playing using active learning and creativity (2017), Accessed 31 July 2020. http://arxiv.org/abs/1706.08560
93. S. Hangl, E. Ugur, S. Szedmak, J. Piater, Robotic playing for hierarchical complex skill learning. Accessed 31 July 2020. https://iis.uibk.ac.at/software/mmr
94. S. Hangl, A. Mennel, J. Piater, A novel skill-based programming paradigm based on autonomous playing and skill-centric testing (2017). Accessed 31 July 2020. http://arxiv.org/abs/1709.06049
95. S. Hangl, E. Ugur, J. Piater, Autonomous robots: potential, advances and future direction. https://doi.org/10.1007/s00502-017-0516-0
96. M.S. Elsayed, N.A. Le-Khac, A.D. Jurcut, InSDN: a novel SDN intrusion dataset. IEEE Access **8**, 165263–165284 (2020). https://doi.org/10.1109/ACCESS.2020.3022633
97. "An Initiative Work on SDN Based Network Intrusion Detection System using Machine Learning|Request PDF. https://www.researchgate.net/publication/341453516_AN_INITIATIVE_WORK_ON_SDN_BASED_NETWORK_INTRUSION_DETECTION_SYSTEM_USING_MACHINE_LEARNING. Accessed 02 June 2021.
98. M. Aakash, R. Shinde, M. Shailesh, P. Bendale, Implications and application of artificial intelligence and machine learning concepts on software defined network and its future prospects. Int. J. Adv. Sci. Technol. **29**(4s), 1142–1152 (2020). Accessed 02 June 2021. http://sersc.org/journals/index.php/IJAST/article/view/6666
99. H. Abraham, A. Offei, R. Agarwal, I. Yunus Akhalwaya, G. Aleksandrowicz, T. Alexander, M. Amy, E. Arbel, Arijit02, A. Asfaw, A. Avkhadiev, C. Azaustre, A. Ngoueya, A. Abhik, et al., Qiskit: an open-source framework for Quantum Computing. https://qiskit.org/. Accessed 02 June 2021.
100. "sklearn.decomposition.TruncatedSVD—scikit-learn 0.24.2 documentation." https://scikit-learn.org/stable/modules/generated/sklearn.decomposition.TruncatedSVD.html. Accessed 02 Jun 2021
101. L. Van Der Maaten, G. Hinton, Visualizing data using t-SNE (2008)
102. "sklearn.manifold.TSNE—scikit-learn 0.24.2 documentation." https://scikit-learn.org/stable/modules/generated/sklearn.manifold.TSNE.html. Accessed 02 June 2021
103. "sklearn.decomposition.PCA— cikit-learn 0.24.2 documentation. https://scikit-learn.org/stable/modules/generated/sklearn.decomposition.PCA.html. Accessed 02 June 2021
104. P.G. Martinsson, V. Rokhlin, M. Tygert, A randomized algorithm for the decomposition of matrices. Appl. Comput. Harmon. Anal. **30**(1), 47–68 (2011). https://doi.org/10.1016/j.acha.2010.02.003
105. N. Halko, P.G. Martinsson, J.A. Tropp, Finding structure with randomness: probabilistic algorithms for constructing approximate matrix decompositions. SIAM Rev. **53**(2), 217–288 (2011). https://doi.org/10.1137/090771806

106. qiskit.aqua.algorithms.VQC—Qiskit 0.26.2 documentation. https://qiskit.org/documentation/stubs/qiskit.aqua.algorithms.VQC.html. Accessed 02 June 2021.
107. qiskit.aqua.algorithms.VQE—Qiskit 0.26.2 documentation. https://qiskit.org/documentation/stubs/qiskit.aqua.algorithms.VQE.html#qiskit.aqua.algorithms.VQE. Accessed 02 June 2021.
108. A. Peruzzo et al., A variational eigenvalue solver on a quantum processor. Nat. Commun. **5** (2013). https://doi.org/10.1038/ncomms5213
109. J.R. McClean, J. Romero, R. Babbush, A. Aspuru-Guzik, The theory of variational hybrid quantum-classical algorithms. New J. Phys. **18**(2) (2015). https://doi.org/10.1088/1367-2630/18/2/023023
110. P.J.J. O'Malley et al., Scalable quantum simulation of molecular energies. Phys. Rev. X **6**(3) (2015). https://doi.org/10.1103/PhysRevX.6.031007
111. N.P.D. Sawaya, M. Smelyanskiy, J.R. McClean, A. Aspuru-Guzik, Error sensitivity to environmental noise in quantum circuits for chemical state preparation. J. Chem. Theory Comput. **12**(7), 3097–3108 (2016). Accessed 02 June 2021. http://arxiv.org/abs/1602.01857
112. B. Bauer, D. Wecker, A.J. Millis, M.B. Hastings, M. Troyer, Hybrid quantum-classical approach to correlated materials. Phys. Rev. X **6**(3) 2015. https://doi.org/10.1103/PhysRevX.6.031045
113. qiskit.aqua.algorithms.QSVM—Qiskit 0.26.2 documentation. https://qiskit.org/documentation/stubs/qiskit.aqua.algorithms.QSVM.html. Accessed 02 June 2021
114. Feature Maps (qiskit.aqua.components.feature_maps)—Qiskit 0.26.2 documentation. https://qiskit.org/documentation/apidoc/qiskit.aqua.components.feature_maps.html. Accessed 02 June 2021.
115. Feature Maps—Qiskit 0.11.1 documentation. https://sooluthomas.github.io/testTranslation/aqua/feature_maps.html. Accessed 02 June 2021.
116. Simulators—Qiskit 0.26.2 documentation. https://qiskit.org/documentation/tutorials/simulators/1_aer_provider.html. Accessed 02 June 2021.

Implications of Deep Circuits in Improving Quality of Quantum Question Answering

Pragya Katyayan and Nisheeth Joshi

Abstract Question Answering (QA) has proved to be an arduous challenge in the area of natural language processing (NLP) and artificial intelligence (AI). Many attempts have been made to develop complete solutions for QA as well as improving significant sub-modules of the QA systems to improve the overall performance through the course of time. Questions are the most important piece of QA, because knowing the question is equivalent to knowing what counts as an answer (Harrah in Philos Sci 28:40–46, 1961, [1]). In this work, we have attempted to understand questions in a better way by using Quantum Machine Learning (QML). The properties of Quantum Computing (QC) have enabled classically intractable data processing. So, in this paper, we have performed question classification on questions from two classes of SelQA (Selection-based Question Answering) dataset using quantum-based classifier algorithms—quantum support vector machine (QSVM) and variational quantum classifier (VQC) from Qiskit (Quantum Information Science toolKIT) for Python. We perform classification with both classifiers in almost similar environments and study the effects of circuit depths while comparing the results of both classifiers. We also use these classification results with our own rule-based QA system and observe significant performance improvement. Hence, this experiment has helped in improving the quality of QA in general.

Keywords Question answering · Question classification · Variational quantum classifier · Quantum support vector machines · Quantum natural language processing · Quantum computing

1 Introduction

NLP is a branch of AI and is arguably the most complex and challenging of all other areas. It has seen some crucial advances in AI. At the same time, QC has grown consistently in terms of hardware and algorithms that are classically intractable

P. Katyayan (✉) · N. Joshi
Department of Computer Science, Banasthali Vidyapith, Banasthali, Rajasthan, India
e-mail: pragya.katyayan@outlook.com

R. Pandey et al. (eds.), *Quantum Computing: A Shift from Bits to Qubits*,
Studies in Computational Intelligence 1085,
https://doi.org/10.1007/978-981-19-9530-9_23

even with reasonable number of resources. This provides extensive opportunities for AI and specifically NLP to group with QC and find solutions to its most severe challenges of all times. Scientists have already proved quantum advantages for NLP tasks including algorithmic speedups and enhanced way to encode complex linguistic structures. QA is an AI-complete problem and we need to explore new dimensions to address it. QC and QML are such areas with the potential of being helpful in solving such problems.

The combination of QC and ML has the potential of changing how we look at previously unsolvable problems. It presents four fusion choices of both worlds as shown in Fig. 1. The approaches are combinations of classical data processed with classical algorithms (cc), classical data processed with quantum algorithms (cq), quantum data processed with classical algorithms (qc) and quantum data processed with quantum algorithms (qq). Since, the repositories of classical data are immense and powerful set of processing capabilities are need-of-the-hour, we have considered processing classical data using quantum algorithms (i.e., cq). Hence, using quantum-enhanced machine learning (or QML) we have tried to utilize quantum algorithms and computers to tap in the extraordinary depths of information processing of classical data. Quantum algorithms are known to see data quite differently than classical ones [2]. If a quantum information processor can produce statistical patterns that are difficult for classical computers, then possibly, they can recognize better patterns in data than classical computers [3].

The most common yet most crucial tasks in text processing these days is classification. Every year, this task gathers significant amount of research work produced by researchers across the world. With this paper we attempt question classification over SelQA dataset that has questions from 10 classes. Question-Answering (QA) is a well-known challenge in the field of NLP and AI. It has three necessary steps: query processing, information retrieval, and response extraction. The query processing step plays a vital role in the whole process because knowing the question is equivalent to knowing what counts as an answer [1]. For instance, we can grab the domain of question [4] and its category [5] through query processing. Question classification

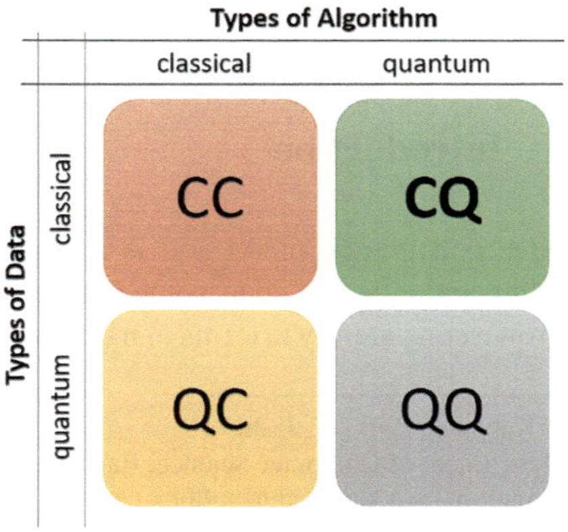

Fig. 1 Four approaches to combine classical and quantum worlds out of which CQ is the most exciting and feasible given the availability of classical data today [2]

provides the domain of possible answers along with the type of answer desired by the question. Example:

Question: How many Despicable Me movies are there?
Domain: Movies.
Category: Count.

These are just two of many attributes possessed by a question. Such information is extracted by using the question classification method. This can be done using two approaches: rule-based [6] or statistics-based (i.e., ML) [5, 7–12]. Since, ML techniques have given some extraordinary results with specific features-set [5, 11], its every researcher's first choice when it comes to tasks like question classification. Studies have shown SVM as the best classifier in this case [8, 9, 11].

We propose the use of QML for question classification. Why do we need QML? The answer to this question actually lies in the Hilbert space. Quantum computation creates exponential number of basis states according to the availability of qubits in the system. For instance, if the quantum computer has 5 qubits, then 25 basis states can be produced by the computation. As the number of qubits increase, we get much more states to work with. The impact of this phenomenon can be realized by the fact that if we have 275 qubits, then the number of states we can represent through them is 2275, which is more the number of particles in the observable universe. Quantum processing gives us more power to represent possible states than classical processing ever can.

Quantum information processing works on qubits that are analogous to classical bits. However, these qubits can exist in the state of superposition that means, as a contrast to the classical bit, qubits can be 0 and 1 at the same time. Following the Dirac notation, we can say, the state of the qubit can be described as $|\psi\rangle = \alpha\,|0\rangle + \beta\,|1\rangle$, where α and β are the complex numbers representing amplitudes of classical states $|0\rangle$ and $|1\rangle$. These complex numbers are subject to normalization condition such that $|\alpha|^2 + |\beta|^2 = 1$. When the state $|\psi\rangle$ is measured, the observation is either $|0\rangle$ or $|1\rangle$ with probability $|\alpha|^2$ or $|\beta|^2$ [13].

Figure 2 describes a rough framework of classical machine learning. In case of ML, we always initiate with data, which is provided to a model that gives us a prediction. We can score the prediction through a cost function and estimate how to update the model's parameters based on gradient-based techniques. In QML, we think of replacing this model, or perhaps some parts of this model with quantum computation that can be executed on a quantum computer while giving us some form of quantum advantage. This would process the classical data on a quantum device by looking at this data from quantum perspective. The involvement of quantum properties would help to find out more patterns in data than classical models.

Studies in QML [3, 13–19] have shown its unique performance with different types of problems and different algorithms. The rise of QML started back in 1995 with the investigation in quantum-based neural network modeling. Then, in the early 2000s, researchers discussed statistical learning theory in quantum settings, which didn't get much attention. Around 2009, a workshop series in quantum computation took off where publications on quantum learning algorithms such as quantum associative

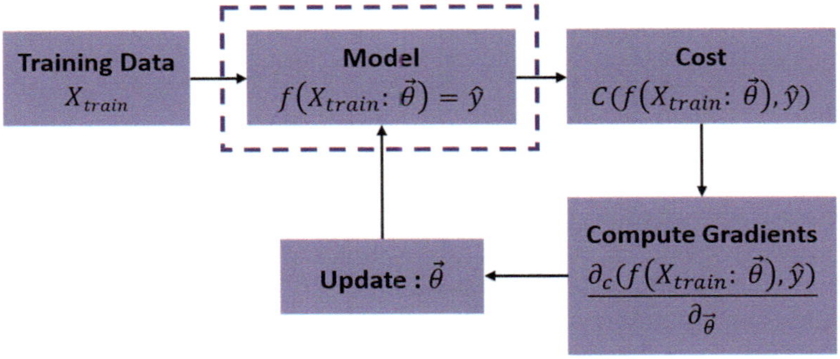

Fig. 2 General framework of machine learning

memory, QBoost, etc. were observed. Around 2013, the term 'quantum machine learning' came into use, and since then, it became popular amongst researchers [2]. QML has many methods and algorithms to support different learning tasks. Under the umbrella of kernel classification methods lies the Quantum Support Vector Machine (QSVM) classifier, which has quite a good reputation in both classical and quantum worlds. In this work, we will use QSVM to classify questions and will analyze its performance.

Question classification is an important step in QA. In this paper, we attempt to improve the quality of QA, by better classification of questions. We perform two major tasks for this. First, we propose binary classification using two quantum classifiers- VQC and QSVM. We describe the experiments and analyze the results of these classifiers. We also present the repercussions of various combinations of hyperparameters for the quantum algorithms and the classification results for each combination with special emphasis on circuit-depth. Secondly, we also show the performance enhancement of our Rule-based Question-Answering (RBQA) system [20] after using the classification results from the first task as a feature.

The main contributions made through this paper are—implementation of quantum binary classification for real-life questions dataset with analysis of how the classifiers behave while classifying questions; practical details of QSVM and VQC configuration; insight about implications of circuit depth on classification; and proof of performance improvement of a QA system with the classification results.

The paper is arranged in the following way: Sect. 2 consists of relevant literature reviews of work done in question classification and quantum machine learning; Sect. 3 explains the dataset used for the experiment; Sect. 4 elaborates on the features extracted; Sects. 5, 6 and 7 explain important concepts and methodology of the research work done; Sect. 8 talks about the experimental setup and Sect. 9 gives the results and analysis. Last, but not the least, Sect. 10concludes the work throws some light on the possible future works.

2 Literature Review

2.1 Question Classification

Hermjakob [5], Hacioglu and Ward [7] has shown the requirement of semantically rich parse trees for better question classification using ML techniques. They have shown that question parsing improves the quality of question classification significantly. Zhang and Lee [6], Metzler and Croft [8] have approached the problem of question classification with ML techniques trained on Bag of Words and Bag of n-grams features. They have observed the performances of Nearest Neighbors, Naïve Bayes, Decision Tree, Sparse Network of Winnows (SNoW), and SVM classifiers. In their paper, they have explained question classification and compared SVM with other ML algorithms and tree kernel and its calculation. Hacioglu and Ward [7], Huang et al. [9] have taken up the ML approach for question classification to replace their previous rule-based classifier. In their paper, they have described the classifier. Metzler and Croft [8], Li et al. [10] have empirically shown that question classification can be done more efficiently by using statistical techniques. They have claimed that rule-based question classification can be too specific at times and can take a lot of effort to be crafted, while machine learning (statistical) techniques can prove suitable as it gets excellent performance with almost negligible efforts.

Huang et al. [9], Liu et al. [11] have emphasized the importance of features for better classification of questions. Instead of taking a rich feature space, they have opted for a compact yet effective feature set. Silva et al. [4], [6] have identified question classification as an essential step for question-answering. They have presented a rule-based question classifier which finds the question headword and then maps it to its target category by using WordNet. They used WordNet to map headwords to categories of questions. They have also provided an empirical analysis of each feature's contribution to the best results. Li et al. [10], [12] have adapted semi-supervised learning to do question classification. They have used a combination of classifiers, i.e., ensemble classifiers. They have compared the ensemble technique with a single classification technique. Liu et al. [11], Hermjakob [5] claim that syntactic feature selection for question classification is computationally very costly. So, they have proposed a hybrid approach for semantic and lexical feature extraction. Gennaro et al. [21] have worked on classifying the intent of a question by using LSTM. They used Glove word embeddings to grab the semantic features of questions.

2.2 Quantum Machine Learning

Abramsky [22] has reported on the high-level methods for quantum computation. He has observed that the current tools are very low-level because there were loads of necessary computations, but the significance of high-level methods was not highlighted. Coecke [23] has reported on the logic of entanglement by exposing, with a theorem, the capabilities of the information flow of pure-bipartite entanglement. He has used this theorem to re-design and analyze popular protocols and has demonstrated the production of new ones. He has explained the extension of these results to the multipartite case. Aïmeur et al. [12], [13] have investigated the collaboration of machine learning and quantum information processing. Their approach is to define new learning tasks related to ML in a quantum mechanical world. Schuld et al. [13], Grant et al. [14] have given a systematic overview of QML along with the technical aspects and approaches. They observed that the researchers worldwide had taken the use of computational costs as well as for translation of stochastic methods in quantum theory logic. Cai et al. [24] have observed the challenge of increasing data and the problems faced by classical computers in managing it. They report on the first entanglement-based two, four, and eight-dimensional vector classification in different clusters. Zeng and Coecke [25] have proposed the applications of quantum algorithms to NLP. They have observed the implementational challenges of the CSC model due to a shortage of classical computational resources. They have shown methods of resolving this issue using quantum computing. Makarov et al. [26] have given a brief overview of quantum logic concerning natural language processing. They have discussed the representation of sentences in quantum logic. They describe sentence representation, similarity analysis, quantum logic in diagrams as well as evaluation on NLP problems. Biamonte et al. [3] have reviewed recent quantum techniques including quantum speedups, classical and quantum machine learning, quantum principal component analysis, HHL algorithm, QSVM, and kernel methods, reading classical and quantum data in quantum machines and quantum deep learning.

Grant et al. [14], Ciliberto et al. [15] have demonstrated that more expressive circuits than the hierarchical structure can be used to achieve better accuracy and are capable of classifying highly entangled quantum states. They have explained in detail the effects of noise on classifiers' performance and deployment of classifiers on a quantum computer. Schuld et al. [27] have pointed out the need for quantum technologies that need fewer qubits and quantum gates and are error-proof. Such an approach is possible through variational circuits, which is very much beneficial for ML as these circuits learn the gates parameters. Their paper explains the circuit-centric classifier, its architecture along with optimization techniques used and graphical representation of gates. They have used hybrid gradient descent for training. They have provided a comprehensive analysis of the results achieved with the experimentation. Ciliberto et al. [15], Kusumoto et al. [16] presented a review on QML with a classical point of view. They have discussed possibilities for both classical and quantum experts. They have focused mainly on the limitations of quantum algorithms and where they stand in comparison to their classical competitions. They have also discussed why quantum is supposed to be a better resource for learning problems. Cárdenas-López et al. [28] have proposed a protocol for performing quantum reinforcement learning. This protocol doesn't require coherent feedback during the learning process. Hence, it can be implemented in various types of quantum systems. Shukla et al. [29] have performed quantum process tomography of all the gates used in IBM processors and have computer gate-error to check the feasibility of complex quantum operations.

They have also compared the quality of these gates with those built using other technologies. They have analyzed with this experiment that technological improvement would be required for achieving the scalability needed for quantum operations that are complex.

Kusumoto et al. [16], Havlíček et al. [17] have exploited the complex dynamics of solid-state nuclear magnetic resonance to enhance machine learning. They have proposed to use Hamiltonian evolution based on inputs to map them to feature space. Havlíček et al. [17], Tacchino et al. [18] have proposed two novel methods that they have implemented on a superconducting processor. In the first method, the quantum variational classifier was built on a variational quantum circuit and similarly classifies the training dataset as the classical SVM. In the second method, they have estimated kernel function and optimized the classifier. They developed artificial data that could be easily and completely be separated by the feature maps. This experiment was completed in two steps- first, the classifier was trained and optimized, and in second, the classifier labels the unlabeled data. Even in the presence of noise, their approach was able to touch a 100% success rate. Tacchino et al. [18], Jurczyk et al. [19] have observed that implementations of artificial neural networks in today's world are hindered by the growing computational complexity, which is a must for training MLP. So, they have introduced an algorithm based on quantum information that implements a binary-valued perceptron on a quantum computer. Mishra et al. [30] have used deep learning and supervised learning to hone quantum techniques by proposing a quantum neural network for cancer detection.

Meichanetzidis et al. [31] performed QA on NISQ device. In their model, the sentences were instantiated as parametrized quantum circuits with meanings encoded as quantum states. They took care of the grammatical structure explicitly by hard-wiring it as entanglement operations. This made this approach NISQ-friendly. Guo [32] introduced a density matrix which modeled a sentence-level attention mechanism. They developed a BiLSTM model based on word-level attention of weak measurement along with sentence-level attention for density matrix. They used this model to perform QA using WikiQA datasets. Song et al. [33] performed rigorous experiments on QA models to check if the relationship between weights of linguistic units and the outputs given by the model were uncertain. They observed that there were few ways where higher weights correlated with the model with a great impact, but usually the data didn't correspond well. This meant that the weight and model prediction were independent of each other and it was possible that different weight distributions produced similar results. Correia et al. [34] developed a theoretical pipeline to perform sentence disambiguation and question-answering which took advantage of quantum features. For the disambiguation task, their contraction scheme dealt with phrases that were syntactically ambiguous. For QA, they extracted a query representation which had all possible answers in equal superposition. Then they implemented Grover's search algorithm to find the correct answer. Zhao et al. [35] found a balance between a model's performance and interpretability while proposed a quantum attention-based language model. Density matrix was used in quantum attention mechanism. The language model was applied in the answer selection module of a typical QA task to achieve effective results.

3 Dataset

The experiments for this research work were performed on a gold-standard benchmark dataset generated by Jurczyk et al. [19], which is popularly known as the Selection-based Question Answering (SelQA) dataset. This dataset has been developed from 486 Wikipedia articles as dumped in August 2014. The corpus has items from 10 most prevalent domains, i.e., Arts, Country, Food, Historical Events, Movies, Music, Science, Sports, Travel, and TV. The original data taken from the websites was broken down in smaller chunks during preprocessing. Each article was segmented into sections with the help of section boundaries available in the raw data. Further, every section was broken into sentences using NLP4J toolkit. The corpus has annotated question-answering examples on the above-mentioned topics. Each instance in the dataset consists of 7 different attributes, out of which the 'question' and its 'type' (i.e., domain) were extracted to make a custom corpus for the question classification task. The dataset was already divided into training data with 5529 sentences and testing data with 1590 sentences. The distribution of each class in the dataset is described in Table 1. Data annotation was done through four sequential tasks on Amazon Mechanical Turk.

Since, we have attempted binary classification using quantum classifiers, we have randomly chosen two classes—'Historical Events' and 'Science' for the experiment. The classes have total 1525 (i.e., 730 and 795, respectively) unique questions. We have balanced the dataset by removing 65 datapoints from science class. Final dataset on which experiments were conducted had 730 questions from each domain, i.e., total 1460 questions out of which 80% datapoints were taken for training and rest 20% for testing purpose.

Table 1 Dataset distribution for each class in SelQA train and test set

Sl. no	Classes in SelQA	Questions in training dataset	Questions in testing dataset	Total
1	Art	467	135	601
2	Country	618	178	791
3	Food	509	147	652
4	*Historical events*	571	164	*730*
5	Movies	574	164	735
6	Music	541	155	677
7	*Science*	621	179	*795*
8	Sport	585	168	741
9	Travel	573	165	734
10	TV	470	135	604
Total		5529	1590	7060

4 Feature Selection

Since we wish to perform binary classification, we take up two questions of two classes from the SelQA dataset, one from 'Historical Events' class and another from 'Science' domain. The two domains are quite different from each other in terms of keywords, which plays a crucial role in accomplishing classification. We extracted eleven distinct features from the questions. Different experiments were performed with different groups of features. For instance, some experiments were performed with four features, some with five, some with seven features while the rest were performed with complete 11 features. The features are mentioned in Table 2, along with features considered in different groups. They are further elaborated in the subsequent sections.

4.1 Content and Non-content Words

Content words are responsible for the important information in a sentence. Non-content words have semantic information where they facilitate anticipation of some feature of the words that follow. They are also called function words which help connect the content words. These are responsible for helping the model better understand the semantics and capture the context. We counted the content words and non-content words of all the questions as two important features. Content words comprised of Nouns, Verbs, Adjectives and Adverbs and other words were counted as non-content words. Examples are given in Table 3.

Table 2 Features extracted for SelQA dataset

Sl. no	Feature name	Feature groups used in experiments			
		4	5	7	11
1	Content words	✓	✓	✓	✓
2	Non-content words	✓	✓	✓	✓
3	Question keywords	✓	✓	✓	✓
4	Wh-words	✓	✓	✓	✓
5	Nouns		✓	✓	✓
6	Verb count			✓	✓
7	1-g Probabilities			✓	✓
8	2-g Probabilities				✓
9	3-g Probabilities				✓
10	4-g Probabilities				✓
11	5-g Probabilities				✓

Table 3 Examples of content and non-content words count from questions

Questions	Content words	Non content words
How many times was the who national fyrd called out between 1046 and 1065?	6	9
When did the germans begin using chlorine gas on the western front?	8	5
As a result of the napoleonic wars the british empire rose in power beginning the historical period known as what?	11	10
What is the natural satellite of earth?	4	4
What two creations of confirmation bias are under study with respect to astrological belief?	8	7

Table 4 Examples of keywords extracted from questions

Questions	Q_keywords
How many times was the who national fyrd called out between 1046 and 1065?	['many', 'time', 'national', 'fyrd', 'call']
When did the germans begin using chlorine gas on the western front?	['german', 'begin', 'use', 'chlorine', 'gas', 'western', 'front']
What is the natural satellite of earth?	['natural', 'satellite', 'earth']
What two creations of confirmation bias are under study with respect to astrological belief?	['two', 'creation', 'confirmation', 'bias', 'study', 'respect', 'astrological', 'belief']
What had faraday concluded based on his electrochemical experiments?	['faraday', 'conclude', 'base', 'electrochemical', 'experiment']

4.2 Question Keywords

We have removed the general English stop-words from the questions dataset and have used the remaining keywords as a feature. These keywords consist of significant words from the individual domains. These are markers that help the classifier in better classification. These keywords are converted into vectors with the help of Bag-of-Words. Examples are given in Table 4.

4.3 Wh-Words

Every question either has a wh-word (how, what, when, where, which, who) or it starts with a verb ending with a question mark (eg. Does he have a gold coin?). We have captured the presence of all the wh-words and have marked the others as 'NA'. This will help in categorizing wh-questions. Since Wh-words list is a closed set of six words, we have extracted only those. Examples are given in Table 5.

Table 5 Examples of wh-words extracted from questions

Questions	Wh-words
How many times was the who national fyrd called out between 1046 and 1065?	How who
When did the germans begin using chlorine gas on the western front?	When
What is the natural satellite of earth?	What
What two creations of confirmation bias are under study with respect to astrological belief?	What
What had faraday concluded based on his electrochemical experiments?	What

Table 6 Examples of nouns extracted from questions

Questions	Nouns
How many times was the who national fyrd called out between 1046 and 1065?	Times fyrd
When did the germans begin using chlorine gas on the western front?	Germans chlorine gas front
What is the natural satellite of earth?	Satellite earth
What two creations of confirmation bias are under study with respect to astrological belief?	Creations confirmation bias study respect belief
What had faraday concluded based on his electrochemical experiments?	Faraday experiments

4.4 Nouns

Nouns in any sentence speak of the significant entities—people, places and things. By identifying Nouns of any domain, we can easily identify the set of significant people, places and things of that particular domain. These are keywords which help in answer generation. Examples are given in Table 6.

4.5 Verb Count

The number of verbs occurring in a sentence displays what kind of domain it is. The count helps us catch the syntactic formation of the system and helps the model learn what the sentence is made up of. It helps in identifying the complexity of sentence. Examples are given in Table 7.

Table 7 Examples of verb count for different questions

Questions	Verb count
How many times was the who national fyrd called out between 1046 and 1065?	2
When did the germans begin using chlorine gas on the western front?	2
What is the natural satellite of earth?	1
What two creations of confirmation bias are under study with respect to astrological belief?	0
What had faraday concluded based on his electrochemical experiments?	3

4.6 N-gram Probabilities

N-grams (where N ∈ {1, 2, ..., 5}) of texts are extensively used in NLP tasks to mine textual data for useful information. These are set of co-occurring words within a particular word-window (e.g., 2-g = 2-word window). Since we are not doing grammatical analysis, n-gram approach is a mechanism of doing the same with statistical learning. We used language modelling to find the occurrences of these n-grams and calculate probabilities of their presence throughout the corpus. We considered probability values up to 5-g as separate features for our dataset.

5 Variational Models

Variational models are quantum circuit-based models which have certain parameters that we can tweak, train, and optimize. Figure 3 shows a variational circuit where U is any set of unitary gates that constructs the feature map. $U(x)$ has parameters x that can be tweaked to get desired results. Next block is $W(\theta)$ which is the model circuit that helps in accomplishing the classification task where θ is the parameter to tweak, train and optimize the circuit. M is the measurement unit. We have used this variational model as a classifier for our classical data to accomplish question classification. The task is to train a quantum circuit $W(\theta)$, on labeled samples of data from $U(x)$, in order to get predictions of labels for unseen data. The first step is to encode the classical data into quantum state so that the quantum circuit can understand it using a feature map. The next step will apply a variational model on the encoded data that will be trained as a classifier. In the third step we will measure the classifier circuit to extract labels and last, but not the least we will use optimization techniques to update the model parameters.

In the above diagram, $W(\theta)$ block is the classifier model. There are several circuits[1] available in Qiskit for such usage viz. *RealAmplitudes, EfficientSU2, TwoLocal, NLocal* etc. For our experiment, we have taken TwoLocal circuit provided by Qiskit's circuit library as the classifier model for training. It is a

[1] https://qiskit.org/documentation/apidoc/circuit_library.html#n-local-circuits.

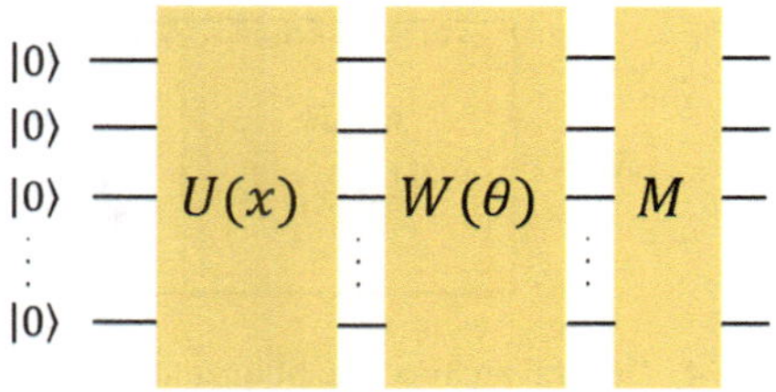

Fig. 3 Representation of a variational model where the quantum circuit consists of parameterized gates that can be changed to get better results

parametrized circuit with alternating rotation and entangled layers. The rotation layers are composed of single qubit gates which are applied on all qubits and the entanglement layer uses two-qubit gates to entangle the qubits. We can provide the combination of gates we wish to use for the rotation layer and entanglement. We used a fully entangled circuit to get the maximum quantum advantage possible.

6 Quantum Support Vector Machines

QSVMs were introduced by Havlíček et al. [17]. Quantum Kernel Estimation (QKE) implements SVM with a quantum kernel using quantum processing twice in the process. Firstly, kernel $K(x_i, x_j)$ is estimated on a Quantum device for each pair of training data $x_i, x_j \in T$. This kernel can later be used in Wolfe-dual SVM to detect the optimal hyperplane. Secondly, the quantum device is again used to estimate the kernel $K(x_i, s)$ for any new datapoint $s \in S$ and the support vectors from the set $x_i \in T$ obtained from the optimization. This is enough to construct the full SVM classifier. Quantum kernel support vector classification algorithm has the following steps:

1. Build the train and test quantum kernel matrices.
    (a) For each pair of datapoints in the training dataset x_i, x_j, apply the feature map and measure the transition probability: $K_{ij} = |\langle 0|U^{\dagger}_{\Phi(x_j)} U_{\Phi(x_i)}|0\rangle|^2$.
    (b) For each training datapoint x_i and testing point y_j, apply the feature map and measure the transition probability: $K_{ij} = |\langle 0|U^{\dagger}_{\Phi(y_j)} U_{\Phi(x_i)}|0\rangle|^2$.
2. Use the train and test quantum kernel matrices in a classical support vector machine classification algorithm (Fig. 4).

The quantum kernel should be hard to estimate classically as compared to their classical counterparts in order to tap into quantum advantage for quantum kernel machine algorithms [17]. We have encoded our datapoints using PauliFeatureMap with full entanglement (explained in the next section) which is hard to achieve classically and that dataset is supplied to the QPU for QKE. The processing and results are tough to simulate classically hence, our QSVM gets the quantum advantage over classical SVM.

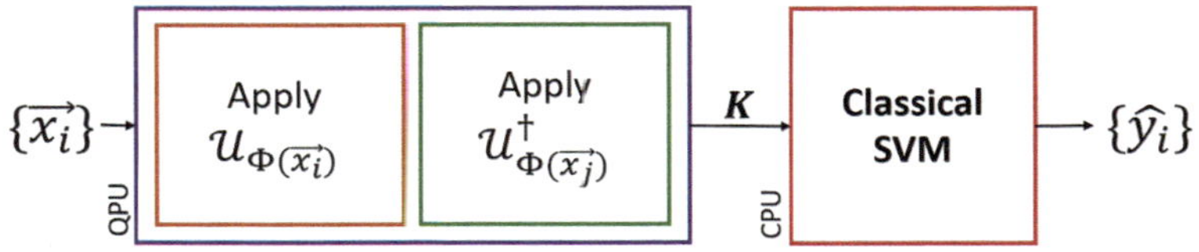

Fig. 4 General architecture of quantum support vector machines as introduced in [16]

7 Data Encoding

In this paper, we propose binary classification of classical data using two quantum classifiers. The classical processes of data encoding are followed during data preprocessing. However, classical data is not quantum readable and hence we need to map the classical data into quantum state space. For classical data, data encoding is usually done using single qubit rotations. The most efficient approach considered is to encode classical data in amplitudes of a superposition which means utilizing N qubits to encode 2 N dimensional data vector. Although, this method is less efficient in terms of space but is very efficient in terms of time because it involves only single-qubit rotations [14]. Here data vectors are re-scaled in an element-wise manner to make them lie between $\left[0, \frac{\pi}{2}\right]$. Next step is to encode every vector element from the qubit using the following equation (Eq. 1):

$$\psi_n^d = cos\left(x_n^d\right)|0\rangle + sin\left(x_n^d\right)|1\rangle \tag{1}$$

where, the classical dataset for binary classification is a set $D = \left\{\left(x^d, y^d\right)\right\}_{d=1}^{D}$, with $x^d \in \mathbb{R}^N$ as N-dimensional input vectors and $y^d \in \{0, 1\}$ are the corresponding data labels. The final data vector is written as $\psi^d = \otimes_{n=1}^{N} \psi_n^d$ and is ready to be used in quantum algorithm [14].

There are several ways to encode classical data to quantum state viz. basis encoding, amplitude encoding, angle encoding and higher order encoding. Since we have a complex dataset to work with, we go for higher order encoding using quantum feature maps. The choice of quantum feature maps depends on the type of data we have, however, there are no standard rules to decide which feature map suits which type of data. If the feature map is hard to simulate classically, that gives us quantum advantage. There are several feature maps provided by the Qiskit library which are capable of higher order encoding viz. $ZFeatureMap$, $ZZFeatureMap$ and $PauliFeatureMap$. Since we are using Pauli feature map to encode our classical data to quantum space, and it is proved to be hard to simulate classically [17], it gives us quantum advantage. Given the complex nature of our data, PauliFeatureMap seems the perfect choice to encode our data as it provides customizable combinations of Pauli Gates which could prove beneficial to map our data on the Hilbert space. We experimentally tried all three feature maps provided by the Qiskit library on small sample datasets carefully taken out from the original SelQA dataset and found

PauliFeatureMap to be giving the best results. PauliFeatureMap when customized with Pauli gates: $P_0 = X$, $P_1 = Y$, $P_2 = ZZ$; can be explained by the following equation (Eq. 2):

$$\mathcal{U}_{\Phi(x)} = \left(\exp\left(i \sum_{jk} \Phi_{\{j,k\}}(x) Z_j \otimes Z_k \right) \exp\left(i \sum_{j} \Phi_{\{j\}}(x) Y_j \right) \exp\left(i \sum_{j} \Phi_{\{j\}}(x) X_j \right) H^{\otimes n} \right)^{d} \quad (2)$$

Each datapoint in our dataset has 11 features representing it, so total qubits we need would be 11. We kept the circuit short-depth with reps = 1 and fully entangled.

8 Experimental Setup

Our dataset has labelled questions from two domains, historical events and science. We first clean and pre-process the dataset from any unwanted punctuations or notations and turn them all in lower-case. We then extract 11 features from the questions. Next, we normalize the features and split the dataset in the ratio of 80:20 where 80% of the dataset would be used for training the classifiers and 20% will be used as the testing set. The dataset distribution is given in Table 8. We encoded the classical data to quantum state using PauliFeatureMap. Further, we used two quantum classifiers to get classification results on the same dataset. First was QSVM and the other was VQC.

QSVM tries to detect a hyperplane in the input feature space that can separate the two classes. There are different kernel methods available that can be used to find out a hyperplane in a high-dimensional input feature space without defining it. Kernels help to project the vectors in the feature space to a higher-dimensional vector space where a linear decision boundary separating two classes is possible. We have used Qiskit Aqua's QSVM module for this experiment. As we have discussed above, we chose PauliFeatureMap for this experiment with a short depth of 1, feature dimension 11 and fully entangled to get quantum advantage. We ran the quantum instance on Qiskit's IBM Qasm Simulator which is a context-aware simulator and can simulate up to 32 qubits. Since, there is no IBM Q device with more than five qubits available on the cloud, we had to run our experiment on the Qasm simulator. We used a Linux-based system with the Ubuntu 20.04 LTS operating system, Quadro RTX 5000 GPU

Table 8 Dataset distribution across training and test set

Labels	Classes	Train	Test	Total
0	Historical Events	534	146	730
1	Science	534	146	730
Total		1168	292	1460

to run this experiment and it took 62 h to complete. The results are discussed in the next section.

VQC, on the other hand, takes another circuit as a model to train as a classifier. As we have discussed above, we have taken TwoLocal circuit of Qiskit library as the model with 'ry' and 'rz' as rotation layer gates and 'cz' as entanglement layer gate. We have used PauliFeatureMap with similar specifications as we used for QSVM. Finally, we have used the COBYLA optimizer to optimize the results with maximum iterations of 100. We executed this experiment on IBM's Qasm simulator and the same hardware as QSVM experiment, and it took about 24 h to complete. The results are discussed in the next section.

8.1 Application of Question Classification on a Pre-developed QA System

We developed a rule-based open-domain question answering (RBQA) system with hand-crafted rules for QA. The system was developed and tested on SQuAD 2.0 dataset [20]; however, the rules were capable of acting on any similar dataset. Figure 5 describes the working of the RBQA system in a nutshell. We tested the RBQA system on questions from the above-mentioned two classes of SelQA dataset. The system gave a satisfactory performance with ~65% accuracy. One notable nature of our system is that it displays exact answers only if it is able to fetch them, in all other cases the system displayed the most relevant sentence from the context that might have the correct answer. Keeping this nature in mind, we applied these classification results as question-feature to our rule-based open-domain question answering (RBQA) system. With this additional feature the QA system gave more exact answers than earlier.

The detailed results of all experiments are given in the next section.

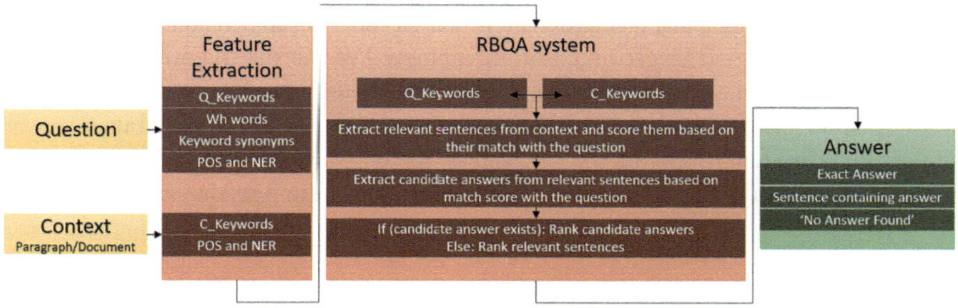

Fig. 5 General working of the RBQA system developed by Katyayan and Joshi [20] explained pictorially

9 Results and Analysis

Quantum machine learning has tremendous potential of solving problems which has been proved by researchers [17]. We chose to use its potential in solving a challenge from NLP domain—Question classification. We had SelQA dataset with questions from 10 different classes. Since we wish to accomplish binary classification only, we segregated two classes—Historical Events and Science from the dataset and balanced it. We extracted 11 features from the dataset and used two classifiers— QSVM and VQC to classify them. We kept the environment similar for both the classifiers, in order to get a comparative result. The results show that VQC is quite faster than QSVM. However, the testing accuracy of QSVM came higher than VQC. QSVM gave 60.61% accuracy while VQC gave 58.21%. This shows that QSVM classified the data more accurately than VQC. There were several challenges to this experiment including the unavailability of bigger IBMQ devices and more computing power. We took a considerably small dataset that took QSVM 60+ hours to process.

The results show a comparison between the QSVM classifier and VQC. These two experiments were executed on the same specifications (e.g., dataset, feature dimension, feature map, optimizers and other hyperparameters and hardware). The execution time of both the classifiers had a big gap. While VQC completed in 24 h, QSVM took 62 h to finish. A possible reason for this would be QSVM's dual processing of the QKE followed by hyperplane estimation while VQC runs just one classifier circuit. Given extra processing of QSVM, the extra time for execution is justified. The results of all the experiments run with both classifiers are presented in Tables 9 and 10.

In case of QSVM (Table 9), we performed 10 different experiments with changes in hyperparameters. Subtle changes in hyperparameters lead to changes in testing accuracies. We started off with just 2 features and lowest feature map depth (i.e., 1) and we got a testing accuracy of 55.98%. Then we increased the depth to 2 and the

Table 9 Results of various experiments run with QSVM (with change in hyperparameters)

Exp no	No. of features	FM	Entanglement	FM depth	FM combination	Testing accuracy
1	2	PauliFM	Full	1	['X', 'Y, 'ZZ']	55.98
2	2	PauliFM	Full	2	['X', 'Y, 'ZZ']	50.71
3	4	PauliFM	Full	1	['X', 'Y, 'ZZ']	53.58
4	4	PauliFM	Full	2	['X', 'Y, 'ZZ']	50.63
5	5	PauliFM	Full	1	['X', 'Y, 'ZZ']	49.50
6	7	PauliFM	Full	1	['X', 'Y, 'ZZ']	53.11
7	7	PauliFM	Full	2	['X', 'Y, 'ZZ']	48.85
8	*11*	*PauliFM*	*Full*	*1*	*['X', 'Y, 'ZZ']*	*60.61*
9	11	PauliFM	Full	2	['X', 'Y, 'ZZ']	54.21
10	11	PauliFM	Full	3	['X', 'Y, 'ZZ']	53.94

Table 10 Results of various experiments run with VQC (with change in variables and hyperparameters)

Exp no	Features	FM	Entanglement	FM depth	FM combination	QC depth	Testing accuracy
1	2	Pauli	Full	1	['X', 'Y, 'ZZ']	3	56.45
2	4	Pauli	Full	1	['X', 'Y, 'ZZ']	4	62.2
3	4	Pauli	Full	2	['X', 'Y, 'ZZ']	3	54.54
4	5	Pauli	Full	1	['X', 'Y, 'ZZ']	1	53.77
5	5	Pauli	Full	2	['X', 'Y, 'ZZ']	2	54.79
6	5	Pauli	Full	3	['X', 'Y, 'ZZ']	3	55.47
7	7	Pauli	Full	1	['X', 'Y, 'ZZ']	1	57.7
8	7	Pauli	Full	2	['X', 'Y, 'ZZ']	2	51.44
9	7	Pauli	Full	2	['ZZ' 'XY' 'ZZ']	2	46.88
10	*11*	*Pauli*	*Full*	*1*	*['X', 'Y, 'ZZ']*	*1*	*58.21*
11	11	Pauli	Full	2	['X', 'Y, 'ZZ']	2	53.76

accuracy fell to 50.11%. Next, we increased the number of features to 4 with depth 1, accuracy increased a little and came up to 53.58%. On increasing the depth to 2 the accuracy again fell to 50.63%. Next, we increased the features to 5 and attempted classification at depth 1, but the accuracy fell to 49.50%. So, we decided to add more features before trying out some more. On increasing the features to 7 at depth 1 accuracy rose to 53.11, but on increasing depth to 2 it again fell to 48.85%. Finally, we added 4 more features to the list and again trained the system on 11 features. With highest depth 3 the accuracy rose to 53.94%, on decreasing depth to 2- accuracy rose to 54.21% and with lowest possible depth 1 the accuracy rose to 60.61%.

In case of VQC, its circuit contains both the feature map as well as a parametrized circuit. There can be numerous possibilities of depth value combinations along with endless combinations for Pauli gates in the Feature map circuit. So, to keep the study diverse as well as finite, we kept the Pauli gate combination set to the default of $X + Y + ZZ$ and depth of both circuits are changed simultaneously to observe the results. Since, VQC performs well on short depth circuits [16], we stopped increasing depths if the accuracy value fell. We started observing with 2 features, FM depth 1 and circuit depth 3 and the accuracy we got was 56.45%. This accuracy was better than that of QSVM's first case with the lowest depth. So, we increased the features to 4 and increased the circuit depth to 4. The testing accuracy jumped to 62.2%. We tried to check if the accuracy fluctuates on increasing FM depth and increased it to 2 while decreasing the circuit depth to 3. Accuracy fell to 54.54%. Next, we increased the features to 5 and kept both the depths to the lowest value 1, the accuracy fell a little to 53.77%. We increased both depth values to 2 and got an accuracy of 54.79%. We again increased the depths to 3 each and got a slight accuracy rise of 55.47%. These were very slight changes in accuracy and so we again added couple features to the list and trained the system. With each depth values set to 1, we got accuracy of 57.7%; each depth value is set to 2, we got a fall in accuracy value to 51.44%. Since the accuracy started falling with increased depths, we tried a different Pauli gates combination for the feature map on depth values 2 for each circuit, but the accuracy fell drastically to 46.88%. So, we understood that changing the Pauli gates combination in this case is probably not suitable, so we used the default combination of X + Y + ZZ gates. We increased the features to 11 and on lowest depths of both circuits we achieved an accuracy of 58.21%. On attempting for a higher depth value of 2 for both circuits, the accuracy fell to 53.76%.

The results in the above tables show the change in accuracy as the number of features and circuit-depth were changed. Since, these algorithms are executed on noisy devices, short-depth circuits perform well as they are compliant with error-mitigation techniques that are capable of reducing the decoherence effect [16]. In case of QSVM, when the number of features were increased gradually the accuracy started falling. However, with 11 features the accuracy reached 60% on a short depth circuit. We tried increasing the depth of the feature map but that worsened the accuracy in every case, so we can say in case of question classification QSVM works best on short-depth circuits. We kept the feature map fully entangled to ensure quantum advantage. In case of VQC, the results get high with 4 features and deeper circuit, however, apart from that the results improve when the circuit depth decreases and at maximum 11 features and shortest possible depth, the accuracy hits 58.21%. If we are to compare QSVM and VQC, then QSVM (exp-8) and VQC (exp-10) match similar hyperparameter values and hence the results are comparable. Table 11 shows few comparable results of both classifiers.

In the above table, it is clearly observed that since QSVM applies only one feature map circuit and VQC requires both feature map and a quantum circuit to work with, the circuit depth of QSVM will always be lower than VQC. QSVM consistently performs poorly than VQC under similar circumstances but with a complete set of 11 features the testing accuracy reaches an all-time high value of 60.61%. Perhaps this particular set of features helped the classifier in finding a better hyperplane. At one instance, VQC surprised us with a 62.2% accuracy at the highest depth of 5. However,

Table 11 Final results of accuracy given by QSVM and VQC for question classification task on same environment variables

Features	QSVM	Testing accuracy (%)	VQC			Testing accuracy (%)
	FM depth		FM depth	QC depth	Total	
4	1	53.58	1	4	5	62.2
	2	50.63	2	3	5	54.54
5	1	49.50	1	1	2	53.77
7	1	53.11	1	1	2	57.7
	2	48.85	2	2	4	51.44
11	*1*	*60.61*	*1*	*1*	2	*58.21*
	2	54.21	2	2	4	53.76

in every other case of VQC the accuracy value fell on increasing depth of the circuit. Also, as compared with QSVM, the testing accuracy in every other comparable case is higher, but with the complete feature set of 11 features, it fell a little behind with 58.21% testing accuracy. So, overall, we understand that for this particular case of question classification given the combination of features we are using, both QSVM and VQC performed well on short-depth circuits and the feature set considered for experiments also affects the accuracy significantly. Stronger the feature, better the classification. Figures 6 and 7 the classifiers' performance pictorially. Figure 6 shows classifier-wise performance based on circuit depth. It can be observed easily that low circuit depths get high accuracy results. Figure 7 highlights the role of features on the increasing accuracy of both the classifiers.

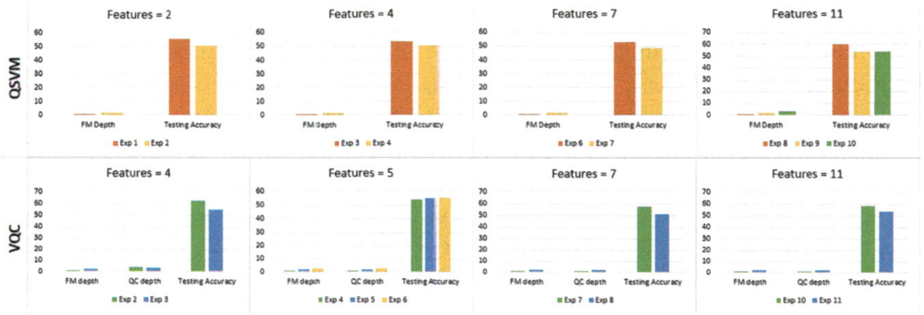

Fig. 6 Graphs showing variations in testing accuracy on increasing circuit depth in different experiments (in case of QSVM, circuit depth = FM depth; in case of VQC, circuit depth = FM depth + QC depth)

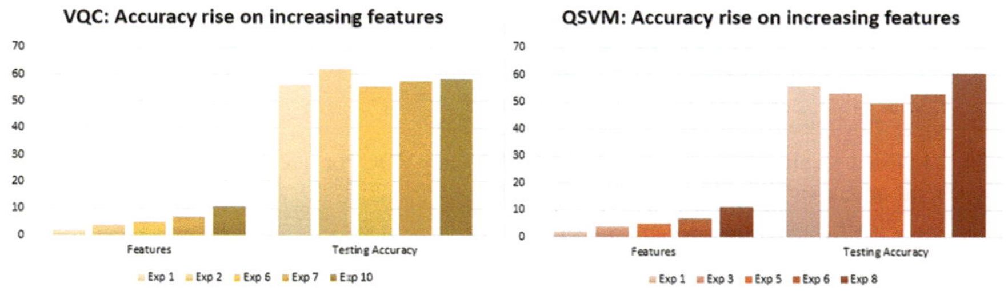

Fig. 7 Graphs showing accuracy rise on increasing features for both VQC and QSVM

9.1 Performance of RBQA System While Using QSVM's Best Classification Results as Features

The second task of our experiment was to apply the best classification results and study the implications on the accuracy of the QA system. First, we established a baseline by testing our RBQA system on the questions from SelQA dataset as defined in Table 8. Our system gave 65% (i.e., 949/1460) correct answers including exact answers and relevant answer sentences. Out of the correctly answered 949 questions, 432 questions got exact answers while 517 got relevant sentences as answers. Next, we used QSVM's best classification result as feature-set for the questions that got correctly answered by our system. We observed that the number of exact answers increased from 432 to 461. This happened because the new feature-set that was introduced, helped the system in finding relevant answers by defining the domain of the question. For instance, if system got confused between multiple candidate answers, the domain helped it in accurately selecting the correct answer. This proves that good quality question classification can help in improving the quality of QA.

10 Conclusion and Future Works

We have experimentally implemented binary question classification using quantum classifiers. We performed several experiments with different hypermetric settings. However, we kept few things constant just to keep the focus on circuit depth and avoid many complex permutations and combinations of all hyperparameters. We selected the best possible options and fixed them. For instance, we used PauliFeatureMaps with default Pauli gates settings in all the cases of both classifiers as they best represent complex data. The experiments show classification results at various depths. Observation of results showed us that in majority of the cases, both the classifiers perform well on short-depth circuits and QSVM slightly outperformed VQC while working under almost similar circumstances. Even on lowest depths, the classifiers' results vary according to different set of features. The performance falls initially on increasing features but when trained on complete 11 features, the testing accuracy rises. This realization enables us to conclude that strong features must be chosen for representing classical data. We had access to few quantum computers but none of them had enough qubits for our experiments. So, the results shown here are simulated using IBM's QASM simulator (supports up to 32 qubits). This also brings to light the vacancy of high-qubit capacity quantum computers for such experiments which involve higher number of features (and hence need more qubits to encode those features). The best classification results achieved by QSVM was used as an additional feature in a QA system which proved to be a performance booster.

Future works may include training the classifiers on a bigger dataset and testing the classifiers' performance on a real quantum computer (NISQ devices). Also, if high-capacity quantum computers are made available in future, it would be interesting to note the performance of these classifiers on a real quantum computer.

References

1. D. Harrah, A logic of questions and answers. Philos. Sci. **28**(1), 40–46 (1961)
2. M. Schuld, F. Petruccione, *Supervised Learning with Quantum Computers*. (Springer, 2018).
3. J. Biamonte, P. Wittek, N. Pancotti, P. Rebentrost, N. Wiebe, S. Lloyd, Quantum machine learning. Nature **549**(7671), 195–202 (2017)
4. J. Silva, L. Coheur, A.C. Mendes, A. Wichert, From symbolic to subsymbolic information in question classification. Artif. Intell. Rev. **35**(2), 137–154 (2011).
5. U. Hermjakob, Parsing and question classification for question answering, in *Proceedings of the ACL 2001 Workshop on Open-Domain Question Answering* (2001)
6. D. Zhang, W.S. Lee, Question classification using support vector machines, in *Proceedings of the 26th Annual International ACM SIGIR Conference on Research and Development in Information Retrieval* (July 2003), pp. 26–32
7. K. Hacioglu, W. Ward, Question classification with support vector machines and error correcting codes, in *Proceedings of the 2003 Conference of the North American Chapter of the Association for Computational Linguistics on Human Language Technology: companion volume of the Proceedings of HLT-NAACL 2003—short papers*, vol. 2 (May 2003), pp. 28–30. (Association for Computational Linguistics)
8. D. Metzler, W.B. Croft, Analysis of statistical question classification for fact-based questions. Inf. Retrieval **8**(3), 481–504 (2005)
9. Z. Huang, M. Thint, Z. Qin, Question classification using head words and their hypernyms, in *Proceedings of the 2008 Conference on Empirical Methods in Natural Language Processing* (Oct 2008), pp. 927–936
10. Y. Li, L. Su, J. Chen, L. Yuan, Semi-supervised learning for question classification in CQA. Nat. Comput. **16**(4), 567–577 (2017)
11. Y. Liu, X. Yi, R. Chen, Z. Zhai, J. Gu, Feature extraction based on information gain and sequential pattern for English question classification. IET Softw. **12**(6), 520–526 (2018)
12. E. Aïmeur, G. Brassard, S. Gambs, Machine learning in a quantum world, in *Advances in Artificial Intelligence: 19th Conference of the Canadian Society for Computational Studies of Intelligence* Canadian AI 2006, Québec City, Québec, Canada, June 7–9, 2006. Proceedings 19 (pp. 431–442) (2006)
13. M. Schuld, I. Sinayskiy, F. Petruccione, An introduction to quantum machine learning. Contemp. Phys. **56**(2), 172–185 (2015)
14. E. Grant, M. Benedetti, S. Cao, A. Hallam, J. Lockhart, V. Stojevic, A.G. Green, S. Severini, Hierarchical quantum classifiers. npj Quantum Inf. **4**(1), 1–8 (2018)
15. C. Ciliberto, M. Herbster, A.D. Ialongo, M. Pontil, A. Rocchetto, S. Severini, L. Wossnig, Quantum machine learning: a classical perspective. Proc. R. Soc. A Math. Phys. Eng. Sci. **474**(2209), 20170551 (2018)
16. T. Kusumoto, K. Mitarai, K. Fujii, M. Kitagawa, M. Negoro, Experimental quantum kernel machine learning with nuclear spins in a solid (2019). arXiv:1911.12021
17. V. Havlíček, A.D. Córcoles, K. Temme, A.W. Harrow, A. Kandala, J.M. Chow, J.M. Gambetta, Supervised learning with quantum-enhanced feature spaces. Nature **567**(7747), 209–212 (2019)
18. F. Tacchino, C. Macchiavello, D. Gerace, D. Bajoni, An artificial neuron implemented on an actual quantum processor. npj Quantum Inf. **5**(1), 1–8 (2019)

19. T. Jurczyk, M. Zhai, J.D. Choi, Selqa: a new benchmark for selection-based question answering, in *2016 IEEE 28th International Conference on Tools with Artificial Intelligence (ICTAI)* (IEEE, Nov. 2016), pp. 820–827.
20. P. Katyayan, N. Joshi, Design and development of rule-based open-domain question-answering system on SQuAD v2. 0 Dataset. Preprint of paper presented, in *3rd International Conference on ICT For Digital, Smart and Sustainable Development (ICIDSSD'22)* (2022), pp. 1–11. arXiv:2204.09659
21. G. Di Gennaro, A. Buonanno, A. Di Girolamo, F.A. Palmieri, Intent classification in question-answering using LSTM architectures (2020). arXiv:2001.09330
22. S. Abramsky, High-level methods for quantum computation and information, in *Proceedings of the 19th Annual IEEE Symposium on Logic in Computer Science, 2004.* (IEEE, July 2004), pp. 410–414
23. B. Coecke, The logic of entanglement (2004). quant-ph/0402014
24. X.D. Cai, D. Wu, Z.E. Su, M.C. Chen, X.L. Wang, L. Li, N.L. Liu, C.Y. Lu, J.W. Pan, Entanglement-based machine learning on a quantum computer. Phys. Rev. Lett. **114**(11), 110504 (2015)
25. W. Zeng, B. Coecke, Quantum algorithms for compositional natural language processing (2016). arXiv:1608.01406
26. I. Makarov, A. Frolenkova, I. Belov, Quantum logic and natural language processing, in *Workshop Computational Linguistics and Language Science* (2017)
27. M. Schuld, A. Bocharov, K. Svore, N. Wiebe, Circuit-centric quantum classifiers (2018). arXiv: 1804.00633
28. F.A. Cárdenas-López, L. Lamata, J.C. Retamal, E. Solano, Multiqubit and multilevel quantum reinforcement learning with quantum technologies. PloS one **13**(7) (2018)
29. A. Shukla, M. Sisodia, A. Pathak, Complete characterization of the directly implementable quantum gates used in the IBM quantum processors (2018). arXiv:1805.07185
30. N. Mishra, A. Bisarya, S. Kumar, B.K. Behera, S. Mukhopadhyay, P.K. Panigrahi, Cancer detection using quantum neural networks: a demonstration on a quantum computer (2019). arXiv:1911.00504
31. K. Meichanetzidis, A. Toumi, G. de Felice, B. Coecke, Grammar-aware question-answering on quantum computers (2020). arXiv:2012.03756
32. P. Guo, Quantum-inspired hierarchical attention mechanism for question answering, in *2021 IEEE 33rd International Conference on Tools with Artificial Intelligence (ICTAI).* (IEEE, Nov. 2021), pp. 943–950
33. S. Song, Y. Hou, G. Liu, The interpretability of quantum-inspired neural network, in *2021 4th International Conference on Artificial Intelligence and Big Data (ICAIBD).* (IEEE, May 2021), pp. 294–298
34. A.D. Correia, M. Moortgat, H.T.C. Stoof, Quantum computations for disambiguation and question answering (2022). arXiv:2106.05299
35. Q Zhao, C. Hou, R. Xu, Quantum attention based language model for answer selection, in *Artificial Intelligence and Mobile Services–AIMS 2021*, ed. by Y. Pan, Z.H. Mao, L. Luo, J. Zeng, L.J. Zhang. Lecture Notes in Computer Science, vol. 12987 (Springer, Cham, 2002)

GPSR Compliance

The European Union's (EU) General Product Safety Regulation (GPSR) is a set of rules that requires consumer products to be safe and our obligations to ensure this.

If you have any concerns about our products, you can contact us on ProductSafety@springernature.com

In case Publisher is established outside the EU, the EU authorized representative is:

Springer Nature Customer Service Center GmbH
Europaplatz 3
69115 Heidelberg, Germany

Batch number: 10726133

Printed by Printforce, the Netherlands